GLOBAL CLIMATE CHANGE

GLOBAL CLIMATE CHANGE

TURNING KNOWLEDGE INTO ACTION

DAVID KITCHEN

University of Richmond

Routledge
Taylor & Francis Group

LONDON AND NEW YORK

First published 2014 by Pearson Education, Inc.

Published 2016 by Routledge
2 Park Square, Milton Park, Abingdon, Oxon OX14 4RN
711 Third Avenue, New York, NY, 10017, USA

Routledge is an imprint of the Taylor & Francis Group, an informa business

Credits and acknowledgments borrowed from other sources and reproduced, with permission, in this textbook appear on pages C-1 to C-8.

Library of Congress Cataloging-in-Publication Data
Kitchen, David, 1956-
   Climate change : turning knowledge into action / David Kitchen.
      pages cm
   Includes bibliographical references and index.
   ISBN-13: 978-0-321-63412-2
   ISBN-10: 0-321-63412-8
 1. Climatic changes–Textbooks. 2. Global warming–Textbooks. I. Title.
   QC903.K562 2013
   551.6–dc23
                                                    2012043914

Cover Image Credit: James Balog / Aurora Photos

ISBN-13: 978-0-321-63412-2 (pbk)

Dedication

This book is dedicated to my children,
Peter, Andrew, Jonathan, Timothy, and Sarah.

Brief Contents

Contents

PART 1   THE EVIDENCE: IS THIS NORMAL?   1

Chapter 1   "So, What's Up with the Weather?"   2

Chapter 2   The Evidence: Observing Climate Change   24

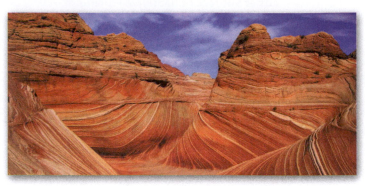

PART 4   IMPACTS OF CLIMATE CHANGE: FROM POLAR BEARS TO POLITICS   208

Chapter 7   The Global Impact of Climate Change   210

A Global Problem   212

Projections of Climate Change   212

Coping with, Adapting to, and Mitigating Climate Change   212

Examples of the Impact of Climate Change on Society   213

PART 5   GLOBAL SOLUTIONS: MANAGING THE CRISIS   **286**

Preface

The danger is that global warming may become self-sustaining, if it has not done so already. The melting of the Arctic and Antarctic ice caps reduces the fraction of solar energy reflected back into space, and so increases the temperature further. Climate change may kill off the Amazon and other rain forests, and so eliminate one of the main ways in which carbon dioxide is removed from the atmosphere. The rise in sea temperature may trigger the release of large quantities of carbon dioxide, trapped as hydrates on the ocean floor. Both these phenomena would increase the greenhouse effect, and so global warming, further. We have to reverse global warming urgently, if we still can.

STEPHEN HAWKING, ABC News interview, Aug. 16, 2006

Every major scientific body in the world now accepts that human-caused global warming is almost certain to cause significant climate change before the end of the 21st century. In 2007, The United Nations Intergovernmental Panel on Climate Change (IPCC) concluded, "Warming of the climate system is unequivocal, as is now evident from observations of increases in global average air and ocean temperatures, widespread melting of snow and ice, and rising global average sea level."

Recent years have only deepened this concern. In 2012 many climate records were shattered, including a new minimum extent for summer ice in the Arctic and the expansive melting of surface ice on Greenland. In the United States, record temperatures started to dominate the eastern two-thirds of the nation by March, and 2012 became the all-time warmest year on record. These high temperatures created drought and wildfires that affected large parts of the nation and the largest hurricane on record hit the northeast coast of the United States late in the season, wreaking havoc in New York and New Jersey. As the year ended, record precipitation in parts of the Pacific Northwest delivered more rain in a few days than normally falls over the entire year. The physical evidence of global climate change is overwhelming, but a vociferous minority still refuses to believe that it has anything to do with human activity. For the average person who wants to understand global climate change and global warming, the debate is very confusing. Many of the facts and figures are complex and hard to understand, and different groups seem to interpret the same data in such different ways. Who should we believe?

The climate change debate has shown us that most scientists lack the skills necessary to communicate a complex and nuanced message to policymakers and the general public, especially when a determined minority is committed to undermining their message. Scientists, industrialists, politicians, and the general public are all valid stakeholders in this important debate, but when extreme views are given unwarranted attention, public confusion and dangerous inaction result.

The world does not have decades to settle outstanding questions about climate change before taking decisive action. Our action (or inaction) today will have very real social, economic, political, and environmental consequences in the future—and our children and grandchildren will hold us accountable.

As an Earth scientist, I understand that Earth's climate and ecosystems are subject to natural changes. The geological evidence is clear that sometime over the next 20,000 years, in the absence of human intervention, we will return to the frozen world that predated modern civilization. Much farther back in time, during the Cretaceous Period, it is equally clear that the world was so warm that deciduous forests stretched almost as far as the poles. Climate change can be natural, but today it is not entirely natural, and for those facing the risk of climate change, the question is almost irrelevant. If the climate is changing, for whatever reason, it places the lives of hundreds of thousands in peril and the welfare of millions more at risk. There is a major humanitarian crisis looming in the near future, and it demands earnest engagement and prudent action.

This book will help you reach an informed decision about global warming and climate change. Your decision will be based on a scientific foundation that separates fact from hypothesis and reason from conjecture. You will not find all the answers in these pages, but you should find yourself prepared to ask more of the right questions.

There is still hope. Your reading and research will illuminate many possible solutions. It is a fascinating journey from science through economics to psychology and politics. It is a path festooned with hyperbolism and speculation, specious conjecture, and professional rivalry, and, at the end, the final destination is still not clear. This is as much a moral, ethical, economic, and political issue as it is a scientific issue, and progress depends on the active engagement of people and governments around the world. We take a serious risk by ignoring the early symptoms of climate change. Wishful inaction has a very poor historical record of success. Whatever the cost—and there will be a cost—we all need to ask ourselves "What are we willing to pay?"

To the Student

This book examines what scientists know about global warming and climate change and considers political and economic solutions that will balance the competing needs of people around the world. It does not answer every question it raises, but it invites you to discover answers for yourself. The text contains brief *Checkpoints* to help you review the material as you read and short *Pause for Thought* sections that ask you to consider some topics in a broader context. By addressing *Critical Thinking* questions at the end of each chapter, you are challenged to think about each problem from the contrasting perspectives of different stakeholders in the debate. This can be achieved through role-play in class and online discussion, where you can examine the arguments proposed by each stakeholder group and analyze their discussions with a professor.

Throughout the book, you are encouraged to learn and apply the scientific method to your study. You are encouraged to think in terms of observing, recording, analyzing, and synthesizing data before developing and testing hypotheses. As with any other scientific debate, you must consider all the available facts about climate change issues before reaching a conclusion.

Climate change is an urgent concern that will impact your life and the lives of your children. I hope you will go out and get involved in the debate after reading this book because, whatever your political opinion, we need informed, engaged, and active citizens who are prepared to take up new positions of leadership in society. The cost of inappropriate action could be measured in trillions of dollars, tens of thousands of lost jobs, and many lost opportunities for economic development. It is equally true that the cost of inaction will be measured by the loss of millions of lives and by a level of environmental destruction and species extinction unseen for millions of years.

To the Teacher

This book is an introductory text for students with a limited background in science, but it has enough content to be suitable for more advanced classes. Unlike most other textbooks on global climate change and global warming, the content does not only focus on the science but also includes extensive coverage of social, economic, political, and environmental aspects of climate change.

This book is optimal for classes where there is time for discussion and debate. Many stakeholders are involved in the climate debate, and many opportunities exist for students to role-play and discuss climate change from different perspectives. Role-playing is a great way to understand why the subject is so divisive, and it encourages students to find answers through further reading, research, and collaborative interaction inside and outside class.

Throughout the book, but especially in Chapters 1 through 6, students are encouraged to learn and apply the scientific method to their study. The overall thrust of the text is to encourage critical thinking and analysis and leave students with a deeper understanding of how climate change will impact all levels of society. The data used in this book are the most up-to date available at time of writing and publication and take into consideration the anticipated conclusions of the IPCC 5th Assessment Report.

Chapter Features

Each chapter contains the following features and tools:

- **Learning Outcomes** at the start of each chapter help students focus on priority concepts and topics.
- **Checkpoint** questions integrated throughout chapter sections help students check their understanding as they read.
- **Pause for Thought** questions throughout the chapters ask students to consider topics in a broader context.
- **Summary** sections revisit the main chapter topics and Learning Outcomes.
- **Why Should We Care?** sections emphasize the most important chapter themes and present brief closing thoughts on the chapter topics.
- **Looking Ahead** sections provide a bridge to and preview of the next chapter's topics and themes.
- **Critical thinking Questions** help students to extend and apply their understanding of chapter topics and themes with higher-order activities that can be done alone or as group work.
- A list of **Key Terms** with references to chapter page numbers reinforce important vocabulary. The Key Terms are also defined in the back of book Glossary.

Chapter Organization

The chapters of this book are grouped into five sections that address specific aspects of the climate change debate.

Part One: The Evidence: Is This Normal?

Is the climate change we observe today part of a natural cycle or due to the emission of heat-trapping greenhouse gases by human activity?

Chapter 1: So What's Up with the Weather? begins with a discussion of the global climate change and global warming debate and introduces some important distinctions between climate and weather. Looking back into the deep history of climate change, it becomes clear that some climate change is normal and natural, but greenhouse gases released into the atmosphere by human activity have driven recent changes.

Chapter 2: The Evidence: Observing Climate Change investigates the physical evidence of global climate change. Data collected by satellite combined with direct measurements from the land and oceans show that the temperature of Earth is rising due to an imbalance between the amount of energy entering and leaving Earth's atmosphere. From this discussion, it becomes clear that we need to explore more of the science behind Earth's climate system if we want to differentiate between the natural and anthropogenic (human-made) factors that drive climate change.

Part Two: Follow the Energy: Atmosphere, Oceans, and Climate

The energy that arrives on Earth from the Sun drives a complex climate machine, where a small change in just one part can have global consequences.

Chapter 3: Earth's Climate System focuses on the physical science of global warming and introduces students to complex interactions between the Sun, atmosphere, hydrosphere, lithosphere, cryosphere, and biosphere that determine how Earth's climate changes. The focus is on the flow of energy through Earth's climate system and how even small changes in the balance between incoming and outgoing energy can be amplified into significant changes in climate.

Chapter 4: Understanding Weather and Climate investigates how these changes in the energy content of Earth's climate system are translated into changes in regional climate and weather through the movement of mass and energy in the atmosphere and oceans.

Part Three: Deep Time: A Long History of Natural Climate Change

The best way to predict the future behavior of any system is to understand how it has behaved in the past.

Chapter 5: Revealing Ancient Climate introduces the tools that scientists use to investigate the history of ancient climate change. These tools range from a simple hand lens that can be used to identify rocks and fossils in the field to highly specialized and expensive analytical equipment that delivers quantitative data on the nature of ancient climate.

Chapter 6: Climate History considers three periods of Earth history when the climate was so extreme that life on Earth nearly came to an end. This chapter also traces the evolution of Earth's climate from the hothouse of the Cretaceous world to the icehouse of today. The chapter covers the last 150,000 years of climate history in more detail, focuses on the origin of long and short-term climate cycles, and investigates the rate at which climate changes.

Part Four: The Impact of Climate Change: From Polar Bears to Politics

Our knowledge of the risks associated with climate change grows each year, but turning this knowledge into effective political action has not been easy.

Chapter 7: The Global Impact of Climate Change investigates the physical and environmental impacts of climate change over the past 150 years and the possible impact of continued warming on future climate. The chapter focuses on the social and economic impact of projected climate change, using case studies from around the world to illustrate the many dimensions of the climate problem.

Chapter 8: Politics and People considers how growing awareness of climate change made global warming an important political issue around the world, culminating in the formation of the United Nations Framework Convention on Climate Change and the Kyoto Protocol. This chapter considers the history of the Kyoto Protocol, reflects on our inability to make further progress with it, and illustrates the role of major stakeholders in the climate debate. The chapter stresses the major social, political, economic, and ethical issues involved in the climate debate and will stimulate discussion about science, society, and the role of the media in determining public opinion.

Part Five: Global Solutions: Managing the Crisis

At a time when global action to prevent climate change is more important than ever before, the world is increasingly distracted by an urgent demand for economic growth in the developing world and by the emergence of new geopolitical rivalries.

Chapter 9: The Energy Crisis introduces the energy crisis that is driven by population growth, and the urgent need to avoid damaging climate change. The chapter identifies energy poverty as a moral and ethical challenge for a world that wants to cut greenhouse gas emissions. Countries such as China and India still lag far behind the developed nations in per capita gross domestic product (GDP), and they need to make use of cheap and abundant coal reserves to generate enough power to support their economic growth. This chapter looks at all the major sources of energy available to meet this rising demand for energy, and considers how different priorities and changing government subsidies could encourage the more rapid development of clean, renewable energy technologies.

Chapter 10: Turning Knowledge into Action looks for ways to balance the competing priorities of economic growth and emissions reduction in a world where rapid population growth is expected to continue well into this century. The chapter considers whether it is possible to minimize greenhouse gas emissions without harming economic development and still prepare the world to adapt to the inevitable climate change that is already locked into Earth's climate system.

There is an immense amount of useful NASA, NOAA, and USGS original data available to students.

Acknowledgments

I want to acknowledge the help and assistance of many people who helped me write this text. To Don Beville formerly of Pearson, for suggesting that I take on the project in the first place. To my dean, James Narduzzi, and colleagues at the University of Richmond, for help and understanding. To Geography/GIS/Meteorology Editor Christian Botting and Senior Project Editor Crissy Dudonis for their humanity, constant encouragement, and help with making this book the best it could be. I would also like to thank their colleagues at Pearson, including Editorial Assistant Bethany Sexton, Assistant Editor Sean Hale, Media Producer Tod Regan, Senior Marketing Manager Maureen McLaughlin, and Senior Marketing Assistant Nicola Houston. I would also like to thank the many production staff at Pearson and elsewhere who helped produce the book, including Managing Editor Gina Cheselka, Production Liaison Connie Long, *International Mapping* Project Manager Kevin Lear, *Element* Associate Director, Full Service Heidi Allgair, and Photo Researcher Christa Tilley.

I am grateful to the following reviewers for their feedback during the book development; they were immensely helpful in focusing and improving the text:

Mark Boardman, *Miami University*;
Wolfgang H. Berger, *University of California: San Diego*
Carsten Braun, *Westfield State College*
Jeffrey Bury, *University of California: Santa Cruz*
Greg Carbone, *University of South Carolina: Columbia*
John Chiang, *University of California: Berkeley*
Dawn Ferris, *The Ohio State University*
Tim G. Frazier, *University of Idaho*
Ryan Zahn Hinrichs, *Drew University*
Peter Jacques, *University of Central Florida*
Bruce R. James, *University of Maryland*
Jean Lynch-Stieglitz, *Georgia Tech*
Scott A. Mandia, *Suffolk County Community College*
Patricia Manley, *Middlebury College*
Heidi Marcum, *Baylor University*
Isabel Montanez, *University of California: Davis*
Dave Robertson, *University of Missouri*
Jame Schaeffer, *Marquette University*
Marshall Shepherd, *University of Georgia*
Richard Snow, *Embry Riddle Aeronautical University*
Robert Turner, *Skidmore University*
Stacey Verardo, *George Mason University*.
A special thanks to Thompson Webb, *Brown University*, for his incredibly helpful accuracy reviews.

But most of all, I want to thank my wife, Michele Cox, and daughter, Sarah, who had to live with me as I spent far too many weekends working inside, when I should have been outside playing with them instead.

David Kitchen
University of Richmond
kitchenclimatebook@gmail.com

About the Author

David Kitchen earned a B.Sc. and Ph.D. in geology from Queen's University–Belfast. After working for two years as a petroleum geologist in the North Sea, he was appointed as lecturer in the University of Ulster where he taught Earth Sciences and worked with research administration and development from 1981 to 2001. He also spent many happy summers working as adjunct professor on field courses for the UK's Open University. In 2001, Dr. Kitchen moved to the United States to work at the University of Richmond in Virginia. As associate dean and associate professor he leads a team developing academic, professional, and lifelong education programs and teaches environmental sciences, with a focus on climate change. He has served as the university representative on the National Council of Environmental Deans and Directors, as environmental fellow for the Associated Colleges of the South, as coordinator of university environmental programs, and as coordinator of the Environmental Studies course team, and he serves on the university's Sustainability Working Group and Environmental Awareness Group. In collaboration with the National Council for Science and the Environment, Dr. Kitchen was awarded a NASA climate change education grant to help develop new online modules in global climate change.

PART 1

THE EVIDENCE: IS THIS NORMAL?

CHAPTER 1: SO WHAT'S UP WITH THE WEATHER?

CHAPTER 2: THE EVIDENCE: OBSERVING CLIMATE CHANGE

Parts of New Jersey and New York City were devastated in late October 2012 by a 4 meter (13 foot) storm surge that was driven ahead of Hurricane Sandy. It is not yet clear if climate change will affect the frequency and intensity of future hurricanes, but increasing global sea level will enhance the risk of all future storm surges.

Earth's climate is changing.

Climate records are being set across the world as heat waves, droughts, floods, snowstorms, tornados, and hurricanes impact the lives of millions of people. There are now clear signs of changing climate on every continent across the globe. As the human and financial costs of extreme weather rise we must understand why global climate is changing and work hard to mitigate its worst impacts. Geological evidence has uncovered a record of continuous and natural climate change, but the extreme weather we observe today is different because none of the usual forces that drive climate change appear to be responsible. Recent climate change appears to be the consequence of a precipitous increase in the emission of heat-trapping greenhouse gases from deforestation, industry and agriculture. Levels of carbon dioxide in the atmosphere have risen by 40% since preindustrial times, and are now higher than at any time during the previous 800,000 years. Computer models project that a level of emissions this high will accelerate climate change until we reduce emissions from all sources. Climate change is a global problem that requires a global solution. Chapter 1 introduces important concepts about weather, climate, and climate change and Chapter 2 presents physical evidence that regional climate is changing rapidly across the globe.

Hurricane Katrina was a powerful Category Five hurricane with winds of 160 mph. Each year hurricanes transfer enormous amount of energy from the tropics to higher latitudes, helping regulate Earth's climate. There may be a link between the intensity of hurricanes and higher ocean temperatures, driven in large part by climate change and global warming.

CHAPTER 1

"So, What's Up with the Weather?"

Introduction

Almost every day, some claim or counterclaim about climate change and global warming hits the headlines. For most of us this is very confusing. Some people consider every storm, drought, heat wave, and record temperature further evidence for global warming; others dismiss these same events as natural climate variation. Isn't everyone looking at the same data? How can the same facts be interpreted in such different ways?

When the interpretation of scientific data makes it imperative that we change the ways we create and use energy, the origin of these data becomes controversial and politically-charged. Before we undertake any action that will have a long-term impact on society, we need to understand how and why global climate changes. The geological record shows us that climate change is a normal part of Earth's history, and while most scientists conclude that recent changes in global temperature and climate are due to the emission of human-made (anthropogenic) heat-trapping **greenhouse gases,** a small but vocal minority disagrees. In order to project the future impact of anthropogenic emissions accurately, we must understand the different factors that have driven climate in the past and how small changes in global temperature will affect Earth's climate today.

Learning Outcomes | *When you finish this chapter you should be able to:*

- Discuss how global temperature has changed over geological time
- Identify some of the principal factors that control global climate over different timescales
- Describe the recent historical temperature record
- Determine possible causes of recent changes in global temperature
- Understand how climate models can be used to project climate change
- Evaluate the possible impact of human activity on the atmosphere

Weather and Climate

Before we begin to discuss climate change it is important to discuss the difference between weather and climate. *Weather* is experienced from day to day. One day can be hot and dry and the next cool and wet. *Climate*, on the other hand, can be related to the statistical probability that any day during the year will be similar to the same day in previous or following years. In Virginia, the winter temperatures are often close to freezing, but the weather on any one day is very unpredictable. The temperature in January can reach as high as 25°C (78°F), but based on weather records, it is very likely that cool air, rain, sleet, and ice will soon return because the climate of Virginia is characterized by cool, wet winters and hot, humid but relatively dry summers. What is the climate like where you live? Is there much variation in weather from year to year?

CHECKPOINT 1.1 ▶ In your own words, describe the difference between weather and climate.

(a)

Is the Climate Changing?

Over the past 100 years, global temperature has been rising at a rate that is slow in terms of a human lifespan but rapid enough to worry climate scientists (**Figure 1.1b**). Climate varies naturally—within limits. Occasional extreme weather events and temperature records do not prove that climate is changing, but scientists have discovered that the past few decades have experienced a statistically significant increase in the number of record heat waves and a decrease in the number of record cold spells (**Figure 1.1a and Figure 1.1b**). (Chapters 2 and 7 will investigate this evidence in more detail.) It is clear that Earth's climate is changing, but we need to find out why and what is likely to happen in the future.

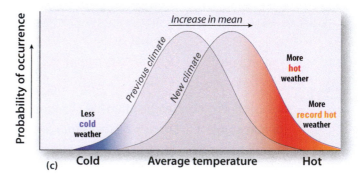

(b)

(c)

How Stable Is the Climate?

Scientists have created a historical temperature record based on tree rings, ice cores, sediment cores, and other **proxy** climate data. This record indicates that global average temperature has been comparatively stable since the last ice sheets retreated from Europe and North America around 9,000 years ago. Following a postglacial temperature maximum around 8,000 years ago, a slow global cooling trend developed that continued until the end of the 19th century

▲ **Figure 1.1:** (a) Forest fires will become more common, at great cost to both human and animal communities. As global temperature rises, evaporation increases and droughts intensify. (b) Annual Average Global Surface Temperature change from 1880–2008 relative to the average temperature of the period 1901–2000 (known as the temperature anomaly). The vertical gray bars indicate the range of uncertainty for each data point. (c) The probability of temperature extremes changes with climate. As the climate shifts to warmer temperatures, there is less chance of a record cold spell and a much greater chance of a record heat wave.

(Figure 1.2). During that time, shifts in the regional pattern of climate still occurred over many parts of the world, such as in Africa, Europe, and North America, and some of these shifts had profound impacts on human settlement. Looking back over the past 5,000 years, we see that modern civilizations have generally grown and prospered under conditions well suited to agriculture, the generation of excess economic capacity, and the growth of commerce and trade, but the threat of climate change was always present. The record is full of examples of human settlements that were devastated by water shortages, famine, and disease.

> **CHECKPOINT 1.2** ▶ Explain how we know that global climate has been relatively stable over the past 8,000 years.

Historical Climate Change

The recent history of climate change is well documented by historical records. A period of warming in parts of the Northern Hemisphere during the 10th to 12th centuries led to prosperity, population growth, and the expansion of settlement. The Vikings arrived in Greenland and North America at this time, and good harvests in Europe led to the expansion of commerce and trade. The records show that this period of prosperity came to end during the 17th and 18th centuries when the climate started to cool, with the devastating human consequences of war, famine, and disease.

There are now many different published reconstructions of global temperature that include this period of time (Figure 1.3). These data, predominantly from measurements on land in the Northern Hemisphere, confirm the existence of a warmer period around 1,000 years ago known as the **Medieval Climate Anomaly (MCA)** and a period of distinct regional cooling that started around 400 years ago known as the **Little Ice Age (LIA).** This is a clear record of natural variation in climate that is not the result of human activity. Even so, the implications are disturbing: If small natural changes in global temperature have such a large impact on society, what can we expect to happen as a consequence of a larger increase in global temperature projected by the end of the 21st century?

> **CHECKPOINT 1.3** ▶ Why is the historical record of climate change alarming to some scientists?

The Hockey Stick One of the most convincing reconstructions of historical climate change came in a groundbreaking paper published in 1998 (Figure 1.4). This paper by climate scientist Michael Mann and colleagues analyzed proxy data from a number of sources in both hemispheres. When plotted on a simple graph, the spatial pattern of these data resembled a hockey stick, with a long shaft reflecting prolonged and gradual global cooling over the past 1,000 years and a pronounced upturned face created by rapid warming in the 20th century. This **hockey stick graph** was controversial when it was first published, because it suggested that the MCA and LIA were transient, hemispheric, or regional phenomena and supported the growing hypothesis that recent warming was due to human activity. There is no longer any scientific basis for this controversy, as the data that produced the first hockey stick graph have been confirmed many times by other researchers using different tools to determine past temperature. It is clear that recent changes in global temperature are unprecedented in the historical record.

(a)

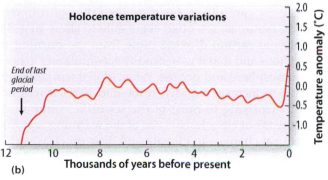

(b)

▲ **Figure 1.2:** (a) The recent and rapid melting of Arctic Sea ice marks the end of a prolonged period of global cooling that lasted for thousands of years. In this enhanced satellite photograph, widespread seasonal (winter) ice is shown in light gray and thicker multiyear ice in white. (b) Temperature change since the end of the last glaciation compared to the mid-20th century average temperature. Note how temperature over the last 8,000 years has trended towards lower global temperatures. This analysis is the average of eight studies using different methods that determine air temperature.

1.1 PAUSE FOR... THOUGHT | *Why is a rise in global temperature of just 1°C so important?*

It may help to think of this in terms of your own body temperature. Similarly to Earth, you gain energy from your internal system and surroundings, and you lose energy from the surface of your skin. You remain healthy as long as you maintain an average core temperature very close to 37°C (98.6°F). What would happen to you if your temperature increased or decreased by just 5°C (9°F)?

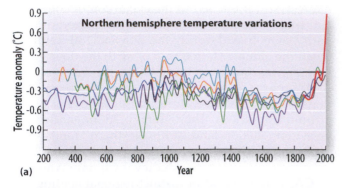

(a)

(b)

(c)

▲ **Figure 1.3:** (a) Temperature change over the last 2,000 years compared to the average temperature in the mid-20th century. The red line at the far right indicates actual 20th century temperature measurements; the other colored lines represent estimates of historical temperature by different authors based on a variety of temperature proxies such as tree rings, corals, and ice cores. These data are biased towards northern hemisphere land temperatures and show evidence of both a Medieval Climate Anomaly when Europe was warm and prosperous and a Little Ice Age when winters in Europe and the United States were much colder on average than today. Data from many studies using different methods to determine past temperature were used to construct this diagram. The red line that starts in the middle of the 19th century is the instrumental temperature record. The wide range of results, particularly for older data, gives some idea of the uncertainties involved in determining past temperature, but there is still remarkable correspondence between most of these data, and the underlying trend is consistent. (b) A scene from Germany where buildings constructed during prosperous medieval times survive to the present day. (c) During the Little Ice Age many rivers in Europe froze so deeply that frost fairs could be held on the thick ice over the winter.

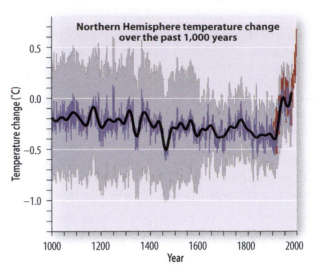

▲ **Figure 1.4:** A more detailed graph of temperature change over the last 1,800 years compared to a 1961–1990 average. This is the infamous "hockey stick curve" that was constructed with data collected from many different sources and is discussed in more detail in Chapter 4. The red line at the right marks the modern instrumental record, the blue line the actual data points, the black line a running average, and the gray areas an estimate of the statistical uncertainty of the data.

CHECKPOINT 1.4 ▶ Why is the evidence for recent rapid climate change no longer in any doubt?

Recent Climate Change

The human impact on global temperature emerged from climate data around the middle of the 20th century, when industrial production reached a peak following the end of World War II. Scientists discovered that carbon dioxide levels in the atmosphere were increasing and understood that this could raise global temperature by trapping more of Earth's outgoing energy. By carefully measuring the ratios of different carbon **isotopes,** scientists were able to show that the additional carbon dioxide came from the burning of fossil fuels and that it was steadily accumulating in both the atmosphere and oceans. When global temperatures started to increase rapidly in the mid-1970s it was not at all surprising to most of the scientific community, but it was not clear how much of the warming was due to natural changes in climate and how much was due to human activity.

CHECKPOINT 1.5 ▶ How did scientists determine that the increase in carbon dioxide in the atmosphere and oceans comes mostly from the burning of fossil fuels?

Temperature Records The instrumental record of temperature change is disappointingly limited. The longest record of surface temperature comes from central England, where a continuous record stretches back to

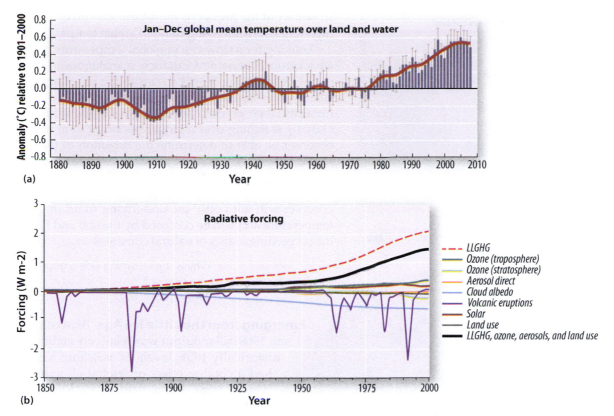

▲ Figure 1.5: Global Temperature and Radiative Forcing (natural and human factors that affect the flux of energy at the top of the atmosphere and act to change global climate) (a) The average temperature measured each year from 1880 compared to the average annual temperatures from 1901 to 2000 (this is known as the temperature anomaly). The vertical gray bars indicate the range of uncertainty for each data point. (b) The impact on radiative forcing (and thus global warming and cooling) of different parameters that affect the intensity of solar radiation and the influence of greenhouse gases (Huber and Knutti, 2011). Compare the impact of carbon dioxide to the other parameters recorded, and note the strong cooling affect of volcanic activity during the Little Ice Age in the mid–19th century.

1659. Reliable regional data only began to be recorded in the 1850s, with the development of modern trade and a little help from the British Empire. Truly global information became available only after the 1970s, with the launch of the first of a number of weather satellites.

Recent Observations on Land Recent data records show that global atmospheric temperature over land and oceans has risen by as much as 0.6°C (1.08°F) on average over the past 100 years (**Figure 1.5a**). In 2005, average land surface temperatures[1] exceeded 1°C (1.8°F) above the 1901–2000 annual average, a general baseline that many climate scientists have adopted. This may not sound like much of a change, but this global figure is amplified within Earth's climate system and translated regionally into much larger changes in temperature. To put this in perspective, a further rise in land–ocean air temperature of just 1°C (1.8°F) would

put atmospheric temperatures at their highest point in 1 million years, but we are now facing a very real possibility that global temperature could rise by as much as 4°C (7.2°F) by the end of the 21st century. A temperature increase of this magnitude would have a devastating impact on global ecosystems and on human society.

Recent Observations in the Oceans The world's oceans have absorbed as much as 90% of the excess heat produced by recent global warming, and a clear warming trend has been identified over the upper 700 meters (~3,000 feet). The average water temperature is now over 0.49°C above the 20th-century average. This may not sound like much of a change, but the oceans are a huge reservoir of heat, and if all this additional energy were suddenly released into the atmosphere, it would be enough to increase air temperature by a much as 22°C (40°F) (**Figure 1.6**).

---

[1] Land surface measurements alone are generally higher than combined land–ocean data because air over the oceans is strongly influenced by ocean temperatures, and these are moderated by ocean circulation and the high heat capacity of water. (It takes more energy to heat 1 gram of water than 1 gram of rock by 1°C.)

CHECKPOINT 1.6 ▶ How much of the additional heat added to Earth's climate system is locked up in the oceans?

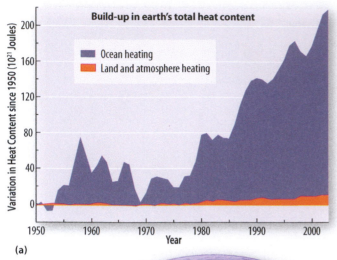

(b)

smooths out the impact of short-term natural cycles and highlights the underlying trend in global temperature.

Looking for a slow rise in global temperature against a background of natural variation is analogous to sitting on the beach trying to spot the incoming tide. Large and small waves lapping against the shore make it difficult for a casual observer to discern how the tide is moving. Only by watching over a longer period of time will the observer be able to determine the direction of the tide. In the same way, a gradual increase in global temperature emerges from the background of natural climate variation only after years of careful observation. A casual observer will not notice the underlying trend in global temperature and will be confused by the ebb and flow of the succeeding waves of natural climate change.

CHECKPOINT 1.7 ▶ Why is the incoming tide a good analogy for recent climate change?

Emerging from the Little Ice Age Between 1880 and 1909, solar output was relatively stable, and historically high levels of volcanic activity had a cooling effect on global climate (see **Figure 1.5b**). From 1910 to 1945, sunspot activity increased and volcanic activity waned; the atmosphere started to warm as Earth finally emerged from the Little Ice Age. It is not completely clear if this warming was entirely free from human influence, but **climate models** can reproduce this warming using only natural factors such as an increase in the intensity of solar radiation and a reduction in the intensity of global volcanism.

Cooling Off Again During the period 1946 to 1975, and especially between the late 1940s and early 1950s, the warming trend from the first half of the 20th century flattened and may have even reversed for a short time. Most scientists attribute this to a combination of **solar cycles,** volcanism, and air pollution from industrial **aerosols.** At the time, the media speculated about whether Earth was about to enter a new ice age, but then, as now, most scientists understood that global temperatures were likely to rise.

CHECKPOINT 1.8 ▶ Why did global temperature stop rising during the middle of the 20th century?

▲ **Figure 1.6:** (a) The global heat content of the oceans has increased since 1955. Note how much energy is stored in the oceans relative to the land and atmosphere. Units of energy are in Joules. (b) Much of the mainstream argument against climate change focuses on air temperature, but this overlooks the overwhelming influence of the vast oceans that cover more than 70% of Earth's surface.

A Century of Warming

Estimates of global atmospheric temperature over land and ocean from 1880 to the present show significant variation from year to year due to the impact of short-term solar, volcanic, atmospheric, and oceanic events (see Figure 1.5a). The temperature curve has a more even profile when averaged over five years because this

A Human Element Emerges Global warming re-emerged after 1975, and for the first time, climate models were not able to reproduce these changes in temperature by using only natural factors. A significant contribution from heat-trapping greenhouse gas emissions was required to match the output of their

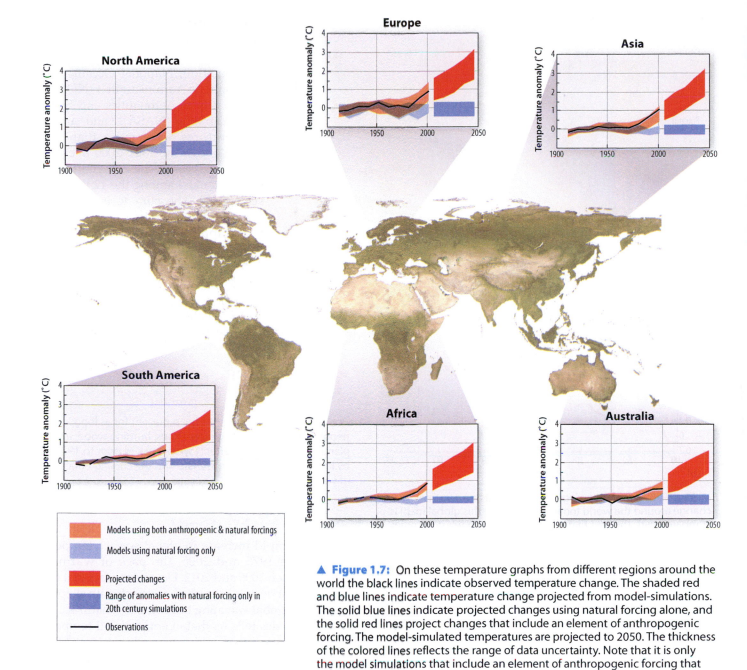

▲ **Figure 1.7:** On these temperature graphs from different regions around the world the black lines indicate observed temperature change. The shaded red and blue lines indicate temperature change projected from model-simulations. The solid blue lines indicate projected changes using natural forcing alone, and the solid red lines project changes that include an element of anthropogenic forcing. The model-simulated temperatures are projected to 2050. The thickness of the colored lines reflects the range of data uncertainty. Note that it is only the model simulations that include an element of anthropogenic forcing that match observed temperature change.

models with observations. This was the first real evidence that the burning of fossil fuels was having a real impact on climate change (**Figure 1.7**).

Subsequent observations, compiled from ground stations, **weather balloons,** and **weather satellites,** confirm that Earth is warming over most of its surface, but warming is uneven. Compared to the average temperature from 1901–2000, surface temperature over the land is increasing much faster than over the oceans (0.85°C vs. 0.37°C) and while much of the Northern Hemisphere at high latitudes is getting warmer (the Arctic is warming at twice the global average rate), parts of Antarctica may be getting colder.

CHECKPOINT 1.9 ▶ What are some of the tools that climate scientists use to understand ancient climate change?

The Hadley Centre Coupled Model, version 3 (HadCM3) climate model shows that the Northern Hemisphere experienced the greatest amount of warming between 1995–2004 (but note the Antarctic Peninsula on **Figure 1.8a**), and projects a further rise in temperature in excess of 5°C–6°C (9.0°–10.8°F) over high northern latitudes between 2070–2100 if heat-trapping greenhouse gas emissions are not curtailed (**Figure 1.9**).

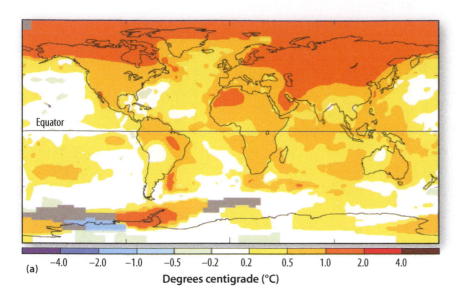

(a)

The Warmest Decade on Record

The year 2010 was the hottest since 1880, according to the **World Meteorological Organization,** National Oceanic and Atmospheric Administration (NOAA), and National Aeronautics and Space Administration (NASA). Land surface temperatures were more than 1°C (1.8°F) above the 1901–2000 average, and every year in the 21st century so far has ranked among the 12 warmest since 1850. The climate is changing rapidly, and scientists suggest that heat-trapping greenhouse gases from industry, agriculture, and deforestation appear to be the most likely cause.

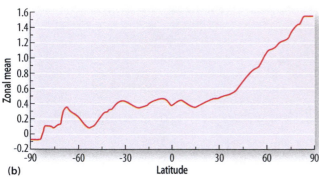

(b)

▲ **Figure 1.8:** (a) When degree of observed temperature change around the world today (illustrated in this global map) is compared with the best fit data from climate models it becomes clear that natural variation in climate alone cannot explain the observed change in global temperature. Note the warming indicated around the Amazon Basin in South America and the Antarctic Peninsula. (b) Temperature is not rising uniformly around the globe. The Northern Hemisphere, and especially the Arctic region, are warming much faster than the Southern Hemisphere.

CHECKPOINT 1.10 ▶ Are the changes in atmospheric temperature observed over the past two decades unusual?

1.2 **PAUSE FOR... THOUGHT** | *Why is there no evidence of anthropogenic warming prior to 1975?*

Anthropogenic warming has been present for as long as people have been cutting down trees, burning wood, and planting crops, but the effects were masked prior to 1975 by large natural swings in climate and the cooling effect of volcanic and industrial aerosols.

Has Warming Stopped?

Following a rapid increase in global surface temperature between 1975 and 2005, the pace of warming slowed between 2005 and 2012. Despite the fact that this is the warmest decade on record, many climate skeptics suggest that global warming has come to a natural end. But basic physics tells us that average global temperature will increase by at least 1.2°C (2.2°F) when we double the level of CO_2 and other greenhouse gases in the atmosphere, and climate models project an actual increase in global temperature by as much as 2°C to 4.5°C (3.6°F to 8.1°F) by the end of the 21st century due to the multiplying effect of changes

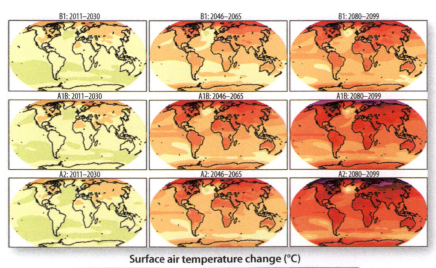

Surface air temperature change (°C)

0.0 0.5 1.0 1.5 2.0 2.5 3.0 3.5 4.0 4.5 5.0 5.5 6.0 6.5 7.0 7.5

◀ **Figure 1.9:** Multi-model projections of annual mean surface air temperature change in °C for different future emissions scenarios that depend on a combination of economic and social factors discussed later in the text. Anomalies are relative to the average of the period 1980 to 1999. The current rate of emissions project a future path between the moderate and high emissions scenarios.

► **Figure 1.10:** The Cretaceous hothouse world. This is an artist's rendering of how large parts of the world looked during the Cretaceous Period. This was a time when the dinosaurs were at the pinnacle of their success, and the landscape was changing as the first flowering trees, plants, and early grasses started to appear. It was a warm world that was very biologically productive with high levels of animal and plant diversity.

in Earth's climate system that are described in detail in Chapters 3 and 4. Regrettably, global warming has not stopped.

CHECKPOINT 1.11 ▶ Why are scientists confident that global warming has not stopped over the past decade?

1.3 **PAUSE FOR... THOUGHT** | *Why do you think that the topic of climate change has become so controversial?*

Is climate change the only controversial area of science? Can you think of any common factors that link climate change with other areas of scientific controversy?

Lessons from the Deep Past

The geological record shows that climate change is a natural part of Earth's history. During the **Cretaceous Period**, over 65 million years ago, fossil, isotope and geochemical evidence suggests that global temperatures were well above current norms. This was a **hothouse world** where vast forests extended toward an ice-free pole in Antarctica, the dinosaurs were at the pinnacle of their success, and the first flowering trees, plants, and early grasses were starting to appear (**Figure 1.10**).

The **icehouse world** we live in today began around 2.5 million years ago in the Northern Hemisphere, and as much as 30 million years ago in Antarctica. Unique rock formations and layers of glacial sediment show us that ice sheets have advanced from the Arctic to reach as far south as New York and the south of Ireland in the recent geological past (**Figure 1.11**). The periodic advance and retreat of these ice sheets has affected the course of human evolution, and the entire history of human civilization has occurred within a geologically brief period of time during the most recent **interglacial stage** when the ice sheets retreated to higher latitudes. This icehouse world is our home, and in the absence of human activity to upset the balance of global climate, we will stay this way for a long time to come.

CHECKPOINT 1.12 ▶ How do we know that the poles were warmer during the Cretaceous Period and that ice has periodically covered New York since the start of the Pleistocene Epoch 2.5 million years ago?

1.4 **PAUSE FOR... THOUGHT** | *Climate has had a major impact on human evolution.*

At the end of the last glacial maximum, modern humans who had evolved in much warmer southern climates were able to displace *Homo neanderthalensis*, a human species much better adapted to cold climates. What changes in human evolution might occur over the next 100,000 years if the world continues to warm?

The long history of global cooling that spans the time between the hothouse of the Cretaceous and the icehouse of today is recorded in the concentration of Oxygen 18 (^{18}O), a heavy isotope of oxygen found in the shells of marine organisms (**Figure 1.12**). These data indicate that ocean temperatures declined steadily from a maximum during the **Eocene Epoch** to a low point during the **Pleistocene Epoch,** with notable periods of very rapidly declining temperature during the Eocene Epoch and late **Miocene Epoch.** We will look again at the geological history of climate change in Chapters 5 and 6 and consider what this can tell us about how Earth's climate is likely to change in the future.

CHECKPOINT 1.13 ▶ According to oxygen isotope data, what happened to global temperature between the Eocene and Pleistocene Epochs?

Major Factors that Affect Climate Change

Many natural factors contribute to climate change and are capable of producing rapid changes in global temperature. The overall process is complex and is explored in more detail in Chapters 5 and 6. Even as global temperatures increase over the next few decades, certain regions may experience short-term cooling as the pattern of circulation in the oceans and atmosphere changes. The following section introduces some of these factors and considers their impact on recent climate change. We will come back to look at them again in much more detail in Chapters 3 and 4.

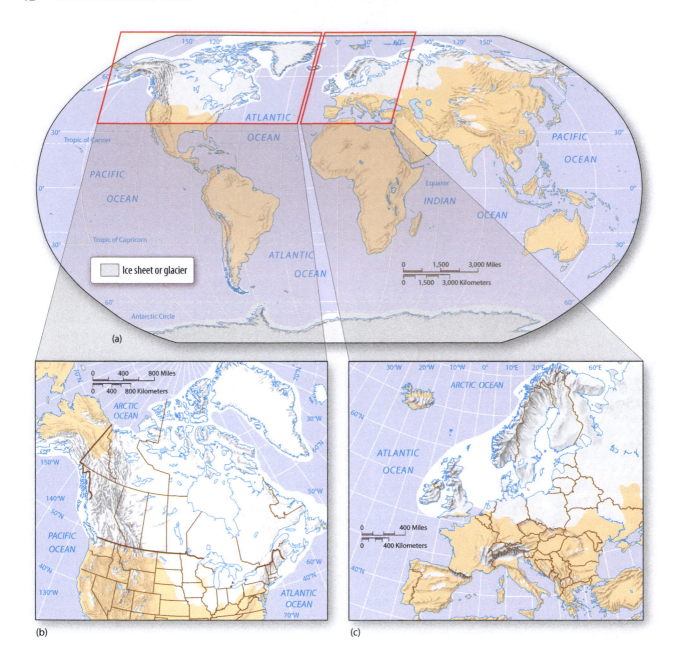

▲ **Figure 1.11:** Ocean temperatures declined steadily from a maximum during the Eocene Epoch to a low point (a) during the Pleistocene Epoch (approximately 26,000 to 19,000 years ago) when ice sheets (b) advanced as far south as New York and (c) almost as far as London.

Radiative Forcing

Radiative forcing refers to an imbalance that develops at the top of the atmosphere between the amount of incoming radiation (energy) from the sun and outgoing radiation (heat) from the Earth. These data are measured by instruments on satellites and show that Earth is very close to radiative balance. This balance is critically important, and it is maintained by many complex processes that move, transform, and radiate energy within Earth's climate system—the atmosphere, hydrosphere, biosphere, cryosphere, and lithosphere (**Figure 1.13**). Even a very small change in the amount of energy moving though this system will have a profound impact on global climate.

The Sun The Sun is by far the most important factor that controls radiative forcing. The radiant energy it produces has been nearly constant for millions of years, although it cycles within well-defined narrow limits due to the growth and decay of sunspots.

Albedo Earth's natural reflectivity (**albedo**) changes with the seasons and plays a critical role in regulating Earth's radiative balance. Snow, ice, clouds, and aerosols

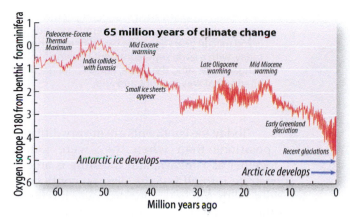

▲ **Figure 1.12:** Temperature change since the end of the Cretaceous Period, estimated from the oxygen isotope chemistry of benthic foraminifera. Ocean temperatures have declined steadily from a maximum during the Eocene Epoch to a low point during the Pleistocene Epoch, with notable periods of very rapidly declining temperature during the Eocene and late Miocene Epochs.

are all very effective at reflecting incoming solar radiation directly back to space, while oceans, forests, and dark rocks and soils are able to absorb most incident radiation.

Radiative Feedback

The net flow of radiation from the top of the atmosphere is determined by a balance between forces that drive climate—such as solar radiation, volcanic activity, and anthropogenic greenhouse gases—and forces that modulate climate such as changes in ice cover, cloud cover, atmospheric water vapor, erosion rates, and the release of greenhouse gases from the land and oceans. Radiative forcing is affected when changes in one part of Earth's climate system are either augmented or diminished by related changes in another part of the system. An example of augmentation (**positive feedback**) is when melting Arctic sea ice with a high albedo is replaced by ocean water with a low albedo. Because the darker water can absorb more of the Sun's incident radiation, the environment warms even further. An example of diminishing (negative) feedback occurs

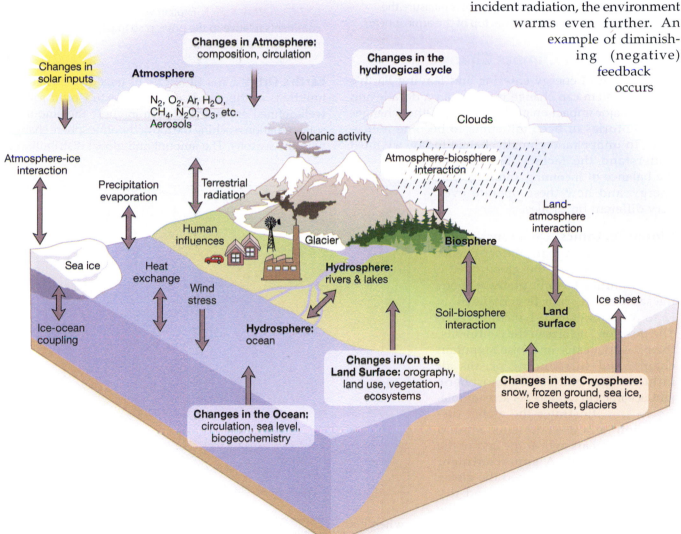

▲ **Figure 1.13:** Some of the main components of Earth's climate system. The radiative balance between incoming radiation from the Sun and outgoing heat from Earth is critically important for life on Earth. This diagram indicates the complexity of the interrelated processes that transform and move energy within Earth's climate system.

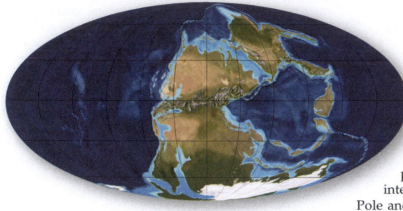

◀ **Figure 1.14:** An artist's rendering of the Permian world from space. The Permian world (~260 million years ago) was very different from today, with a large supercontinent known as Pangaea spanning the globe, and a continental interior that was very hot and arid.

when higher surface temperatures lead to increased evaporation and the formation of certain kinds of clouds that increase the amount of solar radiation reflected out to space and diminish the impact of rising temperature.

CHECKPOINT 1.14 ▶ How do scientists measure the degree of radiative balance at the top of the atmosphere?

Timescales of Climate Change

The amount of energy entering and leaving Earth's climate system can change in less than a decade and have a major impact on global climate. Other changes take centuries or even millennia to become noticeable. To understand climate change today, we must understand the factors that control the balance of incoming and outgoing energy, and how these operate over very different timescales.

Global Tectonics On a timescale of tens of millions of years, the continents drift slowly across the globe on the back of shifting **tectonic plates** at about the same rate as your fingernails grow (2 to 3 centimeters [~1 inch] a year). The position and size of the continents, their vast plateaus, immense mountain chains, and surrounding oceans all influence the pattern of atmospheric and oceanic circulation. As an example, **Figure 1.14** illustrates the global distribution of the continents around 250 million years ago, at the end of the **Permian Period**. A **supercontinent** known as **Pangaea** spanned the globe, and large areas became hot and dry when they became isolated from the oceans. This was a period in Earth's history when life on Earth was very stressed, and the extinction rate was exceptionally high.

CHECKPOINT 1.15 ▶ How many years would it take a continent to move 1,000 kilometers (~600 miles) at a rate typical of the movement of the Earth's tectonic plates?

Today the continents are in very different positions than when they were joined together as Pangaea, and moisture is able to penetrate into all but the deepest continental interiors. With Antarctica located over the South Pole and the North Pole surrounded by landmasses, warm ocean currents are not able to reach high latitudes, and the poles have become exceptionally cold over the past 30 million years (**Figures 1.15** and **1.16**). Continents can build both refrigerators and ovens. At the moment, we are in a refrigerator, and it is not about to change naturally.

CHECKPOINT 1.16 ▶ Explain why the position of the continents relative to the poles is so important in determining global climate.

Earth's Orbit On a timescale of tens of thousands to hundreds of thousands of years, Earth's orbit follows well-defined cycles that determine how the amount of solar radiation reaching the top of the atmosphere changes with the seasons. The amount and global distribution of

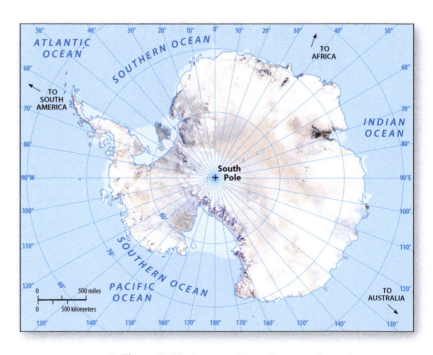

▲ **Figure 1.15:** Snow and ice will accumulate when a continent is sitting on the South Pole. This increase in albedo reflects most summer incident radiation directly out to space, keeping the continent very cold. During the winter, when there is no sunlight, the continent also blocks the flow of warm ocean currents towards the pole, allowing temperatures to plunge even deeper. The position of Antarctica over the South Pole is one of the main reasons Earth is still in the grip of a major ice age.

◄ **Figure 1.16:** Note how the continents of North America, Europe, and Asia surround and isolate the pole from the open ocean and restrict the flow of warm equatorial waters. This helps to keep the pole cold, even during the summer months when there is sunlight 24 hours a day.

The Oceans On a timescale of hundreds of years, deep circulation in the oceans drives a global flow of water known as the **great ocean global conveyor** that buffers and delays the impact of climate change by isolating both heat and greenhouse gases from the atmosphere for long periods of time. The conveyor starts where cold saline waters in the Arctic descend to the deep ocean and then circulate around the globe before returning to high latitudes as warm surface currents that heat the atmosphere and moderate the climate (Figure 1.18).

this energy waxes and wanes with small changes in the tilt of Earth's axis of rotation (**obliquity**), the direction of that tilt (**precession**), and shape of the orbit (**eccentricity**). These changes are often called Milankovitch cycles, after Serbian scientist Milutin Milankovitch, who was one of the first scientists to quantify this effect. Milankovitch cycles can produce major swings in global climate when amplified by Earth's climate system, and they have influenced the advance and retreat of polar ice sheets over the past 2.5 million years (Figure 1.17). Understanding Milankovitch cycles allows us to predict the future of natural climate change. In the absence of human intervention, we can expect ice sheets to expand again sometime within the next 20,000 years.

CHECKPOINT 1.18 ▶ How does deep ocean circulation affect global climate?

Ocean–Atmosphere Interaction On a timescale of years to decades, interactive cycles involving the sun, volcanic activity, the atmosphere, oceans, clouds, aerosols, natural greenhouse gases, and Earth's albedo have a major impact on regional climate—with often disastrous consequences.

The **El Niño–Southern Oscillation (ENSO)** is a natural climate cycle that involves both the atmosphere and oceans. Every three to seven years, warm water that has gradually accumulated to form a deep pool in the eastern Pacific rises to the surface and flows toward

CHECKPOINT 1.17 ▶ In your own words, describe how orbital cycles can influence global climate.

▶ **Figure 1.17:** This diagram illustrates the impact of orbital cycles on global temperature. Small variations in the tilt of Earth's axis (obliquity), the direction of that tilt (precession), and Earth's orbit (eccentricity) change the amount and distribution of energy reaching Earth from the Sun (solar forcing). These act together to produce major swings in the Earth's climate (bottom panel). The numbers in years on the right-hand side of the diagram refer to the observed lengths of particular Milankovitch cycles in thousands of years. Look at the record marked "solar forcing" that indicates how much energy arrives at the top of the atmosphere around 65° North in the summer. This is a critical factor that appears to have driven global climate change for at least the last 2 million years.

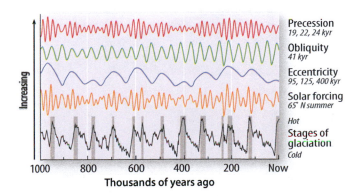

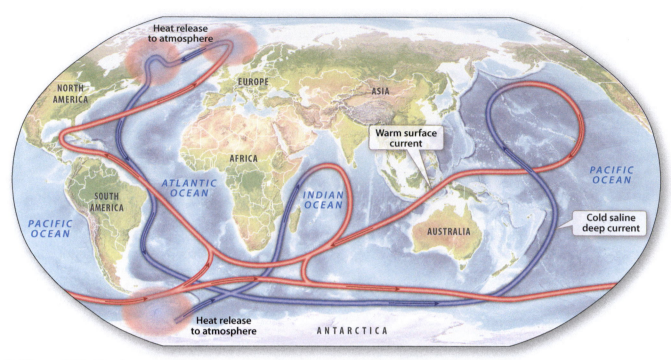

▲ **Figure 1.18:** The deep ocean conveyor. In this simplified diagram, cold saline waters descend into the deep ocean at the poles and then circulate around the globe, returning as warm surface currents that heat the atmosphere above them and moderate the climate in the northern Pacific and Atlantic Oceans.

the east, bringing drought and soaring temperatures to some areas and torrential rain and floods to others. In California, for example, intense storms related to El Niño can lead to destructive flooding and loss of both life and property (Figure 1.19).

The **North Atlantic Oscillation (NAO)** is a natural cycle driven by changes in atmospheric pressure that control the path of strong winds in the upper atmosphere. Changes in the NAO can determine whether winters in the Northern Hemisphere are cold and dry or cool and wet (Figure 1.20).

The eruption of Mount Pinatubo in 1991 increased Earth's albedo and cooled the atmosphere by up to 0.5°C (0.9°F) for nearly five years (Figure 1.5b), as its towering eruption column propelled reflective sulfate aerosols and dust high into the upper troposphere and lower stratosphere (Figure 1.21).

▼ **Figure 1.19:** El Niño events have a major impact on weather across the globe. This flooding in California is a consequence of the intense rainfall that affects parts of the west coast of the United States.

▼ **Figure 1.20:** The North Atlantic Oscillation has a major impact on the weather of the northeastern United States, the North Atlantic, and much of northwestern Europe, including this flooding in England.

▲ **Figure 1.21:** The eruption column from a large volcano ejects dust and gas high into the stratosphere, where it can affect weather for as much as five years after the eruption has ended.

The impact of a single volcanic eruption is certainly dramatic and pronounced (**Figure 1.22**), but a wide range of other natural and anthropogenic activities also produce aerosols that can increase albedo and impact the climate. When combined, all these factors make global climate a complex and chaotic system that is very difficult to predict in the short term.

Working together in a complex and dynamic process of interaction, these factors have controlled how the climate has changed naturally over millions of years. Although much remains to be discovered, especially concerning the shorter cycles and their interaction, these processes are well understood by the scientific community and are reproduced accurately by computer models.

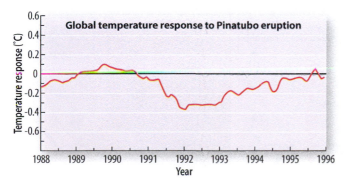

▲ **Figure 1.22:** The temperature of the atmosphere cooled by up to 0.5°C over 5 years following the eruption of Mount Pinatubo (the temperature curve is the five year running mean).

Recent data show that none of the factors that affect natural climate change have changed significantly over the past 10,000 years. The continents have moved only a few centimeters, orbital cycles keep us firmly in an interglacial period, and deep ocean circulation has changed little for the past 8,000 years. The pattern of short-term cycles is less well understood, but as more data on past climates are assembled, and as the resolution of climate models improves, there is no evidence that natural climate change can explain the significant changes in global temperature observed since the mid–20th century.

CHECKPOINT 1.19 ▶ Using the incoming tide as an analogy for recent climate change, how would you represent these decadal and shorter climate cycles?

1.5 PAUSE FOR... THOUGHT *Why would a series of cold winters and cool summers not prove that global warming is coming to an end?*

So many factors affect regional climate that it can be difficult to discriminate the underlying direction of global climate change from natural variation over a short period of time. Recent estimates suggest that a minimum period of at least a decade and preferably between 15 and 30 years is required to be certain of any underlying climate trend.

Greenhouse Gases

Since the start of the Industrial Revolution—and some would say for a lot longer—we have been slowly altering the composition of the atmosphere through the burning of fossil fuels, extensive deforestation, and agricultural practices. The gases produced both directly and indirectly by these activities have the potential to warm the atmosphere and impact the natural cycle of climate change. Heat-trapping greenhouse gases, discussed in depth in Chapters 3 and 4, trap some of Earth's radiant energy close to the surface and keep the troposphere much warmer than it would be if they were not present (**Figure 1.23**). Human activity has knocked the climate system out of equilibrium so that a small but critically important imbalance has developed between the amount of energy entering and leaving the Earth's climate system. This small radiative forcing, when magnified by feedback within the climate system, is enough to generate the global warming observed since the 1970s and cause Earth to warm faster and to higher temperatures than it has experienced in many thousands of years.

CHECKPOINT 1.20 ▶ If all the greenhouse gases were removed from the atmosphere in an instant, what would happen to Earth?

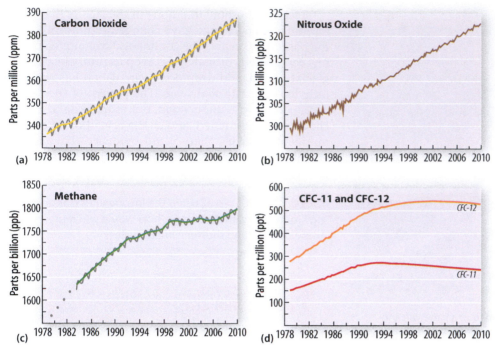

(a) 1978 1982 1986 1990 1994 1998 2002 2006 2010

(b) 1978 1982 1986 1990 1994 1998 2002 2006 2010

(c) 1978 1982 1986 1990 1994 1998 2002 2006 2010

(d) 1978 1982 1986 1990 1994 1998 2002 2006 2010

▲ **Figure 1.23:** The rapid rise in the concentration of important greenhouse gases over the last 35 years has had a major impact on climate. Units are in parts per million, parts per billion, or parts per trillion.

Climate Models

How do we project how much the climate is going to change in the future? Just how much of the recent change in global temperature is due to the impact of anthropogenic greenhouse gases? Climate scientists use climate models as critical tools both to understand present-day climate and project the future of global climate change.

It is important to consider these models in some detail because their projections lie at the core of the climate change debate. Early models were very limited in their ability to capture the true complexity of the climate system, but more recent **global circulation models (GCMs),** such as HadCM3, and the new Coupled Model Intercomparison Project (CMIP5) used by the IPCC are much more effective.

Scientists model climate change using computers that simulate the impact of changes in the flow of energy and mass within Earth's climate system. With over 500,000 lines of computer code and thousands of differential equations, a climate model is highly complex. The accuracy and precision of a model depends on the quality of its constituent algorithms and the data it uses. Many climate models represent the world as a spherical lattice of three-dimensional cells, each 100 kilometers (~60 miles) wide. They calculate how energy and mass flow between the cells and use these data to model patterns of global weather (**Figure 1.24**). As many as 30 layers of cells are stacked vertically to

▶ **Figure 1.24:** Climate models use computers to project how the exchange of mass and energy between different parts of Earth's climate system will affect future climate. By modelling the atmosphere and oceans using cells around 100 kilometers (60 miles) wide they give accurate global projections, but their ability to project changes on smaller scales is more limited.

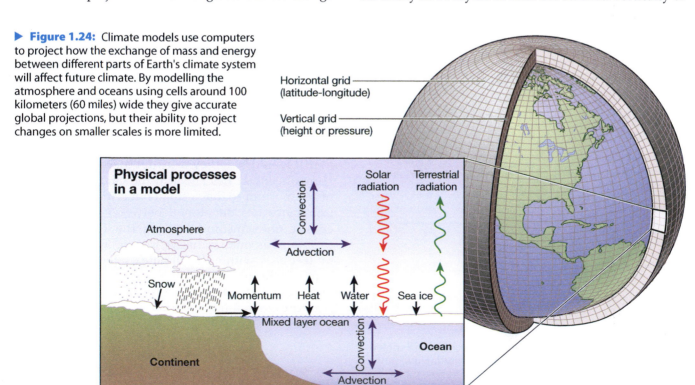

represent the atmosphere, with a further 30 or so layers stacked at depth to represent the oceans. At each step in the process (usually in increments of one hour), the model calculates how energy and mass are exchanged between each cell in the lattice. The size of each cell is limited by the computer power and time required for the calculations. The cells provide excellent resolution at a global level, but at the regional level—below 200 to 300 km^2 (~116 mi^2)—resolution degrades rapidly, like a pixelated photograph. **Regional climate models (RCMs)** have been developed that are able to focus computational power on areas as small as 50 km^2 (~20 mi^2) over a short period of time, but true global resolution at these scales is not yet available.

CHECKPOINT 1.21 ▶ Why are climate models less accurate at a regional level than at a global level?

Dealing with Model Uncertainty   There are three sources of uncertainty in climate models:

- *Internal variability:* The first source of uncertainty is the intrinsic internal variability that is built into the climate system. Even in the absence of radiative forcing, natural cycles in the Sun, atmosphere, and oceans will cause short-term variation in climate.
- *Model uncertainty:* There are many different climate models, and each one assigns slightly different weighting to the different parts of the climate system. Statistical problems associated with scaling and zonation effects related to the size and location of individual cells in climate models introduce further errors. Even a small difference in the amount of cloud cover, for example, could have a major impact on the modeled temperature of the troposphere.
- *Scenario uncertainty*: This type of uncertainty comes from the difficulty of projecting the future level of emissions from economic models.

On the timescale of one or two decades, internal variability and model uncertainly dominate, but move beyond 50 years, and scenario and model uncertainty start to dominate. Beyond 80 to 100 years, scenario uncertainty becomes strongly dominant, as economic and technological forecasts become highly speculative.

Given these factors it is clear why Earth's future climate is difficult to predict. Global climate is a complex, nonlinear system that is very difficult to predict in the short term (1 to 30 years). Over the medium term (30 to 100 years) these natural irregularities tend to even out statistically and allow much more accurate climate projections. Long-term projections (more than 100 years) depend on the future level of greenhouse gas emissions.

CHECKPOINT 1.22 ▶ What do scientists mean when they talk about uncertainty, and how is this different from a lack of knowledge?

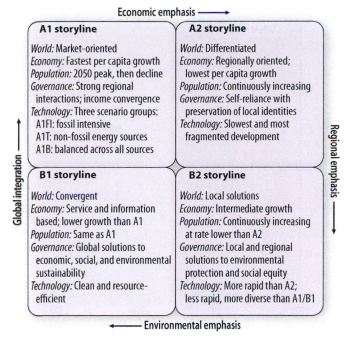

▲ **Figure 1.25:** The different social, environmental, and economic criteria used to define the alternative storylines of the IPCC *Special Report: Emissions Scenarios* are used by economists to estimate the level of future greenhouse gas emissions.

How Accurate Are Economic Projections?

Climate models depend on accurate projections of future greenhouse gas emissions, and these in turn depend on economic forecasts. The future rate of economic growth, population growth, technological development, and the organization of global society will all affect the future production of heat-trapping greenhouse gases. The influential **United Nations Intergovernmental Panel on Climate Change (IPCC)** projects how greenhouse gas emissions may change in the future. The IPCC makes these projections by using economic models to illustrate a number of contrasting "storylines" that it outlines in its **Special Report: Emissions Scenarios** (**Figure 1.25**):

- *A1:* The A1 storyline and scenario family is based on a future world of very rapid economic growth and a global population that peaks in mid-century but declines thereafter, and foresees the rapid introduction of new and more efficient technologies.
- *A2:* The A2 storyline and scenario family is based on a very heterogeneous world where global population continues to increase, and economic growth is more regionally oriented, more fragmented, and slower to develop than in other storylines.
- *B1:* The B1 storyline and scenario family is based on a more convergent world where global population grows as rapidly as in the A1 storyline, but where rapid changes in economic structure focus on the

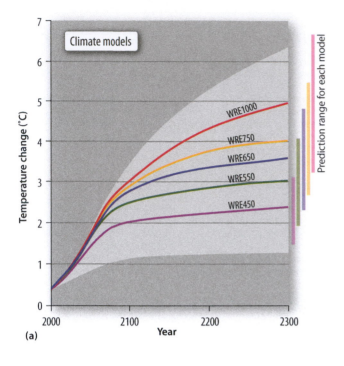

(a)

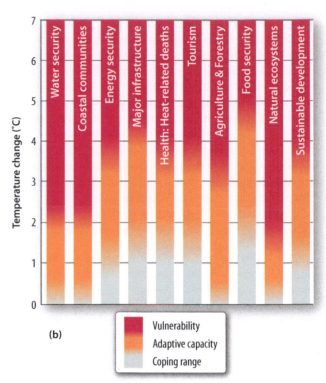

(b)

▲ **Figure 1.26:** (a) Climate models project how global temperature will change according to the different economic and emission models outlined in the *IPCC-SRES* report. (b) Climate change will impact many aspects of our economy and society. The blue areas of this graph indicate temperature change that society can cope with using current resources, the orange areas indicate a level of temperature change that will require significant adaptation, and the red areas indicate a level at which temperature change would be damaging to both human and natural environments.

development of a service and information-based economy that uses less raw materials and encourages the introduction of clean and resource-efficient technologies.

- *B2:* The B2 storyline and scenario family is based on a less likely world where there is an intermediate level of economic development, global population grows more slowly, and there is an emphasis on local solutions to economic, social, and environmental problems.

The IPCC will soon start to use a new set of emissions models to project future climate change based on a set of long term scenarios that extend as far as 2300 and utilize the full range of emission stabilization, mitigation and baseline emissions scenarios available in the scientific literature. The climate modelling community needs more detailed information than SRES provides, because there is an increasing interest in scenarios that explore the impact of different climate policies. Such scenarios allow evaluation of the "costs" and "benefits" of long-term climate goals.

Four new Representative Concentration Pathways (RCPs) are proposed:

- The **RCP3_PD** model assumes that radiative forcing peaks by 2050 at ca. 3 Wm^{-2} and decreases to 2.6 Wm^{-2} by 2100.

- The **RCP4.5** model assumes we are close to stabilizing radiative forcing at 4.5 Wm^{-2} in 2100 and that carbon dioxide concentrations, and radiative forcing are held constant after 2100.
- The **RCP6** model assumes that radiative forcing reaches 6 Wm^{-2} by 2100 before declining to stabilize at 4.5 Wm^{-2} .
- The **RCP8.5** model assumes that radiative forcing is still increasing and that emissions are still high. This results in very high radiative forcing of 16 Wm^{-2} and a concentration of carbon dioxide in the atmosphere as high as 3,000 parts per million.

The RCP models are not forecasts and are not policy prescriptive. Like the SRES scenarios, they simply describe a set of possible futures with different levels of emissions that are based on current literature and provide important information for decision-making.

The actual path of global emissions may be different from any of these that the IPCC has projected. However, the rapid rise of new economic superpowers such as China and India, and their increasing demand for fossil fuels, makes it very likely that the "worst-case" A2 or RCP6 scenarios might become the "best case" we can actually hope for.

Climate modelers run the same computer program many times using these different emissions scenarios and present their output as a range of possible futures that depend on the actions we take today (Figure 1.26). Each time a model is run with the same input data the answer can be slightly different due the nonlinearity of the climate system. After many runs, however, the answers all cluster round a statistical mean that is the most likely outcome. It is these data that define the risk of climate change for policy makers who must plan legislation to limit the emission of greenhouse gases.

CHECKPOINT 1.23 ▶ Describe the global factors that determine which of the *Special Report: Emissions Scenarios* storylines we are following most closely today.

Should Climate Models Be Used to Guide Policy?

Many important decisions hang on the accuracy of climate model projections. There is therefore a lot of concern about their reliability. Their use of mathematical equations from physics, chemistry, and biology is robust and well tested. They have proven capable of accurately replicating past climate with precision ("hindcasts"), a fact that builds confidence in their ability to project the future. The absence of regional resolution is a problem in a political world where "all politics are local," but the global view of climate models has convinced many policymakers of the need to take immediate action to prevent the worst model projections from ever becoming reality. Despite the initial reservations of many scientists and some outstanding technical uncertainties, climate models have evolved to become powerful tools that should be used with growing confidence by all policy makers as they determine the future of climate change legislation.

1.6 PAUSE FOR... THOUGHT | *How can a small change in the concentration of a trace gas have such a large impact on Earth's climate?*

One way to answer this is to think of adding just 8 drops of food dye to 1 liter (just over 2 pints) of water. This is equivalent to 400 parts per million (ppm) and is quite enough to turn the water opaque. Or think of hydrogen sulfide, the gas that produces a "rotten eggs" smell; your nose can detect hydrogen sulfide at just 0.00047 ppm. At 350 ppm of hydrogen sulfide, which is lower than the level of CO_2 in the atmosphere today, you would be dead. It is not the amount of a gas that matters; it is the effect that the gas has on the system in question.

The Next Decade?

We still need to discover a lot about Earth's climate system before we can use computer models to accurately project temperature change over periods of less than a decade. Short-term relationships between the Sun, atmosphere, biosphere, cryosphere, hydrosphere, and lithosphere are complex, nonlinear, and difficult to model. It is highly likely that climate models have accurately projected how global temperature will change over the next 50 years, but natural changes in Earth's climate system can mask this effect in the short term. The underlying radiative imbalance that is driving climate change has not gone away, and the oceans and atmosphere have continued to accumulate more energy. As the natural forces that slowed the pace of global warming to climate change over the previous decade weaken, the pace of global warming will increase. The year 2012 was the warmest year on record in the United States, Arctic sea ice reached a new record minimum, and violent storms and wildfires continue to wreak havoc across the world (Figure 1.27). This is exactly what the climate models projected.

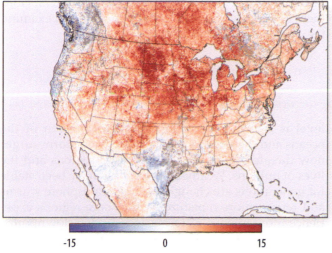

Land Surface Temperature Anomaly (°C)

◀ **Figure 1.27:** March 2012 temperatures compared to the average of the same eight day period of March from 2000–2011. Areas with warmer than average temperatures are shown in red; near-normal temperatures are white; and areas that were cooler than the 2000–2011 base period are blue. The unseasonable warmth broke temperature records in more than 1,054 locations between March 13–19.

Summary

- Climate change is both natural and normal, and it is driven by many different factors acting over different timescales. Gradual changes in orbital configuration over the past 8,000 years have reduced the amount of solar radiation reaching high latitudes, and global temperature has fallen slowly due to positive (for cooling) feedback in the climate system.
- Over this time, small shifts in the pattern of winds and ocean circulation have led to major, and sometimes rapid, changes in global and regional climate. It is therefore important to determine whether the changes we observe today are part of this natural cycle or driven by human activity.
- So far, the evidence from field observations, the historical temperature record, and climate models suggests that recent climate change is a consequence of both human and natural forcing.
- Before 1975, much of the warming appears to have been due to natural variation in climate due to solar forcing. Post-1975, climate models are unable to reproduce the observed changes in climate without including the effect of anthropogenic heat-trapping greenhouse gases.
- The burning of fossil fuels has altered the composition of the atmosphere, changed the balance of radiative forcing at the top of the atmosphere, and left an isotopic signature in every living thing on Earth.
- Projecting the future of climate change is difficult. Earth's climate system is complex, with a high degree of interconnectivity between its different components.
- Climate models have excellent accuracy and precision at a global level and are becoming more effective at a regional level. Their ability to successfully replicate past climate change builds confidence in their ability to project the future, but their projections depend on the economic data that feed them.
- Until we decide what to do about greenhouse gas emissions, climate models can only present us with contrasting scenarios of future climate. Whatever we do, the immediate prognosis is not good.

Why Should We Care?

There is strong evidence based on sound data that the climate is changing, but a few scientists and a significant percentage of the general public in the United States and United Kingdom still doubt that most of this change is due to human activity. At a time of prolonged global economic recession, they argue that we should not divert precious resources to stop change that will be mild and natural. But what if they are wrong? What if the climate models are correct? As we will see in Chapters 2 and 7, the consequences of inaction could be catastrophic. We need to answer some important questions as we consider climate change:

- How much change do we observe?
- How much is natural?
- How much is related to human activity (anthropogenic)?
- What has happened in the past?
- What is going to happen in the future?
- What can we do about it?
- What should we do about it?
- What is the cost of unnecessary action?
- What is the cost of inaction?

We will address these questions in other chapters. Hold them in mind as you read through the text and examine the data in more detail.

Looking Ahead . . .

Chapter 2 presents physical evidence from around the world that rising global temperature is changing Earth's climate. The signs of climate change are visible around the world from the poles to the equator. The impact is greatest at northern latitudes, especially in Arctic regions, where sea ice, ice sheets, tundra, and glaciers are melting rapidly, but is also evident in Antarctica where ice shelves that existed for thousands of years have become unstable and broken apart. Global sea level is rising due to the thermal expansion of the oceans and the addition of meltwater, and storm surges now threaten many low lying coastal regions and the lives of millions. Weather patterns that have been stable for generations are changing as the atmosphere warms and retains more moisture. While some regions experience more intensive storms and sudden downpours, others suffer from record droughts and wildfires.

Critical Thinking Questions

1. Describe how an older friend or relative remembers what the weather was like when they were younger. Older folks will often tell you that the weather has changed. Do you think this is because of normal variation in the weather, or does it signify a change in climate? How can you tell the difference?

2. Why is climate change always bad for someone?

3. In what ways could one argue that global climate was more suitable for life during the Cretaceous Period?

4. Can you think of times over the past 2,000 years when climate made a difference to world history? How would the modern world have changed if the climate during that time had been different?

5. When does "good greenhouse gas" become "bad greenhouse gas"? How would you convince someone that the Sun is not responsible for recent climate change?

6. Why has the temperature of the land surface increased much faster than surface ocean temperatures?

7. How much confidence do you have that climate models are able to give us an accurate forecast of Earth's future climate? What additional information is needed to increase confidence in these forecasts?

8. How can an understanding of uncertainty help formulate an appropriate response to climate change?

Key Terms

Make sure you are familiar with the key terms introduced and highlighted in this chapter. A full glossary is available at the end of the book.

Aerosols, p. 8
Albedo, p. 12
Climate model, p. 8
Cretaceous Period, p. 11
Eccentricity, p. 15
El Niño–Southern Oscillation (ENSO), p. 15
Eocene Epoch, p. 11
Global circulation model (GCM), p. 18
Great ocean global conveyor, p. 15

Greenhouse gases, p. 3
Hockey stick graph, p. 5
Hothouse world, p. 11
Icehouse world, p. 11
Interglacial, p. 11
Isotope, p. 6
Little Ice Age (LIA), p. 5
Medieval Climate Anomaly (MCA), p. 5
Miocene Epoch, p. 11
North Atlantic Oscillation (NAO), p. 16

Obliquity, p. 15
Pangaea, p. 14
Permian Period, p. 14
Pleistocene Epoch, p. 11
Positive Feedback, p. 13
Precession, p. 15
Proxy, p. 4
Radiative forcing, p. 12
Regional climate models (RCM), p. 19

Solar cycle, p. 8
Special Report: Emissions Scenarios, p. 19
Supercontinent, p. 14
Tectonic plates, p. 14
United Nations Intergovernmental Panel on Climate Change (IPCC), p. 19
Weather balloon, p. 9
Weather satellite, p. 9
World Meteorological Organization, p. 10

A farmer in Logan, Kansas sifts through arid topsoil and crops devastated by recent drought. As climate changes, many parts of the world are expected to experience less rainfall and higher rates of surface evaporation leading to drought, crop failure, and the threat of famine.

CHAPTER 2

The Evidence: Observing Climate Change

Earth's climate is changing. The level of the oceans is rising, the cryosphere is melting, and changes in the pattern of precipitation are bringing intense rainfall and flooding to some areas and devastating drought to others. The physical evidence for climate change is all around us. It is particularly evident around the Arctic regions of the Northern Hemisphere, but the Southern Hemisphere is also warming, especially around the Antarctic Peninsula and the West Antarctic Ice Sheet. Higher summer temperatures in hot, arid regions of the world have increased the rate of evaporation, and deserts are advancing to claim land that was once arable. In temperate regions the physical evidence of climate change is less evident. Winter nights are not as cold, there are fewer heavy frosts, spring comes a little earlier, rainfall is more intense, and there are more record-breaking summer heat waves. On the other hand, people living on small islands in the Pacific or tending farms on some of the world's great deltas are under increasing pressure from the effects of rising sea level.

As global average temperature changes at an unprecedented rate and greenhouse gas emissions continue unabated, climate models project that global temperature will continue to rise and trigger significant climate change. Climate change is real and dangerous, and millions of people live in fear of its immediate consequences.

Learning Outcomes | *When you finish this chapter you should be able to:*

- Explore the evidence for global climate change
- Understand the meaning of *global temperature* and how it is determined
- Distinguish between the main factors that control global sea level
- Explain how sea level has changed continuously throughout geological time
- Predict how the cryosphere will respond to further global warming
- Discuss the global impact of retreating glaciers and ice sheets
- Reflect on the impact of melting permafrost on the Arctic ecosystem
- Discuss how the addition of more energy and water vapor to the atmosphere will affect global weather patterns

Global Temperature

Chapter 1 described how global temperature has increased rapidly over the past century, but there are many questions that still need to be answered. How is it possible to talk about global temperature when there is so much seasonal and regional variation in climate around the world? Where do the temperature data come from, and why are they usually expressed in the form of temperature anomalies and not actual physical temperatures? What does global temperature actually mean in a world with so many climate zones? How will a change in global temperature affect the African Sahara, the Amazon forests, the tropical forests of Indonesia, the African savannah, the marginal deserts of the Sahel, the temperate forests of Europe, the corn belt of the United States, the Arctic taiga, the high Himalayas, or the Siberian tundra (Figure 2.1)?

For millions of years, Earth has maintained a balance between the amount of the Sun's energy reaching the top of the atmosphere and the amount of infrared radiation emitted out into space. Even small changes in this balance can have a large impact on climate. However, not all changes in regional climate indicate that the balance has changed. Both regional warming and cooling can occur because of cyclical or even random (**stochastic**) changes associated with the redistribution of existing energy within Earth's climate system. It is quite possible for some areas of Earth to cool at the same time as global average temperature is rising.

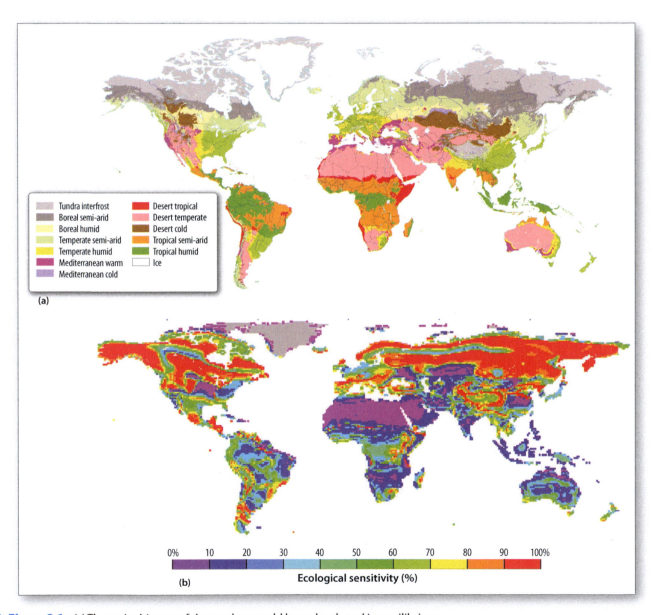

▲ **Figure 2.1:** (a) The major biomes of the modern world have developed in equilibrium with the physical environment. When climate changes due to global warming, many of these biomes, and especially those that are already under threat from human activity, will be damaged. (b) 21st-century ecological sensitivity. This map shows the predicted percentage of the ecological landscape being driven toward changes in plant species as a result of projected human-induced climate change by 2100.

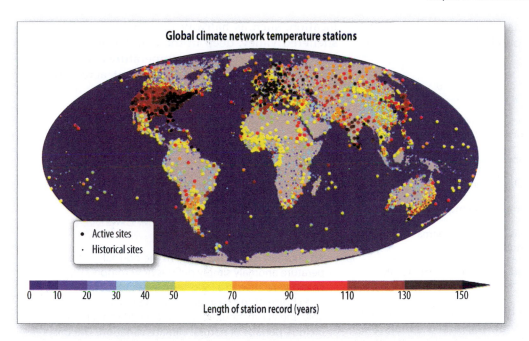

Global climate network temperature stations

- Active sites
- Historical sites

0   10   20   30   40   50   70   90   110   130   150
Length of station record (years)

▲ **Figure 2.2:** This map shows the 7,280 fixed land-based temperature stations in the Global Historical Climatology Network catalog, color coded by the length of the available record. The 2,277 sites that are actively updated in the database are marked as "active" and shown in large symbols; other sites are marked as "historical" and shown in small symbols. Global temperature also uses marine sea surface temperature data to estimate global temperature.

How can we tell if the climate change we observe today is a result of natural internal variability or is the result of a global energy imbalance due to an increase in solar insolation and/or the impact of greenhouse gas emissions? The best way would be to measure the balance between incoming and outgoing radiation at the top of the atmosphere. If they are in balance (equilibrium), then the climate change we observe today is most likely due to internal variability. On the other hand, a positive imbalance would lead us to expect global warming. Satellites can measure these data (see Chapter 3), but the difference is around 0.75 Wm^{-2}, almost too small for the instruments to resolve. We need some other way of measuring the temperature of the whole Earth that can help us determine if there is an imbalance between incoming and outgoing energy.

CHECKPOINT 2.1 ▶ If satellites can measure the energy imbalance at the top of the atmosphere, why do you think we still measure global temperature by using surface thermometers?

Estimating Global Temperature

To calculate global temperature using physical thermometers, scientists divide the surface of Earth into a 5° × 5° grid, and they determine the average temperature in each box in that grid from all the data available. A global average temperature is then obtained by adding (**stacking**) average temperatures measured in each 5° × 5° box around the globe and correcting for the fact that, due to Earth's geometry,

the boxes are smaller near the poles and larger near the equator. When this process is complete, it is possible to compare the average temperature of Earth today with temperatures in periods from the historic record. This is not an easy task; even today there are many parts of Earth where few data are available. Most weather stations on land are clustered around population centers, leaving other areas, especially near the poles and in the middle of deserts and tropical forests, poorly represented in the climate record. Direct observation of ocean air temperature is even less complete and is confined to coastal stations, weather buoys, and shipping lanes (**Figure 2.2**).

CHECKPOINT 2.2 ▶ Look at Figure 2.2 again. Note which parts of the world are best represented in the data record and which are least well represented. Why is this so?

Why Use Temperature Anomalies?

Temperature anomalies measure how the temperature at any location has changed relative to a **base period** that covers a much longer period of time at the same location (usually more than 15 and preferably more than 30 years). Selection of the most appropriate base period is very important. When chosen correctly, it can even out the impact of short-term cycles such as warm El Niño events and the impact of volcanic activity. For this reason, a minimum of 30 years of data is normally recommended. When chosen incorrectly, the base period can skew the data and suggest either greater or lesser warming than is actually occurring.

These are the base periods most commonly used to calculate global climate anomalies and the organizations that use them:

Base Period	Organization
1951–1980	Goddard Institute for Space Studies (GISS)
1961–1990	Climate Research Unit (CRU)
1971–2000	NOAA (general)
1979–1998	Remote Sensing Systems (RSS)
1981–2010	University of Alabama at Huntsville
The entire 20th century	National Climate Data Center

Each organization uses a baseline that best suits the data available. The University of Alabama at Huntsville (UAH) publishes NASA satellite data using a baseline that is very different from most others. It excludes temperature data prior to the warmest decades on record. This does not invalidate the data, but it does make anomalies appear much smaller in comparison to most other published data.

Organizations' uses of different base periods means that their numerical results are different, but this does not affect the relative change of temperature measured from year to year or the overall magnitude of global warming measured over the past 150 years.

There are two reasons most climate scientists use temperature anomalies and not absolute temperature to measure global, regional, and local climate change. The first reason is that a careful study of global temperature data has shown that while surface temperatures vary significantly across a region, temperature anomalies are remarkably uniform up to a distance of 1,200 kilometers (746 miles) around each data point. Using temperature anomalies to measure global temperature *change* allows data from remote locations, such as the deep Arctic, to be estimated from data points over 1,000 kilometers (621 miles) away.

CHECKPOINT 2.3 ▶ In your own words, describe why most scientists use temperature anomalies and not physical temperature to track global climate change.

The second reason anomalies are used is to overcome problems with data accuracy and the geographical location of data stations. All records are affected by factors such as how far off the ground the measurements were taken, at what altitude, how close to buildings and urban centers, over what kind of surface (such as grass, concrete, or tarmacadam), and at what time of day the measurements were taken. These varying factors can make it very difficult to quantitatively estimate the actual regional temperature. The physical temperature measured at any weather station depends on its location, such as halfway up a mountain, in a deep sheltered valley, deep in the countryside, or near an airport runway. However, temperature anomalies have proven to be remarkably consistent across weather stations from the same region.

CHECKPOINT 2.4 ▶ Why is it not good practice to choose a base period of less than 30 years when calculating temperature anomalies?

2.1 PAUSE FOR... THOUGHT | *Why do scientists and others use different periods of time to define temperature anomalies?*

An anomaly is any change from the normal, so a temperature anomaly simply defines how temperature has changed relative to a long-term average (base period). The choice of the base period depends on the data available. Ideally, it should cover at least 30 years in order to include periods of natural climate variation. The base period 1901–2000 is commonly used in the literature, but other periods can be used, depending on the data available. Anomalies based on data that are too limited to account for natural changes in climate or that start during a period of unusual climate (such as an unusually warm El Niño period) will give misleading results and even suggest that the world is cooling when it is, in fact, only recovering from an unusual spike in temperature.

The Temperature Data

Nothing is quite as misunderstood and misrepresented in climate science as temperature data, and yet nothing is more important. Climate scientists measure global temperature by combining data from land and the oceans, and this remains the best available method to determine how much extra energy is being added to Earth's oceans and atmosphere.

Land Surface Temperature

Thousands of weather stations around the globe have recorded surface land temperatures using simple thermometers for hundreds of years. Many government and university research centers record and publish temperature data, including NOAA's National Climate Data Center, NASA's Goddard Institute of Space Studies, the Japanese Meteorological Agency, the Berkeley Earth data set, and the combined UK Meteorological Office's **Hadley Centre and Climate Research Unit at the University of East Anglia (HadCRUT).**

Ocean Air and Sea Surface Temperature

Ocean air temperature is measured on board commercial and naval vessels, but coverage is confined to shipping lanes, and daytime measurements are strongly affected by the location of the thermometers. Nighttime temperatures are less affected by the location of thermometers, but the data are incomplete. To overcome this challenge, scientists prefer to use **sea surface temperature (SST)** as the preferred measure of temperature over the oceans. Of course, ocean and air temperatures are not exactly the same, but they are closely related, and any temperature anomaly measured at the ocean surface is very likely to be reflected in the air immediately above.

To calculate ocean temperature anomalies, NOAA and other research agencies use an **Extended Reconstructed Sea Surface Temperature (ERSST) analysis**, along with the **International Comprehensive Ocean–Atmosphere Data Set (ICOADS)**, the most comprehensive record of data on sea surface temperature collected from moored and drifting buoys, coastal stations, and ships. These data range from the 17th century to the present day, and more recent data have sufficient geographical coverage to complement the existing land data and almost cover the entire globe. In the most recent version of ERSST (version 3), satellites are also used to collect data in less well-covered regions, especially in the Southern Ocean.

CHECKPOINT 2.5 ▶ Why do climate scientists prefer to use sea surface temperature and not surface air temperature to measure the temperature of the atmosphere above the oceans?

Air Balloons

In order to determine the impact of greenhouse gases on the atmosphere it is necessary to measure how air temperature changes with height. Air balloons measure air temperature, humidity, and atmospheric pressure at more than 900 upper-air observation stations around the globe, mostly in the Northern Hemisphere, using **small radiosondes** that later fall back to Earth by parachute. These data are used to determine how the middle and upper layers of the atmosphere are responding to global warming.

CHECKPOINT 2.6 ▶ What information do air balloons provide to help scientists understand how the atmosphere is responding to Earth's energy imbalance?

Satellites

Data from weather and research satellites operated by NASA and NOAA have been used to complement, extend, and enhance the land- and ocean-based record since 1978. Raw satellite data are available for many groups to analyze, but the data sets prepared by the **University of Alabama at Huntsville (UAH)** and **Remote Sensing Systems (RSS)** are most prevalent.

Satellites measure air temperature at different levels in the atmosphere by using instruments called **Microwave Sounding Units (MSUs)** that measure the temperature-dependent microwave flux generated by oxygen molecules in the atmosphere. This radiation has a much longer wavelength than infrared (heat) radiation and travels through a broad vertical layer of the atmosphere. The data must be corrected for many factors, including orbital drift, sensor variation, and time of day. Different methodologies for analysis result in small differences between the UAH and RSS data sets, but there is generally good correspondence between the RSS version 3.3 warming trend of +0.148°C/decade, the UAH version 5.4 trend of +0.140°C/decade, and the current rates of global temperature change derived from the surface temperature record of +0.17°C/decade.

Satellites and Sea Surface Temperature

Since 1967, climate scientists have been measuring sea surface temperature using satellites by examining thermal infrared radiation emitted by the oceans. Both the amount of radiation and the wavelength of maximum radiation depend on temperature, so scientists can use these data to determine temperature. Today's satellites use **Advanced Very High Resolution Radiometer (AVHRR)** instruments to provide global coverage of sea surface temperature, but land surface temperature is more difficult to determine by satellite due to the impact of topography on the reflected signal.

CHECKPOINT 2.7 ▶ Satellites are powerful tools for measuring the temperature of the atmosphere. What corrections have to be made before satellite data can be used? How accurate are these measurements?

Global Historic Climatology Network

The largest international data set of global temperature is maintained at NOAA's **National Climate Data Center (NCDC)** as part of the **Global Historic Climatology Network (GHCN)**. The GHCN has 3,832 records spanning over 50 years, 1,656 records spanning over 100 years, and 226 records spanning over 150 years. The longest continuous record in the collection began in Berlin in 1701.

Although the temperature records from older thermometers are often less accurate than those from newer ones, most of the data have a high degree of precision (that is, they gave the same answer each time under similar conditions). This precision allows climate scientists to use temperature anomalies as a more accurate measure of global warming than absolute temperature. The NCDC is using the most recent data set (GHCN-M version 3) along with the ERSST data set to form a merged land and ocean surface temperature data set to calculate global average temperature anomalies from 1880 to the present.

Missing Data

Even after climate scientists combine all the available data sets, there are still large areas of the Arctic and Antarctic that remain incompletely covered. Observations clearly show that the Arctic is warming faster than the rest of the planet and must be included in any global temperature analysis. Climate scientists use two contrasting ways to compensate for the missing data:

- *Climate Research Unit (CRU) model:* The CRU model assumes that the temperature in the Arctic changes in proportion to the rest of the Northern Hemisphere; this model almost certainly underestimates the magnitude of global warming.
- *Goddard Institute for Space Studies (GISS) model:* The GISS model estimates temperature in the Arctic by extrapolating data from weather stations up to 1,200 kilometers (746 miles) away. The complexity of temperature change across this region may lead to errors in this model.

The CRU model is more scientifically conservative, but the GISS model is built on well-established research and is more likely to produce an accurate analysis of the global temperature anomaly.

CHECKPOINT 2.8 ▶ In your own words, describe the important difference between the meanings of *accuracy* and *precision*.

Changes in Sea Level

The ancient historical record is full of stories about devastating floods. Today, nothing else has gripped the imagination of the general public quite as much as the threat of rising sea levels. The most recent predictions from the IPCC (**Figure 2.3**) are conservative and suggest that we can expect sea level to rise between 18 and 76 centimeters (7 and 30 inches) over the next 100 years. This is not much more than we have experienced over the past 100 years, but it would continue to present significant problems in places such as the Maldives, Venice, and Bangladesh.

This IPCC report would be reassuring if we actually understood how ice sheets melt. Recent research suggests that ice sheets do not melt like a block of ice, as many of the early climate models assumed, but can instead rapidly collapse into the sea. Many ocean scientists believe we are likely to see a rise in sea level of more than 1 meter (3.3 feet), and maybe as much as 2 meters (6.6 feet), by the end of the century. There is still a possibility that sometime over the next 100 to 200 years, the Greenland Ice Sheet and/or the West Antarctic Ice Sheet could start to collapse, eventually resulting in a rapid rise in sea level of up to 6 to 12 meters (20 to 40 feet). Such a rise would inundate many coastal cities and low-lying areas at a rate that would make it difficult for many natural coastal ecosystems to adapt. This is not the most likely scenario, as there is evidence that parts of the Greenland Ice Sheet have been stable in the past when temperatures were warmer than they are today. However, we do not have room for complacency, and much more research is required.

CHECKPOINT 2.9 ▶ Why are the IPCC's conservative estimates of rising sea level not reassuring?

Is Changing Sea Level Normal?

Sea level is constantly changing. On a time scale of millions of years, it is controlled by the expansion and contraction of vast subsea mountain chains called

▲ **Figure 2.3:** (a) A graph of recent and predicted sea level rise. (b) A flooded street in Cartagena, Columbia, due to rising sea level, November 15, 2007.

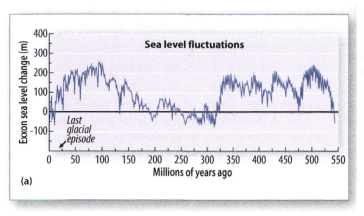

(b)

(a)

▲ **Figure 2.4:** (a) Sea level change over geological time, determined from a detailed analysis of sedimentary basins from around the world by Exxon. Note the difference in scale between this figure and Figure 2.3. (b) Fossils of marine shells encased in undisturbed sedimentary rock found many meters above sea level remind us that sea level has been changing constantly over geological time.

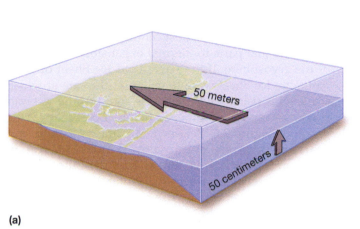

(a)

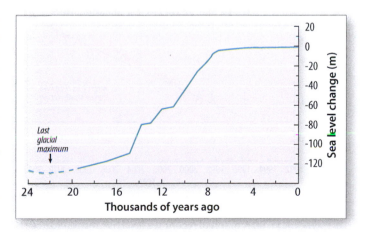

(b)

▲ **Figure 2.5:** (a) A diagram that illustrates how coastal plains can flood rapidly, with only a small increase in sea level. (b) Areas of salt marsh and tidal flats around our coasts will become flooded within decades if sea level continues to rise at the current rate.

mid-ocean ridges that form where Earth's giant tectonic plates are moving slowly apart. During the Cretaceous Period, over 65 million years ago, the Pacific Ocean mid-ocean ridge was especially active. An immense chain of volcanoes rose up from the deep ocean floor and displaced billions of tonnes (a tonne is a metric system unit of mass equal to 1,000 kilograms (2,204.6 pounds) of water onto the shallow continental shelves around the globe. At the high point of this process, sea level stood well over 100 meters (328 feet) above current levels (**Figure 2.4**).

On a timescale of thousands of years, the advance and retreat of ice sheets over the Northern Hemisphere has been marked by great swings in sea level. Only 21,000 years ago, at the end of the last major ice advance, sea level reached as low as the margin of the continental shelf, some 120 meters (394 feet) lower than today, and the shoreline lay up to 100 kilometers (62 miles) off the current East Coast of the United States. These flat coastal plains around

▲ **Figure 2.6:** A graph of postglacial rise in sea level.

the globe would have flooded rapidly with only a small rise in sea level, so we shouldn't be surprised to find such strong folk memories of global floods (**Figures 2.5** and **2.6**).

At regional and local levels, sea level can also depend on other factors, such as tectonics, sediment supply, and subsidence. Southeastern England is slowly subsiding into the North Sea due to the impact of **regional tectonic stress.** In the United States, the Mississippi Delta is slowly subsiding under its own weight, while we also restrict the supply of new sediment by channeling river flow and create additional subsidence by extracting vast volumes of oil and gas from the subsurface.

Global sea level started changing more rapidly again in the 19th century, after a period of stability that lasted 2,000 to 3,000 years. Sea level rose around 170 millimeters (6.7 inches) during the 20th century, roughly the same as the previous century. But as more heat has been added to the oceans, the rate of sea level rise has increased from 1.8 millimeters per year to 3.6 millimeters per year (0.07 to 0.14 inches). (The upper limit may reflect a component of natural **decadal** cyclicity.) Much of this change is due to **thermal expansion**—the way that water expands when heated.

Sea level is not rising equally around the globe. Both the eastern Pacific Ocean and the western Indian Ocean are actually experiencing a fall in sea level due to the impact of natural decadal cycles such as El Niño–Southern Oscillation (ENSO) and the North Atlantic Oscillation (NAO). A sudden increase in the rate of precipitation over land can also lead to a slight decrease in sea level, as long as the water is retained in surface and groundwater reservoirs. The pattern is not simple, and the interrelationships are at times complex and confusing, but the overall picture is still clear: Sea level is rising around the globe and can be expected to rise at least a further 0.18 to 0.76 meters (7.1 inches to 30.0 inches) by the end of the 21st century (**Figure 2.7**).

CHECKPOINT 2.10 ▸ What are the main factors that control sea level, and over what timescale do they operate?

How Fast Is the Sea Rising?

The rate at which sea level rises will have a major impact on the ability of society and the environment to adapt. A slow increase in sea level is not a major threat to the environment, as **ecosystems** have evolved over millennia to adapt to such changes. The Chesapeake Bay in Virginia (**Figure 2.8**) is a good example of an ancient river valley that flooded as ice sheets melted and sea level rose at the end of the last glaciation. An observer at the time might have considered this flooding an ecological disaster, but the ecosystem adapted and evolved to meet the demands of a new environment. A gradual rise in sea level of less than 1 meter (3.3 feet) does not raise social or economic concerns either, as new technologies and increasing wealth in the developing world offer an array of options for climate mitigation and adaptation.

But sea level is now rising at a rate of around 3 millimeters (0.12 inches) per year, faster than it has at any other time over the past century. This rise is due in roughly equal parts to thermal expansion and melting of the cryosphere. If the Greenland Ice Sheet or West Antarctic Ice Sheet were to collapse, the rate at which sea level is rising would increase significantly. A rapid rise in sea level has happened before. Look again at the graph in Figure 2.6. Between 14.2 and 14.7 thousand years ago (kya), sea level rose by 25 meters (82 feet) in less than 500 years, probably as a result of the collapse of an ice sheet over present-day Antarctica.

If the sea level were to rise much more than 1 meter (3.3 feet) in 100 years, many coastal ecosystems would

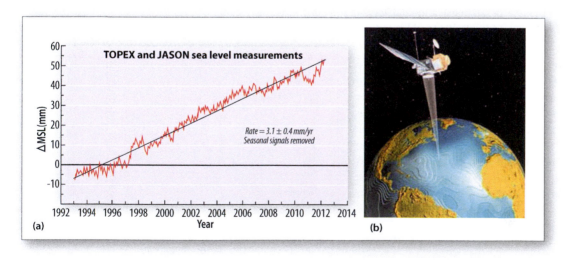

(a) TOPEX and JASON sea level measurements

Rate = 3.1 ± 0.4 mm/yr
Seasonal signals removed

(b)

▲ **Figure 2.7:** (a) A graph of recent sea level rise, measured from NASA's TOPEX and JASON satellite instruments, showing changes in sea level averaged over a 60-day period (blue line) and the overall trend (black line) of 3.1 millimeters per year (0.12 inches). Note the small decrease in sea level between 2010 and 2012. This developed after unusually wet weather globally increased the amount of rainwater stored on land in rivers, lakes, reservoirs, and groundwater. This temporary decrease in sea level soon recovered as all the extra water made its way back slowly to the oceans. (b) Launched in 1992, the TOPEX/POSEIDON satellite was the first mission to analyze sea surface topography with precision. It has since been replaced by the JASON-2 satellite launched in 2008.

change to a level where many species can no longer easily adapt (see Chapter 7).

Why Is Sea Level Rising?

Meltwater from glaciers and ice sheets has the greatest potential to impact sea level over the next 200 years. However, meltwater has not been responsible for most of the observed rise in sea level over the past 200 years.

The ability of water to store large amounts of energy with only a small increase in temperature—that is, to have a high **heat capacity**—has allowed the oceans to absorb as much as 90% of the additional energy added to Earth's climate system since 1955. As water expands when it is heated (thermal expansion), even a small increase in volume of as little as 0.1% could cause sea level to rise as much as 1 meter (3.28 feet).

Thousands of measurements of ocean temperature by ship and satellite now show that part of the oceans have warmed to a depth of at least 3,000 meters (9,843 feet), with widespread warming over the upper 700 meters (2,297 feet) (see Chapter 3). Consequently, as much as 57% of the rise in sea level since the 1950s has been due to thermal expansion. The contribution from glaciers and ice sheets is certainly increasing, but thermal expansion will continue to be important well into the next century.

CHECKPOINT 2.11 ▶ Explain the contribution that thermal expansion has made to changing global sea level over the past 100 years.

▲ **Figure 2.8:** A NASA satellite photograph of a river valley and its tributaries that was drowned as sea level rose at the end of the last glaciation to form the Chesapeake Bay.

find it hard to adapt, leaving damaged ecosystems that would take millennia to come to equilibrium with their new physical environment (**Figure 2.9**). A rapid rise in sea level would also create a social and economic disaster with greatest impact in developing countries where poverty leads to overcrowding near the coasts and communities lack the basic physical and economic resources they need to build adequate sea wall defenses.

It is very important to note that climate change and sea level change are natural and normal. The primary concern today is not the inevitability of that change but the *rate at which such change is taking place*. Ecosystems have evolved to adapt and even exploit natural climate change, but we may be accelerating the natural rate of

2.2 **PAUSE FOR...** | *When will the cities flood?*
            THOUGHT

Many movies and documentaries show disturbing images of cities flooding due to climate change. Is this really going to happen? It is clear that sea level will rise as the oceans warm and ice sheets melt, but it is less clear how quickly this will happen. It is possible that sea level could rise 1 meter (3 feet) or more before the end of the century, but it might take hundreds of years before Florida is flooded, as predicted in Al Gore's movie *An Inconvenient Truth*. Scientists need to understand more about the mechanisms that control the collapse of ice sheets before they can answer this question with certainty.

The Melting Cryosphere

The fact that permanent snow and ice still cover 10% of Earth's land area and as much as 7% of the oceans is an ever-present reminder of how recently we emerged from a deep ice age. By mid-winter, nearly 49% of land in the Northern Hemisphere is deeply covered in snow and ice, with almost 75% of the world's fresh water locked up in glaciers and ice sheets.

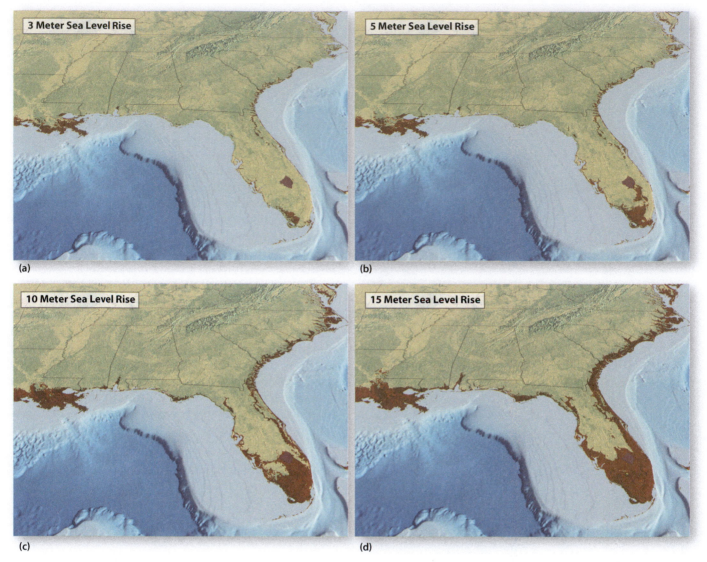

3 Meter Sea Level Rise

(a)

5 Meter Sea Level Rise

(b)

10 Meter Sea Level Rise

(c)

15 Meter Sea Level Rise

(d)

▲ **Figure 2.9:** Flat lying areas around the coast of the United States such as Florida, the Outer Banks and parts of the Gulf Coast will be among the first to notice the impact of rising sea level. With so many people and so much invested in coastal property, rising sea level is a major social and economic hazard. The four maps illustrate the impact of changing sea level by 3 meters (10 feet), 5 meters (16 feet), 10 meters (33 feet), and 15 meters (50 feet) on the geography of the southeast United States. These scenarios are realistic in the long-term, but it is unlikely that sea level will rise more than 1 meter by the end of this century.

CHECKPOINT 2.12 ▶ If the ice sheets were not melting at the moment, why would sea level still continue to rise?

As discussed earlier in this chapter, the rapid warming of the atmosphere over Greenland, the Antarctic Peninsula, and many mountain glaciers is adding to sea level by increasing the amount of summer meltwater. The observed melting of land-based ice may account for over 40% of current sea level rise. The processes of ice sheet **denudation** and glacier retreat are still not fully understood. As the atmosphere warms near the poles, the air is able to hold more moisture that falls as snow, and a steady increase in snowfall has been observed since the end of the last ice age. Over the high central parts of the ice sheets in east Antarctica and Greenland, this enhanced

precipitation might even offset some of the loss of ice at the margins and slow the rate of rising sea level.

CHECKPOINT 2.13 ▶ What factors that control global climate determine that the world is not yet free from the grip of an ice age?

Polar Sea Ice

Sea ice is a layer of ice that forms on the surface of the ocean when water temperatures drop lower than the $-1.9°C$ ($28.6°F$) necessary to freeze salt water. During the Antarctic winter, some 18 million km^2 (7 million sq mi) of thin seasonal sea ice covers an area that effectively doubles the area of the Antarctic continent. This ice is so thin that it is not durable, and it rapidly melts over

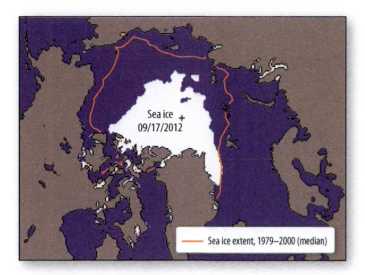

◄ **Figure 2.10:** A diagram showing the recent decline in the minimum extent of Arctic sea ice. The boundary lines mark the limits of ice during the remarkable ice minima of 2012. Compare these boundaries with the 1979–2000 median ice minimum

the summer. In contrast, some of the 12 million km² (4.6 million sq mi) of sea ice that forms during the Arctic winter is able to last year round, building a **multiyear ice layer** that is 3 to 4 meters (9.8 to 13.12 feet) thick.

Sea Ice Area   Arctic sea ice declined on average by 2.7% per decade from 1978 until 1996 (the first period of accurate satellite records), and climate models now foresee an accelerated decline of as much as 20% by 2050 (**Figure 2.10**). September 2002 saw the start of a continuing and rapid acceleration in the decline in the extent of sea ice in the Arctic, with a record minimum extent of summer ice. A small recovery in 2003 was followed in 2005 by a new record minimum and in 2007 by a drastic fall in sea ice area to a new record low of 4.28 million km² (1.65 sq mi), a figure 39% below the 1976–2000 average and maybe as much as 50% below the average of the 1950s (**Figure 2.11**). The summer of 2012 set a new record low in ice area of 3.6 million km² (1.39 million sq mi). This degree of melting is faster than predicted by most climate models and suggests that further drastic reductions may lead to ice-free summers in the Arctic by 2030, for the first time in over 1 million years.

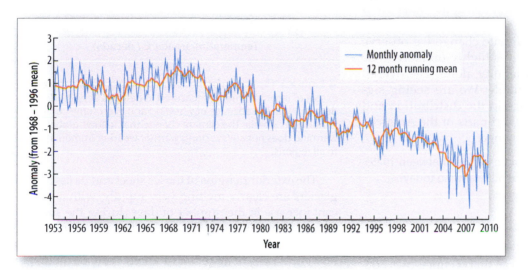

▲ **Figure 2.11:** Arctic sea ice extent standard anomalies, January 1953–September 2010. Until the 1970s, the extent of Arctic sea ice was relatively constant, but it has been decreasing since the 1980s.

Sea Ice Volume   A loss of sea ice volume in the Arctic is even more indicative of climate change than the decrease in sea ice area. There are two major kinds of ice in the Arctic. Annual sea ice forms each winter and can reach up to 1 meter (3.3 feet) thick before melting the following summer. Multiyear ice survives the summer melt and can grow up to 5 to 7 meters (16.4 to 23 feet) thick over subsequent winters.

Two satellites (**ICESat and CryoSat**) have been measuring sea ice volume since 2003, and their data show a rapid decline in multiyear ice volume from 62% of the total in 2003 to 32% in 2008 (**Figure 2.12**). Some of these changes are due to cyclical effects of wind and ocean currents that transport multiyear ice away from the Arctic, but most are due to melting, and much of this melting appears to occur underneath the ice as the temperature of the Arctic Ocean rises.

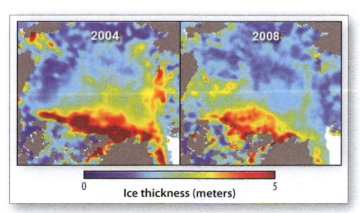

▲ **Figure 2.12:** ICESat measurements of the distribution of winter sea ice thickness over the Arctic Ocean in 2004 and 2008. Note the significant loss of thick ice (in red) over 3 to 5 meters (16.4 feet) thick between 2004 and 2008.

CHECKPOINT 2.14 ▶ Why is sea ice volume more indicative of warming in the Arctic than sea ice area?

Historical Change   As we will see with many other aspects of climate science, there are many uncertainties due to our lack of historical data. Declassified data on ice thickness from submarines goes back as far as 1958, but specialized passive microwave satellites that give reliable data from across the Arctic were not launched until 1979. There is undeniable evidence of substantial natural variation in temperature that must be taken into account. The period from 1925 to 1945, for example, was unusually warm in the Arctic, and regional sea ice volumes were reduced at that time. There is, however, an important difference between that decrease and the current one. There is no evidence that the Arctic warming of the 1930s was anything more than a regional event. Climate variability around the world is not at all unusual, and regions such as the Arctic are expected to change naturally on a decadal or shorter scale due to natural cycles in the oceans and atmosphere.

In contrast, the global climate change we observe today is very unusual, and it is highly unlikely to be due to natural decadal variability alone. The overall conclusion today is undeniable: We are seeing significant changes taking place in the distribution and thickness of ice in the Arctic and parts of the Antarctic as a result of global climate change.

CHECKPOINT 2.15 ▶ If the Arctic warmed between 1925 and 1945, why are we worried about warming today?

The Greenland Ice Sheet

Greenland is more than four times as large as the state of Texas, and it has a massive ice cap that occupies over 80% of the island. Formed by the accumulation of over 2.85 million km^3 (0.68 cu mi) of ice and snow, and covering a surface area of 1.7 km^2 (0.66 sq mi) to a depth of over 2 kilometers (1.2 miles), much of the ice on Greenland is at least 110,000 years old. This is a very old and cold place. Mean temperatures in the island's interior reach as low as –31°C (-23.8 °F). After Antarctica, Greenland is the second largest body of ice in the world.

Slowly Melting   The mean air temperature over Greenland has risen by as much as 3°C (5.4 °F) over the past century (**Figure 2.13**). In 2007, the IPCC concluded that the Greenland Ice Sheet may have lost as much as 50 to 100 gigatonnes per year (around 80 km^3 [(19.2 cu mi)] per year) between 1993 and 2003. (A **gigatonne** is 10^9 metric tonnes, and 1 metric tonne is 1,000 kilograms [1.1t(US)].) With accelerating warming in the Arctic, some estimates now place this higher than 220 gigatonnes per year (240 gigatons(US)) and accelerating as fast as 30 gigatonnes per year (33 gigatons(US)).

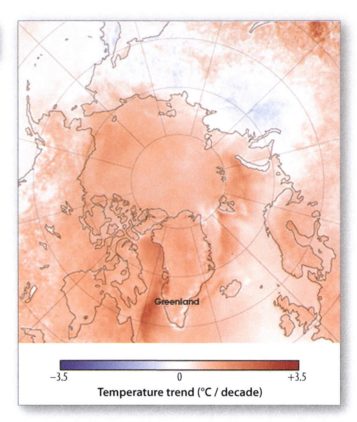

▲ **Figure 2.13:** A map of the Arctic, showing the widespread extent of recent warming, measured in degrees Centigrade per decade. Note the absence of warming and even some cooling in northern Siberia. The climate change that follows global warming does not always bring warmer weather immediately to all regions.

The exterior margins of the ice sheet and its associated glaciers are clearly melting very rapidly, with a significant increase in the number of days during the summer on which meltwater is flowing on the surface of the ice sheet. Glaciers draining the ice sheet at the margins are flowing at rates of more than 1 kilometer (0.62 miles) per year, with significant acceleration since 1996. The fastest-flowing outlet glacier on Greenland is the amazing Jakobshavn Glacier (**Figure 2.14**), which has a surface velocity of up to 7 kilometers (4.4 miles) per year!

Ice flow rates in the interior of the Greenland Ice Sheet are much slower than at the margins. In the interior, ice moves at just a few meters per year, especially at higher elevations in the south, indicating that the ice sheet here is actually quite stable. Close monitoring by NASA's **Gravity Recovery and Climate Experiment (GRACE)** and other satellites has even shown a slight increase in the volume of ice in the interior, as warmer air blowing in from the surrounding ocean holds more moisture that falls as snow (**Figure 2.15**), but this effect is far too small to offset the massive loss of ice mass from the margins of the ice sheet.

CHECKPOINT 2.16 ▶ Why is a small increase in ice volume in the interior of the Greenland Ice Sheet consistent with a warming atmosphere?

Figure 2.14: The location of the end (terminus) of the Jakobshavn Glacier has been recorded since 1913 to produce this historical record (the coast is beyond the lower-left corner of this image). If repeated along the Greenland coast, this loss of glacial ice would signal the collapse of the entire ice sheet.

If Greenland were to melt, the result would be a 7.2-meter (24-foot) rise in sea level. Most models suggest that if this happens, it will be a slow process, adding only about 1 meter (3 feet) to sea level after 1,000 years, and it will take many thousands of years to complete (**Figure 2.16**). Other scientists, such as James Hanson, who heads NASA's GISS, are deeply concerned that multiple **positive feedbacks** in the climate system could lead to much more rapid ice sheet disintegration (Figure 2.16). There is some observational evidence to support this, as the rate of flow of coastal glaciers is exceeding model predictions, but other evidence from ice cores suggests that parts of the ice sheet have remained stable during past interglacial periods when the temperature was much warmer than it is today.

It is becoming clear that current models of the cryosphere are not very good at predicting the response of glaciers and ice sheets to a rapid and prolonged rise in temperature and that we still have a lot to understand. There is a possibility that melting will be slower than predicted. There is a much greater chance that models have underestimated the rate at which these massive ice sheets will collapse.

The Antarctic Ice Sheets

The continent of Antarctica is a remote and foreboding place. It is isolated from the rest of the climate system by deep circulating vortices in the atmosphere and ocean, and the temperature can plummet to as low as –90°C (-130°F) in the winter and struggle to reach 10°C (50°F) around the coast in the summer. Antarctica is nearly 1.4 times the size of Europe, and most of the continental interior is more than 3 kilometers (1.9 miles) above sea level and covered by an ice sheet that is on average 1.6 kilometers (1 mile) thick and up to 4.8 kilometers (3 miles) thick in the center (**Figure 2.17**).

The total volume of ice is more than 30 million km^3 (7.2 million cu mi), and it contains 90% of all the fresh water on Earth. Around the margins of the continent, fast-flowing, deep rivers of ice called ice streams drain the continental interior (**Figure 2.18**). Where the ice streams meet the sea, they start to float and form bounding ice shelves that actually hold back the flow of ice from the interior. Icebergs frequently break away from the outer margins of these great ice shelves and drift in the surrounding ocean currents. If all the ice on Antarctica were to melt, global sea level would rise

| Thinning | No change | Thickening |

Change in elevation

Figure 2.15: Changes in elevation of the Greenland Ice Sheet, showing the loss of ice from the ice sheet margins but some thickening over parts of the interior (the color white indicates no change in thickness). This is not unexpected because warmer air holds more water vapor that falls as snow in the interior of the island.

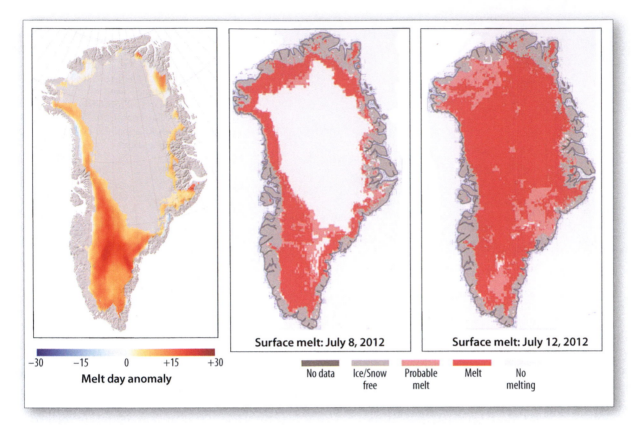

Surface melt: July 8, 2012

Surface melt: July 12, 2012

−30 −15 0 +15 +30
Melt day anomaly

No data | Ice/Snow free | Probable melt | Melt | No melting

▲ **Figure 2.16:** With higher Arctic summer temperatures, the number of summer-melt days on the Greenland Ice Sheet has risen dramatically. The greatest degree of melting is to the south and west, but in 2012 an unusually prolonged period of high temperature appears to have melted the entire surface of the ice sheet.

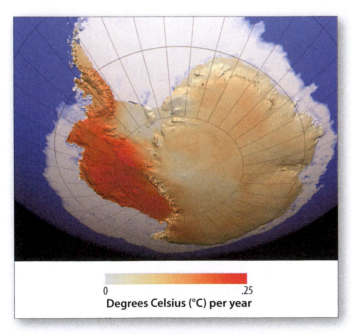

0 .25
Degrees Celsius (°C) per year

▲ **Figure 2.17:** A map of Antarctica, showing recent surface temperature anomalies, measured in degrees Celsius per year. Note the most rapid change in temperature on the Antarctic Peninsula and the West Antarctic Ice Sheet.

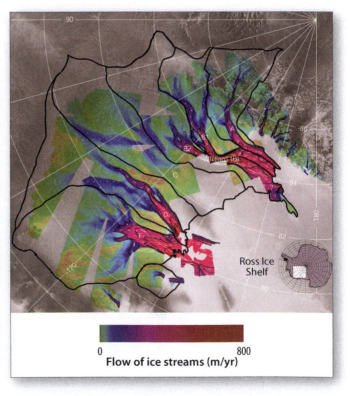

Ross Ice Shelf

0 800
Flow of ice streams (m/yr)

▲ **Figure 2.18:** A map of Antarctica, illustrating the flow of ice streams inside the West Antarctic Ice Sheet, as measured by satellite. Flow rates increase from green through purple to red.

over 61 meters (200 feet), but Antarctica has three very different regions that are responding in different ways to global warming.

The East Antarctic Ice Sheet The thick **East Antarctic Ice Sheet (EAIS)** contains 83% of Antarctica's ice and rests on an ancient continental landmass. Recent satellite observations suggest that the center of the ice sheet may be slowly growing due to a slight increase in precipitation, possibly related to elevated atmospheric temperatures. In contrast with Greenland, this is more than enough to offset melting around the margins of the ice sheet. This massive ice sheet is at least 30 million years old and appears to be very stable. Only an extreme and very unlikely rise in global temperature could cause any significant melting to take place.

The West Antarctic Ice Sheet The **West Antarctic Ice Sheet (WAIS)** contains 13% of Antarctica's ice and has developed over a submerged volcanic archipelago and reaches as deep as 2,500 meters (8,202 feet) below sea level. Grounded on the ocean floor, this ice sheet has become unstable during previous interglacial warm periods, and it has the potential to melt today, leading to a rise in sea level of around 5 to 6 meters (16 to 20 feet). As in Greenland, however, it is uncertain how long this process might take. Instability will occur if there is a rise in sea level that causes the ice sheet to lift off the ocean floor and/or if rising ocean temperatures start to erode the ice sheet from beneath. While most climate models predict that the WAIS will remain fairly stable, there are worrying signs that some models may not reflect what is actually happening on the ground. A catastrophic collapse of the WAIS is not probable; a gradual response occurring over thousands of years is still much more likely, but once initiated, the melting of this ice sheet could have a major effect on sea level beyond the year 2100.

The Antarctic Peninsula The Antarctic Peninsula is a long neck of land stretching from Antarctica toward South America. This is the only part of Antarctica that is showing significant effects of global warming. Over the past 50 years, temperatures have risen over 2.5°C (4.5°F), at an astonishing rate of 0.5°C (0.9°F) per decade. The 12,000-year-old Larsen B Ice Shelf on the northeast margin of the peninsula became the focus of international climate research when, in 2002, it suddenly collapsed over a period of just a few days (Figures 2.19 and 2.20). Larsen B had a surface area of 3,250 km² (1,254 sq mi) and was over 120 meters (394 feet) thick; its shattering released 720 billion tonnes (794 t(US)) of ice into the Weddell Sea.

The warming ocean currents that undermined the Larsen B Ice Shelf continue to melt ice from the margins of the Antarctic Peninsula. Since 1974, there has been a gradual decline in ice shelf volume in the region of over 13,500 km² (5,212 sq mi). The collapse of the Larsen B Ice Sheet had a huge impact on climate science. Scientists once thought that ice sheets would disintegrate slowly, but it became apparent that rapid collapse can occur as an ice shelf is undermined by warmer water, lifted by rising sea level, or impacted by changing

▼ **Figure 2.19:** Satellite image of the Larsen B Ice Shelf prior to breakup. The light-blue lines on the surface of the ice show where it was starting to melt.

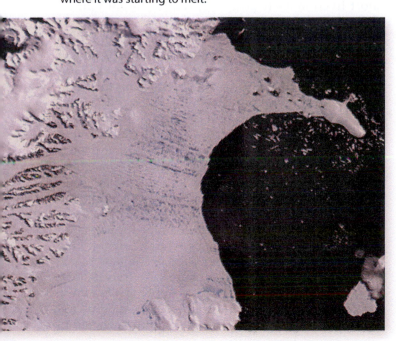

▼ **Figure 2.20:** The Larsen B Ice Shelf just a few days later. The shelf has completely fragmented. The light blue areas are a mixture of seawater, ice slush, and icebergs.

winds. Once ice shelves are gone, there is nothing to keep ice from flowing from the interior of the ice sheet, and ice streams can accelerate. The entire WAIS could collapse rapidly in this manner.

CHECKPOINT 2.17 ▶ Describe how warming ocean currents can have a greater impact on the ice than can a warming atmosphere.

2.3 PAUSE FOR... THOUGHT | *Would the world be a better place if all the ice sheets melted?*

The answer to this question is highly subjective. The last time Earth had no ice sheets was during the Cretaceous Period and the early part of the Paleogene circa 65 million years ago. Sea level was more than 100 meters (300 feet) higher than today, with shallow oceans covering much of the interior of the larger continents. Life was abundant and diverse both on land and in the oceans, and the climate was relatively stable. Global warming and melting ice sheets are not themselves a threat to the global environment. It is possible that the world would be a better place if barren tundra and ice sheets were replaced by diverse temperate ecosystems. But the problem today is not climate change alone; it is the rate at which such change is taking place, and the threat this creates for many vulnerable communities across the world.

Permafrost

Permafrost is permanently frozen ground that acts like an insulator, trapping moisture, heat, and trillions of tonnes of biogenic methane deep under the surface. All over the Arctic region, permafrost is starting to thaw more deeply and more widely than ever before, and although the full impact of this melting is uncertain, it is bound to accelerate the rate of climate change and radically change the nature of Arctic ecosystems.

Deep permafrost underlies vast swaths of **tundra** in Arctic Canada, Alaska, Siberia, and the Tibetan Plateau (**Figures 2.21** and **2.22**). In total, 20% of Earth's landmass is affected by seasonal and permanent permafrost, and there are clear signs that some areas are starting to warm and melt (**Figures 2.23** and **2.24**). Since 1900, the maximum area covered by frozen ground has decreased by 8% (7 million km² or 2.7 sq mi) during the Northern Hemisphere winter and by up to 15% during the spring (Figure 2.24).

The actual depth of permafrost can vary from a few meters to more than 1,000 meters (3,281 feet), but at the top of the permafrost layer, temperatures have increased by as much as 3°C (5.4°F) in recent years. Some places in Alaska, Canada, and Siberia have seen buildings and roads subside as the surface of the permafrost has thawed deeply, creating thousands of

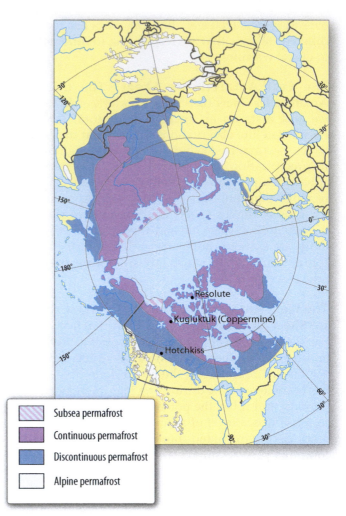

Subsea permafrost
Continuous permafrost
Discontinuous permafrost
Alpine permafrost

▲ **Figure 2.21:** A map of the area of the globe covered by permafrost.

short-lived lakes (**Figure 2.25**) that eventually drain as the ice beneath them finally melts.

The snow that covers permafrost for much of the year has a high **albedo,** reflecting as much as 90% of the Sun's energy back into the atmosphere. This has been very effective at protecting the permafrost from the warming effects of the Sun for much of the year. With the recent earlier onset of spring-like temperatures, however, freshly exposed soils that have a very low albedo of around 10% absorb energy that further heats the surface and accelerates the process of melting.

Melting also releases trapped greenhouse gases from beneath and within permafrost. The cold, wet, boggy lakes and pools that remain as permafrost melts are starved of oxygen and an ideal home for **methanogenic** bacteria, which release methane (CH_4) into the atmosphere (**Figure 2.26**). In addition, abundant organic material once frozen in the permafrost is exposed at the surface and oxidized in the air, producing even more carbon dioxide (CO_2). This enhanced release of natural greenhouse gases can intensify global warming,

▲ **Figure 2.22:** Tundra forms a barren and forbidding landscape. Recent surface melting of surface permafrost produces a lot of standing surface water that eventually drains as the permafrost melts to greater depths.

forming a *positive climate feedback loop*, as more warming melts more permafrost and results in the release of even more CH_4 and CO_2.

CHECKPOINT 2.18 ▶ How does melting permafrost contribute to global climate change?

Mountain Glaciers

All around the globe, many glaciers are retreating at an astonishing rate. Widespread melting of glaciers appears to be a direct result of atmospheric warming, but there is still a lot of confusion surrounding the fact that some glaciers are still advancing. Why is this? The answer lies in the fact that glaciers are complex rivers of ice that ebb and flow, depending on a number of competing factors.

If a glacier is to keep advancing, the amount of snow that falls during the winter must equal or exceed the amount of summer melt. In addition, a combination of other factors—such as the degree of slope, changing summer temperatures, temperature at the base of the ice, and variable surface reflectivity—make each glacier unique. There are more than 160,000 glaciers around the globe, and most of them are not well documented. But thanks to satellites, it is possible to monitor many more glaciers today than in the past, and the results are unequivocal: Glaciers around the world—in the Arctic and Antarctic, in the European Alps and the Himalayas, in the American Northwest, the Andes, and the Antipodes—are melting at an unprecedented rate; but glaciers are not melting everywhere. Climate change can still produce

▼ **Figure 2.23:** The permafrost landscape on the shore of the Yana River in Yakutia in northeastern Russia.

▶ **Figure 2.24:** (a) The mean annual temperature at the permafrost surface and (b) projected change in mean annual temperature at the permafrost surface.

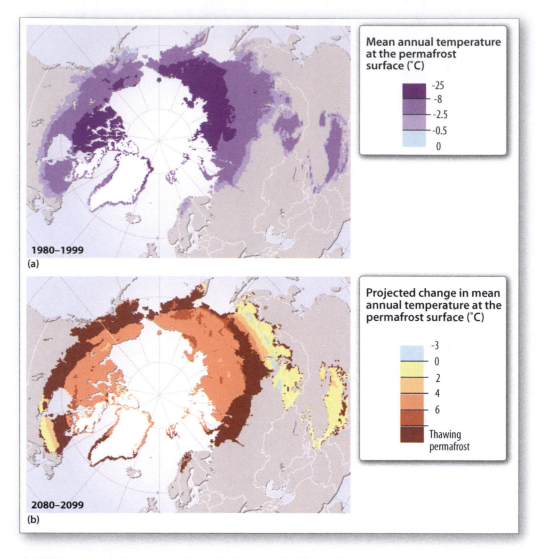

Mean annual temperature at the permafrost surface (˚C)

- -25
- -8
- -2.5
- -0.5
- 0

1980–1999
(a)

Projected change in mean annual temperature at the permafrost surface (˚C)

- -3
- 0
- 2
- 4
- 6
- Thawing permafrost

2080–2099
(b)

▼ **Figure 2.25:** Lakes form on the surface of tundra as permafrost melts.

▲ **Figure 2.26:** A researcher setting methane trapped under ice ablaze.

some surprises, including regional cooling and more winter precipitation that can cause some glaciers to advance. Individual glaciers in some parts of the world are advancing, but they are exceptions; most glaciers are rapidly losing mass and retreating toward their source (**Figure 2.27**).

CHECKPOINT 2.19 ▶ What factors control whether a glacier will advance or retreat?

   Ice loss from retreating glaciers and small, mountain ice caps is estimated to have been responsible for as much as 27% of recent sea level rise (**Figure 2.28**). It is difficult to determine the last time that glaciers were in similar full retreat because historical data is sparse. The recent discovery of the frozen body of an "ice man" beneath melting ice in the European Alps certainly suggests that this particular mountain pass had been frozen for at least the past 5,000 years. Remote-sensing data from satellites such as NASA's **Advanced Spaceborne Thermal Emission and Reflection Radiometer (ASTER) satellite** help to quantify this loss of mass

from mountain glaciers. Some estimates suggest that the European Alps may have lost as much as 30% to 40% of surface area and as much as 50% of volume since the 1850s. Chapter 7 will show how the retreat of glaciers in the Himalayas poses a serious regional threat to local communities and water resources in southern and eastern Asia.

CHECKPOINT 2.20 ▶ Why is our knowledge of the rate of glacier retreat much more limited prior to the 1970s than from the 1970s to the present?

Changing Patterns of Climate

The slow increase in global temperature since 1880 is having a profound impact on weather around the globe. There are now more intense heat waves, a longer growing season, fewer frost days, an apparent increase in the frequency and severity of floods and storms, and the first days of spring arrive earlier than they used to (**Figure 2.29**). All these phenomena are affected strongly by natural variations in regional climate. Any single event can be due to natural causes, but as energy is added to the atmosphere and oceans, a global pattern of climate change is emerging from the background of natural variation.

Changes in Diurnal Temperature

Both daytime and nighttime warm temperature extremes have increased, but nighttime temperatures are rising faster (**Figure 2.30**). Why is this? The more rapid increase in average nighttime temperature is a clear response to increasing amounts of the greenhouse gases water vapor and CO_2 in the atmosphere that act to slow radiative heat loss. This is direct evidence that greenhouse gas emissions are changing the balance of energy in the troposphere.

CHECKPOINT 2.21 ▶ What impact will an increase in water vapor and carbon dioxide in the atmosphere have on the nighttime ground temperature of the desert in Nevada?

Changes in Precipitation

The amount of water vapor that air can hold increases rapidly with temperature, at a rate of about 7% per 1°C, and as part of a natural water cycle, this will inevitably lead to an increase in rainfall. To some degree, this is exactly what has been observed (**Figure 2.31**), but the details are complex. The warming atmosphere and oceans also impact many interrelated natural climate cycles (see Chapter 3), and these in turn have a significant impact on regional climate.

   Although Earth is warming on average, some parts of the planet will still get cooler, some wetter, some hotter, and some much drier. It is still difficult to prove that an increase in precipitation in any one region is due to

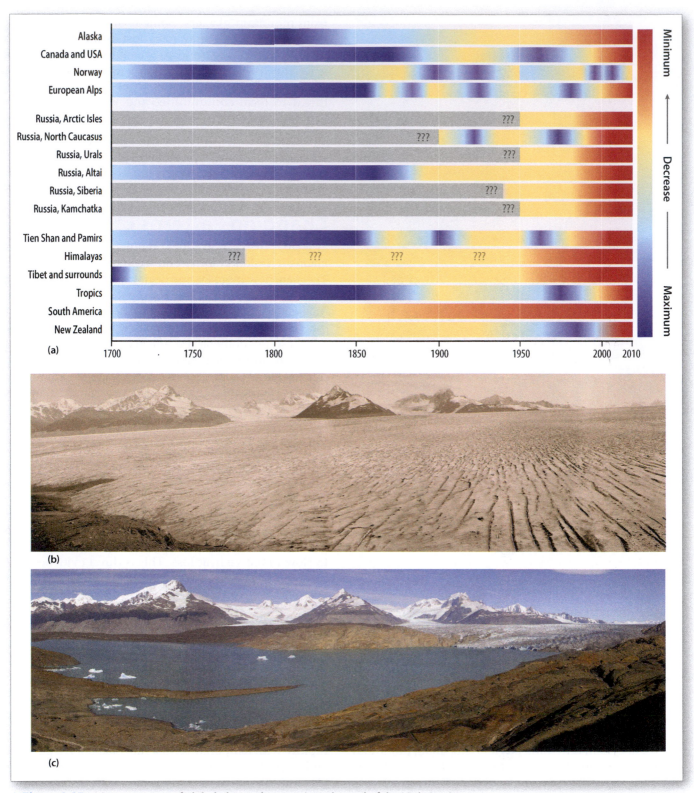

▲ **Figure 2.27:** (a) An overview of global glacier changes since the end of the Little Ice Age. Deepening blue colors mark periods of glacial retreat, and deepening red marks periods of glacial advance. Question marks note areas of uncertainty. The before (b) and after (c) pictures of the Upsala Glacier, Patagonia, Argentina, show why the retreat of glaciers since 1850 has become an increasingly important climatic issue.

climate change and not just a function of natural climate variability; there are just too many historical and physical unknowns to have certainty. But recent studies appear to suggest a strong connection, and around the globe, there is an emerging pattern of many regions experiencing an increase in the intensity of precipitation (heavy rain) and in the frequency of intense precipitation events (the number of really severe downpours), although not

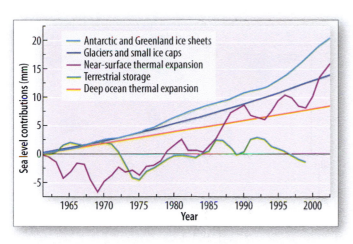

▲ **Figure 2.28:** This graph shows the variable contribution to sea level rise from different sources since the early 1960s.

example, the frequency of precipitation has increased by as much as 20% since 1900, and the number of days with precipitation has increased by 50% over the past century. Some of this change is undoubtedly related to regional decadal variation in atmospheric and oceanic circulation, but not all of it. The signal is currently weak, but the physics of air saturation is well established. If warmer air comes in contact with water, it will become more humid, and this will lead to more precipitation.

The immediate impact of more intense precipitation is an increase in the number of extreme flooding events. Although changes in regional atmospheric circulation are also a major factor, major floods in Europe in 2007 were widely blamed on global warming, as were catastrophic floods in parts of West Africa, Uganda, Sudan, Kenya and Ethiopia, Australia, and Pakistan in 2010 and 2011.

The impact of flooding is certainly greatest in the poorer nations of the world, where people are crowded onto vulnerable flood plains or packed into shantytowns that cling precariously to unstable slopes around many of the world's large cities. Developed nations can afford to meet the rising challenges with new or

necessarily an increase in the general frequency or overall amount of precipitation (Figure 2.32).

In the United States, there has been a general increase in precipitation of 10% since 1910, but there is significant regional variation. In the upper Midwest, for

▶ **Figure 2.29:** Graphs of historical and predicted change in the number of frost days (a) and heat waves (b) and in the length of the growing season (c) predicted by multimodel simulations from three global climate models using emissions data from the IPCC Special Report on Emission Scenarios (see Chapter 1). The maps show how the spatial patterns of these observations are projected to change between two 20-year means (2080–2099 minus 1980–1999) for the A1B scenario. Note how the length of the growing season and the number of frost days (which are closely related) has changed in the Northern Hemisphere.

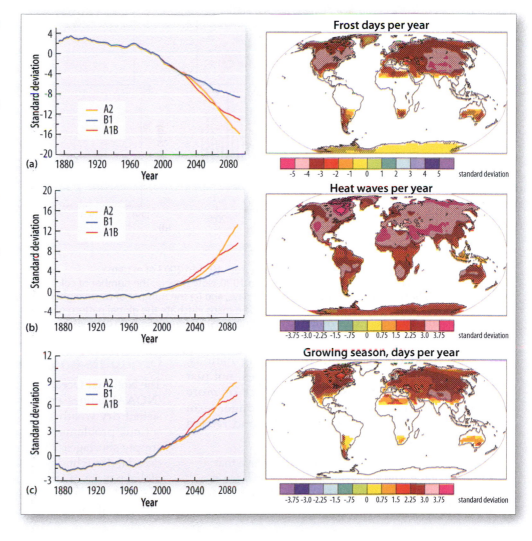

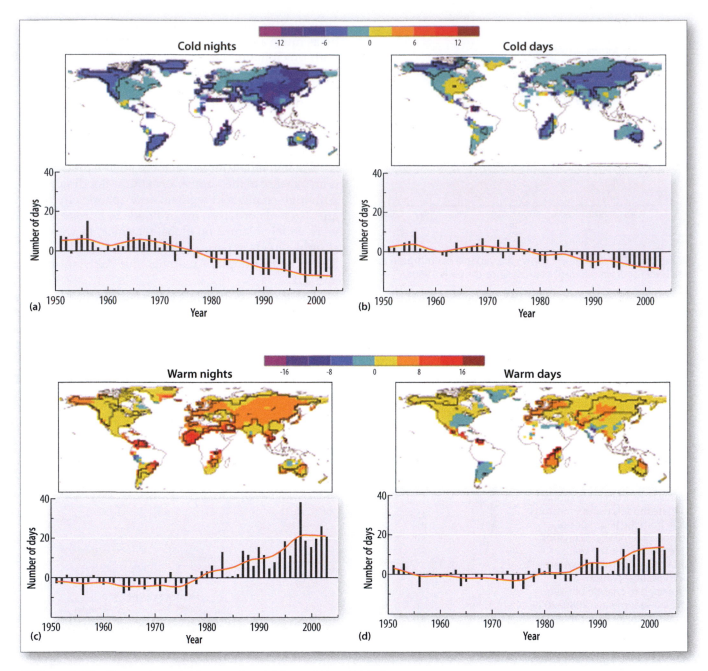

▲ **Figure 2.30:** Observed trends (days per decade) from 1951 to 2003 of extreme temperatures, based on average values from 1961 to 1990. Note how (a) the number of cold nights decreased faster than (b) the number of cold days, and (c) the number of warm nights increased faster than (d) the number of warm days. The maps illustrate the global pattern of climate change and the graphs the observed data (gray bars) and trend lines (in orange).

upgraded flood defenses, but developing nations, especially in Africa, face major challenges, not just from the critical and immediate impact of more frequent floods but from the resulting longer-term health risks due to contaminated water and related diseases (see Chapter 7).

CHECKPOINT 2.22 ▶ If climate scientists believe that many areas of the world may not see a large increase in the total amount of rainfall, why is there still reason to be concerned?

Droughts and Fires

The **National Oceanic and Atmospheric Administration (NOAA)** defines a *drought* as "a period of unusually persistent dry weather that persists long enough to cause serious problems such as crop damage and/or water supply shortages." The frequency and severity of multidecadal drought depends on many factors, including the global pattern of atmospheric and oceanic circulation and regional patterns of temperature, precipitation, evaporation, soil moisture, and wind speed. Together these

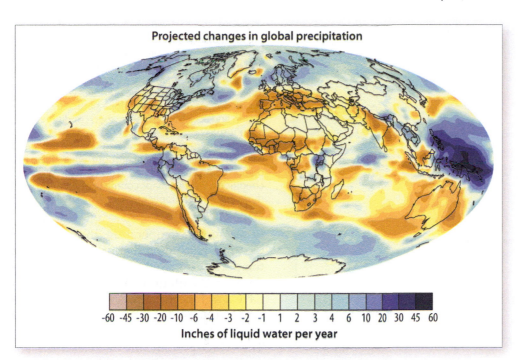

Projected changes in global precipitation

Inches of liquid water per year

◀ **Figure 2.31:** Expected changes in global precipitation by the end of the 21st century, expressed as inches of liquid water per year.

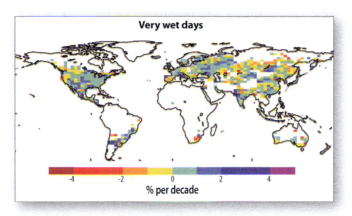

Very wet days

% per decade

▲ **Figure 2.32:** The change in the number of very wet days from 1951 to 2003, expressed as a percentage change per decade from areas where there is a reliable record. A very wet day is defined as a day with ≥1 mm of rain that exceeds the 95th percentile in the climate record. The colored squares indicate areas where data are available and the white areas where the climate record is poor or incomplete.

factors make a critical difference in the ability of an area to withstand even small changes in climate (Figure 2.33).

During periods of drought, forested areas and dry scrubland are at greatly increased risk of fire, and even the thick, wet rainforests of the Amazon Basin proved vulnerable to fire following an extensive drought in 2005, when thousands of square kilometers of forest burned. In California an increase in the number of wildfires—such as the huge brush fire in the Angeles National Forest off San Gabriel Canyon in 2012—is responsible for the loss of millions of dollars and puts many lives at risk. Agriculture also suffers badly, with a significant decrease in productivity and increase in crop disease as exemplified in 2012 by the lowest corn yield in the United States in 17 years and record corn prices. Droughts bring hunger,

disease, and conflict to people already living on the edge of survival and increasing hardship for those lucky enough to have the means to survive.

It is not possible to state with statistical certainty that recent droughts and wildfires are the consequence of global warming, but these observations are consistent with computer models that project that the greatest impact of future drought will be the interior of continents and at mid latitudes. Around the world, from the central and southwestern United States through parts of northern Mexico and the Amazon Basin, to the Mediterranean, the Sahel, southern Africa, western Asia, and southern Australia, many people and natural ecosystems are expected to suffer as the climate changes.

CHECKPOINT 2.23 ▶ What are the different factors that control the severity of drought, and how is an increase in the frequency of drought likely to affect the Amazon forests?

Hurricane Frequency and Intensity

Following the Hurricane Katrina disaster in 2005 (Figure 2.34) a movie on climate change by Vice President Al Gore sparked controversy among climate scientists and skeptics alike by suggesting (indirectly) that Katrina was a consequence of global warming. His film certainly heightened awareness of the potential risks associated with an increase in hurricane intensity, and this stimulated further research into possible links between hurricanes and global warming.

The frequency of hurricane and typhoon activity and the paths these storms follow are determined by many factors that involve complex interactions between the oceans and atmosphere (Figures 2.35 and 2.36). Some research models predict that storms may become more frequent, but this depends on many factors in addition

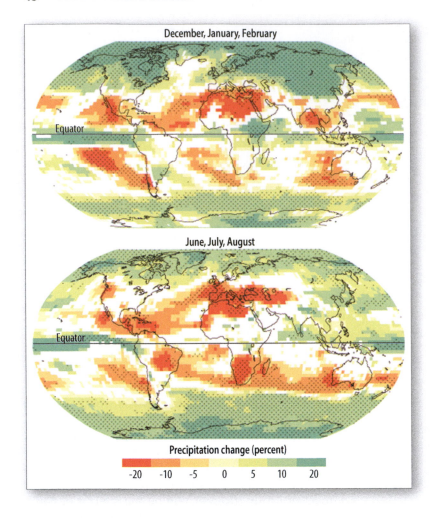

Precipitation change (percent)
-20 -10 -5 0 5 10 20

◀ **Figure 2.33:** Southern Europe faces the prospect of increased desertification (desert expansion) as annual precipitation decreases and evaporation rates increase with rising global temperature. This map uses a Drought Severity Index, which assigns positive numbers when conditions are unusually wet for a particular region and negative numbers when conditions are unusually dry. A reading of –4 or below is considered extreme drought. Regions that are green will likely be at lower risk of drought, while those in orange could face more unusually extreme drought conditions. The two maps contrast conditions during (a) winter (upper map) and (b) summer (lower map).

to sea surface temperature. In the Atlantic Ocean, a natural oscillation in sea surface temperature called the **Atlantic Multi-decadal Oscillation (AMO)** is known to have a significant impact on the frequency of hurricanes. The spawning ground for Atlantic hurricanes is the tropical Atlantic. The AMO periodically increases the sea surface temperature in this area, which affects the frequency of hurricanes. The presence of a strong El Niño is also known to decrease storm intensity by creating strong upper-level winds that tear a growing hurricane apart before it has time to form. El Niño is still a subject of very active research, and a lot remains to be discovered. Hurricanes and the factors that control them are discussed in more detail in Chapter 4.

▼ **Figure 2.34:** Category 5 Hurricane Katrina over the Gulf Coast, August 2005.

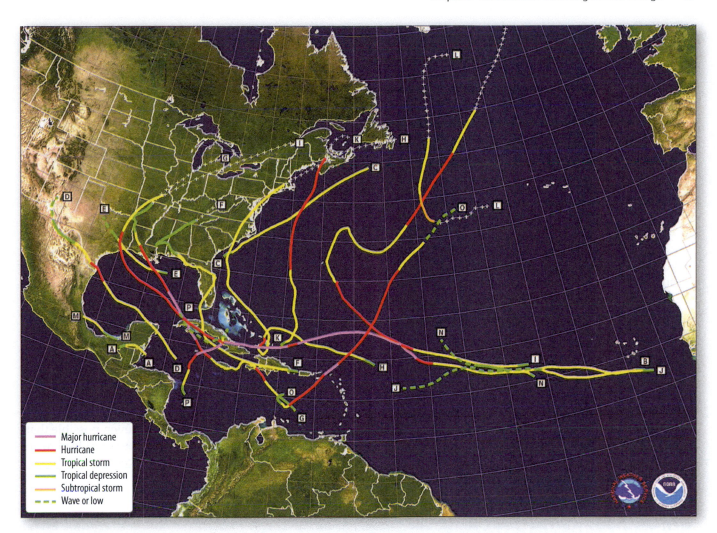

▲ **Figure 2.35:** Tropical storm and hurricane tracks in the Atlantic Ocean during the 2008 season. As each tropical storm develops it is assigned a name based on the letters of the alphabet. The letters (A through P) indicate the relative chronology of the storms in that year.

Some scientists have concluded that an increase in hurricane intensity by as much as 2% to 11% by 2100 is possible as the surface temperature of the ocean increases. Warmer sea surface temperatures allow more water vapor to evaporate into rising air over the oceans. When this water later condenses to form rain, it releases latent heat that feeds the intensity of growing storms. This simple relationship suggests a plausible link between sea surface temperature and storm intensity, but this is still unproven.

CHECKPOINT 2.24 ▶ Did Hurricane Katrina have anything to do with global climate change? Explain your answer.

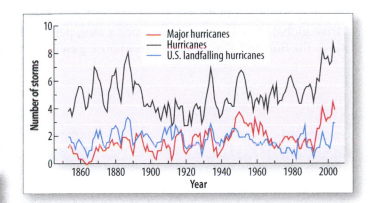

▲ **Figure 2.36:** Atlantic hurricane counts, 1851–2006 (five-year running mean). The increase in the number of events after the year 2000 may be related to climate change but also reflects natural cyclical changes in ocean temperature in the North Atlantic.

Summary

- In many places around the world, there are signs that the oceans and atmosphere are warming and that global climate is changing. Global sea level is rising. There are signs of increased melting throughout the cryosphere, especially in northern Polar regions, where regional warming is well above the global average.
- Global surface temperature as measured by satellite and ground stations around the world is increasing on land faster than the oceans, but the large heat capacity of water means that a vast amount of energy is accumulating in the oceans.
- Global sea level is rising steadily due to a combination of thermal expansion of the oceans and the addition of water from the melting of ice sheets and glaciers.
- Sea level is constantly changing with global temperature and the growth and melting of ice sheets. During the last glacial maximum, sea level was as much as 100 meters (300 feet) lower than today, and the geological record of sea level change shows that sea level fluctuates slowly and continuously with the growth and decay of mid-oceanic ridges.
- The cryosphere in the northern hemisphere is reacting rapidly to climate change with a significant reduction in the area, thickness, and age of sea ice, the retreat of glaciers, extended melt seasons, and the melting of permafrost. Positive feedback into the climate system from changing albedo and the release of biogenic carbon dioxide and methane is adding to the impact of warming and accelerating climate change.
- While the rate of melting of Arctic sea ice is an important indication of regional climate change, it adds nothing to sea level as this ice is already floating on water, but meltwater from melting glaciers and ice sheets in the Arctic and Antarctica is adding water to the oceans and increasing sea level.
- The melting of permafrost is leading to an increase in the rate of coastal erosion and to the release of large quantities of biogenic carbon dioxide and methane into the atmosphere. These greenhouse gases are accelerating the rate of global warming.
- Patterns of precipitation are changing. Extreme climate events—including extreme heat waves and more intense precipitation events—are more common today than in the past. It is even possible that elevated sea surface temperatures could lead to an increase in the intensity of hurricanes.

Why Should We Care?

Climate is changing around the globe, and there is clear evidence of significant warming in both the atmosphere and the oceans. Natural climate cycles are important but cannot explain the sustained global change observed for over 100 years. Only two factors have the power to drive global climate change over such a long period of time: the Sun and atmospheric greenhouse gases. Both have the ability to alter the flow of energy at the top of the atmosphere, and both are highly likely to have played at least some role in changing climate over the past century. How can we tell which of these two factors is dominant? To answer this question, Chapter 3 will examine the historical record of both solar activity and greenhouse gas emissions.

Looking Ahead . . .

Chapter 3 investigates how energy from the Sun is received, distributed, transferred, stored, and released within Earth's climate system. Earth's temperature is determined by a delicate balance between the amount of energy arriving from the Sun and the amount of energy leaving Earth from the top of the atmosphere. The addition of greenhouse gases to the atmosphere has upset this balance. How much global temperature will change as a result depends on competing factors that can both augment and diminish the direct impact of greenhouse gas emissions.

Critical Thinking Questions

1. Was it an ecological disaster when the Chesapeake River valley flooded at the end of the last glaciation? Explain your answer.

2. The Arctic region was warm during the 1930s. What is different about the warming that is taking place at the moment that makes us so sure it is not just a repeat of that natural event?

3. The Greenland Ice Sheet is thought to have lost 80 km³ (19.2 cu mi) of ice per year between 1993 and 2003. Find out how many cubic kilometers of ice make up the entire ice sheet. (Wikipedia has a reasonable estimate.) Calculate how long it would take the entire ice sheet to melt at that rate. What if the rate doubled? Do you consider this an immediate threat?

4. Do you think that air temperatures or sea temperatures are a more important indicator of the likely instability of the West Antarctic Ice Sheet?

5. As permafrost melts, it releases methane, a strong greenhouse gas, into the atmosphere. Do you consider this a natural process, or is it anthropogenic? Why?

6. Why do environmentalists seldom talk about some of the advantages of early global warming? Is this silence justified, given the controversial nature of the debate? Explain your answer.

7. Is it ever right to overstate your cause and make specious connections between natural disasters and climate change when there is little or no scientific evidence to support these connections? Do the ends ever justify the means?

Key Terms

Please make sure you are familiar with the key terms introduced and highlighted in this chapter. A full glossary is available at the end of the book.

Albedo, p. 40
Atlantic Multidecadal Oscillation (AMO), p. 48
Advanced Spaceborne Thermal Emission and Reflection Radiometer (ASTER) satellite, p. 43
Advanced Very High Resolution Radiometer (AVHRR), p. 29
Base period, p. 27
Decadal, p. 32
Denudation, p. 34
East Antarctic Ice Sheet (EAIS), p. 39

Ecosystems, p. 32
Extended Reconstructed Sea Surface Temperature (ERSST) analysis, p. 29
Gigatonne, p. 36
Global Historic Climatology Network (GHCN), p. 29
Gravity Recovery and Climate Experiment (GRACE), p. 36
Hadley Centre and Climate Research Unit at the University of East Anglia (HadCRUT), p. 28
Heat capacity, p. 33

ICESat and CryoSat, p. 35
International Comprehensive Ocean–Atmosphere Data Set (ICOADS), p. 29
Methanogenic, p. 40
Microwave sounding units (MSUs), p. 29
Multiyear ice layer, p. 35
National Climate Data Center (NCDC), p. 29
National Oceanic and Atmospheric Administration (NOAA), p. 46
Positive feedback, p. 37

Radiosondes, p. 29
Regional tectonic stress, p. 32
Remote Sensing Systems (RSS), p. 29
Sea ice volume, p. 37
Sea surface temperature (SST), p. 29
Stacking, p. 27
Stochastic, p. 26
Thermal expansion, p. 32
Tundra, p. 40
University of Alabama at Huntsville (UAH), p. 29
West Antarctic Ice Sheet (WAIS), p. 39

PART 2

FOLLOW THE ENERGY: ATMOSPHERE, OCEANS, AND CLIMATE

CHAPTER 3: ENERGY AND EARTH'S CLIMATE

CHAPTER 4: UNDERSTANDING WEATHER AND CLIMATE

Small changes in the balance of mass and energy between the atmosphere, biosphere, cryosphere, oceans, and lithosphere control the direction of global climate change. In this image from Antarctica, an iceberg drifts out to sea before melting in the Southern Ocean

To project the future of climate change and its impact on mankind and the environment, we must understand how small changes in radiative forcing, both natural and anthropogenic, can affect Earth's climate system. To do this, we must understand how both mass and energy are exchanged between the atmosphere, oceans, land, cryosphere, and biosphere. Only then will it be possible to project how much of the climate change we observe today is natural and how much is driven by heat-trapping greenhouse gases. Chapter 3 investigates the flow of energy though Earth's atmosphere and oceans, and Chapter 4 shows how this flow translates into patterns of global weather and ocean circulation.

Viewed at this angle from space, the thin blue veneer of Earth's atmosphere appears tenuous and vulnerable. Visible light is able to reach the camera from the surface, but infrared radiation is trapped within the atmosphere and keeps the planet warm enough to support liquid water and abundant life.

CHAPTER 3

Energy and Earth's Climate

Earth's climate system can be defined as the sum of all exchanges of energy and mass between the atmosphere, hydrosphere, cryosphere, biosphere, and lithosphere that act together to determine the state of global and regional climate. Different factors affect how energy travels though the climate system. Some help to retain energy and warm the planet, while others enhance the flow of energy back to space and cool the planet. Changes in average global temperature are a measure of how closely balanced these competing factors are, and recent changes in global average temperature indicate that slightly more energy is being retained than lost.

On a scale of thousands of years, changing solar intensity is by far the most important factor controlling global climate, but over recent decades small changes in the sun are more likely to have caused global cooling rather than warming. Volcanic eruptions can also have a major impact on global climate, but they tend to cool the atmosphere and the level of global volcanic activity is relatively low at the moment. Natural variations in the atmosphere and oceans can produce large swings in climate from year to year like those we observe today, but are not able to produce the steady underlying trend of increasing global temperature observed over recent decades. The impact of man-made heat trapping greenhouse gases is the only factor that can reasonably explain recent changes in global temperature.

Learning Outcomes | *When you finish this chapter you should be able to:*

- Describe and explain the roles that the Sun and orbital cycles play in controlling Earth's climate
- Understand how energy is received at the top of the atmosphere and then cycled through Earth's climate system
- Recognize that complex interrelationships exist between different parts of the climate system and that small changes in just one part can have a global impact
- Identify major factors that influence radiative forcing
- Compare the nature and origin of both natural and enhanced greenhouse gas effects
- Distinguish between natural and anthropogenic radiative forcing and demonstrate why anthropogenic forcing is dominant
- Understand the meaning of climate sensitivity and why this has proved difficult to quantify.
- Describe how positive and negative climate feedbacks add to uncertainty when projecting the exact impact of increasing concentrations of greenhouse gases

The Source of All Energy: The Sun

Without the Sun, there would be no climate. The slow flow of **geothermal** heat from Earth's interior is not enough to keep the surface warm, and without a constant flow of energy from the Sun, it would soon freeze, and life could exist only deep under the surface, where geothermal energy could sustain conditions necessary for survival. Solar energy is the fuel that drives Earth's climate machine, and any discussion of climate change must first investigate whether changes in the output of solar energy could be responsible for recent climate change.

Deep within the core of the Sun, 150 million kilometers (93 million miles) away, is a nuclear furnace where the temperature reaches as high as 15 million degrees Kelvin (27 million degrees Fahrenheit). Under such intense temperature and pressure, hydrogen has been converted into helium by nuclear fusion for over 4.5 billion years (**Figure 3.1**). After a journey within the radiation zone that can last thousands of years, this energy, some 174 petawatts (PW; 1 PW = 10^{15} W), is radiated into space from the photosphere, the visible surface of the Sun, at a temperature of around 6,000K, and then takes just over 8 minutes to reach Earth. The photosphere (**Figure 3.2**) is in constant motion, driven by a deeper convecting layer, where the complex rotation of the Sun causes intense **magnetic fields** to twist, fold, tighten, and eventually break and shoot out massive **solar flares**, or "prominences," deep into space, beyond the Sun's tenuous corona of hot plasma. (**Figure 3.3**).

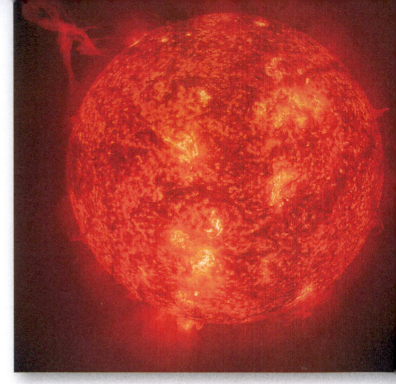

▲ **Figure 3.2:** Seen in detail, the Sun is in constant turbulent and violent motion.

CHECKPOINT 3.1 ▶ Is the surface of the Sun hotter than Earth's core? What is the temperature of each?

By the time this energy reaches as far as Earth's orbit, 150 million kilometers (93 million miles) away, it has spread out over a vast area and the energy flux (the flow of energy per square meter) is much weaker, equivalent to the output from five or six household light bulbs (measured

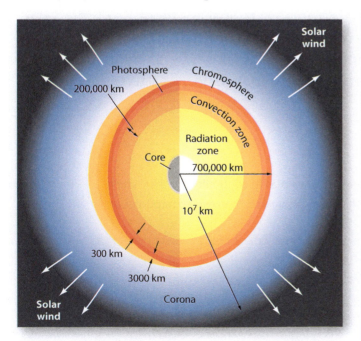

▲ **Figure 3.1:** The internal structure of the Sun is complex and hidden beneath the visible surface of the photosphere. Between the photosphere and the core a zone of intense convection overlies the radiation zone where photons generated in the interior can be trapped for thousands of years before they finally escape at the surface. The chromosphere is a tenuous layer that is only weakly visible during a solar eclipse.

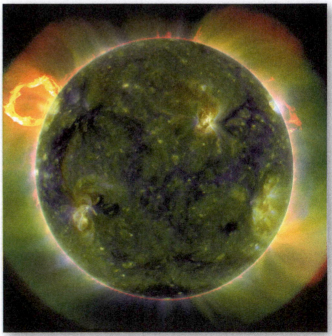

▲ **Figure 3.3:** The Sun is a furnace of nuclear fusion that is 4.5 billion years old. This is a full-disk multiwavelength extreme ultraviolet image of the Sun taken on March 30, 2010. False colors trace different gas temperatures. Reds are relatively cool (about 60,000 Kelvin, or 107,540°F); blues and greens are hotter (greater than 1 million Kelvin, or 1,799,540°F).

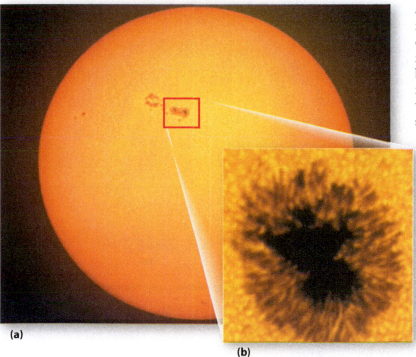

(a)

(b)

▲ **Figure 3.4:** (a) Sunspots are darker, cooler areas of the Sun's surface where intense magnetic fields reduce the flow of energy to the photosphere. (b) An umbra or "shadow" forms the dark center of each sunspot, surrounded by a slightly lighter region called the penumbra.

in Watts) for every 1 square meter (10.8 sq ft) of Earth's surface. The amount of radiation received at the top of the atmosphere from the Sun is called the **total solar irradiance (TSI)**, and was first measured directly by satellite in 1978 as 1,370 watts per square meter (Wm^{-2}). From this total, we receive *on average* only 342 Wm^{-2} of radiation at the top of the atmosphere because of the combined effect of Earth's curvature and rotation. (Remember that half the time is night, when there is no sunlight!)

TSI is not static, and even small changes in solar irradiance can have an impact on global climate. Scientists have shown that TSI varies naturally by as much as 0.5 to 0.75 Wm^{-2} due to sunspot cycle activity and by as much as 3.4% (46 Wm^{-2}) due to changes in the patterns of Earth's orbit.

CHECKPOINT 3.2 ▶ Why is the amount of solar radiation measured at the top of the atmosphere so much less, on average, than the amount of solar radiation measured by satellite?

Sunspots

The surface of the Sun is not featureless. Early astronomers in China around 800 B.C. were among the first to record dark spots on the surface. After many years of careful observation, German astronomer Heinrich Schwabe published an article in 1843 describing how these "sunspots" wax and wane in a rhythmic cycle that is now called the **Schwabe cycle**.

Sunspots are large (up to 50,000 kilometers [31,068 miles]), slightly depressed areas in the photosphere that

are around 1,500K cooler than the surrounding Sun (**Figure 3.4**). A sunspot marks a point of intersection where tightly coiled magnetic fields penetrate the photosphere and partially block radiation from deeper in the Sun from reaching the surface. They come in pairs: one where the coiled magnetic field exists on the surface of the photosphere and the other where it returns beneath the photosphere. Sunspots are darker and cooler than the surrounding Sun but are surrounded by bright areas of irradiance called **faculae**. Faculae actually emit energy with greater-than-average intensity, so although sunspots appear to make the Sun darker and cooler at the surface, they actually increase the total amount of solar radiation it emits.

The Solar Cycle

Sunspot intensity waxes and wanes over an approximately 11-year solar cycle. Each new cycle is marked by a reversal in the north–south polarity of the Sun's magnetic field, and these cycles have been carefully recorded since 1755. The recent end of cycle 23 and beginning of cycle 24 (**Figure 3.5**) was an unusually quiet period for the Sun, with very little sunspot activity, and represented a 12-year low in solar irradiance.

Many other solar cycles have been identified. These include a 22-year Hale cycle that is linked to the Schwabe cycle and additional cycles with periods of 87, 216, 2,300, and 6,000 years. There is even an unproven suggestion that a 1,500-year solar cycle may have driven climate changes in the North Atlantic toward the end of the last glaciation; we will look at this again in Chapter 6.

CHECKPOINT 3.3 ▶ How and why do Schwabe cycles affect global temperature?

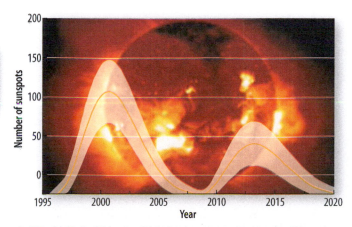

▲ **Figure 3.5:** The transition between sunspot cycles 23 and 24 saw an unusually prolonged period of lowered TSI. This figure illustrates how the intensity of sunspots varies over time with pronounced peaks and troughs of solar activity. Earth tends to warm slightly when activity is high.

Changes in TSI Measured by Satellite

Satellite observations confirm that changes in sunspot intensity and cycle duration can **modulate** the average amount of radiation reaching Earth by as much as 0.1% both above and below background levels. There was a slight increase in TSI over the industrial era (ca. 0.04 Wm^2) but since 1978, over the last three solar cycles, TSI has decreased to ca. −0.04 Wm^2. This figure alone is much too small to be a major factor in the recent global warming. Therefore, if the Sun was responsible for some early global warming and climate change, there must be other factors that amplified this effect.

Could the Sun Be Responsible for Recent Climate Change?

While most scientists believe that the impact of the Sun on recent climate change is limited, some propose that solar activity could have a greater influence on climate than is generally accepted. **Figure 3.6** shows that the Sun was very active during the 20th century, with more sunspot activity than at any other time over the past 8,000 years. This could explain some of the global warming that took place before 1978, but since that time, TSI has actually decreased, while global temperatures have continued to rise (**Figure 3.7**). Climate models can explain and reproduce these observations only if there is an additional component to recent warming, and this is very likely to be the recent rapid rise in the level of carbon dioxide and other greenhouse gases in the atmosphere.

Changes in sunspot activity also affect the strength of Earth's protective magnetic field. When sunspot activity is low, the magnetic field is weaker, and more high-energy cosmic rays are able to penetrate deep into the atmosphere. Some scientists suggest that this could lead to an increase in cloudiness by creating more cloud condensation nuclei (small particles that allow water to

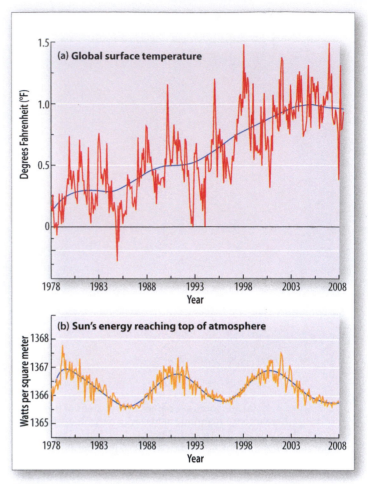

▲ **Figure 3.7:** The upper graph (a) traces the actual path (red line) and averaged path (blue line) of global temperature change since 1978. The lower graph (b) traces the actual path (yellow line) and averaged path (blue line) of solar energy output as measured by satellite at the top of the atmosphere since 1978. These data clearly show that small changes in global temperature may be related to the sunspot cycle, but that overall changes in energy received from the Sun cannot explain the observed long-term warming trend in global surface temperature. It is therefore not likely that recent global warming is due to changes in solar intensity. Note that the vertical temperature axis on this diagram is in degrees Fahrenheit.

condense more readily). This, in turn, would have a major impact on Earth's albedo and lead to global cooling. This is an interesting hypothesis, but there is currently no consistent evidence to suggest that it is correct. It is especially difficult to promote this hypothesis when the recent decline in sunspot activity has coincided with the hottest decade in the historical record.

> **CHECKPOINT 3.4 ▶** Is there any evidence that the Sun is responsible for recent changes in global temperature? Explain your answer.

The Sun clearly plays a critical role in determining the temperature of Earth, and recent observations and climate models have confirmed this effect. However, the absence of any significant upward trend in solar activity since 1978 suggests that the Sun is not the major factor driving recent global warming.

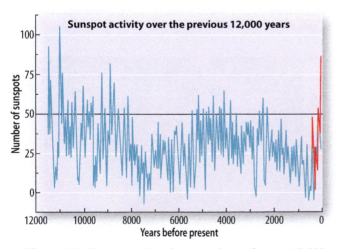

▲ **Figure 3.6:** Sunspot activity has varied over the past 12,000 years. The data in this chart are estimated from tree ring data and carbon isotopes and show that the level of recent solar activity is unprecedented in the past 8,000 years. The red line to the right adds recent data from direct observations.

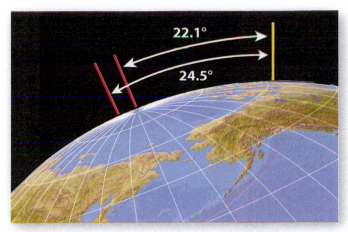

▲ **Figure 3.8:** Earth's axis is tilted—this is its obliquity—and the degree of tilt varies over time. This does not change the total amount of energy arriving from the Sun, but it does change the intensity of solar energy at different latitudes.

Orbital Cycles

As discussed in Chapter 1, the intensity of radiation from the Sun arriving at the top of Earth's atmosphere varies with cyclical changes in the shape of Earth's orbit and the tilt of Earth's axis known as Milankovitch cycles. This certainly appears to have been an important factor in determining how Earth's climate shifts between glacial and interglacial phases, although the actual changes in radiation (heat) intensity are so small that they must be amplified by feedback in Earth's climate system to have this effect. Three main factors are involved:

- The tilt, or **obliquity**, of Earth's axis
- The way this axis "wobbles"[1] over time, called **precession**
- The shape of Earth's orbit changes over time, called its **eccentricity**

Could these cycles be responsible for recent climate change?

Obliquity

Earth revolves around an axis that is inclined at an angle of around 23.4° to the plane of its orbit (**Figure 3.8**). Over time, the axis cycles back and forth between 22.1° and 24.5°, with a period of around about 41,000 years. It is the tilt of Earth that determines the passage of the seasons. If Earth's axis were not inclined, every part of Earth would receive a constant amount of energy from the Sun all year, with most at the equator and less at the poles.

When the axis is inclined, Earth still receives the same total amount of energy from the Sun, but that energy is distributed very differently around the globe. As Earth moves in orbit around the Sun, it spends some time with the Northern Hemisphere pointing away from the Sun (Northern Hemisphere winter), some of the time with the Northern Hemisphere pointing toward the Sun (Northern Hemisphere summer), and twice a year with both poles equidistant from the Sun (the spring and fall **equinoxes**).

When the Northern Hemisphere points more directly toward the Sun in the summer, it heats quickly, and ice that formed the previous winter is more likely to melt. As the tilt of Earth increases, the amount of solar radiation reaching latitudes north of 65° increases by more than 1% for every 1° increase in obliquity. The difference over the cycle at 65°N can be as much as 50 Wm^{-2} during the summer and 18 Wm^{-2} when averaged over the entire year. This appears to be a critical factor in controlling the advance and retreat of ice sheets. It is not so much the amount of snow that falls in the winter that is important as how much of this snow melts during the following spring and summer. The overall effect is most noticeable in the Northern Hemisphere because this is where most of the landmass of Earth is concentrated.

At the moment, the tilt is decreasing from its current angle of 23.4°, and it will reach a new minimum in around 8,000 years. This will make the difference between seasons in the Northern Hemisphere less intense, resulting in a much shorter summer melting season and new ice accumulation. That could signal the start of the next major period of ice advance.

CHECKPOINT 3.5 ▶ How much and over what period of time does the inclination of Earth's axis change? What impact does this have on global temperature?

Eccentricity

Earth's orbit around the Sun is an ellipse, not a perfect circle. The amount that the orbit departs from a true circle is known as its eccentricity (e), and this varies from $e = 0.005$ (an almost perfect circle) to $e = 0.06$

---

[1] Put your elbow on a table and point your index finger straight up. Now by only moving your finger, trace a circle with the tip of your finger. This is the way Earth's axis "wobbles" over time. The size of the circle depends on how much you tilt your finger as it moves, much as the inclination of Earth's axis modulates the impact of precession.

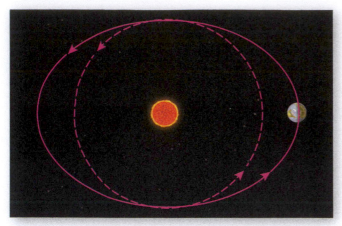

▲ **Figure 3.9:** Earth's orbit varies from circular to very slightly elliptical over time. When the orbit is more elliptical, Earth passes much closer to the Sun during parts of its orbit, increasing the total amount of solar radiation reaching the surface.

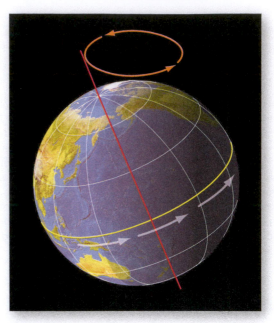

▲ **Figure 3.10:** Earth's axis is not only inclined but also rotates, or "wobbles," around a fixed point relative to the position of the stars in a precession cycle that lasts around 23,000 years.

(eccentric but still very circular). At the moment, the eccentricity of Earth's orbit is around $e = 0.017$.

Earth's orbit cycles between these states over periods of approximately 100,000 and 400,000 years (**Figure 3.9**). When the orbit is more elliptical, Earth spends part of the year farther from the Sun and part closer to the Sun than it would if the orbit were fully circular. The point in an elliptical orbit where Earth is closest to the Sun is known as perihelion, and the point where it is furthest away is called aphelion. As a result, the total amount of energy Earth receives from the Sun varies by around 0.2%. This is not enough to make a significant difference to Earth's climate, but in combination with the obliquity cycle, the seasonal difference can vary by around 100 Wm^{-2} (>20%) at high latitudes. This *is* enough to have a major impact on climate. Just 11,000 years ago, when the ice sheets of the last glacial advance were melting, perihelion occurred in July, when Earth's northern axis was pointing toward the Sun and polar heating in the Northern Hemisphere was at a maximum. At the moment, perihelion (closest to the Sun) occurs on January 3 and aphelion (farthest from the Sun) on July 4, a situation more likely to promote Northern Hemisphere ice sheet advance than global warming. Warmer winters result in enhanced precipitation near the poles, while cooler summers mean less ice melts.

CHECKPOINT 3.6 ▶ Over which months of the year is Earth currently closest to the Sun due to orbital eccentricity? What impact does this have on climate?

Precession

Earth's axis is not only inclined but also rotates, or "wobbles," around a fixed point relative to the position of the stars in a cycle that lasts around 23,000 years. This effect, called precession, is similar to what happens

when the axis on a child's top rotates around a fixed point as it spins (**Figure 3.10**). Precession also has an important influence on the amount of energy received from the Sun. When Earth's orbit is eccentric, especially when the axis is tilted at a high angle, precession determines whether the North Pole or South Pole is inclined toward the Sun at **perihelion** and therefore how much radiation is received from the Sun during the Northern Hemisphere summer (**Figure 3.11**).

Over time, all the different Milankovitch cycles interact to control the amount of solar radiation reaching Earth's atmosphere. The Northern Hemisphere has prolonged cool summers when the orbit is at its most eccentric, the tilt is at its smallest (less energy is gained at high latitudes and more energy is gained at tropical latitudes), and **aphelion** occurs in summer (when Earth is farthest from the Sun). This combination leads to ice advance. The opposite is true when the orbit is more eccentric, the tilt is large, and perihelion occurs during Northern Hemisphere summer. This combination leads to ice retreat. Most of the time, individual cycles act to both enhance and diminish TSI, and the net effect varies.

Over the past 800,000 years, the advance and retreat of the ice sheets appears to have been determined by the eccentricity of Earth's orbit, but prior to this, the obliquity cycle was more dominant. It is unclear why this transition occurred or even if it is likely to occur again. **Figure 3.12** shows how the different cycles combine to affect the amount of solar radiation received at 65° north of the equator. You can see the individual obliquity and precession cycles, but also notice how they are modulated (made larger or smaller) by the longer 100,000-year cycles of eccentricity. Solar radiation can be

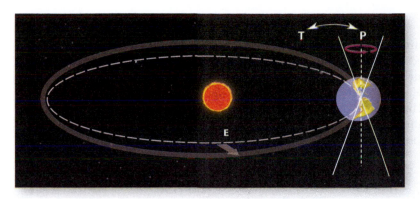

◀ **Figure 3.11:** The precession cycle determines whether the North Pole or South Pole is inclined toward the Sun at perihelion, the period when Earth is closest to the Sun as part of an eccentric orbit.

Could Orbital Cycles Be Responsible for Recent Climate Change?

Because Milankovitch cycles are regular, they can be used to predict the future. The eccentricity of Earth's orbit is currently low, as part of a larger 400,000-year cycle, and should stay that way for another 100,000 years. A decrease in obliquity has resulted in a gradual cooling of the Northern Hemisphere over the past 6,000 years that was discussed in Chapter 1 (**Figures** 3.12 and **3.13**). These data make it clear that orbital cycles are cooling Earth's climate and are not likely to be responsible for recent climate change.

CHECKPOINT 3.8 ▶ What is the evidence that Milankovitch cycles are unlikely to be the cause of recent climate change?

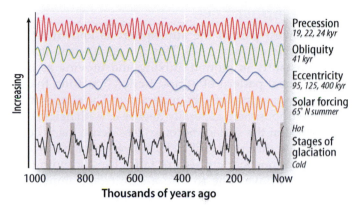

▲ **Figure 3.12:** Changes in the obliquity, eccentricity, and precession of Earth's orbit affect the critical amount of solar insolation received in the Northern Hemisphere and seem to control the advance and retreat of glaciers and ice sheets. Interglacial periods are indicated by the gray bars.

seen to arrive in slow pulses or packets of energy over time, and it is these that drive the major climate cycles.

CHECKPOINT 3.7 ▶ How does axial precession affect global climate?

3.2 **PAUSE FOR... THOUGHT** | *Putting natural cycles to rest.*

Obliquity, eccentricity, and precession are all pushing Earth toward a colder climate, and solar output is lower than it has been for decades. There are no other natural sources of radiative forcing, apart from the impact of greenhouse gas emissions, that can explain the underlying trend of global warming.

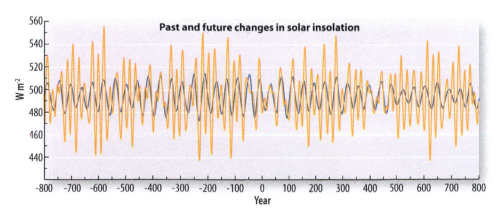

▲ **Figure 3.13:** Climate models can project the future impact of Milankovitch cycles, and trends suggest that the current interglacial period may be longer than usual, as part of a 400,000-year eccentricity cycle. The low eccentricity of the current orbit reduces the contrast between seasons and lowers the probability of ice sheets advancing. The orange line marks changes in solar insolation at high latitudes. Note the strong influence of precession that produces large swings in insolation. The green line marks the contribution to these changes from the obliquity cycle.

The Atmosphere

Energy from the Sun must pass through the atmosphere on its way to Earth's surface, and energy emitted by Earth must pass though the atmosphere on its way out to space. We must understand what controls this flow of energy if we want to understand recent climate change. This section describes the structure and composition of the atmosphere and discusses the important role that greenhouse gases play in regulating global surface temperature.

Structure and Composition of the Atmosphere

The atmosphere is a complex mixture of gases that evolved slowly following the formation of Earth (Figure 3.14). Much of the atmosphere is derived from degassing of the mantle, but continual bombardment by meteorites and comets for millions of years after Earth formed also contributed to its overall composition. Today, the atmosphere is composed of 78.08% nitrogen (N_2), 20.95% oxygen (O_2), and a variable amount of water vapor (1% to 3%), with traces of other gases, including carbon dioxide (~390 ppm) and methane (~1.8 ppm). The presence of oxygen is entirely due to the evolution of life and the emergence of photosynthesis around 3,500 million years ago.

The atmosphere is very thin relative to the diameter of Earth (Figure 3.15). If Earth were the size of a soccer ball, the atmosphere would be about as thick as a coat of varnish on its surface. Most of the mass of the atmosphere is concentrated close to the surface. Around 50% lies within 5 kilometers (3 miles) of the surface, and as much as 99% within 25 kilometers (15.5 miles). It really is a thin, and in many ways vulnerable, veneer of gas that sustains all life at the surface.

> **CHECKPOINT 3.9 ▶** What is the origin of all the oxygen in Earth's atmosphere?

The Troposphere

The lowest part of the atmosphere is known as the **troposphere** (Figure 3.16). This is a zone where the air is constantly in motion, mixing and moving as part of a global pattern of atmospheric circulation that is complex and imperfectly understood. The temperature of the troposphere generally decreases with height. Air warmed at the surface becomes less dense than cooler air above, then rises rapidly by convection and expands as the atmospheric pressure falls. The ascending air does not have time to lose a significant amount of heat to its surroundings by radiation or mixing, but its temperature falls because it is expanding and must use some of its energy to push against the surrounding air (**adiabatic expansion**). The temperature of the surrounding atmosphere also cools with altitude (the **environmental lapse rate**), but it is not uncommon to find large temperature inversions (reversals) due to the mixing of contrasting air masses, nighttime heat loss, and/or the impact of ground temperature. Once the convecting air reaches the same temperature as the surrounding

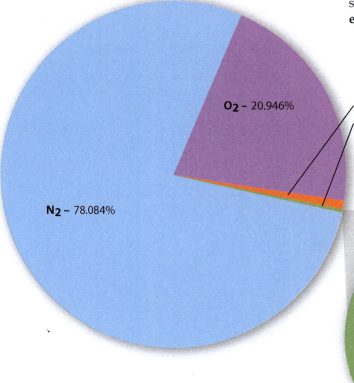

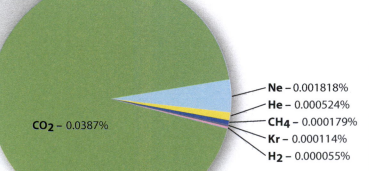

▲ **Figure 3.14:** Composition of Earth's atmosphere. Note how the concentration of greenhouse gases is very small. Water vapor, although an important greenhouse gas that constitutes up to 3% of the lower troposphere, is not present in this diagram because it has a very short residence time in the atmosphere.

◀ **Figure 3.15:** A view of Earth's atmosphere from high above the troposphere. The atmosphere may seem large but it is only a thin gaseous mantle held loosely to the surface of Earth by the force of gravity. Without the atmosphere, Earth's surface would resemble that of the distant Moon.

air it stops rising, cools by radiating energy, becomes denser than the air below, and descends towards the surface. The energy the convecting air lost on ascent to push aside the surrounding air is now recovered as the descending air is warmed by compression due to the increasing weight of the overlying atmosphere (adiabatic compression). The rate at which the convecting air warms and cools is known as the **adiabatic lapse rate** and it varies from around 9.8 °C per kilometer (5.38 °F per 1,000 feet) when the air is dry (the dry adiabatic lapse rate) to 5 °C per kilometer (2.7 °F per 1,000 feet) when the air becomes saturated and condensation starts to release latent heat (the moist adiabatic lapse rate).

If the troposphere were allowed to evolve adiabatically from an initial state without disturbance it would eventually achieve an ideal state of equilibrium with neutral buoyancy (no air rising or falling) where the vertical temperature profile follows the adiabatic lapse rate. In this way, the temperature of air at the top of the atmosphere is directly linked by adiabatic gradients to the temperature at the ground surface. Why is this important? If the effect of greenhouse gases is to warm air near the top of the troposphere, the atmosphere will adjust by convection and radiation to establish new lapse rates that lead to surface warming. This lesson is important. What we do at ground level affects the top of the troposphere. And what happens at the top of the troposphere has a direct impact on us at the surface.

You can experience adiabatic cooling by just blowing hard on the back of your hand. The air coming out of your mouth is warm but under pressure, so it expands and cools as it pushes the surrounding air out of the away. You can experience adiabatic warming by pumping up a bicycle tire and feeling the pump warm at the point of attachment.

CHECKPOINT 3.10 ▶ Why does the troposphere get cooler with height, despite being closer to the Sun?

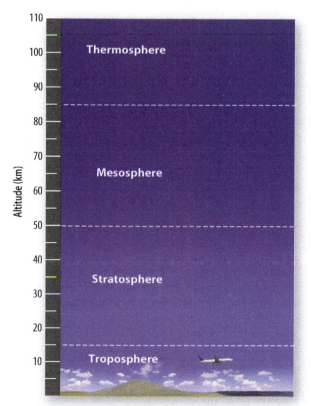

▲ **Figure 3.16:** Structure of Earth's atmosphere. Approximately 90% of the mass of the atmosphere lies in the troposphere.

A transition zone called the **tropopause** marks the boundary between the troposphere below and the stratosphere above. There is no fundamental difference between the troposphere and stratosphere, except that an increasing concentration of ozone, a greenhouse gas, above the tropopause puts an effective lid on convection by increasing the ambient temperature so that air rising from below is no longer buoyant (**Figure 3.17**).

CHECKPOINT 3.11 ▶ What is the primary distinction between the troposphere and the stratosphere?

The height of the tropopause depends on the temperature of the column of air beneath. Air above the tropics is much warmer than air at the poles, and has expanded so that the troposphere near the tropics can extend as far as 16 kilometers (9.9 miles) from the surface. At the poles, the air contracts because it is cold and much denser, and the top of the troposphere may be as shallow as just 8 kilometers (5 miles) above the surface. This temperature-driven change in the height of the top of the troposphere from equator to pole is very important because it sustains upper-level winds that can drive weather at the surface of Earth. Above the stratosphere, the air is thin and tenuous and has little direct impact on climate.

CHECKPOINT 3.12 ▶ Why is the troposphere so much deeper in the tropics than at the poles?

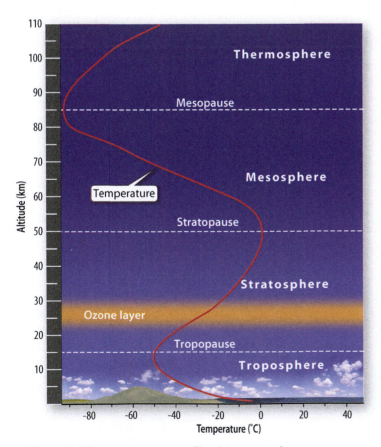

▲ **Figure 3.17:** A temperature profile of the atmosphere. Notice that the presence of ozone, a greenhouse gas, reverses the cooling trend of the troposphere. This puts an effective lid on convection in the troposphere.

What Is a Greenhouse Gas?

The Sun and orbital cycles are important factors that control Earth's climate, but there is no evidence that they are driving global warming. In Chapter 2 we concluded that the only other mechanism that could drive global warming is an increase in the level of greenhouse gases in the atmosphere. But what are these heat-trapping greenhouse gases?

The wavelengths of **infrared radiation** emitted from Earth's surface and atmosphere closely match the molecular dimensions of the gases water vapor (H_2O), carbon dioxide (CO_2), methane (CH_4), and nitrous oxide (N_2O) in the atmosphere. In the same way that a tuning fork absorbs energy and is set ringing by the matching (resonant) frequency of sound, greenhouse gases absorb infrared radiation and convert this electromagnetic energy into kinetic energy. You can think of the bonds that join the atoms in these molecules as if they were springs that start to flex, bend, and turn as they absorb different parts of the infrared spectrum.

Of course, as they absorb this energy, they also heat up, and as they heat up, they release infrared radiation back into the atmosphere at different frequencies that depend on their temperature. This radiation spreads out equally in all directions. Some travels up toward the top of the troposphere, some spreads out laterally, and some heads back toward the surface. Much of this radiation is absorbed by other greenhouse gases and emitted many times before it either reaches the surface of Earth or finally escapes to space.

CHECKPOINT 3.13 ▶ Why is water vapor a greenhouse gas, but nitrogen gas is not?

Water Water vapor is the most abundant greenhouse gas in the atmosphere and is responsible for as much as 36% to 70% of the total greenhouse gas effect. The exact amount in the atmosphere varies with elevation and temperature. The near-surface atmosphere can hold as much as 3% by mass in warm tropical regions, where it is often more than 10,000 times more abundant than other natural greenhouse gases. Despite these impressive data, however, water vapor does not drive climate change. The reason is that water vapor has a very short **residence time** in the atmosphere. The average molecule of water stays only a few hours or days at most in the atmosphere before falling as precipitation. If other greenhouse gases with a much longer residency time were not present to keep the atmosphere warm, most water vapor would soon precipitate, and the atmosphere would become very cold and dry.

As the concentration of other greenhouse gases increases, however, even more water vapor can be retained in the atmosphere and its potent greenhouse properties increase the temperature of the troposphere. Increasing levels of water vapor in the atmosphere are thus a good example of a positive feedback loop arising from the radiative forcing of the other greenhouse gases.

> **CHECKPOINT 3.14 ▶** Which of the greenhouse gases is most abundant? Why is this gas not considered to be a pollutant by the Environmental Protection Agency?

Carbon Dioxide Volcanoes and other geological processes produce carbon dioxide (CO_2) as a natural and very important part of Earth's atmosphere. As the building block for photosynthesis, CO_2 is essential for life on Earth, and as a greenhouse gas, it prevents Earth from cooling to the point where all surface water would freeze.

As we will see in Chapter 6, the level of CO_2 in the atmosphere varies naturally over time. CO_2 is a soluble gas and readily dissolves in seawater, especially when the water is cool and when the level of CO_2 in the atmosphere is high. As the oceans have warmed and cooled over the millennia, they have acted as an important sink and source of CO_2 for the atmosphere. When Earth warms, the oceans emit CO_2 into the atmosphere, where it helps to accelerate warming. When Earth cools, the oceans absorb more CO_2 from the atmosphere and help to accelerate the rate of cooling.

Before industrial times, the atmosphere contained around 280 ppm of CO_2, a level that was in equilibrium with the oceans, atmosphere, and global temperature. As ocean temperatures increased over the past century, the amount of CO_2 released by the oceans increased as expected, but when human activity started adding CO_2 directly into the atmosphere, the net movement of CO_2 reversed, and the oceans started to absorb more CO_2 from the atmosphere than they emit.

Of the 10 gigatonnes of carbon emitted by human activity as CO_2 each year, about 45% remains in the atmosphere and 30% is absorbed in the oceans. The rest is locked away in the terrestrial biosphere. The level of CO_2 in the atmosphere has reached as high as 400 ppm. This is an increase of nearly 40% over background levels, and it is unprecedented in Earth's recent geological history.

> **CHECKPOINT 3.15 ▶** What factors controlled the preindustrial level of carbon dioxide in the atmosphere?

David Charles Keeling, working at the Scripps Institute of Oceanography, was the first person to

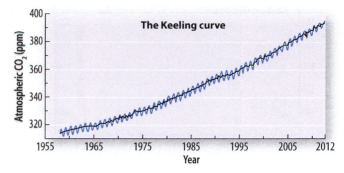

▲ **Figure 3.18:** The Keeling curve chart of rising CO_2. Since 1958, the level of CO_2 in the atmosphere has been measured above the summit of Mount Mauna Loa in Hawaii.

document the slow and steady rise in the concentration of CO_2 in the atmosphere. During the International Geophysical Year of 1958, when scientists from around the world attempted to coordinate global measurements of the physical environment, Keeling started his groundbreaking project to sample pollution-free air high on the flanks of Mauna Loa in Hawaii. He charted the rise in CO_2, from a base level of 315 parts per million (ppm) in 1958 (equivalent to 0.0315%), to 370 ppm in 2002, and over 395 ppm in 2013. Keeling's graph of CO_2 over time, now generally known as the **Keeling curve,** shows clear seasonal variation but an overall steady trend of rising CO_2 that appears to be accelerating (**Figure 3.18**).

> **CHECKPOINT 3.16 ▶** Why has the Keeling curve become such an icon of climate change science?

Unlike water, CO_2 has a long residence time in the atmosphere, which means that even if our emissions were to stop tomorrow, half of the anthropogenic gas now in the atmosphere would still be there after 30 years. A further 30% could be removed by the year 2300, but as much as 20% would stay in the atmosphere for millennia. Our actions today will have impacts on the atmosphere surrounding many future generations.

Methane Methane (CH_4) is a very efficient greenhouse gas but present at very low concentrations in the atmosphere (**Figure 3.19**). It has a much shorter residence time in the atmosphere than CO_2, lasting only 10 to 12 years before it oxidizes to CO_2 and H_2O.

One way to assess the strength of a greenhouse gas is to compare its relative impact on the atmosphere with CO_2. By calculating the impact of 1 kilogram (2.2 pounds) of CH_4 over 100 years with the impact of the same mass of CO_2, it turns out that CH_4 is at least 20 times more effective at warming the atmosphere.

Methane is produced naturally by **anaerobic bacteria** in places as diverse as wetlands, termite mounds, and the ocean floor. Trillions of kilograms of biogenic CH_4 are also stored in the form of icy compounds called methane hydrates, or clathrates, that form in the deep ocean and under the permafrost of the Arctic.

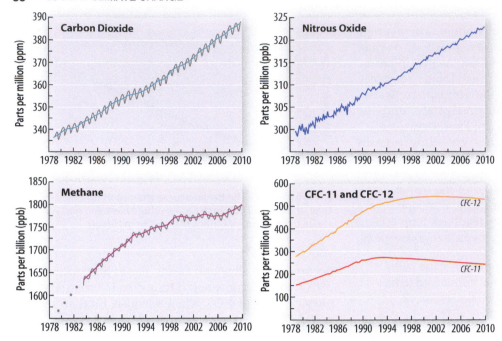

▲ Figure 3.19: Graphs illustrating the rapid rise in the concentration of some important greenhouse gases over the past 30 years. CFC-11 and CFC-12 are varieties of chlorofluorocarbon.

Over the standard 100-year reference period, 1 kilogram (2.2 pounds) of N_2O in the atmosphere will have over 310 times the impact of 1 kilogram (2.2 pounds) of CO_2. The level of N_2O in the atmosphere is now rising at a rate of 0.26% per year, and the current level is around 18% greater than the 270 ppb of preindustrial times.

Ozone Ozone (O_3) is a highly reactive gas that is concentrated in parts of the **stratosphere** and the lower troposphere. As an anthropogenic greenhouse gas, ozone has greatest impact in the lower troposphere, where it is a by-product of photochemical reactions with nitrogen and carbon monoxide, themselves the products of industry, internal combustion, and burning vegetation.

Concentrations of tropospheric O_3 have increased some 36% since the industrial revolution and almost doubled since 1800. The residence time of O_3 in the troposphere is as low as 25 days, but it is a potent greenhouse gas that causes 3% to 7% of observed global warming, placing it fourth among the important greenhouse gases.

Stratospheric ozone is important to us all, as it protects the surface of Earth from high-energy **ultraviolet radiation.** The widespread production of industrial chlorofluorocarbons (CFCs) in the 1970s and their release into the atmosphere created a major environmental problem as high-altitude photochemical reactions depleted stratospheric ozone. This depletion allowed much more damaging ultraviolet radiation to reach the ground, but the depletion of O_3 also removed a strong greenhouse gas, and the overall impact was to effect a slight cooling (by -0.1 Wm^{-2}) of the stratosphere. The appearance of a large hole in the ozone layer over the Antarctic during the boreal winter promoted public demand for legislation to limit the release of these ozone-damaging chemicals (**Figure 3.20**).

The result of this public pressure was the **Montreal Protocol,** an international treaty that was signed in 1987 and now limits further damage to the ozone layer. Although ozone is still depleted by around 4% below the 1964–1980 average, it is on the way to recovery. Ironically, the recovery of stratospheric ozone, an efficient heat-trapping greenhouse gas, may exacerbate the problem of climate change by warming Antarctica.

Human sources of CH_4 include paddy fields, coal mining, natural gas production, waste dumps, animal slurry, biomass burning, and the intestines of ruminants raised for meat. Together these sources account for over 60% of global methane production.

The concentration of CH_4 has already increased by over 150% since 1750 and appears to be rising again today, after a period of relative stability that lasted for over a decade. The draining of wetlands for agriculture and improved practices in the development of natural gas reserves may have started to slow the pace of growth, but the recent increase in the rate at which CH_4 is escaping into the atmosphere is probably related to melting of permafrost and methane hydrate ice beneath the seafloor. A rising concentration of CH_4 is bad news. As sea level continues to rise, it will flood coastal areas and create enlarged wetlands that will produce even more biogenic methane. The continued warming of oceans and permafrost will also release much more methane from the further breakdown of the icy methane hydrates.

CHECKPOINT 3.17 ▶ What factors controlled the preindustrial level of methane in the atmosphere?

Nitrous Oxide Nitrous oxide (N_2O) is the third largest contributor to the atmospheric greenhouse gas effect. Present in only very small concentrations (324 ppb), it is a powerful greenhouse gas with a residence time in the atmosphere of over 150 years. Most N_2O in the atmosphere is of natural origin, mostly from bacterial activity in tropical and temperate soils and from biological activity in the oceans. Roughly 40% of atmospheric N_2O, however, is the result of human activity, mostly from agriculture, biomass burning, industry, and livestock.

CHECKPOINT 3.18 ▶ Why do we want ozone in some parts of the atmosphere but hope to avoid it in others?

The Halocarbons Industrial processes have released a formidable array of human-made chemicals into the atmosphere over the past 20 years. Many are known to make significant contributions to the greenhouse gas effect. The chlorofluorocarbons (CFC), perfluorocarbons (PFC), hydrofluorocarbons (HFC), sulfur hexafluoride, and a list of more than 34 other specific chemicals highlighted by the IPCC are present in minute concentrations (parts per trillion [ppt]) compared to CO_2 but have great potential for global warming (**Figure 3.21**). On average, these chemicals are 2,000 to 3,000 times more effective than CO_2 at causing global warming—and some are up to 23,900 more effective than CO_2 but are present in such small quantities that they have relatively little impact on climate change.

CHECKPOINT 3.19 ▶ Why is one of the greenhouse gases more responsible for driving climate change than the others?

Could Greenhouse Gases Be Responsible for Recent Climate Change?

There is no doubt that greenhouse gases heat the atmosphere. The physics of this process is well understood. A fairly simple calculation shows us that the temperature of the atmosphere will eventually increase by at least 1.2°C if the level of CO_2 doubles from preindustrial levels to 560 ppm without any additional climate

feedback. To understand the potential for further enhanced warming, we need to understand how energy is transformed and transported though Earth's atmosphere and oceans.

CHECKPOINT 3.20 ▶ In the absence of any feedback in Earth's climate system, how much do we expect global temperature to increase after doubling the concentration of carbon dioxide in the atmosphere from preindustrial times?

3.4 PAUSE FOR... THOUGHT | *Are greenhouse gases evenly dispersed throughout the atmosphere?*

This is a matter of perspective. Convection (vertical movement) and advection (lateral movement) in the atmosphere are very effective at mixing the atmosphere both vertically and laterally over time. The uneven distribution of human and natural sources of greenhouse gases, however, can produce significant local variation in their geographical and seasonal concentration and distribution.

Heating the Atmosphere

The following sections investigate the flow of energy from the top of the atmosphere to the depths of the oceans. You will learn why greenhouse gases warm the atmosphere and test the hypothesis that they are responsible for recent climate change.

By the time radiation from the Sun reaches Earth, it has been dispersed over a vast area of space. The energy covers a range of wavelengths, with greatest intensity in the ultraviolet, visible, and infrared parts of the spectrum (**Figure 3.22**).

If Earth were flat and facing the Sun, it would receive around 1,366 Wm^{-2} at the top of the atmosphere, roughly the same as suspending 14 household 100-watt light bulbs directly above a large dining table. Because Earth is a sphere, the energy is actually wrapped over the illuminated surface of the globe that is facing the Sun ($2\pi r^2$), reducing the incident power to around 680 Wm^{-2}. Finally, because at any one time the far side of Earth is in total darkness, the average energy Earth actually receives is only half of this value, around 342 Wm^{-2} (**Figure 3.23**).

The Sun's energy is not evenly distributed around the globe (**Figure 3.24**). Earth is nearly spherical, so the Sun's energy illuminates the equator from directly overhead, but toward the poles, it strikes at an oblique angle, and the energy is spread over a much larger area. In a final twist, the tilt of Earth brings darkness to both poles during their respective winters.

As Earth is warmed by radiation from the Sun, it simultaneously emits radiation back out to space from all

▲ **Figure 3.20:** A computer-enhanced false color image of the ozone "hole" that appears over Antarctica during the winter. The extent of ozone loss is illustrated by colors changing from green (least) to purple (greatest).

Area comparison

SOUTH AMERICA

ANTARCTICA

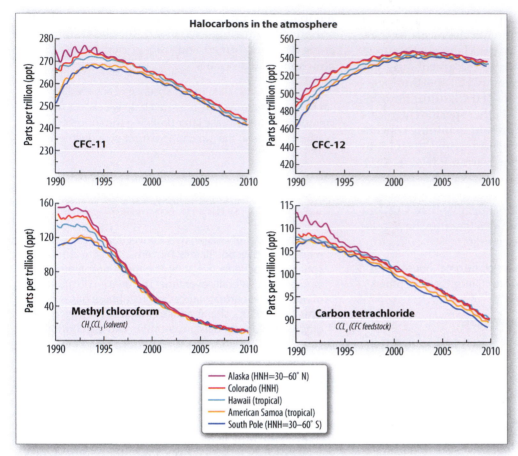

▲ **Figure 3.21:** The atmospheric concentrations of most heat-trapping halocarbons fell rapidly after the Montreal Protocol came into force in 1989.

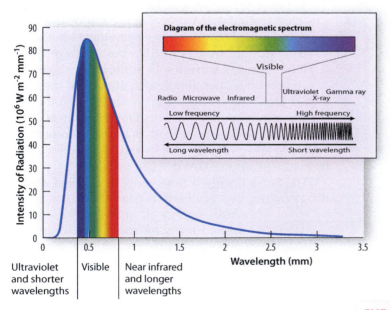

▲ **Figure 3.22:** The spectrum of sunlight received at the top of the atmosphere. The greatest intensity of radiation ranges from the ultraviolet through visible to near infrared parts of the spectrum (see inset).

surfaces, both day and night (**Figure 3.25**). This terrestrial radiation is all in the infrared part of the spectrum, and it is 1 million times less intense than radiation from the Sun's photosphere (**Figure 3.26**). The exact amount of energy leaving Earth's atmosphere varies based on season, latitude, and longitude, but an overall equilibrium (balance) develops between the net amount of incoming solar radiation and outgoing terrestrial radiation *when measured at the top of the troposphere.*

Within the troposphere, the uneven distribution of incoming solar radiation creates an important energy imbalance that "fuels" our climate. Net cooling of the poles and warming at the equator create a strong latitudinal (equator to pole) temperature gradient that is strongly reinforced at the poles because ice reflects much more of the Sun's already dispersed energy directly out to space (**Figure 3.27**).

CHECKPOINT 3.21 ▶ How does a difference between the total amount of energy arriving at the surface of the equator and at the poles drive Earth's climate system?

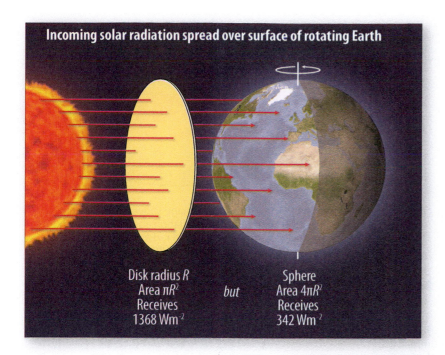

Incoming solar radiation spread over surface of rotating Earth

Disk radius R
Area πR^2
Receives
1368 Wm^{-2}

but

Sphere
Area $4\pi R^2$
Receives
342 Wm^{-2}

◀ **Figure 3.23:** Energy from the Sun is wrapped over the surface of Earth, reducing the overall intensity by a factor of 4, the ratio of the surface area of the disc to the surface area of the sphere that is illuminated by the Sun ($4\pi r^2/\pi r^2$).

The Energy Budget

Satellites are now able to measure the energy flow at the top of the troposphere in great detail. Seen from space, Earth shines with both visible and infrared energy (**Figure 3.28**). Satellite photographs of the **short-wavelength flux** (energy from the Sun) clearly show the high albedo of deserts, cloud tops, tropical forests, and polar ice. Photographs of the long-wavelength flux (radiation emitted by Earth) show areas of intense tropical heating partly obscured by cooler patches associated with radiation from very high cloud tops above large tropical storms.

▶ **Figure 3.24:** Energy from the Sun is unevenly distributed over the surface of Earth, with much more energy arriving at the equator than at the poles.

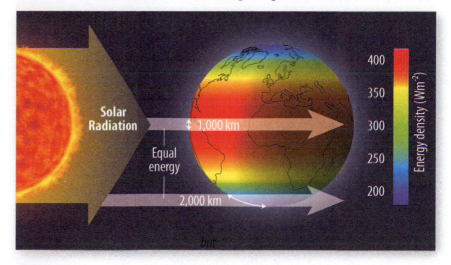

Solar Radiation

1,000 km

Equal energy

2,000 km

Energy density (Wm^{-2})

400
350
300
250
200

but

▼ **Figure 3.25:** Outgoing heat (long-wave radiation). Earth is only heated where it is facing the Sun, but it emits radiation from all surfaces, both day and night.

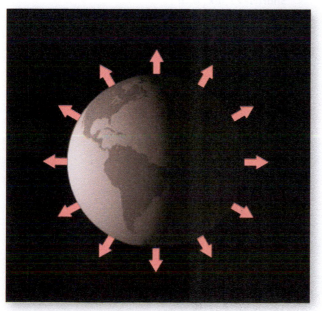

Cold Comfort

If Earth had an atmosphere with no greenhouse gases, some of the Sun's energy would still be reflected directly out to space from high-albedo surfaces at ground level, but most would be absorbed to heat Earth's surface. When that heat was later released from the surface of Earth as infrared radiation, there would be nothing to stop it from escaping directly to space (**Figure 3.29**).

Under such conditions, Earth would be a cold planet with an average temperature of around −18°C (−0.4°F), depending on Earth's albedo. Tiny quantities of greenhouse gases such as carbon dioxide, methane, and nitrous oxide in the atmosphere make the difference between this cold, dead planet and our current living world, where the average temperature is a much more comfortable 16°C (60.8°F) and most water is in the liquid state.

CHECKPOINT 3.22 ▶ What would be the temperature of Earth's atmosphere in the absence of greenhouse gases?

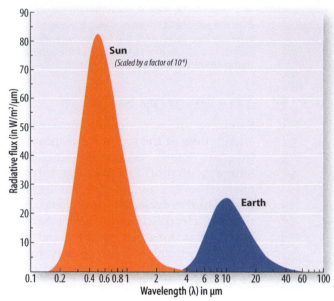

▲ Figure 3.26: Radiation curves of the Sun and Earth. Earth only emits radiation in the infrared parts of the spectrum, and with much less intensity than the radiation it receives from the Sun.

▲ Figure 3.27: This satellite image shows the difference between the amount of solar radiation absorbed by Earth's surface and the amount of infrared radiation emitted back into space during the period 1985–1986. Notice the net gain in radiative energy close to the equator (increasing from red to purple) and net loss of energy at high latitudes (increasing from green to blue).

The relationship between the temperature of a body and the amount of energy it radiates is very important. As Earth warms, it radiates infrared energy out to space with increasing intensity, in proportion to the fourth power of temperature ($e \sim T^4$)—known as the Stefan–Boltzmann relationship. This means that the intensity of radiation released will increase by a factor of 16 every time the temperature doubles ($2^4 = 16$).

CHECKPOINT 3.23 ▶ What is the approximate relationship between Earth's surface temperature and the intensity of infrared energy Earth emits?

The Distribution of Greenhouse Gases

Greenhouse gases are not evenly distributed throughout the atmosphere. In the lower troposphere,

▶ Figure 3.28: (a) The top image shows the amount of radiation reflected from the Sun (short wavelength flux- increasing from dark green through white colors) and (b) the lower image shows the amount of radiation emitted by Earth (long wavelength infrared (IR) flux- increasing from dark blue through red to yellow colors). Note the high albedo from deserts, tropical forests, and high altitude clouds in the upper image and the strong IR radiation from the hot tropics and subtropics in the lower image. Note also how the cooler tropical forests and cold high cloud tops emit IR radiation much less intensely.

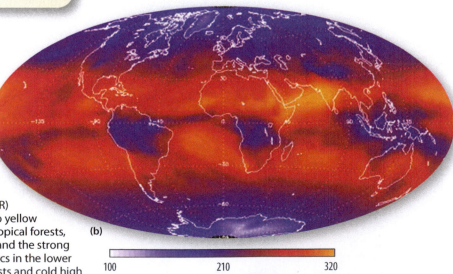

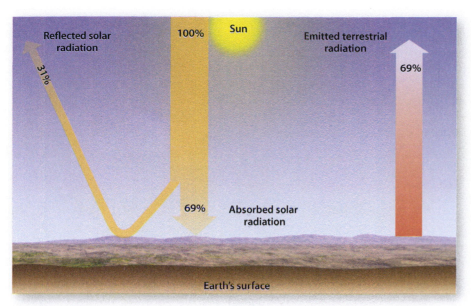

▲ **Figure 3.29:** A simple radiation model to determine the temperature of Earth if there were no greenhouse gases in the atmosphere.

water vapor is the dominant greenhouse gas, and it is so abundant that the lower troposphere is almost opaque to infrared radiation. The average photon released by a molecule of H_2O or CO_2 at this level in the atmosphere will not travel far before being captured again by another greenhouse gas molecule. Higher in the troposphere, water vapor is progressively lost through condensation and precipitation, and its importance as a greenhouse gas diminishes. This leaves the middle and upper levels of the troposphere and the stratosphere much more susceptible to the impact of anthropogenic greenhouse gases (**Figures 3.30** and **3.31**).

CHECKPOINT 3.24 ▶ Why does the relative importance of water vapor as a greenhouse gas decrease with height in the troposphere?

Measuring the Flow of Energy

The average flow of energy in and out of Earth's atmosphere has been measured and is illustrated in **Figure 3.32**. This is a simplified guide to the general flow of energy through the atmosphere. From day to day, actual values are affected considerably by factors such as season, latitude, and cloudiness.

Of the 342 Wm^{-2} of direct solar energy that enters Earth's atmosphere on average each day:

- 77 Wm^{-2} (22.5%) is reflected by the atmosphere
- 30 Wm^{-2} (8.8%) is reflected by Earth's surface
- 67 Wm^{-2} (19.6%) is absorbed in the atmosphere
- 168 Wm^{-2} (49.1%) is absorbed at Earth's surface

If we add these figures together, we see that 235 Wm^{-2} (67 Wm^{-2} +168 Wm^{-2}) is absorbed into the climate system, and the rest (77 Wm^{-2} +30 Wm^{-2}) is lost to space.

In return, each day at the top of the troposphere, Earth emits:

- 165 Wm^{-2} of infrared energy from greenhouse gases in the atmosphere
- 30 Wm^{-2} directly from clouds
- 40 Wm^{-2} directly from the surface via an "atmospheric window"

In total, then, 235 Wm^{-2} (165 Wm^{-2} +30 Wm^{-2} +40 Wm^{-2}) of energy leaves the climate system, and an equilibrium exists between incoming and outgoing radiation.

The Natural Greenhouse Gas Effect

In the lower and middle troposphere, the effects of greenhouse gases are especially dramatic. Of the 235 Wm^{-2} of infrared energy emitted by Earth's surface and atmosphere, only 40 Wm^{-2} is able to escape directly to space. This energy is emitted over a relatively short interval of wavelengths, known as the **atmospheric window,** where infrared photons can evade capture. Greenhouse gases in the troposphere, such as water, carbon dioxide, methane, and nitrous oxide, absorb and emit the remaining 195 Wm^{-2} of infrared energy.

CHECKPOINT 3.25 ▶ What is the atmospheric window?

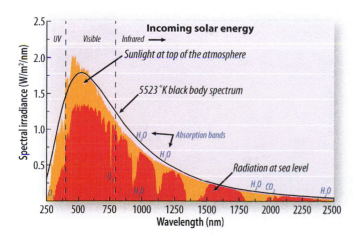

◀ **Figure 3.30:** This diagram shows the intensity of radiation (irradiance) received from the Sun at Earth's surface. Compare the ideal spectrum from the Sun (continuous gray line) with the observed spectrum (orange) at the top of the atmosphere and the observed spectrum (red) at the bottom of the atmosphere. The ideal and observed curves are very similar at the top of the atmosphere, but large "bites" of energy are missing from the spectrum observed from the surface. These bites correspond with the 67 Wm^{-2} of solar radiation that are absorbed by ozone and other greenhouse gases in the stratosphere and troposphere. The energy missing from the surface has been absorbed in the atmosphere.

▶ **Figure 3.31:** This satellite image taken in July 2008 by an atmospheric infrared sounder on the NASA AQUA satellite shows how the distribution of CO_2 in the atmosphere varies at one moment in time. The intensity and distribution of CO_2 changes with the seasons, wind direction, and the location of major sources from industry, agriculture, and deforestation.

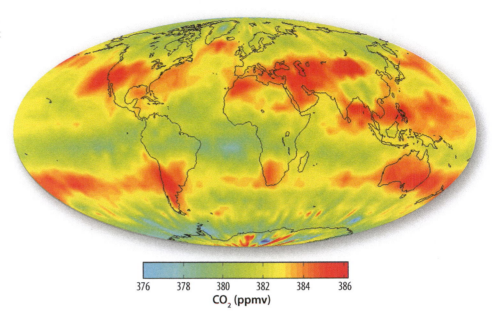

376   378   380   382   384   386
CO_2 (ppmv)

This is a fascinating and very dynamic process. Greenhouse gas molecules in the atmosphere absorb infrared energy, but they also emit infrared energy with changing intensity that depends on the ambient (surrounding) temperature. This constant flux of energy flows equally in all directions, but on average, half travels upward toward the top of the atmosphere, and the rest travels downward toward the surface. On this journey, each photon is captured, and a new **photon** is emitted multiple times before energy is either finally relayed out to space or arrives back at Earth's surface.

The impact of this energy "trapped" near the surface of Earth is very large. An additional 324 Wm^{-2} of infrared radiation emitted by greenhouse gases is added to the 168 Wm^{-2} of incoming energy from the Sun absorbed at Earth's surface. It is this natural greenhouse gas effect that gives the world an average comfortable temperature of 16°C (60.8°F) as opposed to a freezing –18°C (–0.4°F).

▼ **Figure 3.32:** A flow diagram that illustrates global heat flow: how energy from the Sun is cycled through Earth's atmosphere. Figures are average values for all of Earth, and actual values will vary with factors such as season, latitude, and cloudiness. See the text for a full discussion.

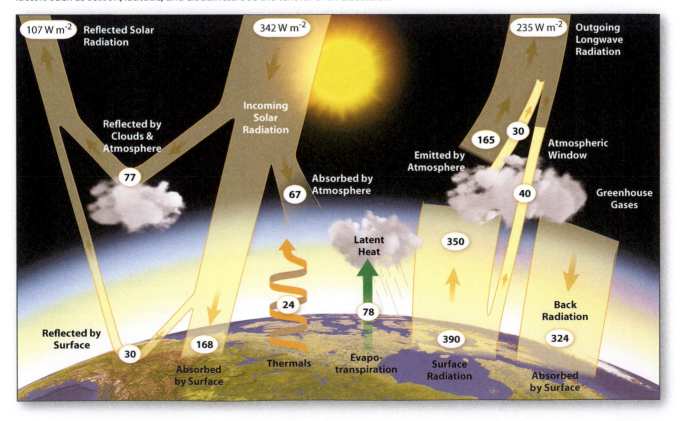

It is only toward the top of the troposphere, where the air is very cold and dry, that greenhouse gas concentrations are low enough that most infrared heat can escape to space, thus cooling Earth. It is as if Earth has a skin at this altitude, and loses most of its heat from this surface.

Think of it this way: If you could fly to this altitude in the troposphere and look around with glasses that could only "see" at these infrared wavelengths, it would appear as if you were immersed in a glowing fog. With altitude, the fog would clear until you found yourself in the stratosphere, looking down on a shifting surface below. Unless you changed your glasses for a new pair that allowed you to see via the infrared atmospheric window that allows 40 Wm^{-2} to escape to space, you would never see the surface of Earth. From your perspective, most of the energy Earth emits would come from an opaque glowing skin near the top of the troposphere.

The greenhouse gases at the altitude of this "skin" in the troposphere are cold (around −18°C [−0.4°F]) and emit energy much more weakly than at the surface. Even so, at this **skin temperature** they still emit the 235 Wm^{-2} necessary to balance the total amount of incoming energy from the Sun.

Satellites are now able to measure the actual spectrum of infrared energy leaving Earth, and **Figure 3.33** compares this measured spectrum with theoretical spectra based on greenhouse gases being emitted at different temperatures in the atmosphere. These data can be used to estimate the height from which infrared energy finally escapes to space (given an average lapse rate of around 6.5°C km^{-1} [3.5°F/1,000 feet]).

CHECKPOINT 3.26 ▶ In your own words, explain the meaning of the term *skin temperature*.

The diagram charts changing wavenumber (the inverse of wavelength) of infrared radiation along the x-axis against the intensity of radiation emitted at each wavelength along the y-axis (a quantity that depends on temperature).

Radiation emitted from wavenumbers 400 through 580 is known to come from water vapor, and the spectrum indicates that it was emitted at temperatures typically found between 2 and 5 kilometers (6,561 and 16,404 feet) from the surface, where most water vapor is concentrated.

Radiation emitted from wavenumbers 600 to 800 represents emission from CO_2, with an intensity that indicates an origin from cold gas near the top of the troposphere. This is direct physical evidence that CO_2 in the atmosphere prevents infrared radiation at these wavelengths from escaping from the lower troposphere, and it is only at this altitude that the radiation can finally escape to space.

At shorter infrared wavelengths (higher wavenumbers to the right on the diagram), there is a strong signal emitted at close to surface temperatures. This is the atmospheric window where radiation emitted at or close to Earth's surface is able to escape to space without capture by greenhouse gases.

Note a hole in this window at around wavenumber 1,100 that is caused by ozone emission at around 250K and close to 5 kilometers (16,404 feet) from the surface. Still further to the right, at even shorter wavelengths, is infrared energy emitted by water vapor over an interval from 5 to 12 kilometers (16,404 to 39,370 feet) above the surface.

The area under this entire curve (the integral of all radiation emitted over this range of wavelengths) is an estimate of the total amount of radiation emitted by Earth and the atmosphere. The clear difference between the observed radiation curve with all the "bites" removed by greenhouse gases and the much smoother theoretical curve is a measure of how much radiation is actually trapped by the greenhouse gas effect.

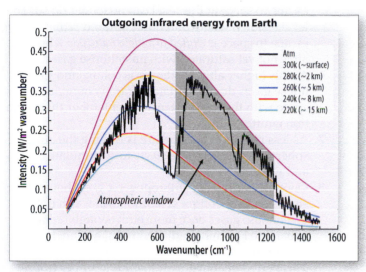

▲ **Figure 3.33:** The shape and intensity of the radiation curve emitted from any surface depends on its temperature. This graph shows emission curves (solid lines) that would be observed at the top of the atmosphere if radiation were emitted by ideal surfaces over a range of different temperatures. These temperatures range from the 300°K (0°C = 273.15 K) typical of Earth's surface to the 220°K typical of the top of the troposphere. By looking at the actual emission curve that is observed at the top of Earth's atmosphere (red line) and comparing it to the ideal curves for surfaces at different temperatures, it is possible to tell from what level in the troposphere the observed radiation leaving Earth is coming. (Remember that the temperature of the troposphere falls steadily from the surface toward the top of the troposphere.) These data show that radiation across the infrared spectrum of wavelengths is able to escape directly to space from different levels between the surface and the top of the troposphere in varying amounts. Radiation between wave numbers 600 and 700 is only able to escape at an altitude very close to the top of the troposphere. This is because infrared radiation at these wavelengths is trapped by CO_2 gas in the troposphere, and it is only near the tropopause, where the air is very cold and the concentration of CO_2 low, that radiation at these wavelengths can finally escape directly to space. On this diagram, by convention, radiation intensity is plotted against wavenumber. Wavenumber is a measure of the number of wave cycles per centimeter and is the reciprocal of wavelength (1/λ). In other words, higher frequencies occur on the right side of the diagram.

3.5 PAUSE FOR... THOUGHT | *Will Earth ever experience a runaway greenhouse gas effect like Venus?*

Earth and Venus are very similar. Volcanic activity on both planets has pumped vast amounts of carbon dioxide into their atmospheres. Although Venus may have had large oceans for as much as 2 billion years of its geological history, a runaway greenhouse effect finally led to the evaporation of all liquid water. On Earth, marine biological activity, weathering, and tectonics continually lock away carbon dioxide and prevent a runaway greenhouse effect from developing. It is interesting to speculate how important the evolution of life and biological activity on Earth has been in maintaining an atmosphere cool enough to prevent Earth turning into another hostile world like Venus.

The Enhanced Greenhouse Gas Effect

The greenhouse gas effect is a natural atmospheric phenomenon that maintains and sustains life on Earth. Our planet would be a permanent snowball and quite possibly devoid of life if we lost these gases. Over geological time the concentration of these gases in the atmosphere has changed slowly, mostly due to the impact of volcanism, and Earth's temperature has changed accordingly. The critical balance between greenhouse gas production and natural sequestration was upset as soon as mankind started to cut down trees and used fire to control the landscape for farming.

The Lower Troposphere

The **enhanced greenhouse gas effect** occurs as additional greenhouse gases are added to the atmosphere by industrial activities, agriculture, and deforestation. Adding more greenhouse gas to the lower troposphere has little direct effect, as the air at this level is already opaque over much of the infrared spectrum. But at higher levels in the atmosphere, even a small increase can have a significant impact.

Warming the Middle and Upper Troposphere

The atmosphere is well mixed by turbulence throughout the troposphere so that the concentration of CO_2 and other minor greenhouse gases in the middle and upper troposphere remains relatively constant. In contrast, the concentration of water vapor, the most abundant greenhouse gas, decreases rapidly with altitude, leaving the middle troposphere less opaque to infrared radiation. The impact of adding anthropogenic greenhouse gases at the altitude of 5 to 8 kilometers (16,404 to 26,246 feet) is critical, as this is the level where most infrared radiation finally escapes to space.

The process is continuous but can be considered as a series of steps:

1. Anthropogenic greenhouse gases make the atmosphere more opaque to infrared radiation, and the troposphere warms.
2. As air descends toward the surface it is warmed further due to adiabatic compression. This increases surface temperature and enhances evaporation.
3. Enhanced evaporation allows more water vapor to rise convectively toward the middle of the troposphere.
4. As the concentration of water vapor and other greenhouse gases in the middle troposphere increases, the air at this altitude gradually becomes opaque to infrared radiation and warms even further (especially in the tropics).
5. When air at 5 to 8 kilometers above the surface (16,404 to 26,246 feet) becomes saturated with greenhouse gases, the infrared radiation that used to escape to space at this level can no longer do so. The level ("skin") where outbound infrared radiation *can* escape to space is *shifted to a higher altitude* where the air is not yet saturated with greenhouse gases.
6. At this higher altitude the air is colder, and greenhouse gases do not radiate infrared energy out to space with the same intensity (remember that radiation emitted $\sim T^4$).
7. As less radiation than before is leaving the atmosphere, the troposphere starts to accumulate energy and warms.
8. Once the addition of anthropogenic greenhouse gases stops, the temperature at altitude will warm until greenhouse gases once more emit enough radiation to balance incoming radiation from the Sun. Energy equilibrium is restored and global warming will stop.
9. This period of adjustment in middle-upper tropospheric temperature appears to be a fairly rapid process, but if greenhouse gases are added continually, as they have been over the past century, the atmosphere will stay in a constant state of disequilibrium.

CHECKPOINT 3.27 ▶ Explain why the enhanced greenhouse effect is important for life on Earth.

Back to the Surface

The changes in temperature near the top of the troposphere have an important impact on the temperature at Earth's surface. Warmer air near the top of the troposphere will transfer this energy to the surface as it descends and warms adiabatically (**Figure 3.34**).

Figure 3.34a shows the unperturbed atmosphere. The surface temperature is elevated by the natural greenhouse gas effect, and the average height from which infrared radiation escapes to space is around 15 kilometers (50,000 feet). The temperature is approximately −18°C (255K), and the flow of energy leaving the atmosphere at 240 Wm^{-2} (200 Wm^{-2} from near the top

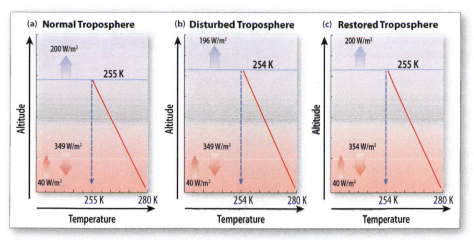

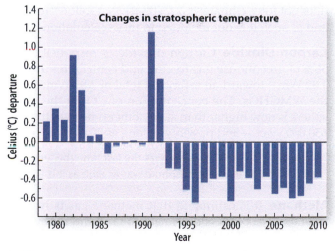

▲ Figure 3.34: This diagram illustrates how the enhanced greenhouse gas effect works. (a) The unperturbed atmosphere. The surface temperature (280K) is elevated by the natural greenhouse effect, and the average height from which infrared radiation escapes to space is around 15 kilometers (50,000 feet). The temperature is approximately –18°C (255K), and the flow of energy at 240 Wm^{-2} (200 Wm^{-2} from the top of the troposphere and 40 Wm^{-2} via the atmospheric window) balances the level of incoming radiation from the Sun. (b) What happens when heat-trapping greenhouse gases are added to the troposphere? As the troposphere becomes more opaque to infrared radiation, the height at which radiation can escape to space increases to a level where the atmosphere is colder (254K). These colder gases can only emit radiation at 196 Wm^{-2}, much less than the 200 Wm^{-2} needed to balance incoming radiation from the Sun. The balance of radiation is thus disturbed, and the atmosphere starts to accumulate additional energy. (c) What happens as the atmosphere adjusts to these changes? The high-altitude gases warm to 255K until they also start to emit radiation at 200 Wm^{-2}, and the lower troposphere and surface warm to 281K as more infrared radiation (354 Wm^{-2}) is returned to the surface.

of the troposphere and 40 Wm^{-2} via the atmospheric window) balances the level of incoming radiation from the Sun. Figure 3.34b shows what happens when greenhouse gases are added to the troposphere. As the troposphere becomes more opaque to infrared radiation, the height at which radiation can escape to space increases, and colder, higher gas becomes the new source for much of the radiation escaping to space. These colder gases can emit radiation only at 196 Wm^{-2}, much less than the 200 Wm^{-2} needed to balance incoming radiation from the Sun. The balance of radiation is disturbed, and the atmosphere starts to accumulate additional energy. Figure 3.34c shows what happens as the atmosphere adjusts to these changes. The high-altitude gases warm until they also start to emit radiation at 200 Wm^{-2}, and the lower troposphere and surface warm as more infrared radiation is returned to Earth's surface.

CHECKPOINT 3.28 ▶ Why is the middle troposphere (especially in the tropics) expected to exhibit more *relative* warming than Earth's surface?

Up to the Stratosphere

As the **opacity** of the troposphere increases with the addition of anthropogenic greenhouse gases, the upward flux of energy (heat) to the lower stratosphere is reduced. At the same time, greenhouse gases within the stratosphere continue to emit infrared radiation even though they receive less energy from the troposphere below. The overall impact is that the stratosphere cools while the troposphere warms, a change already observed by satellite (**Figure 3.35**). If the Sun were responsible for the recent rise in global temperature, it would warm the entire atmosphere, including the stratosphere. The observed cooling of the stratosphere and warming of the troposphere is strongly indicative of warming by greenhouse gases.

▲ Figure 3.35: This figure shows how the temperature of the stratosphere has changed from the base period 1979–1998. Unusually high temperatures in 1991 and 1992 due to the eruption of Mount Pinatubo interrupt a cooling trend that started around 1985.

3.6 PAUSE FOR... THOUGHT | *Is it true that Earth loses much of its energy from close to the top of the troposphere?*

Common sense suggests that most energy must escape to space from Earth's warm surface. It is helpful to think of Earth as being surrounded by an opaque outer gaseous skin that emits just enough energy from its cold surface to balance the rate at which energy is added to the interior and surface by the Sun. Beneath this "skin," the atmosphere would appear in the infrared like a dense glowing fog that thickens progressively and radiates more strongly toward Earth's surface.

Radiative Forcing

The previous sections show that the Sun and greenhouse gases are the two primary factors that drive climate change on a scale of decades, but they are not the only factors that affect the climate. This section introduces more factors that control the flow of energy through the atmosphere and investigates why some of these factors tend to cool the planet while others enhance warming. The sum of all these competing warming and cooling factors will determine exactly how much Earth will warm as a result of changes in both solar energy and greenhouse gas concentrations.

Radiative forcing can be defined as the instantaneous change in the net, downward minus upward, irradiance (expressed in Wm^{-2}) at the tropopause due to a change in an external driver of climate change such as a change in the concentration of greenhouse gases or the energy output of the Sun. Radiative forcing is similar to **climate forcing**, but it does not account for rapid adjustment of some components of the atmosphere. **Adjusted forcing** reflects additional factors such as the impact of anthropogenic aerosols on clouds and snow cover and is a better indicator of the overall impact of radiative forcing on climate. The best current estimate of radiative forcing is around $2.4 W m^{-2}$ but the adjusted forcing is lower at around $2.2 W m^{-2}$. It is not easy to estimate the current level of radiative forcing. It should be possible to measure this from space, but the level is so low that it is almost impossible to detect with current satellite instruments. Part of the problem is that there are many competing factors. We know that greenhouse gases and the Sun can make a positive contribution to warming, but cooling factors such as aerosols and volcanic eruptions must also be considered.

This section considers these competing forces and how they act together to determine the temperature of Earth. The sensitivity of Earth to changes in radiative forcing is a matter of keen debate. As you will see, there are many different factors, and some are still not clearly understood.

CHECKPOINT 3.29 ▶ In your own words, explain the term *radiative forcing*.

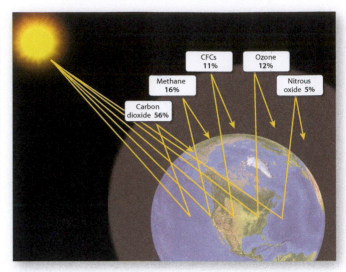

▲ **Figure 3.36:** The main greenhouse gases that drive global warming are carbon dioxide, methane, nitrous oxide, ozone, and human-made chlorofluorocarbons. Water is also an important greenhouse gas, but without the others, it would soon be lost from the atmosphere. Numbers are their relative percentage contribution to the overall greenhouse gas effect.

The Sun

TSI over the last three solar cycles has decreased at a rate of around $-0.04 Wm^{-2}$ but the contribution over the entire industrial era remains small and positive at around $0.04 Wm^{-2}$.

Greenhouse Gas Forcing

The current radiative forcing from greenhouse gases is rising "at least six times faster than at any time during the two millennia before the Industrial Era, the period for which accurate records exist." The total impact of anthropogenic radiative forcing from all well-mixed greenhouse gases (WMGHG) in the troposphere is around $2.83 Wm^{-2}$ (**Figure 3.36**). Remember that water vapor does not force climate; it is feedback that follows forcing, and as such it is not included in these calculations.

Carbon Dioxide Carbon dioxide is responsible for 56% of greenhouse gas forcing and contributes around $1.82 Wm^{-2}$ toward an estimated total of $2.83 Wm^{-2}$ from the WMGHGs. The concentration of CO_2 in the atmosphere is now higher than at any other time over the past 650,000 years—and possibly as long as 20 million years.

CHECKPOINT 3.30 ▶ When was the last time the level of carbon dioxide in the atmosphere was as high as it is today?

Methane It is estimated that methane gas is responsible for 16% of greenhouse gas forcing and that it contributes around $0.48 Wm^{-2}$ toward the total greenhouse gas forcing of $2.83 Wm^{-2}$ from the long-lived greenhouse gases (excluding water vapor).

Nitrous Oxide Nitrous oxide, at over 319 ppb, is considered to be responsible for around 5% of greenhouse gas forcing and contributes around 0.17 Wm^{-2} toward the total greenhouse gas forcing of 2.83 Wm^{-2}.

Ozone Ozone contributes 0.40 Wm^{-2} toward greenhouse gas forcing in the troposphere, and ozone loss due to CFCs contributes $-.10$ Wm^{-2} towards cooling of the stratosphere.

CHECKPOINT 3.31 ▶ What is the cumulative radiative forcing from the addition of all well-mixed anthropogenic greenhouse gases?

Other Sources of Radiative Forcing in the Atmosphere

Greenhouse gases play a major role in maintaining Earth's atmospheric temperature, but tiny particles of carbon, acidic aerosols, and clouds also play critical roles by changing the amount of solar energy that is absorbed and reflected.

Black Carbon and Organic Carbon Black carbon (soot) and **organic carbon** (organic particles) are mostly the products of the incomplete combustion of fossil fuels and the burning of biomass. These particles are unequally distributed around the globe and form complex aerosols high in the atmosphere. Black carbon and organic carbon from the burning of fossil fuel can both absorb and reflect energy from the Sun. The best current estimate is that the overall contribution from black carbon is small and positive (warming), at 0.40 Wm^{-2}, and even smaller and negative (cooling) from organic carbon, at -0.04 Wm^{-2}.

When black carbon particles fall from the atmosphere and land on ice and snow, they darken the surface and cause the surface to absorb more energy. It is estimated that this may contribute an overall warming impact of 0.04 Wm^{-2} toward total adjusted radiative forcing of 2.2 Wm^{-2}.

CHECKPOINT 3.32 ▶ What is black carbon, and what impact does it have on radiative forcing?

Aerosols and Global Dimming Not all the gases and pollutants we add to the atmosphere cause global warming. We have already seen that the addition of CFCs to the atmosphere had a mixed impact by warming the troposphere but cooling the stratosphere by depleting ozone. Indeed, one of the questions raised by scientists studying early climate models is Why has Earth not been warming more than we observe? To answer this question, further research has highlighted the significant cooling effect of industrial pollution and volcanic activity on the atmosphere.

Aerosols produced by industrial and volcanic processes are tiny particles of solid and liquid that become suspended in the atmosphere and act to partially block the passage of sunlight. By increasing the aerosol optical thickness of the atmosphere (AOT) they decrease

▲ **Figure 3.37:** Human-made aerosols over India, China, Russia, and other industrialized countries that depend on fossil fuels for energy cause significant dimming of solar radiation at the surface and delay the impact of global warming. This image of the Bay of Bengal gives a good impression of the extent of the aerosols problem.

the amount of radiation reaching the ground. This aerosol radiation interaction reduces adjusted radiative forcing by around -0.4 Wm^2 and is sometimes referred to as global dimming; anyone who has taken a window seat on an airplane will be familiar with the thick haze that can often obstruct a clear view of the ground. These aerosols can both reflect and absorb incoming radiation from the Sun, and they also absorb infrared radiation emitted by Earth. In other words, they contribute components of both cooling and heating at the same time, competing factors that make analyzing their impact very difficult.

NASA **MODIS (Moderate-resolution Imaging Spectroradiometer)** satellite photographs over northern India show skies filled with anthropogenic aerosol particles rich in sulfates, nitrates, organic carbon, black carbon, and fly ash aerosol that can increase global dimming (**Figure 3.37**). Note how the high mountains of the Himalayas and the Tibetan Plateau rise above the level of pollution.

In addition to their reflective and absorptive properties, many aerosols are also **hydrophilic** (water attracting) and contribute to the formation of tiny cloud droplets that make clouds more reflective. They can also inhibit precipitation so that the clouds last longer, reflect more sunlight, and cool the atmosphere even

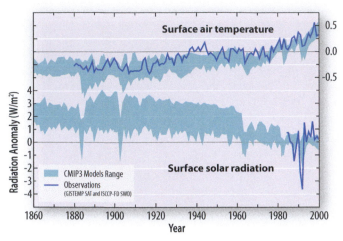

▲ Figure 3.38: A graph of surface air temperature and surface solar radiation. This graph shows how solar insolation and surface temperature have changed relative to base levels from 1860 to 2000. Note the impact of Mount Pinatubo in 1991. Can you see the impact of earlier volcanic eruptions? The thin blue lines represent real-world data; the thick turquoise lines represent data from computer models.

further. This balance between competing warming and cooling factors determines the overall contribution of individual aerosols to climate change.

The impact of global dimming was first discovered in experiments set up during the International Geophysical Year in 1957. One of the key observations involved experimental data from Israel and Estonia that indicated a recent 22% drop in levels of direct sunshine. These simple, but very effective, controlled experiments measured the rate at which water evaporated from open pans. The more direct sunlight, the more evaporation occurred.

Although the impact of aerosols is often greatest at regional level, scientists estimate that worldwide solar radiation at the surface dropped by at least 1.3% per decade between the 1950s and 1990s due to the growth of heavy industry and the burning of undergrowth

▶ Figure 3.39: The Mount Pinatubo eruption and other volcanic eruptions create aerosols that penetrate to the stratosphere and can cause global dimming and cooling for years following an eruption.

following deforestation. The demise of industry in former Communist countries and the enactment of clean air legislation in the United States and Europe helped to reduce the effect towards the end of the 1990s, but observations by satellite have confirmed that renewed growth in aerosol emissions has occurred due to rapid economic development in China and India.

It is generally accepted that the short-term reversal of 20th-century atmospheric warming that occurred between 1945 and 1978 is, at least in part, due to the cooling effect of sulfate aerosols. It is likely, therefore, that the recent increase in the rate of emission of aerosols from the developing nations will soon affect radiative forcing by more than the current -0.4 Wm^2.

Although generally good news for human health, a decrease in global dimming will probably act to increase the rate of global warming. Some global climate models even estimate that temperatures could rise by as much as $6°C$ (10.8°F) by the end of the century as the underlying growth in greenhouse gas emissions is gradually unmasked.

> **CHECKPOINT 3.33 ▶** In your own words, describe the concept of global dimming and explain how it could have a very important impact on radiative forcing if clean-air legislation were enacted worldwide.

Even as solar radiation at Earth's surface was decreasing due to atmospheric dimming, some global warming was still taking place. **Figure 3.38** is a graph that illustrates the relationship between global surface air temperature and the amount of short-wave radiation (radiation from the Sun) received at the surface of Earth between 1860 and 2000. This graph shows how the atmosphere continued to warm, even as the amount of radiation reaching the surface decreased due to air pollution. The sharp anomaly in the surface solar radiation curve in 1991 is due to the cooling effect of the eruption cloud from Mount Pinatubo (**Figure 3.39**

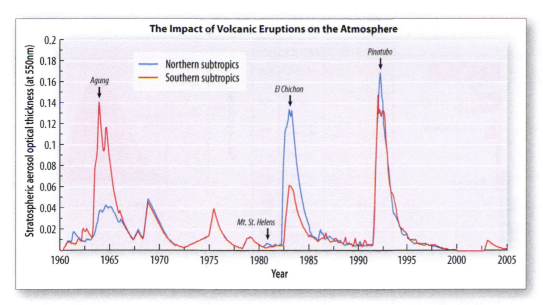

▲ **Figure 3.40:** The impact of some recent volcanic eruptions on stratospheric aerosol optical thickness (AOT).

and **Figure 3.40**). Climate models have been able to successfully reproduce the global impact of this event, a notable achievement that adds to their credibility (**Figure 3.41**). As clean-air legislation continues to reduce aerosol concentrations in the atmosphere, it is very likely that the atmosphere will start to warm even faster in the near future.

Figure 3.40 illustrates in more detail the effect on the atmosphere of the 1991 Mount Pinatubo eruption in the Philippines and the 1982 eruption of El Chichón in Mexico. You can see very clear peaks in **optical thickness**—a measure of how much incident light is scattered or absorbed—associated with the eruptions. In addition, you can see the general trend of decreasing optical thickness associated with cleaner air starting around 1991.

3.7 **PAUSE FOR...** | *Is volcanic activity likely* **THOUGHT** | *to increase in the future?*

The recent history of volcanic activity is fairly well known from the distribution of lava and ash. The precise dating of eruptions is more difficult in the distant past, but it appears as if volcanic activity has had only a transient impact on global climate for the past 55 million years. Volcanic activity can only cool the atmosphere if it occurs at fairly low latitudes and is sustained and powerful enough to eject aerosols high into the stratosphere, where they can reside for a long time and reflect the Sun's energy. Volcanic activity can only warm the atmosphere if the rate of emission of CO_2 exceeds the rate at which it can be absorbed by chemical weathering and by the oceans. The pattern of global tectonics today makes it very unlikely it will happen in the foreseeable future.

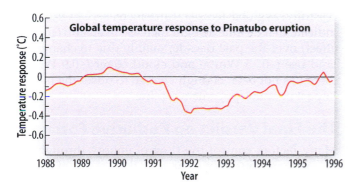

▲ **Figure 3.41:** The impact of the 1991 eruption of Mount Pinatubo on global temperature lasted up to five years. At its maximum, it reduced global temperature by as much as 0.5°C. The ability of climate models to reproduce this effect is evidence that they are accurate and correctly estimate climate sensitivity to small changes in radiative forcing.

Although sulfate aerosols absorb some infrared radiation, they have a net effect of cooling the atmosphere by reflecting more of the Sun's radiation back into space. It is estimated that as much as -0.4 Wm^{-2} of incoming radiation is reflected back, with contributions from the burning of biomass (-0.2 Wm^{-2}) and fossil fuels (-0.1 Wm^{-2}) having the greatest impact. In addition to their direct impact, sulfate aerosols are also estimated to contribute a further -0.07 Wm^2, which would cool Earth by increasing the reflectivity of clouds.

CHECKPOINT 3.34 ▶ What impact do large volcanic eruptions have on radiative forcing, and how long do these effects last?

Figure 3.42 is an IPCC summary of the total global mean radiative forcing now affecting the climate

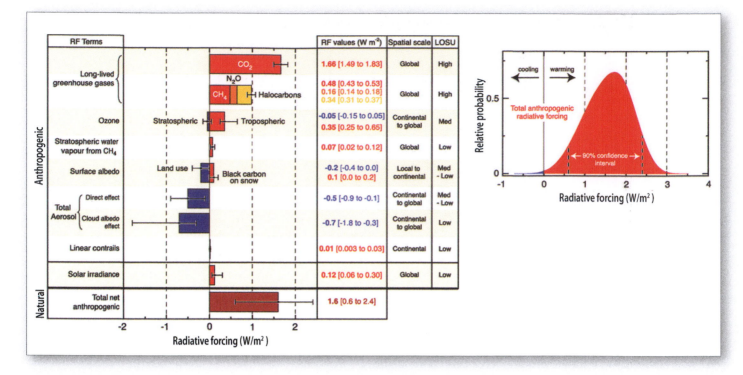

▲ **Figure 3.42:** This diagram from the 2007 IPCC *Fourth Assessment Report* is a summary of the different factors involved in forcing climate change. The graph of relative probability of radiative forcing on the right-hand side of this figure indicates that there is 90% confidence that Earth's climate is receiving on average 0.6 to 2.4 Wm^{-2} of radiative forcing at the top of the atmosphere.

system. The error bars remind us of the relative uncertainty of some of these data, but most possible errors end up producing even more warming. The impact of all greenhouse gases warming the planet is around 2.83 Wm^{-2}, but after the effect of the Sun and cooling feedback from clouds and aerosols are taken into account, the net total of anthropogenic radiative forcing after adjustments for feedback is around 2.2 Wm^{-2}. The natural forcing of greenhouse gases is offset by feedback in the climate system that cools Earth and reduces the overall impact of greenhouse gases.

3.8 PAUSE FOR... THOUGHT | *Do we really understand aerosols?*

The role of aerosols in driving climate change is an area of active research. Aerosols have the ability to both warm and cool the climate, depending on their location, size, and composition, but most estimates suggest that the overall effect is cooling at around −0.70 Wm^{-2} to −0.9 Wm^{-2}. More research is required to quantify their effect in order to reach a more precise estimate of climate sensitivity.

The Role of Albedo and Clouds

Albedo is a measure of how strongly sunlight is reflected by Earth's surface and atmosphere. In general, clouds, snow, ice, and deserts have higher albedo than forests,

lakes, oceans, tundra, and agricultural land. Earth reflects around 30% of incident energy from the Sun, or just over 100 Wm^{-2}, back out into space. Compared with the estimate of just 2.83 Wm^{-2} for the combined greenhouse gas radiative forcing, it is clear that even a small change in albedo of just 5% could have a very significant impact on global temperature. Our incomplete knowledge of the impact of changing cloud cover on albedo remains one of the most significant uncertainties in climate science. Cloud formation can be enhanced by both the presence of aerosols and other particles in the atmosphere and as a direct feedback from changes in regional climate (and therefore difficult to model successfully) as global temperatures increase. Observations indicate that despite a slight increase in albedo (cooling effect) over the past decade, mainly due to changes in land use (−0.15 Wm^{-2}) and cloud cover (−0.9 Wm^{-2}), there has not been enough change to offset the overall impact of anthropogenic warming.

The Final Verdict on Radiative Forcing

Many factors control the flow of energy through the atmosphere, but it is clear that the effects of greenhouse gases are real and intensifying. Total instantaneous radiative forcing is around 2.4 Wm^{-2} but when the rapid adjustment of some components is taken into account, adjusted radiative forcing is around 2.2 Wm^{-2}. This is about the same energy flux as a few holiday tree lights for every square meter of Earth's surface. The atmosphere is responding to this change by warming. But by how much is it warming? As discussed earlier, the maximum increase in temperature following the doubling of CO_2 from preindustrial levels is projected to be around 1.2°C in the absence of climate-related feedback.

> **CHECKPOINT 3.35 ▶** Explain why it is important to understand if there have been major changes in cloud cover over recent decades.

The Question of Climate Sensitivity

This section investigates how positive and negative feedbacks in Earth's climate system can amplify and attenuate the overall impact of radiative forcing. It will become clear that global climate is sensitive to even small changes in the overall balance of energy.

By combining all the factors discussed so far in this chapter, it is clear that instantaneous radiative forcing is currently positive and growing from around 2.4 Wm^{-2}. How is this going to change global climate over the next century and beyond? **Climate sensitivity** is a measure of how much mean global surface temperature will change as energy is added to or subtracted from the climate system. This is most commonly expressed as the increase in temperature expected after doubling the level of preindustrial CO_2 in the atmosphere from 280 ppm to 560 ppm.

Basic physics suggests that doubling the level of preindustrial CO_2 in the atmosphere will increase global temperature by just over 1°C (1.8°F). But there is more to climate sensitivity than simple physics. Feedback within the climate system can both amplify and attenuate this sensitivity, and global temperature has already exceeded the 1°C limit. Melting of ice in the Arctic, the increase in methane emissions from permafrost, and the increase in water vapor in the middle troposphere all have potential to increase radiative forcing well beyond 2.4 Wm^{-2}. On the other hand, an increase of just 3% reflectivity from cloud cover could balance most, if not all, of the radiative forcing from greenhouse gases. All these factors have to be included in global climate models that estimate climate sensitivity.

Most current climate models estimate sensitivity at somewhere between 1.5°C and 4.5°C. These estimates have long tails of uncertainly, but the longest tails "wag" in the direction of increased sensitivity rather than decreased sensitivity. The most probable value is 3°C for a doubling of CO_2, but it could be as low

as 1°C or as high as 6°C. The lowest limit is reassuring, but the upper limit would be catastrophic, and the unfortunate answer is that we don't know for sure which is correct. The sensitivity of climate change to the balance of energy at the top of the atmosphere is difficult to determine with precision, and some factors remain poorly resolved, but climate models and geological evidence suggest that doubling of the level of carbon dioxide in the atmosphere will result in a global rise in temperature of at least 3°C. This is close to the level where some models suggest that inherent, chaotic instability in the climate system (tipping points) could lead to very rapid, damaging, and permanent climate change. Complex systems like Earth's climate have multiple levels of feedback and so much background "noise" that they often exhibit highly unpredictable nonlinear and even stochastic (random) behavior. It is possible that our climate could experience exceptionally rapid change if the climate system is disturbed significantly and rapidly from equilibrium. As we shall see in Chapter 6, it has happened before.

To understand the full complexity of this question, and why it is so difficult to answer, we need to understand more about how Earth distributes energy throughout the climate system and how sensitive Earth has been to changes in solar intensity and greenhouse gases in the past.

> **CHECKPOINT 3.36 ▶** Why is climate sensitivity at the core of the climate change debate?

3.9 PAUSE FOR... THOUGHT | *The important question of sensitivity.*

When I first started teaching and writing about climate change, opponents of climate mitigation were adamant that the world was not warming and that greenhouse gases were a healthy addition to the atmosphere. By focusing on uncertainty, they hoped to delay any action that might reduce our use of fossil fuels. Over the past 10 years, their arguments have changed, as more and more data confirm that anthropogenic greenhouse gases are driving global climate change. Today, the evidence has compelled most skeptics to accept that the world is warming due to anthropogenic greenhouse gases, but they are still critical of IPCC estimates of climate sensitivity, the only remaining area of uncertainty. It is possible that climate sensitivity lies toward the low end of IPCC projections as skeptics propose, but most recent data suggest that the IPCC midrange estimates are correct.

Summary

- The amount of energy from the Sun that reaches the top of Earth's atmosphere varies over time due to internal processes and small changes in the eccentricity of Earth's orbit. Changes in the obliquity and precession of Earth's axis then vary the distribution of this energy at different latitudes.

- The energy received from the sun is then distributed throughout Earth's climate system by the circulation of mass and energy within the oceans and atmosphere, processes that have maintained average global temperature within a limited range for billions of years.

- Many different factors affect the balance at the top of the atmosphere. Some, such as the amount of energy from the sun, anthropogenic greenhouse gases, black carbon, and aerosols, act to force climate change away from a state of quasi-equilibrium. Others, such as clouds, albedo, sea ice, and the natural release of natural greenhouse gases, are a response to climate forcing and can act to reinforce climate change (positive feedback) or ameliorate the impact of climate change (negative feedback).

- The natural greenhouse effect maintains global surface temperatures within limits necessary to maintain the presence of complex life on Earth. The enhanced greenhouse effect from the addition of anthropogenic greenhouse gases to the atmosphere traps more energy within the troposphere and leads to pronounced cooling of the stratosphere and warming the troposphere until the amount of energy leaving Earth once more balances the amount of energy arriving from the sun.

- Positive feedback from water vapor, natural greenhouse gases, changes in albedo, and some kinds of cloud cover enhance the effect. The negative forcing from aerosols, natural aerosols, black carbon, and other kinds of cloud cover ameliorate the impact, but there is still a net imbalance of radiation at the top of the atmosphere that is growing and is enough to increase the risks associated with climate change over the following century.

- Climate models and geological evidence suggest that doubling the level of carbon dioxide in the atmosphere will result in a global rise in temperature of at least 3°C. This is close to the level where some models suggest that inherent, chaotic instability in the climate system (tipping points) could lead to very rapid, damaging, and permanent climate change.

- Complex systems like Earth's climate have multiple levels of feedback and so much background "noise" that they often exhibit highly unpredictable nonlinear and even stochastic (random) behavior. We must consider the real possibility that our climate could experience exceptionally rapid change if the climate system is disturbed significantly and rapidly from equilibrium.

Why Should We Care?

An increase in solar radiation and major volcanic eruptions can explain most of the observed changes in atmospheric temperature over the first part of the 20th century. The same is not true of the past 30 years. Global temperature has continued to increase, even as solar intensity fell toward the end of sunspot cycle 23. Climate models are not able to reproduce this trend without including a substantial component of anthropogenic warming. There is growing satellite evidence that anthropogenic greenhouse gases have changed the amount of infrared radiation leaving the troposphere, and all observations of warming in the atmosphere, on land, and in the oceans are consistent with theoretical projections. *It is clear that human activity is a major force driving climate change.*

Looking Ahead . . .

Chapter 4 will look at the flow of energy through the atmosphere and oceans and examine how interaction between the atmosphere and oceans will be critical in determining the path of climate change. You will discover that all parts of Earth's climate system are deeply interconnected so that a change in the flow of energy in one region is communicated rapidly to affect weather in another region thousands of kilometers away, sometimes with devastating consequences.

Critical Thinking Questions

1. You are a climatologist in the 22nd century. Sunspot activity has been falling over the past three 11-year cycles, and there has been a resurgence in explosive volcanism. You are asked to write a report on the impact of these changes on Earth's climate. Summarize your conclusions and explain why these changes are not likely to have a long-term impact on global climate.

2. Future planetary scientists notice that the climate on Mars is warming during a period of intense sunspot activity. On Earth, however, the Arctic continues to cool, following a trend established over many decades. Use your knowledge of Milankovitch cycles and sunspots to suggest why the Arctic does not show any significant warming.

3. Mars has large volumes of frozen water at the poles and widespread permafrost at lower latitudes. Using your knowledge of greenhouse gases, describe how future Mars colonies could turn this desert planet back to a water planet.

4. Imagine that greenhouse gases reduce the amount of energy leaving the atmosphere by 2 Wm^{-2}, while at the same time there is a 2% increase in the amount of solar radiation reflected by clouds. How would you determine if Earth is likely to warm or cool?

5. At a public meeting in your town, a coal company makes a case for opening a new mine. It claims that coal is good for the environment because sulfate emissions create aerosols that help to reflect solar radiation and keep Earth cool. You rise to counter this argument. What do you say?

6. The rate of global temperature change depends on Earth's climate sensitivity to radiative forcing. Explain why this sensitivity is difficult to determine with confidence based on modern climate data alone. If current estimates are incorrect, is the correct answer more likely to be lower or higher than first estimated?

Key Terms

Please make sure you are familiar with the key terms introduced and highlighted in this chapter. A full glossary is available at the end of the book.

Adiabatic expansion, p. 62
Adiabatic lapse rate, p. 63
Adjusted forcing, p. 76
Aerosols, p. 77
Anaerobic bacteria, p. 65
Aphelion, p. 60
Atmospheric window, p. 71
Black carbon, p. 77
Climate forcing, p. 76
Climate sensitivity, p. 81
Eccentricity, p. 59

Enhanced greenhouse gas effect, p. 74
Equinoxes, p. 59
Environmental lapse rate, p. 62
Faculae, p. 57
Geothermal, p. 56
Global dimming, p. 77
Hydrophilic, p. 77
Infrared radiation, p. 64
Keeling curve, p. 65
Magnetic field, p. 56

MODIS, p. 77
Modulate, p. 58
Montreal Protocol, p. 66
Obliquity, p. 59
Opacity, p. 75
Optical thickness, p. 79
Organic carbon, p. 77
Perihelion, p. 60
Photon, p. 72
Precession, p. 59
Radiative forcing, p. 76
Residence time, p. 65

Schwabe cycle, p. 57
Short-wavelength flux, p. 69
Skin temperature, p. 73
Solar flare, p. 56
Stratosphere, p. 66
Sunspots, p. 57
Total solar irradiance (TSI), p. 57
Tropopause, p. 63
Troposphere, p. 62
Ultraviolet radiation, p. 66
Water vapor, p. 64

Rising sea level increases the probability of destructive storm surges that damage more property and take more lives than any other aspect of a major storm. These New Jersey taxis were left languishing in flood water following Hurricane Sandy in 2012.

CHAPTER 4

Understanding Weather and Climate

Introduction

Monsoon rains, hurricanes, deserts, and frozen tundra are all part of a global climate machine that takes electromagnetic energy from the Sun and turns it into the ceaseless motion of the atmosphere and oceans. In this way, vast amounts of heat energy are transferred from the tropics to the poles. This global transfer of energy is critically important to our understanding of the future of climate change. Earth's climate is unimaginably complex and interconnected so that small changes in temperature in one part of the world can have an almost immediate impact on weather in another. To project the future of global climate change it is important that climate scientists understand how natural processes in the atmosphere and oceans redistribute the net excess of energy that arrives at low latitudes and net loss of energy that occurs at high latitudes. The processes in the atmosphere and oceans that shift mass and energy around the globe are like interconnected cogs meshed together and moving synchronously. Any change in one part of the system, large or small, will affect the entire machine. Climate scientists have identified all the major "cogs" in this global machine, but new unexpected phenomena still emerge, and many of these appear to increase the rate and risk of climate change.

Learning Outcomes | *When you finish this chapter you should be able to:*

- Identify and describe the origin and nature of major natural climate cycles
- Describe the basic structure of the atmosphere and explain the difference between the troposphere and stratosphere
- Discuss how convection and advection in the atmosphere transport heat from the tropics to the poles
- Describe and explain how cyclonic frontal systems transport energy across the polar front to higher latitudes
- Discuss how the oceans transport heat from the tropics to the poles and why the oceans have the capacity to retain vast amounts of heat over many decades
- Explain why major climate cycles involving both the atmosphere and oceans determine the path of climate change on a scale of decades
- Explain why climatologists prefer to use a period of time of at least 15 and preferably up to 30 years as a base period to determine whether there has been a true underlying shift in climate

Global Heat Transfer

In order to project the future of climate change, we must understand the processes that transfer energy between different parts of Earth's climate system today. Only then can we hope to determine how any future changes in the energy content of the atmosphere and oceans will affect global climate. The transfer of energy within and between the different parts of Earth's climate system (the atmosphere, hydrosphere, lithosphere, cryosphere, and biosphere) is a dynamic process that operates over different time scales, and involves energy in many forms (solar radiation, thermal energy, infrared radiation, potential energy latent heat, kinetic energy and electrical energy).

The Source of Energy

The source of nearly all the energy in the atmosphere and oceans is electromagnetic energy from the Sun. As solar energy arrives, it is rapidly converted into other forms of energy and then distributed around the globe by a combination of radiation, convection, conduction, and advection. This global flow of energy is discernible through the ceaseless movement of the atmosphere and oceans, and through the emergence of both short and long-term climate cycles that influence global weather.

The Global Distribution of Energy

When energy arrives from the Sun it is not evenly distributed over Earth's surface. The latitude where the sun is directly overhead, and solar intensity greatest, varies seasonally with Earth's orbit between the Tropics of Cancer (23.4°N) and Capricorn (23.4°S) due to the tilt of Earth's axis. The intensity of radiation also varies with the curvature of Earth's surface, because incoming solar energy is spread over an increasingly larger area at high (polar) latitudes than it is at lower latitudes. The net impact is that Earth gains more energy from the Sun than it emits back out into space at low latitudes up to 40° north and south of the equator, but loses more energy than it receives at high latitudes from 40° to 90° toward the poles (**Figure 4.2**). This energy imbalance sets up a temperature gradient between the tropics and poles, which sets the atmosphere and oceans in motion to equalize the imbalance. If there were no motion in the atmosphere or oceans to redistribute the excess energy received at low latitudes, the tropics would simply continue to heat, and the poles would continue to cool, creating a hostile environment for any form of life.

Despite the surplus of energy received at low latitudes and the net loss of energy lost at high latitudes, the flow of energy in the atmosphere and oceans allows Earth to reach a state of equilibrium, where the poles stay cold but get no

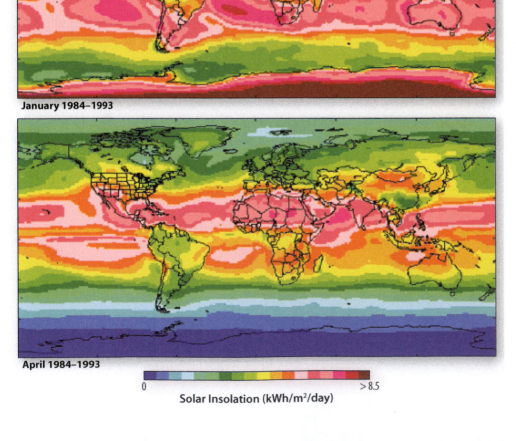

January 1984–1993

April 1984–1993

Solar Insolation (kWh/m²/day) 0 >8.5

◀ **Figure 4.1:** False-color images showing the average solar insolation over the years 1984–1993 for the months of January and April. Note how the band of maximum insolation moves north of the equator with the advent of spring in the Northern Hemisphere.

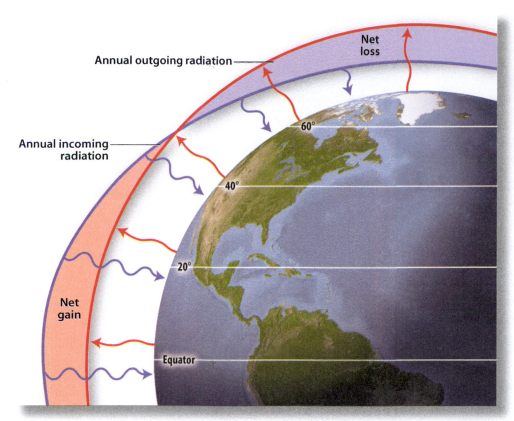

◀ **Figure 4.2:** This plot of net incoming solar radiation and net outgoing infrared radiation shows how latitudes higher than 40°N and 40°S of the equator lose more infrared radiation from Earth than they gain from the sun. This energy imbalance sets up motion in the atmosphere and oceans that acts to counter the growth of large global temperature gradients.

colder and the tropics stay warm but get no warmer. Any change in the amount of energy entering or leaving Earth's climate system will force the climate to adjust in ways that eventually balance the perturbation and restore Earth to a new state of thermal equilibrium.

Understanding Atmospheric Circulation

On a smooth nonrotating planet, the pattern of atmospheric circulation would be simple. Air heated at the equator would rise toward the top of the troposphere, move toward the poles, cool, descend, and flow back along the surface toward the equator (**Figure 4.3**). However, this is not what we observe with our planet. Atmospheric circulation is made complex by the effect of Earth spinning on its axis, at over 1,600 km hr^{-1} (994 mph) near the equator, and the presence of oceans, plateaus, and mountains that interrupt the flow of air.

Despite this complexity, years of data give us an accurate picture of the pattern of atmospheric circulates over time. These data reveal complex patterns of vertical convection and lateral advection in the troposphere that are effective at transferring energy from lower to higher latitudes and maintain an overall balance in global temperature (**Figure 4.4**).

Convection in the Tropics: Hadley Cells

Hadley Cells are large scale atmospheric convection cells that develop close to the tropics, where the Sun's

energy is most intense and surface air temperatures are very high. High evaporation rates over the oceans charge the surface air with as much as 3% by weight of water vapor and create a distinctive atmospheric boundary layer that stores an immense amount of energy in water vapor in the form of the latent heat of vaporization (the energy absorbed from its surroundings when water changes from a liquid to gas phase).

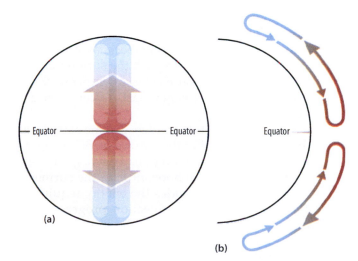

(a)

(b)

▲ **Figure 4.3:** (a) The idealized pattern of circulation that would develop on a uniform, smooth planet that is not rotating on its axis. (b) Warm air rises at the equator and flows uninterrupted to the poles where it cools, descends, and returns to the equator along the surface.

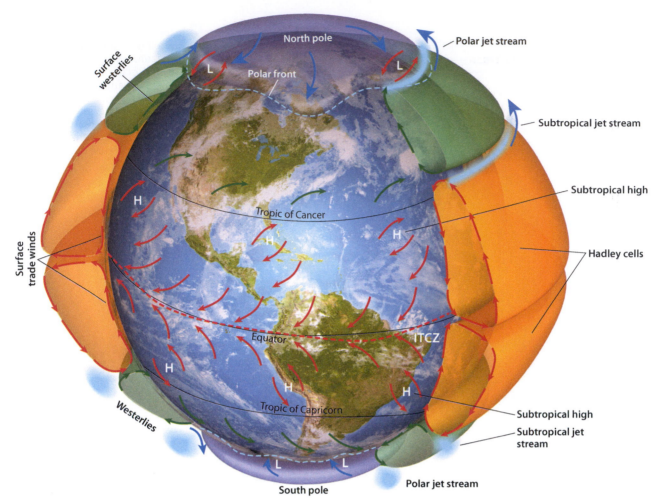

▲ **Figure 4.4:** An idealized pattern of global circulation in the atmosphere, showing the major atmospheric convection cells discussed in the text and their associated surface winds. ITCZ is the Intertropical Convergence Zone. H and L mark areas of consistently high and low atmospheric pressure.

Ascent to the Stratosphere The process that generates Hadley Cell convection starts when surface air is warmed at Earth's surface under the intense tropical Sun. As the air warms it expands, becomes less dense, and starts to rise into the troposphere. As it expands it cools adiabatically (this happens because it must use some of its internal energy to push the surrounding air out of they way) and soon reaches a temperature (the dew point) where water vapor starts to condense and form clouds.

As water vapor starts to condense, some of the latent heat that was absorbed at the surface during evaporation is released into the rising air. This helps to keep the air warmer and more buoyant than the surrounding troposphere, and provides the energy required to maintain convection. As convection progresses, the buoyant air rises rapidly towards the stratosphere and generates a band of large thunderstorms that are clearly visible from space (**Figure 4.5**).

CHECKPOINT 4.1 ▶ Why does surface air start to cool rapidly as it rises through the atmosphere and expands?

▼ **Figure 4.5:** Earth from space. Look for the band of clouds close to the equator over the Pacific Ocean that marks the position of an intertropical convergence zone (ITCZ).

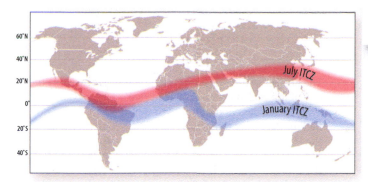

▲ **Figure 4.6:** The average position of the ITCZ in January and July. As the continents warm faster than the surrounding oceans, they deflect the position of the ITCZ to higher latitudes than occurs over the oceans.

The Intertropical Convergence Zone   Beneath these rising storms, surface air flows in from both north and south to replace the air rising by convection and creates a zone of surface convergence. This is called the **intertropical convergence zone (ITCZ),** and its position depends on the latitude where solar heating is at a maximum. In many ways, the ITCZ, represents the "thermal equator," the line of maximum average temperature at any time during the year. It is not fixed in place, but moves back and forth across the equator with the shifting seasons and is also diverted from a simple equatorial path by warm ocean currents and land surface topography (**Figure 4.6**).

CHECKPOINT 4.2 ▶ What factors determine the shifting position of the ITCZ?

Descent from the Stratosphere   During its rapid ascent to the upper troposphere, air from the tropics loses most of its moisture as precipitation (**Figure 4.7**). Convection eventually slows, and stops around 16 kilometers (10 miles) above the surface, where air temperature starts to rise due to the presence of the natural greenhouse gas ozone in the stratosphere. At this point the rising air is no longer warmer than the surrounding troposphere and loses its buoyancy. The warming of the stratosphere by ozone puts an effective "lid" on convection in the troposphere and defines a boundary layer, the tropopause, between the convecting troposphere below and the more stable stratosphere above.

CHECKPOINT 4.3 ▶ Why do the towering thunderclouds that pump mass and energy into Hadley cell circulation finally come to rest at the tropopause?

The addition of more convecting air to this boundary layer creates a zone of convergence as air "piles up" beneath the stratosphere. As it can no longer move vertically, it is displaced laterally, and splits into two divergent air streams that flow northwards and southwards toward the poles and mark the top of the Hadley Cell. By the time the air reaches

▼ **Figure 4.7:** A towering cumulonimbus storm cloud. Note how the top of the cloud spreads out along the base of the stratosphere. Storms like this take heat from the surface of the ocean and release it high in the troposphere and even into the stratosphere when the storms have enough momentum to penetrate through the tropopause.

latitudes around 25° to 30°, North and South of the equator, it has finally cooled to the point where it has lost its buoyancy and starts to descend back towards the surface as the descending limb of a Hadley cell. The descending air regains much of the heat it lost on ascent as it is compressed by the growing weight of the atmosphere above (air warms when it is compressed). By the time it reaches the surface, it is once more hot but exceptionally dry. Sailors named these latitudes the **doldrums** because their ships were often becalmed under descending air for weeks, waiting for a breeze to move them along.

Mid-latitude Deserts

Some of the world's largest deserts occur where air moving within a Hadley Cell returns to the surface (**Figure 4.8**). These most often occur at mid-latitudes, approximately 30° north and south of the equator, where the descending air converges on the surface to build large areas of high pressure that keep moisture from surrounding weather systems at a distance (**Figure 4.9**). This is why we have the Sahara in Africa, the Atacama in Chile, the deserts of Central Australia, and the western deserts of the United States.

> **CHECKPOINT 4.4 ▶** What are the main factors that control the position of Earth's largest hot deserts?

Mid-latitude Westerlies: Ferrel Cells

In the same way that convergence drives the flow of air trapped at the top of the troposphere, surface air at "desert" latitudes is displaced laterally by the convergent flow of air descending from the top of the troposphere. Some flows back toward the equator as easterly winds, or trade winds, that complete the cycle of Hadley Cell convection. However, some flows toward the poles as the surface limb of a **Ferrel Cell** (**Figure 4.10**). This easterly directed flow of warm air (the name "Westerlies" refers to the direction from which the wind blows) from lower latitudes is strongly affected by mixing along a boundary where it meets cold air descending from higher latitudes. There is a large difference in potential energy between the vertically expanded warm air and the vertically contracted cold air. Low-pressure cells or cyclones that develop along this boundary turn this potential energy into kinetic energy that pumps the warm air north to mix with the cold air coming south. The physical properties of the westerlies are strongly affected by Earth's surface. Over land, they remain hot and dry, but over the oceans, the westerlies absorb moisture quickly and become humid.

> **CHECKPOINT 4.5 ▶** What major atmospheric phenomenon controls airflow in the atmosphere in the mid-latitudes between the Hadley Cells and Polar Fronts?

The Polar Fronts

More heat is lost at high latitudes through infrared radiation from Earth than is received from the Sun (Figure 4.2). This leaves the air so intensely cold, dense, and dry that deep convection is unable to develop, and the tropopause lies only 8 kilometers (5 miles) above the surface. This frigid air sinks toward the surface and creates a broad area of convergent high pressure that forces cold polar surface air to flow towards the equator. When these winds blow over the open ocean, they can pick up moisture and thus energy, but when they blow over sea ice or the wastelands of northern Canada and Siberia, they are among of the coldest and driest on Earth.

> **CHECKPOINT 4.6 ▶** Why does cold air descend toward the surface over the poles?

As this cold air flows away from the poles towards lower latitudes it collides with air moving toward the poles from the **subtropics** along the **polar fronts** that develop 30° to 60° around the Arctic and between 55°S and 65°S around Antarctica. The complex **cyclonic** and **anticyclonic weather patterns** that develop along this front are responsible for most of the weather we observe in the United States and Western Europe, as well as the intense storms that form around Antarctica.

The boundary between cold polar and warm subtropical air masses at the top of the troposphere is marked by a steep drop in the height of the tropopause and decrease in air temperature (**Figure 4.10**). Over a short distance, the altitude of the tropopause can drop from as much as 16 kilometers above the surface to as little as 8 kilometers (10 to 5 miles), and very strong high altitude winds blow across the boundary. These

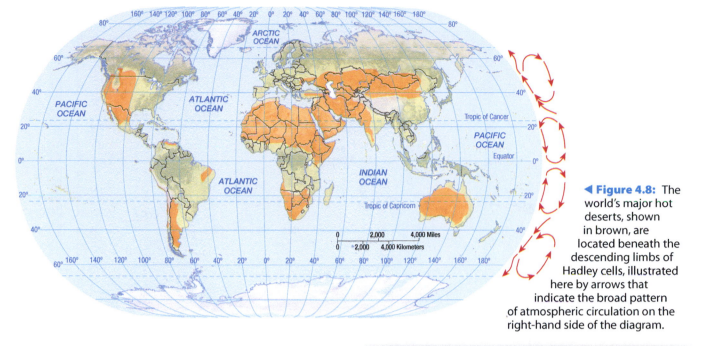

◀ **Figure 4.8:** The world's major hot deserts, shown in brown, are located beneath the descending limbs of Hadley cells, illustrated here by arrows that indicate the broad pattern of atmospheric circulation on the right-hand side of the diagram.

winds do not flow directly down the gradient of the tropopause as might be expected, but are diverted by an apparent force created by the rotation of Earth. This is called the **Coriolis effect,** and it acts on all fluids (air and water) that move over the surface of Earth. As a rule of thumb, both winds and water currents (as well as airplanes) are diverted to the right in the Northern Hemisphere and to the left in the Southern Hemisphere. The Coriolis effect diverts the strong upper tropospheric winds that form along the polar front into relatively narrow streams of air known as the *polar jet streams.* They are typically a few hundred kilometers wide and up to 3 kilometers (2 miles) deep. As fast rivers of air, they ebb and flow around Earth above the polar front with wind speeds of up to 400 kilometers (250 miles) per hour. Small instabilities in this flow develop quickly into large atmospheric waves known as **Rossby waves** (**Figure 4.11**) that form shifting meanders in the jet streams. When upper-level air within the Rossby waves **converges,** it descends toward the surface, creating an area of high pressure. When upper-level air within the jet stream **diverges,** surface air is drawn high into the troposphere, creating an area of low pressure at the surface (**Figure 4.12**). These shifting disturbances of surface atmospheric pressure are closely linked to the movement of contrasting **air masses** along boundaries known as *fronts.* Air masses are large volumes of air that cover a large surface area and acquire uniform properties of temperature and humidity in equilibrium with the land or ocean surface beneath. Over the tropical oceans, for example, the air in the summer is often uniformly warm and humid, while over the North Atlantic it is humid and cool. Over the American interior, the air mass in the summer is often hot and dry, and over the Arctic, the air mass is cold and very dry.

4.2 **PAUSE FOR... THOUGHT** | *Do the jet streams really form a continuous stream of air around each pole, as shown in many textbooks?*

The jet streams are constantly shifting, fragmenting, reforming, strengthening, and weakening as they encircle the globe. While this flow may be statistically continuous, on any one day, the flow is quite complex.

CHECKPOINT 4.7 ▶ How can adjacent air masses acquire such different physical properties?

Cyclonic Frontal Systems: Heating the Poles

The movement of the jet streams is closely linked to changing patterns of surface pressure and the movement of contrasting air masses that form rapidly moving cyclonic **frontal systems.** This is especially so in the Northern Hemisphere, where landmasses, ice sheets, mountains, and oceans help disturb the regular flow of the **jet stream.**

Typically over 1,000 kilometers (600 miles) across, these frontal, low pressure systems move rapidly across Earth's surface from west to east. At the trailing edge of each low pressure system cold surface polar air flows towards lower latitudes, forming an advancing cold front while warm subtropical surface air moves to higher latitudes along a developing warm front. Because warm humid air is forced to rise over cold air along both fronts extensive bands of clouds may form, and some of these may produce rain or snow depending on the season (**Figure 4.13**).

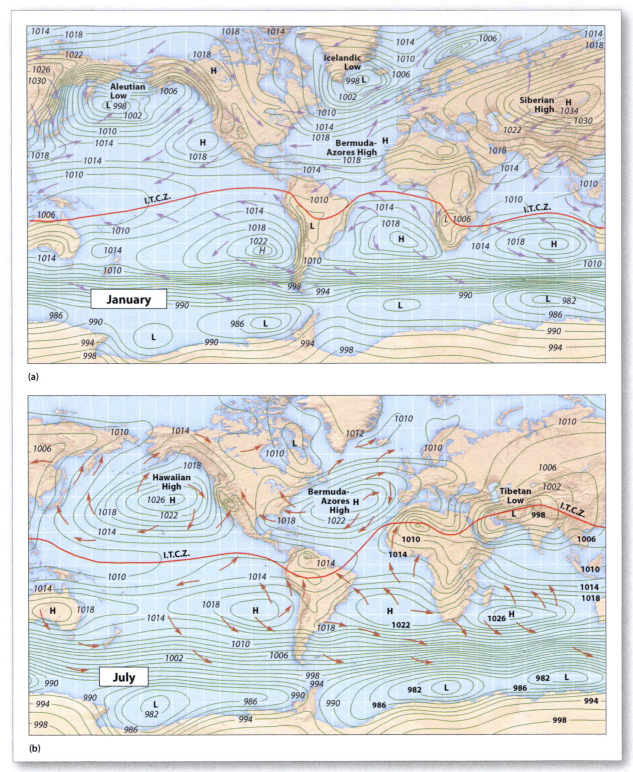

▲ **Figure 4.9:** The average global atmospheric pressure, in millibars, changes with the seasons. Note the large high-pressure belts (anticyclones) in the mid-latitudes that form beneath the air descending from Hadley cells. Note the changing position of the ITCZ.

As each system develops, the cold front usually overtakes the more slowly moving warm front and lifts the warmer air to higher altitudes to form an occluded front, where it cools, condenses, and releases latent heat into the surrounding atmosphere.

Embedded within these dynamic air masses and frontal systems are streams of air known as **conveyor belts.** One of these, the warm conveyor belt, is a critical conduit that moves heat and moisture across the polar front. The warm conveyor belt rises rapidly from sea level in the warm sector of the frontal system in advance of the cold front, crosses over the warm front at altitude, and turns to join the flow of the jet stream. These conveyor belts are often easily seen on satellite

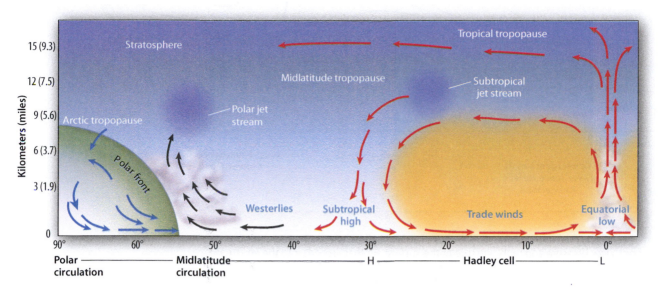

▲ **Figure 4.10:** A slice of Earth's lower atmosphere from the equator to the North Pole showing the idealized position of the major atmospheric cells and the position of the subtropical and polar jet streams at the upper boundaries between these cells. The actual position of the Polar Front changes with the seasons. The red arrows indicate warm tropical air, the black arrows cool air from mid latitudes, and the blue arrows cold polar air.

photographs, where they are marked by their high moisture content and heavy precipitation (**Figure 4.14**).

Condensation on this scale pumps a vast amount of energy in the form of latent heat into the atmosphere and is one of the most important ways in which atmospheric circulation equalizes the imbalance of temperature around the globe. Recent research suggests that the atmosphere is responsible for as much as 75% of

the **meridional** (north–south) transfer of heat from the equator to the poles, and the greatest transfer of heat appears to occur at latitudes where the polar and subtropical air masses converge (**Figure 4.15**). Associated with these fast moving low-pressure systems are small, transient areas of high pressure (Figure 4.12) that move with the jet stream, but large, strong and persistent high pressure areas known as anticyclones that spread

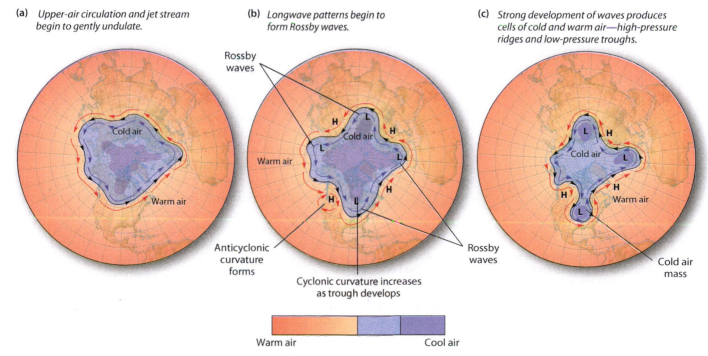

▲ **Figure 4.11:** Along the northern polar front, the jet stream always flows from west to east. Large-amplitude Rossby waves that develop in this airflow determine the position of the jet streams and influence the development of frontal systems lower in the troposphere.

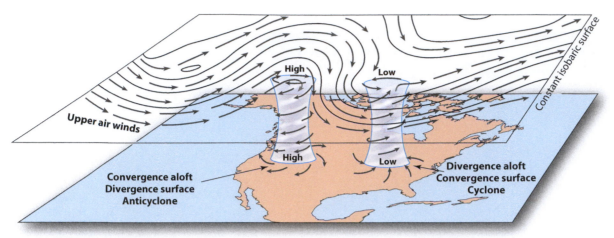

▲ **Figure 4.12:** The convergence and divergence of air within Rossby waves creates shifting centers of both low pressure and high pressure at the surface. Air that ascends into the jet stream around low pressure areas is deflected by the Coriolis effect into a counterclockwise (cyclonic) flow of air. Air that descends from the jet stream towards the surface around high pressure areas is deflected into a clockwise (anticyclonic) flow of air. These shifting areas of low and high pressure move with the jet stream and help to move energy from lower to higher latitudes.

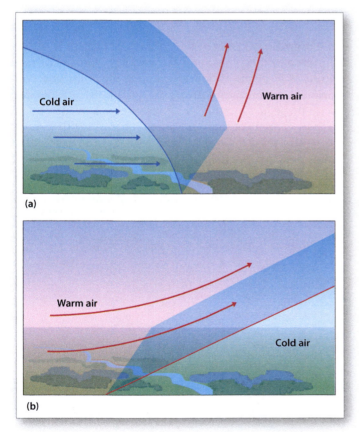

▲ **Figure 4.13:** Frontal low pressure systems are atmospheric disturbances of varying strength that develop along the Polar front and move in the direction of the jet streams. Within each low pressure system, (a) cold polar air (upper figure) advances from the rear of the disturbance by displacing warm air that is forced to rise ahead of an advancing cold front. At the leading edge of the low pressure system, (b) warm air advances along a warm front by flowing over the colder air beneath. Because warm and often humid air is forced to rise over cold air along both fronts, extensive bands of clouds may form, and some of these may produce rain or snow depending on the season.

over 3,000 kilometers (1,800 miles) in all directions can also develop. These are common over Siberia during the winter and over the Atlantic in the spring and have a large impact on weather because they block and disrupt the west to east flow of the jet stream. When they appear over North America during the winter they often direct the flow of upper air into a strong meridional (North-South) flow that pumps warm air into the Arctic and delivers an icy blast of Arctic air to the Deep South. In this way anticyclones also play a critical role in the transfer of heat from low to high latitudes.

CHECKPOINT 4.8 ▶ What role do atmospheric conveyor belts play in controlling the flow of energy toward the poles?

The Oceans

The impact of rapid temperature change in the atmosphere is immediately noticeable, but it is the deep oceans, which can take decades or centuries to respond to changes in global temperature, that will finally direct the future path of global climate change.

The oceans cover 71% of Earth's surface and are kept in continual motion by the combined impact of temperature, wind, gravity, and Earth's rotation. Large-scale circulation patterns move all the water in the oceans from the surface to great depth and back again in a cycle that lasts hundreds of years.

Water is so much denser than air that even the fastest ocean currents are slow compared to movement in the atmosphere, but surface ocean currents are still able to deliver as much as 35% of the heat needed to maintain thermal balance between the equator and the poles. We must understand how the oceans gain, store, transport, and release energy if we are to be able to predict future climate change.

▼ **Figure 4.14:** (a) The warm conveyor (orange arrow) belt rises rapidly from sea level in the warm sector of the frontal system in advance of the cold front (blue line on the map), crosses over the warm front (red line on the map), and turns to join the west-to-east flow of the jet stream (not shown). (b) Both the warm and cold (blue arrow) conveyor belts are associated with extensive cloud cover and the release of latent heat into the troposphere. The dry conveyor belt (green arrows) is an area of relatively dry, cloud-free air that descends towards the surface behind the advancing cold front.

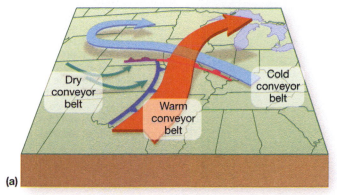

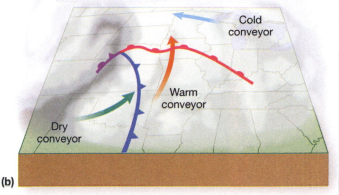

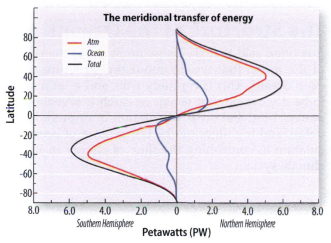

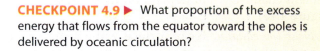

▲ **Figure 4.15:** This diagram compares the average annual contribution of the oceans and atmosphere to the rate at which energy is transferred from the equator to the poles (in petawatts; 1 pW = 10^{15} W). The atmosphere is responsible for as much as 75% of the meridional (north–south) transfer of heat. The greatest transfer of heat occurs at latitudes where polar and subtropical air masses converge, and where Earth's overall heat budget transitions from a net gain of energy to a net loss of energy.

CHECKPOINT 4.9 ▶ What proportion of the excess energy that flows from the equator toward the poles is delivered by oceanic circulation?

The overall **heat content** of the ocean is vastly greater than that of the atmosphere due to the high heat capacity of water, and it has been growing rapidly since the late 1970s (**Figure 4.16**). The Arctic and parts of the Antarctic are warming faster than anywhere else on Earth (see Chapter 2). Currents descending into the deep ocean from the polar regions take this additional heat and store it in a vast deep reservoir of thermal energy. This energy is conserved at depth for hundreds of years before global circulation finally releases it to warm the atmosphere.

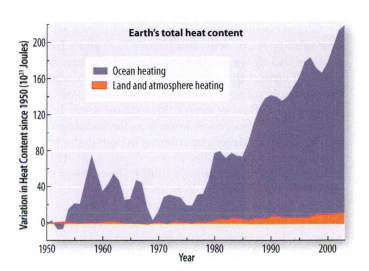

▲ **Figure 4.16:** Earth's total heat content. Heat content is rising much faster in the oceans than in the atmosphere. This "stored" energy in the oceans will have a significant effect on future climate.

The Circulation of Water in the Oceans

The oceans contain many different water masses of contrasting temperature and salinity that reflect their regional geography and climate. The global circulation of these water masses is driven by a combination of meridional (north–south) and zonal (east–west) flow that regulates the transfer of momentum, heat, and mass (moisture) between the oceans and the atmosphere. Surface currents are driven by wind and gravity and directed by the Coriolis effect. Deep currents are driven by density gradients created by regional changes in evaporation, precipitation, temperature, or salinity, and their course is greatly affected by the complex topography of the sea floor.

◀ **Figure 4.17:** This enhanced satellite photograph of the Arctic taken during the winter (note the extent of ice and snow cover) illustrates how the North Pole is surrounded by land that restricts the flow of water and ice and, thus, the exchange of energy with the surrounding oceans. This helps keep the polar oceans cold, and a ready source of cold water flows out at depth into the Atlantic Ocean. Note the large area of the Arctic Ocean that is covered by winter sea ice. This isolates the atmosphere from the warming effects of the oceans.

Global circulation begins where cold, dense, relatively low-salinity Arctic seawater (**Figure 4.17**) sinks from the surface and flows southward along the ocean floor at depth. It crosses the equator (in contrast to circulation in the atmosphere) and flows toward the South Pole, where is joined by a flow of even colder, dense water from the Antarctic. This process is known as *thermohaline circulation* because it is driven by changes in both water temperature (*thermo*) and salinity (*haline*; **Figure 4.18**). This flow of deep cold water continues until it slowly rises toward the surface and warms in both the Pacific and Indian Oceans. Surface currents then complete the global circulation by transferring this warm water back toward the poles. In this way, the tropics are cooled and the poles are warmed, and the energy balance of the climate system is maintained.

CHECKPOINT 4.10 ▶ In your own words, describe how thermohaline circulation develops in the Arctic and Antarctic.

To understand more about this critical cycle of energy, we need to discover much more about the detailed structure of the oceans.

4.3 PAUSE FOR... THOUGHT | *Do we really understand the pattern of global ocean circulation?*

Our understanding of ocean circulation has increased considerably over recent decades. The pattern of surface flow is now well recorded, but it is much more difficult to know what is happening with deeper ocean currents. The generalized models shown here are adequate to understand the overall pattern of energy flow, but it is important to understand that the actual pattern of deep-ocean circulation is much more complex than illustrated.

The Structure of the Oceans

It is often said that we know more about the surface of the moon that we do about the structure of the deep oceans. This is certainly true, and a lot remains to be discovered, but recent research is revealing the pattern of global ocean circulation in unprecedented detail, and highlighting the importance of ocean circulation in controlling the flow of energy in Earth's climate system.

The Surface

All water spends some time at the surface as part of the slow cycle of **ocean convection** and takes on distinctive physical properties that are climate dependent. The top 100 to 400 meters (300 to 1,300 feet) of the oceans is known as the mixed layer because the water is thoroughly mixed by the **turbulent action** of wind and waves (**Figure 4.19**). Each near-surface water mass acquires uniform and unique characteristics of temperature, **salinity,** oxygen content, and trace element chemistry that depend on its point of origin (Figure 4.18).

- *Near the equator:* Near-surface waters are warm and relatively fresh due to intense tropical rainfall.
- *At subtropical latitudes:* Increased evaporation and little precipitation create more saline near-surface water.
- *At mid-latitudes:* Near-surface waters are less saline due to the increased rainfall.
- *Near major rivers:* Surface waters are fresher and have very distinctive trace element chemistry that reflects the geology of the adjacent river basin.
- *At high latitudes:* Salinity is highly variable due to many competing factors:

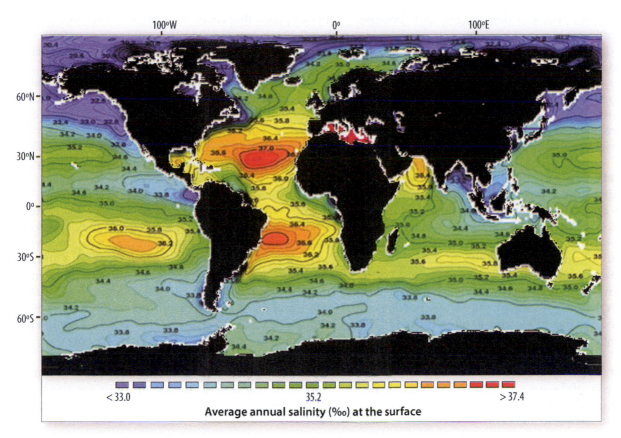

▲ **Figure 4.18:** The salinity of the ocean is not constant. Notice how the salinity of the poles is much less than at the equator. Why is this? High evaporation in the subtropics leads to high salinity in these waters, and the presence and seasonal melting of freshwater sea ice around the poles decreases salinity in those areas. Note the very high salinity of the Mediterranean Sea that arises from high rates of evaporation and limited exchange of water with the open ocean. Figures are in practical salinity units (PSU), a dimensionless quantity based on water conductivity. One PSU is approximately equivalent to one gram of salt per liter of water.

- Factors that decrease salinity include input from rivers at peak flow during the Arctic spring and the melting of seasonal sea ice.
- Factors that increase salinity include strong seasonal winds that increase surface evaporation and the formation of brines rejected as sea ice forms during the winter.

Clearly, the details are complex, but these properties are invaluable as diagnostic tools in the study of ocean convection and circulation.

The Transition Zone   A **transition zone** extends beneath the mixed layer between 400m and 1000m (1,300 and 3,300 feet) deep. This is often called a **pycnocline**

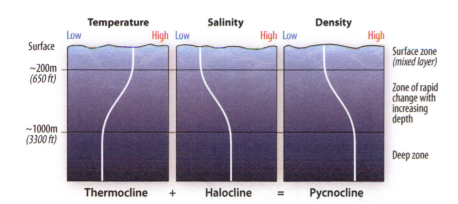

◀ **Figure 4.19:** A transition zone extends beneath a mixed layer between 400 and 1,000 meters (1,300 and 3,300 feet) below the ocean surface. This is called a pycnocline because the density of water increases rapidly with depth and is often marked by both a thermocline reflecting a drop in temperature and a halocline due to an increase in salinity.

from the Greek for "thick or density gradient" because the density of water increases rapidly with depth and is often marked by both a **thermocline** or "heat gradient," reflecting a drop in temperature, and a **halocline** or "salt gradient," due to an increase in salinity. This density stratification inhibits vertical circulation over much of the open ocean apart from areas where surface divergence draws deeper water towards the surface, or where cold dense polar water sinks to the bottom of the ocean (Figure 4.19).

Intermediate Water Intermediate water forms between active surface currents and deep water flows. Much of the intermediate water around the world originates close to the Antarctic, where surface **convergence** causes cold, relatively fresh surface water to sink below the surface to intermediate levels before flowing across the equator as far as the North Atlantic and Pacific Oceans. In the eastern North Atlantic, this Antarctic intermediate water (AAIW) mixes with warm, salty water flowing out from the Mediterranean Sea and is often drawn to the surface by regional surface water divergence.

Deep Water Beneath the transition zone, the temperature and density of the oceans decreases only gradually. The average depth of the flat and relatively featureless abyssal plains that cover most of the ocean floor is more than 3,000 meters (9,843 feet), but this plunges to around 11,000 meters (36,000 feet) in the deep-ocean trenches. Here, the water is cold and flows very slowly at a rate of only a few centimeters per second. Typical **deep-ocean water** has a temperature of about 3°C (37.4°F) and a salinity measuring about 34–35 practical salinity units (psu), which is equivalent to 3.4%–3.5%. The most important mass of deep water is the North Atlantic Deep Water (NADW), which forms in the North Atlantic in the Norwegian, Greenland, and Labrador Seas.

Bottom Water The very coldest and densest **bottom-ocean water** in the world is restricted to the deep-ocean floor around the Arctic and Antarctic Oceans. In the Arctic, this very cold, dense water is trapped within the Greenland, Iceland, and Norwegian Basins and cannot escape. In contrast, intensely cold water (–1.9°C [28.6°F]) with a salinity of 34.62% that forms on the surface of the Weddell Sea and Ross Sea on the Antarctic continental shelf is able to spill onto the deep-ocean floor.

The Guiding Hand of the Ocean Floor The most prominent features on the ocean floor are the major **ocean ridges** (**Figure 4.20**). These volcanic mountain

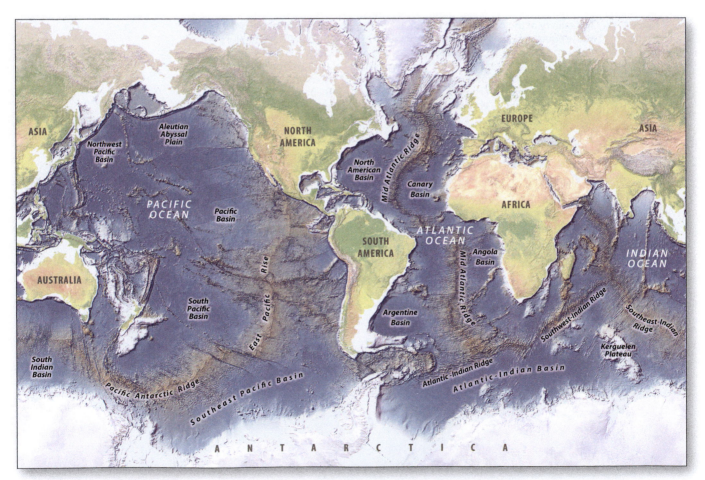

▲ **Figure 4.20:** The world's major ocean basins, showing ocean ridges, seamounts, trenches, and abyssal plains. All these features have a major impact on the direction of flow of deep ocean currents.

chains transect the deep-ocean floor and mark the boundaries between tectonic plates where new ocean floor is created. At over 2 kilometers (6,500 feet) high and over 1,000 kilometers (3,000 feet) across, ocean ridges create formidable barriers to the flow of deep-ocean water. **Ridge axes** are offset by fracture zones in the crust called **transform faults** that can displace the axis by many hundreds of kilometers and can act as conduits that allow deeper water to spill over into adjacent ocean basins. As new ocean floor spreads away from the ocean ridges, it cools, thickens, and subsides. Over time, thickening layers of sediment blanket the underlying topography to form flat, featureless **abyssal plains** that cover much of the deep-ocean floor.

CHECKPOINT 4.11 ▶ What major features on the ocean floor direct and deflect deep-ocean currents?

Thermohaline Circulation

There is still much to be learned about the incredible cycle that takes as much as 90% of water in the oceans on a continuing journey from the surface to abyssal depths and back again. Known as the **Great Ocean Conveyor** (**Figure 4.21**), this global pattern of **thermohaline circulation** starts with cold, dense water around the poles that sinks to the depths of the ocean. Once it reaches the ocean floor, it flows slowly on a global journey that may take hundreds of years before it rises back to the surface in the subtropics, where it is

warmed before returning to the poles as near-surface or surface currents.

The deep-water leg of this journey is important for two main reasons. First, it controls the transport of heat from the tropics to the poles. As water sinks to the ocean floor, it creates a mass deficit that is filled by warm surface water drawn in from the subtropics. Second, as the Arctic Ocean warms, some of this energy is stored in these deep-ocean currents, where it may stay for hundreds of years before returning to the surface. The previous 30 to 40 years of warming in the Arctic has already stored a great deal of energy in these high **heat capacity** currents.

The North Atlantic Deep Water (NADW)   The **North Atlantic Deep Water (NADW)** is the primary current that drives thermohaline circulation. It forms in the North Atlantic in the Norwegian, Greenland, and Labrador Seas. Two main factors combine to set this flow in motion. The first is evaporative cooling by strong winds that blow across the surface. Evaporation removes heat from the water and leaves the surface water slightly saltier, cooler, and denser (this is the latent heat of evaporation). The second factor occurs because ice cannot accommodate salt within its crystal structure, and the salt is continuously expelled into the surrounding ocean as an ice pack forms.

This cold (2–3C° [35.6–37.4 F°]), salty, dense water sinks to the sea floor and pools at the bottom of the Nordic Basin before spilling out through crevasses that cut through topographic highs that connect Greenland,

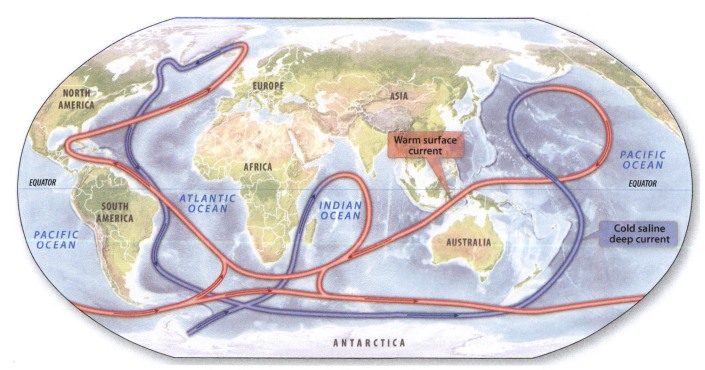

▲ **Figure 4.21:** A simplified map of the ocean conveyor that transfers tropical heat to polar oceans. Note how the conveyor crosses the equator, unlike atmospheric circulation. The red (warm) and blue (cool) colors indicate the general temperature of the water.

Iceland, and northwestern Europe (Figure 4.22). The combined flow rate from these different sources is very large. A sverdrup is the unit used widely in oceanography to denote the rate at which currents flow; 1 **sverdrup** is equal to 1 million meters³ per second, approximately equivalent to the combined flow rate of all of the world's rivers. The flow rate of cold, dense water into the North Atlantic is as high as approximately 20 sverdrups.

Once it reaches the Atlantic sea floor, this large volume of very distinctive water starts to flow on its momentous journey around the world's oceans, a journey that will not be completed for hundreds of years. The current starts by flowing slowly southward along the ocean floor, bound to the west by the American continental shelf and to the east by the high Mid-Atlantic Ridge. The current continues on its journey south along the abyssal plain, at a depth of around 4,000 meters (13,000 feet), until around 60° south, where its flow is interrupted by colder bottom water flowing north from the Antarctic (Figures 4.24 and 4.25).

The Antarctic Bottom Water (AABW)

The **Antarctic Bottom Water (AABW)** is the coldest and most dense seawater on Earth (Figure 4.23). It forms beneath the Antarctic Circumpolar Current and spills north through fracture zones into the Atlantic, Pacific, and Indian Oceans. This water is even colder (as low

as −1.9°C [28.6°F]) than the NADW, and even though it is a little less salty than the NADW, it is so cold that it sinks to the depths of the oceans around the Antarctic. As it flows off the continental shelf, it partially mixes with deep and intermediate water masses before finally flowing outward toward the equator (Figure 4.24). This mixed water flow is slightly warmer, with a characteristic temperature of around 0.9°C (33.6°F) that allows it to be identified as far as 45° north in the Atlantic Ocean. The rest of the water is so dense that it becomes trapped close to the Antarctic. How this partially mixed AABW water eventually rises to the surface is still unclear. Strong-density stratification in the ocean limits the opportunity for the vertical flow of water in the absence of surface divergence. One possibility is that these deep currents are destabilized by seamounts and mid-ocean ridges that cause sufficient turbulence to bring this water to the surface (Figure 4.25).

CHECKPOINT 4.12 ▶ Why does the NADW flow over the AABW, when it is actually more saline?

Wind-Driven Circulation

Surface-water circulation involves only 10% of the volume of the ocean but has a major impact on both weather and climate. A small part of surface circulation is generated directly by the thermohaline circulation, but most is due to the effects of wind, gravity, and the Coriolis effect.

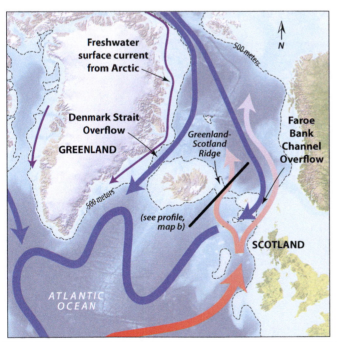

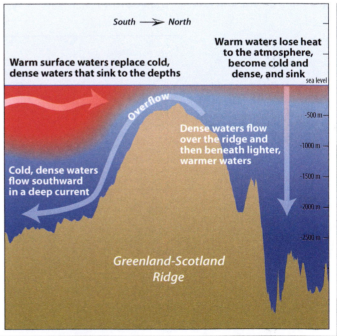

▲ **Figure 4.22:** (a) Cold Arctic water spills out through crevasses that cut through topographic highs that connect Greenland, Iceland, and northwestern Europe as cooling water from the North Atlantic Current takes its place, transferring energy poleward. (b) A vertical section to show how topographic highs such as the Greenland-Scotland Ridge direct and control the flow of water from the Arctic Ocean to lower latitudes and limit the flow of warmer water into the Arctic.

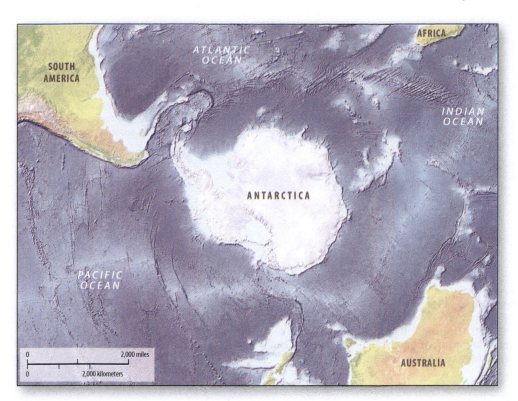

◀ **Figure 4.23:** The continent of Antarctica is isolated at the South Pole. As AABW flows at depth to higher latitudes it encounters physical barriers on the sea floor. Look at the seafloor topography and try to analyze where the greatest barriers to the movement of the AABW are located.

The Marine Atmosphere Boundary Layer (MABL)

The interface between the oceans and the atmosphere is known as the **marine atmosphere boundary layer (MABL).** Here the atmosphere exchanges momentum, thermal energy, water, oxygen, carbon dioxide, and other gases with the ocean. Wind blowing over the ocean is slowed by turbulence and friction and sets the surface water in motion. The Coriolis effect starts working immediately when the water moves, deflecting currents to the right in the Northern Hemisphere and to the left in the Southern Hemisphere.

CHECKPOINT 4.13 ▶ What is the MABL, and why is it important?

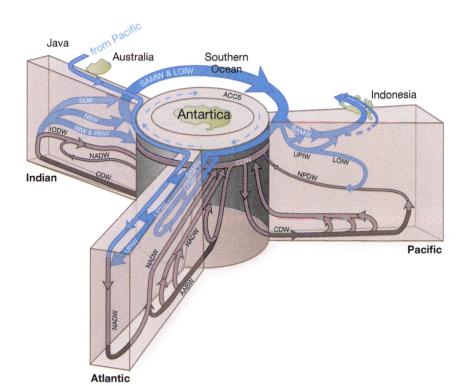

◀ **Figure 4.24:** The AABW forms beneath the Antarctic Circumpolar Current and spills north through fracture zones into the Atlantic, Pacific, and Indian Oceans. This diagram illustrates the complexity of the three-dimensional circulation of water within the Atlantic, Indian, and Pacific Oceans. AABW, Antarctic Bottom Water; ACCS, Antarctic Circumpolar Current System; BIW, Banda Sea Intermediate Water; CDW, Circumpolar Deep Water; IODW, Indian Ocean Deep Water; LOIW, Lower Intermediate Water; NADW, North Alantic Deep Water; NIIW, North Indian Intermediate Water; NPDW, North Pacific Deep Water; RSW, Red Sea Water; SAMW, Subantarctic Mode Water; SLW, Surface Layer Water; UPIW, Upper Intermediate Water.

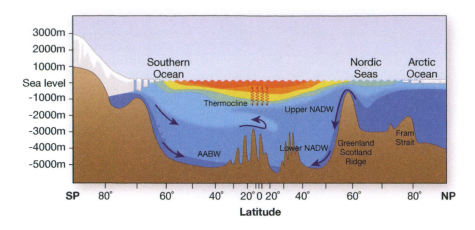

◀ **Figure 4.25:** The different bodies of water in the Atlantic are stratified due to temperature and salinity. The AABW is so much colder than the NADW that it hugs the sea floor. Turbulence and convergence around barriers such as ocean ridges and seamounts may force deep water to finally rise toward the surface to warm again.

▶ **Figure 4.26:** As the surface layers of the ocean move, they in turn apply shear stress to ever-deeper layers, in a process that continues until the stress is fully dissipated at a depth of around 100 to 150 meters (330 to 490 feet). The overall impact of this complexity is that the direction of transport of water in the upper 100 to 150 meters (330 to 490 feet) of water is at right angles to the wind direction.

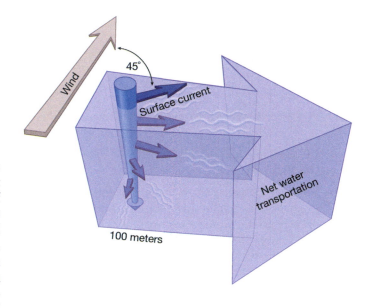

Ekman Transport

As the surface layers of the ocean move, they in turn apply shear stress to ever-deeper layers, in a process that continues until the stress is fully dissipated at a depth of around 100 to 150 meters (330 to 490 feet). This process is easiest to understand by thinking of an ocean divided into discrete layers. As the upper layer starts to move, the Coriolis effect deflects it so that it starts to flow at an angle of around 45° to the wind. Surface water flowing in this new direction then applies stress to the water underneath, which starts to flow and is likewise deflected by the Coriolis effect so that it flows at an even greater angle to the surface flow. As each successive layer starts

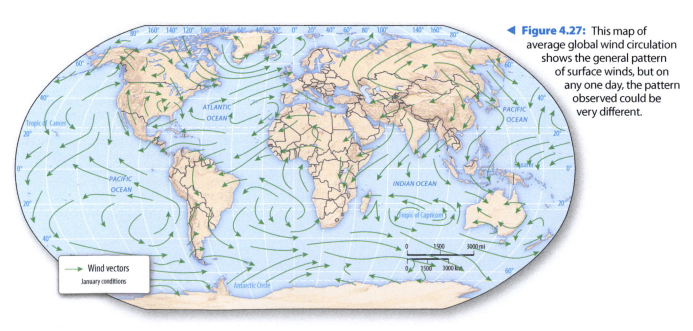

◀ **Figure 4.27:** This map of average global wind circulation shows the general pattern of surface winds, but on any one day, the pattern observed could be very different.

to move, it is likewise deflected, and over the depth of the water column, this will lead to a progressive spiral of deflection from the original surface current, known as an Ekman spiral (**Figure 4.26**).

The net impact of all this movement is that the overall direction of transport of water in the upper 100 to 150 meters (330 to 490 feet) of water—the **Ekman transport direction**—is at right angles to the wind direction: to the left in the Southern Hemisphere and to the right in the Northern Hemisphere. This has a profound impact on the interaction between the atmosphere and the oceans.

CHECKPOINT 4.14 ▶ What is the impact of the Coriolis effect on the upper 100 to 150 meters (330 to 490 feet) of the oceans?

4.4 PAUSE FOR... THOUGHT | *Are the Coriolis effect and the Ekman transport confusing?*

A full explanation is beyond the scope of this book, but students can manage if they learn the simple rule that winds are deflected to the right in the Northern Hemisphere and to the left in the Southern Hemisphere (no matter what direction they are coming from). The equivalent rule of thumb for Ekman transport is that the net flow of water is at 90° to the right of the wind direction in the Northern Hemisphere and 90° to the left in the Southern Hemisphere.

Winds and Oceans

The global pattern of surface wind direction and speed (**Figure 4.27**) is determined by the position of the Hadley, Ferrel, and polar cells (Figure 4.9). Acting on

the surface layers of the oceans, these winds control the flow of major surface currents that play a critical role in heat transport from the tropics to the poles (**Figure 4.28**).

Tropical Divergence

As surface winds from the Hadley cells converge along the ITCZ they blow over the ocean and transfer energy to the upper layers, causing the water to flow. The impact of Ekman transport on this surface flow is a net divergent transport of surface water away from the ITCZ in both the Northern and Southern Hemispheres (**Figures 4.29** and **4.30**). This **tropical divergence** draws deeper water toward the surface and plays an important role in the vertical circulation of the ocean at low latitudes. The impact of this divergence can be seen in the Atlantic and eastern Pacific Oceans (**Figure 4.31**). A clear band of cooler water marks the site of **divergence,** and because this is nutrient-rich water, it is also a zone of high biological productivity (**Figure 4.32**).

CHECKPOINT 4.15 ▶ Why is there a band of cooler, more nutrient-rich ocean water along the equator?

Subtropical Convergence and Gyres

Tropical and subtropical ocean currents are dominated by the flow of winds blowing around semi-permanent anticyclones (high pressure cells) that develop beneath the descending air of a Hadley cell. As this air meets the ocean surface, some starts to flow back toward the equator, and some toward the poles, but this flow is deflected under the influence of the Coriolis effect to form large stationary anticyclones in both the Northern and Southern Hemispheres.

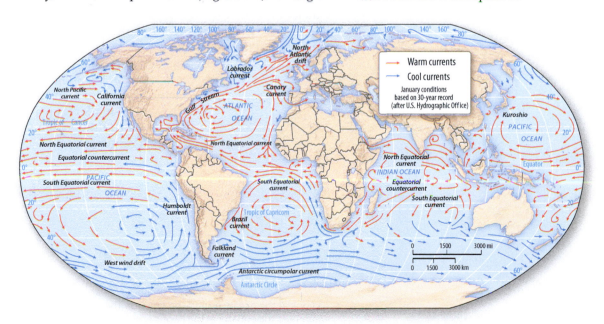

▲ **Figure 4.28:** This map of average global ocean circulation shows the general pattern of surface ocean currents, but on any one day, the pattern observed could be very different.

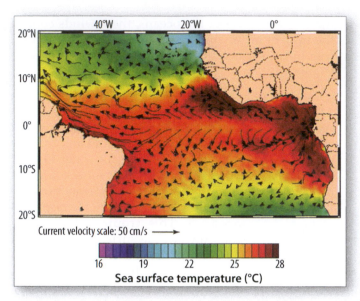

▲ **Figure 4.29:** This map of sea surface temperature and surface ocean currents in the equatorial Atlantic Ocean shows how currents diverge from the equator, allowing slightly cooler water to rise to the surface. This movement is driven by Ekman transport associated with winds (not shown) that converge from the northeast and southeast toward the equator.

As anticyclonic winds blow over the ocean at around 20° to 30° north and south of the equator, they force the surface water beneath to flow, and the Coriolis effect deflects this at an angle slightly to the right of the prevailing wind direction in the Northern Hemisphere and to the left in the Southern Hemisphere. As this surface flow is propagated to greater depths, Ekman transport acts over the top 100 to 300 meters (300 to 1,000 feet) to direct the net flow toward the center of anticyclonic wind circulation, forming basin-wide rotating ocean currents called **gyres** (**Figure 4.33**).

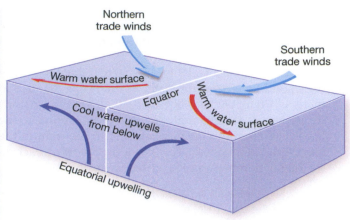

▲ **Figure 4.30:** Sea surface temperature and currents in the equatorial Atlantic Ocean. Note that currents diverge from the equator. This movement is driven by Ekman transport associated with winds that converge on the equator.

CHECKPOINT 4.16 ▶ How do the positions of the Hadley, Ferrel, and polar cells in the atmosphere impact the circulation of surface water in the oceans?

The rotating currents in a Gyre develop because Ekman transport elevates sea level toward the center by as much as 1 meter (3 feet) (**Figure 4.34**). To counter this effect, water beneath a gyre flows away from the elevated center under the force of gravity, and the upper 800 to 1,000 meters (2,600 to 3,000 feet) of this flow is diverted by the Coriolis effect to form a current that flows in a clockwise direction around a gyre in the Northern Hemisphere and in a counterclockwise direction in the Southern Hemisphere (**Figure 4.35**). This balance between the effect of Ekman transport (flow in) the pressure gradient created by gravity (flow out) and the Coriolis effect is known as *geostrophic flow*.

There are five major subtropical gyres in the world's oceans—two in the Atlantic, two in the Pacific, and a fifth in the Indian Ocean—with smaller gyres that direct surface flow in subpolar waters (Figure 4.35).

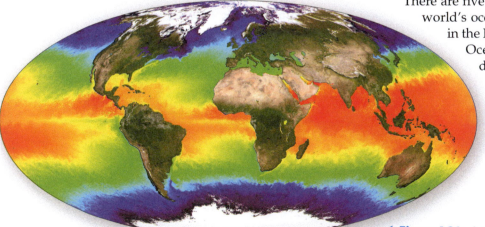

◀ **Figure 4.31:** A map of global ocean temperature. Look at the ocean around the equator. The thin band of cooler water near the equator in the Pacific and Atlantic oceans develops where divergence at the surface allows deeper, cooler, more nutrient-rich water to well up to the surface.

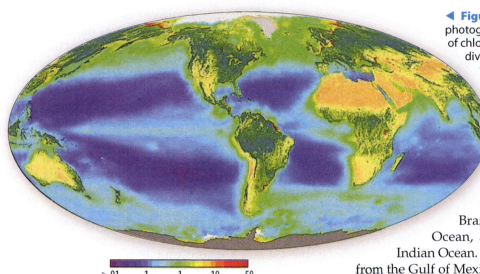

◀ **Figure 4.32:** An enhanced satellite photograph showing a greater concentration of chlorophyll along the equator, where divergence allows nutrient-rich waters to well up toward the surface. (Compare with Figure 4.31.)

>.01   .1   1   10   50

Chlorophyll *a* concentration (mg/m³)

western boundary currents are typical of gyres in both the Northern and Southern Hemispheres. Equivalent currents are the Pacific Ocean Kuroshio Current (Figure 4.33 and 4.36), the Brazil Current in the South Atlantic Ocean, and the Aghulas Current in the Indian Ocean. As the Atlantic current flows north from the Gulf of Mexico, it forms the warm Gulf Stream that flows up the east coast of North America, moving offshore around Cape Hatteras in North Carolina to the North Atlantic Current (Figure 4.37).

These currents play a critical role delivering heat from the tropics across the polar front and well beyond the Arctic Circle. The rotation of Earth and the impact of the Coriolis effect always shift the point of highest elevation in a gyre toward its western boundary. This creates an asymmetry that confines the flow of water in a smaller area and generates some of the fastest, strongest, deepest surface ocean currents in the world. The flow of water on the western boundary of the North Atlantic gyre current is astounding, reaching velocities as high as 2 to 4 ms⁻¹, at flow rates as large as 30 sverdrups around the Gulf and as high as 80–100 sverdrups as it flows past Cape Hatteras off the coast of North Carolina.

CHECKPOINT 4.17 ▶ Why is the ocean surface at the center of a gyre higher than the surrounding ocean? What impact does this have on surface water circulation?

The North Atlantic can be used to illustrate a typical pattern of surface circulation in gyres and the important impact they have on the meridional transport of heat. On the southern limb of the gyre (Figure 4.36), the western directed current meets the American continent and is deflected to form a **western boundary current** that mixes with the flow of very warm water emerging from the Gulf of Mexico and the surrounding ocean. These

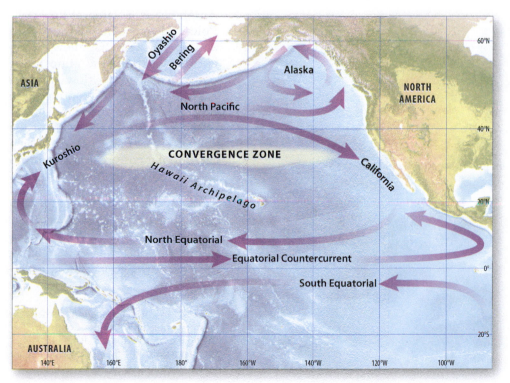

◀ **Figure 4.33:** Gyres are basin-wide rotating ocean currents that are generated by wind friction. This map shows the location of the major gyre of the Pacific Basin and indicates a major area of surface convergence.

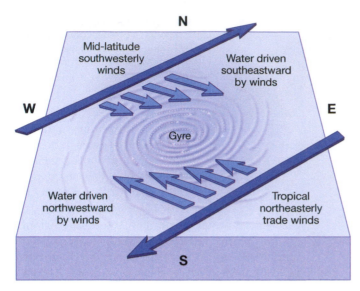

▲ **Figure 4.34:** The net impact of Ekman transport is to move water over the top 100 to 300 meters (300 to 1,000 feet) of the ocean toward the center of rotation. This builds up a large, shallow mound of water up to 1 meter (3.3 feet) higher than the surrounding sea level.

As this phenomenal current of warm water flows around the north of the gyre, a part of the flow breaks off and heads out across the Atlantic to form the North Atlantic Current (**Figure 4.38**). This current reaches deep within the Arctic Circle, releasing energy acquired in the tropics into the polar atmosphere, and as the polar air warms it is able to evaporate more water and transfer additional energy even further north, in the form of latent heat. As the North Atlantic Current slows and cools, it eventually becomes so dense that it starts to sink along with other polar waters to the depths of the ocean, as part of the global thermohaline circulation (Figure 4.38).

The southerly return flow of cold surface water from the poles on the eastern boundary of the gyre is known as the Canary Current and is much broader, more shallow, and slower than the Gulf Stream, with current velocities around 10 centimeters (0.33 feet) per second and a flow rate of around 12 Sverdrups.

A detailed study of the other major subtropical gyres is beyond the scope of this text, but Figure 4.36 illustrates typical circulation flow rates around these gyres and points to the important role they play in directing the transport of heat from the subtropics to the poles.

▼ **Figure 4.35:** The locations of the major subtropical gyres.

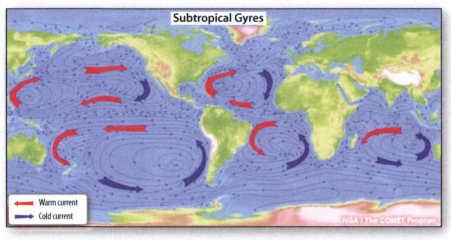

CHECKPOINT 4.18 ▶ What are western boundary currents? What role do they play in the meridional (north–south) transfer of energy?

The Antarctic Circumpolar Current

Antarctica is isolated at the South Pole and surrounded by sea ice and open water that allow strong currents in both the atmosphere and the oceans to flow uninterrupted around the continent. In the troposphere, frontal systems transfer heat into the heart of the Antarctic, but in the oceans, a deep circumpolar current acts as a formidable barrier to the transfer of heat.

In the oceans surrounding Antarctica, the predominant upper level winds are the Southern Hemisphere westerlies (**Figure 4.39**). Deflected by the Coriolis effect, these winds develop a strong clockwise rotating

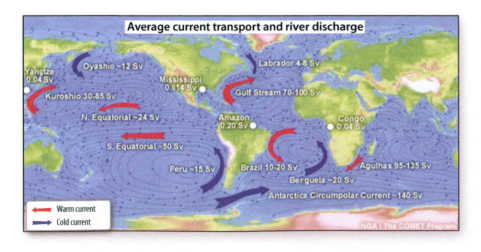

◀ **Figure 4.36:** The locations of major western boundary currents. Compare the flow from the Amazon River (0.20 sverdrups) with the flow of the Gulf Stream (70 to 100 sverdrups).

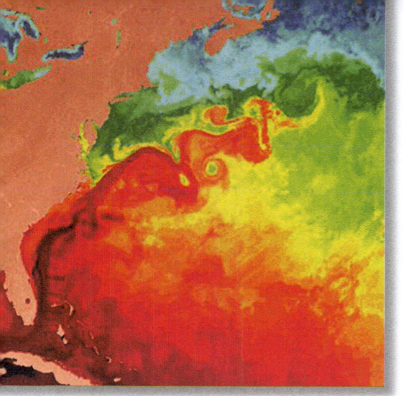

▲ **Figure 4.37:** Thermal image of the Gulf Stream taken by a geostationary satellite positioned over the western Atlantic. Warmer temperatures are indicated by yellow to red colors and colder temperatures by greens to dark blue.

flow that encircles the continent. In the oceans below, the net Ekman transport from this flow is to the north, away from the continent. This creates a sloping ocean surface that gains in height away from the pole. Under the force of gravity, the water "piled up" by Ekman transport starts to flow back "downhill" toward the continent but is deflected by the Coriolis effect into the clockwise-directed **Antarctic Circumpolar Current** (**Figure 4.40**). This strong, deep current has the largest flow rate of any ocean

current on Earth and flows in excess of 100 sverdrups. Strong vertical mixing of water around Antarctica means that there is no pycnocline to prevent this current from reaching down to the sea floor, where it acts as a formidable barrier to prevent warmer waters from reaching the continent.

Over the continent, cold, dry, dense air descends and flows northward over the surface of the immediately surrounding ocean. These winds are deflected by the Coriolis effect to form a counterclockwise flowing current that generates net Ekman transport toward the continent. This is the **Antarctic Coastal Current,** and the relationship between this and the Antarctic Circumpolar Current is very important. The net Ekman transport imposed by these two currents creates flow in opposite directions. The Circumpolar Current pushes water to the north. The Coastal Current pushes water to the south. Between the two currents is an area where surface waters are diverging and where deep water flowing toward Antarctica from the north is finally drawn up toward the surface, where it helps to warm the atmosphere. Here, the NADW finally mixes with Antarctic waters and becomes mixed with water in the Circumpolar Current.

CHECKPOINT 4.19 ▶ Why is there no equivalent to the Antarctic Circumpolar Current in the Northern Hemisphere?

Sea Ice

Seasonal polar sea ice plays an important role in modulating the exchange of heat between atmosphere and oceans around the poles. Ice reflects more of the Sun's energy and prevents direct transfer of heat by conduction between

◀ **Figure 4.38:** The North Atlantic Current, an extension of the Gulf Stream, takes energy from the tropics and releases it high into the Arctic, while cold Arctic water returns to the Atlantic at depth. (Also see Figure 4.22.)

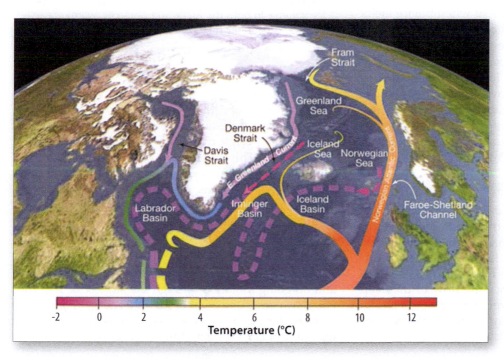

Temperature (°C)

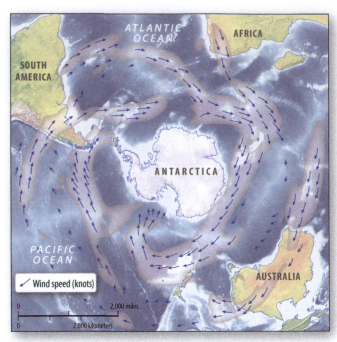

▲ **Figure 4.39:** The Antarctic polar jet stream (gray) develops in association with strong frontal systems that transfer heat into the heart of the Antarctic. On this diagram, the length of each arrow is proportional to the wind speed (in knots).

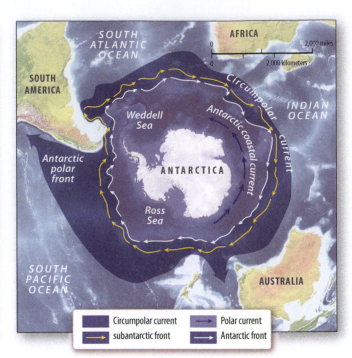

▲ **Figure 4.40:** The strong, deep Antarctic Circumpolar Current encircles the continent in a clockwise direction and is the largest ocean current on Earth, with flows in excess of 100 sverdrups.

the oceans and atmosphere. Ice also isolates polar winds from the oceans and prevents the exchange of latent heat by evaporation.

Seasonal Ice

As water turns to ice it releases energy to the surrounding environment in the form of the latent heat of freezing. When sea ice is stationary, this represents an important annual flow of energy, but has no global impact, because the same amount of energy is removed from the surrounding environment when the ice melts again in the spring (the latent heat of melting).

Studies have shown, however, that the ice often drifts out from the land toward the open ocean before melting. This means that energy released to the environment as ice formed close to the poles is recovered from the environment where ice melts some distance from the poles. The net effect is the transport of heat from outside to inside the ice shelf, and this icy heat pump is considered to be an important component of Antarctic polar heat flow.

The Oceans as a Heat Sink

This study of ocean circulation explains why the oceans are more important than the atmosphere in determining the future path of climate change. Thermohaline circulation turns the ocean into a large **thermal capacitor** that can drive climate change for hundreds of years. Surface currents help to balance the flow of the deep currents and are important over shorter timescales in maintaining the balance of energy between the equator and the poles.

CHECKPOINT 4.20 ▶ How does the seasonal formation and movement of sea ice contribute to the transfer of energy to the South Pole?

4.5 PAUSE FOR... THOUGHT | *Hidden secrets.*

While the media and climate skeptics tend to focus on the area of the ocean covered by ice each winter, a better measure of the long-term behavior of sea ice is the total ice volume. In both the Arctic and Antarctic, warm ocean currents are rapidly melting the ice from below.

Atmosphere–Ocean Interaction

If the vast reservoir of energy in the deep oceans will determine the long-term direction of climate change, how is that energy delivered from the oceans to the atmosphere? This section explores this question and examines how Earth's major convection systems, the atmosphere and oceans, are deeply interconnected. We need to understand these processes if we hope to project the future of climate change.

Complex Interconnected Systems

Earth maintains a dynamic balance between incoming and outgoing radiation through the large-scale meridional (north–south) transfer of thermal energy (Figure 4.2). Some of these processes take place almost entirely within the atmosphere or oceans, but many rely on the constant interchange of mass, **momentum,** and thermal energy between the two. These processes are still not fully understood. They operate over very different **spatial and temporal scales,** convolving in complex ways that cause Earth's climate to change naturally, periodically, and substantially over the short term.

Systems like this are described as **nonlinear systems.** Their future behavior depends on small variations in starting conditions that make their short-term evolution hard to predict. As a consequence, Earth's climate suffers from a lot of random variability, known as **stochastic variability.** It may seem that six hot summers in a row are a sign of global warming, but in reality, this record tells us no more about climate change than six cold winters would tell us that the problem has gone away.

CHECKPOINT 4.21 ▶ What are nonlinear systems? What makes their behavior so difficult to predict?

Climate change is seldom gradual and more often involves jumps as the climate system adjusts to quite small changes in internal energy. It is not at all clear that the relatively moderate climate that has served our civilization so well over the past 5,000 years will be stable in the event of even moderate climate forcing. Some research has even suggested that when random "noise" exceeds some threshold, it can "flip" the climate from one semi-stable state to another, a property known as **stochastic resonance.**

The major ocean–atmosphere cycles are well documented. They work over different timescales, ranging from seasonal to decadal. Their impact on climate is well understood and reproduced accurately by climate models. Even though individual cycles may be confined to one hemisphere, or even just one ocean basin, their impact can be global in a system as interconnected as Earth's climate. The ability of such small changes on one side of the globe to impact climate on another is known as *teleconnection.* There are many examples, but the El Niño–Southern Oscillation (ENSO) is probably the best known beyond climate science because it plays such an important role in regulating the heat budget of Earth.

Tropical Cyclones

Tropical cyclones are large, thermally driven, short lived, infrequent episodes of intense tropospheric convection and precipitation that are collectively responsible for the transfer of an enormous amount of heat and momentum from the tropics toward the poles.

Tropical cyclones can form only under restricted seasonal conditions where sea surface temperatures are greater than 27°C (80.6°F) and winds in the middle and upper troposphere support the development of strong and deep convection. Each year in the tropics, many atmospheric disturbances develop as poorly organized, transient areas of low pressure but never develop the internal organization of a tropical cyclone because

▲ **Figure 4.41:** A large tropical cyclone showing a well-developed eye at the center of the storm.

they dissipate quickly, before they have time to merge, organize, and develop the complex structure that can maintain the essential flow of energy needed to feed these gigantic storms (**Figure 4.41**).

CHECKPOINT 4.22 ▶ Why do relatively few tropical cyclones develop each year, even when sea surface temperature is warmer than necessary for their formation?

Small clusters of thunderstorms known as **tropical disturbances** can develop into tropical cyclones in all oceans under the right conditions. Many occur each year around the globe; in the Atlantic Ocean they are known as hurricanes, in the Western Pacific and China Sea they are called typhoons, and in the Indian Ocean and Australia they are known as cyclones.

The energy that drives a tropical cyclone comes from the release of latent heat from warm, moist air that was in prolonged contact with very warm water before rising into the storm. The amount of energy transferred from the oceans to the atmosphere in this way is enormous—on the order of 600×10^{12} watts (Joules of energy per second), or equivalent to 200 times the world's daily production of electrical power. Those who have experienced the destructive power of a hurricane have some idea how this feels, but most of this energy is actually released into the upper troposphere as air spills out from the top of the hurricane.

Tropical cyclones may be low-frequency events, but at up to 800 kilometers (500 miles) across, they are so wide and the amount of energy released is so large that they play a very important role in the meridional transport of heat, especially in the Northern Hemisphere (**Figure 4.42**).

It is still not clear what the impact of global warming will have on tropical cyclones. There is some evidence that the average intensity of these storms may increase as more latent heat is captured

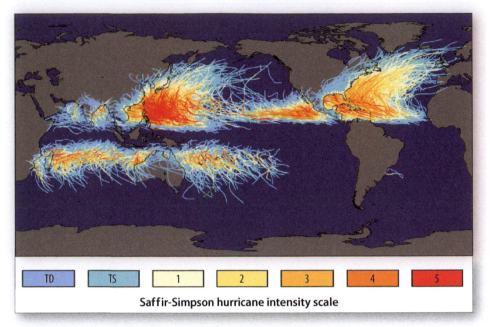

▲ **Figure 4.42:** This map traces the recorded paths of tropical cyclones measured on the Saffir-Simpson scale of hurricane intensity over 150 years. Note how much of Earth is affected by these massive storms that transfer vast amounts of energy from the tropics towards higher latitudes. TD is short for Tropical Depression, TS for Tropical Storm and the numbers for increasing hurricane magnitude.

by increased evaporation as the oceans warm, but it is not at all clear that their frequency will increase. Too many factors are still unclear. Warm, El Niño ENSO episodes, for example, decrease the frequency of North Atlantic hurricanes, and the very passage of intense tropical cyclones disturbs the upper 100 meters (330 feet) of the ocean, bringing to the surface colder water that might inhibit the development or dampen the intensity of new storms in the immediate future. Changes in the state of the Atlantic Multidecadal Oscillation (see page 121) also have an impact. **Figure 4.43** shows how the frequency of hurricanes in the North Atlantic increases during the warm phase of this oscillation.

Monsoons

Monsoons are areas of deep atmospheric convection and precipitation that occur outside the immediate vicinity of the equator. They develop over land surfaces during the summer when heating from the Sun is most intense and migrate over the adjacent oceans during the winter (**Figure 4.44**).

This happens because rock and soil have a low heat capacity (the amount of energy it takes to raise the temperature of a substance by 1°C). They heat quickly to shallow depths but cool just as quickly when heating stops. Water, in contrast, has a high heat capacity, heats slowly to greater depth, and retains this temperature for much longer when heating stops.

As the surface of the land warms during the summer, it transfers energy to the atmospheric boundary layer. Convection soon generates low pressure at the surface and this draws in humid air from the surrounding oceans. The intense precipitation and deep convection that result from this flow release latent heat high into the troposphere that helps to intensify and drive the monsoon circulation (**Figure 4.45**).

Toward the end of summer, the land surface cools rapidly and starts to radiate more energy than it receives from the Sun. The oceans also cool, but they cool more slowly than the land due to their much higher heat capacity. Soon, the oceans are warmer than the surrounding land, and the location of atmospheric convection shifts from the land toward the oceans, reversing the direction of the monsoon winds. This leads to deepening low pressure and more precipitation over the oceans.

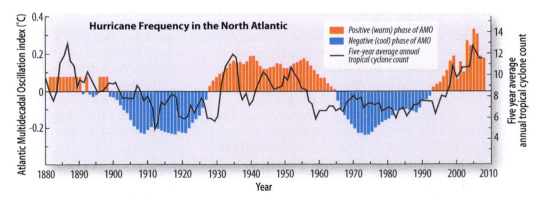

▲ **Figure 4.43:** The frequency of hurricanes in the North Atlantic increases during the warm phase of the Atlantic Multidecadal Oscillation.

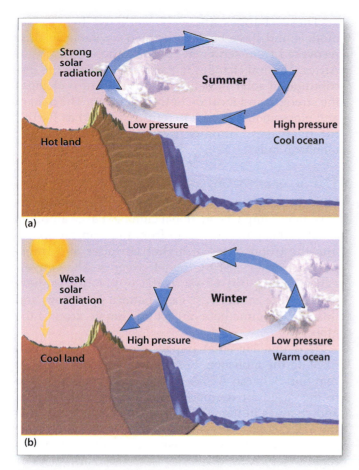

(a)

(b)

▲ **Figure 4.44:** Rapid heating of thin air over the Tibetan Plateau (a) drives the development of summer monsoons over Asia as air rises and draws in moist air from the surrounding oceans. (b) During winter, the air over the Tibetan Plateau cools rapidly, but the sea stays warmer due to its high heat capacity. This generates convection over the oceans that draw winds from the land toward the sea.

Monsoons develop in many places across the world, but the most intense and most studied systems are the extensive monsoons that develop over southern Asia and West Africa. The summer monsoons in these regions are especially important because they are a vital source of water for agriculture and human consumption.

The Asian Monsoon

The southern Asian monsoon develops high on the Tibetan Plateau, where the thin surface air is heated, becomes buoyant, and rises to create low pressure near the surface and high pressure at altitude. Air converging near the top of the troposphere disrupts zonal (west–east) circulation in the Northern Hemisphere and diverts the jet stream from just south of the Himalayas to north of the Tibetan Plateau. This shift in the jet steam draws in moist air from the surrounding Arabian Sea and Bay of Bengal, and as it is forced to rise over the land, it produces the intense precipitation that is so characteristic of the monsoon season (**Figure 4.46**).

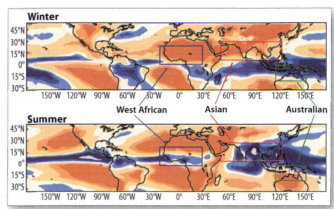

▲ **Figure 4.45:** Seasonal rainfall patterns; winter and summer. Monsoons occur with different intensity in a number of locations around the world where there is a consistent reversal in seasonal wind patterns. Can you spot the development of monsoons in Central America, West Africa, and southern Asia?

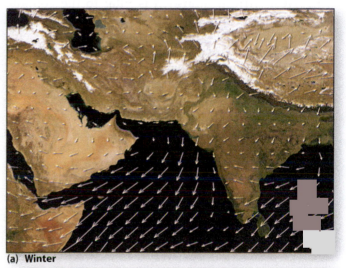

(a) Winter

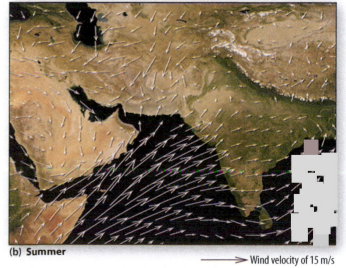

(b) Summer

Wind velocity of 15 m/s

▲ **Figure 4.46:** This map of Asia from NASA illustrates how the southern Asian Monsoon affects both wind strength and direction in (a) winter and (b) summer. Note the very strong winds over the Arabian Sea.

The West African Monsoon

The West African monsoon develops in West Africa and Sub-Saharan Africa as the ITCZ moves north over the summer. Before the seasonal arrival of the ITCZ, regional winds blow over the continent from the northeast and are hot and dry. As the seasons shift, the ITCZ arrives and disrupts the strong flow of air from the northeast (the surface Hadley cell winds) and establishes strong convection over northwest Africa that is able to draw in moist air off the Atlantic Ocean (**Figure 4.47**).

CHECKPOINT 4.23 ▶ What factors would affect the intensity of the monsoon in West Africa if the obliquity (tilt) of Earth started to increase again?

Monsoons and Milankovitch Cycles

The West African monsoon is vital to the continuing survival of much of Sub-Saharan Africa. Recent research has shown that the current monsoon is only a weak remnant of a much-expanded and stronger West African monsoon that affected this region around 8,000 years ago. During that time, Milankovitch cycles determined that Earth was slightly more inclined on its axis, and that the axis was pointing towards the Sun during the summer months, so that solar radiation was more intense at mid-latitudes. As a result, the ITCZ moved to higher latitudes during the summer, and sedimentary and fossil evidence indicate that the intensified monsoon delivered rain over such a large area that it turned the Sahara from desert to savanna, with large lakes, flowing rivers, and abundant plants and animals.

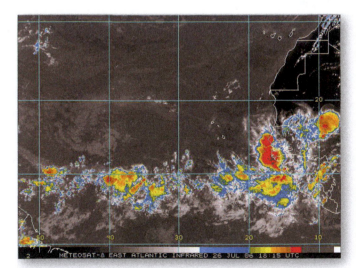

▲ **Figure 4.47:** As the ITCZ moves north during the summer, it reverses the flow of air over parts of northwest Africa and allows moist air to penetrate into the continental interior. This satellite image shows a number of large storms affecting northwest Africa in July. Guided by upper-level airflow, these storms move west into the Central Atlantic and are capable of generating hurricanes.

These changes are well predicted by climate models and have been confirmed by Saharan dust recovered from sediment cores off the west coast of Africa. Natural changes in Earth's tilt change the distribution of solar radiation, and this in turn changes the strength and location of the monsoon. The Sahara will blossom again, only to return to desert as Earth's tilt decreases, summer insolation falls, and the monsoon fails.

The Impact of Monsoons

Both the southern Asian and West African monsoons have major impacts on global climate. Large amounts of latent heat are drawn up from the tropical oceans into the troposphere at higher latitudes and transported further east by the jet streams. The convergence (piling up) of air in the upper troposphere disrupts the zonal (west–east) pattern of airflow, and this has an immediate impact on climate around the globe.

It is unclear what impact global warming will have on the monsoon. Some climate models suggest that Sub-Saharan Africa will become more arid due to increased air temperatures, surface evaporation, and changes in the ITCZ. Other models suggest that the ITCZ could intensify, and more intense convection could actually increase precipitation over the region.

Interannual and Longer-Term Natural Variations

Monsoons and tropical cyclones illustrate how seasonal interaction between the atmosphere and oceans can affect climate. Both monsoons and tropical cyclones exert a strong regional impact on climate and influence global climate through the meridional (N-S) transfer of energy. On a broader geographical and temporal (time) scale, however, a number of larger natural cycles exert an even more important influence on climate. Many of these larger cycles have been recognized for years and are well documented because of the impact they have on society. Others have been discovered more recently and are less well documented. All are still objects of intense research, as they hold vital clues to climate sensitivity and can help us project a much more accurate view of the future.

The North Atlantic Oscillation (NAO)

The **North Atlantic Oscillation (NAO)** was introduced in Chapter 1 and has a significant impact on the weather of North America and Western Europe, because it controls how heat and moisture are distributed throughout the troposphere and how heat and moisture are exchanged with the oceans.

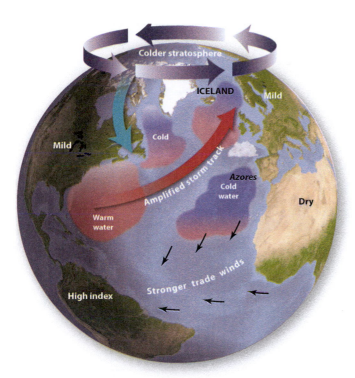

▲ **Figure 4.48:** When low pressure over Iceland is deep and strong high pressure develops over the Azores, the NAO is said to be in a positive phase, pulling the polar jet stream to the north. The light blue arrow illustrates how cold air spills from the Arctic to lower latitudes, and the red arrow, how warmer air in North Atlantic storms takes a northerly path towards Scandinavia. The line of arrows at high latitudes is symbolic of a strong polar jet stream.

In the North Atlantic Ocean, low pressure over Iceland and high pressure over the Azores control the path of the polar jet stream and thus the path of frontal systems as they move across the Atlantic. The NAO is said to be in a positive phase when deep low pressure develops over Iceland and strong high pressure develops over the Azores (**Figure 4.48**). This pulls the polar jet stream to the north and keeps cold polar air confined to northern Europe and the Arctic. Atlantic storms tend to be more severe during this phase of the NAO, but they track further north and leave southern Europe very dry.

The NAO is said to be in a negative phase when both the Icelandic low pressure and the Azores high pressure weaken (**Figure 4.49**). When this happens, the polar jet stream tracks further south, allowing cold polar air to spill southward behind it. Atlantic storms tend to be less severe during these times, but winters in parts of the United States and northern Europe are much colder, while parts of the Arctic and Greenland can be much warmer, as during the winters of 2010 and 2011.

CHECKPOINT 4.24 ▶ Why does a strong negative phase of the NAO bring colder weather to northwestern Europe, wetter weather to the Mediterranean, and warmer weather to parts of the Arctic?

4.6 PAUSE FOR... THOUGHT | *Let it snow, let it snow, let it snow . . .*

The influence of the NAO on the weather of the North Atlantic became very clear during the winters of 2009–2010 and 2010–2011, when an unusually strong negative NAO allowed cold Arctic air behind the jet stream to spill out across western North America and Western Europe, while warmer air was drawn northward into the Arctic. Paradoxically, the decline in Arctic ice may be partly responsible for this change in weather. During the fall and early winter, a more ice-free ocean has time to transfer further energy to the overlying air. The presence of this "bubble" of warmer, more humid air may have played a role in shifting the patterns of atmospheric pressure around Greenland and Iceland that forced the jet stream along a southerly path. The very mild winter of 2012 that followed is a clear reminder that periodic weather extremes, either warm or cold, tell us nothing about long-term change in climate until a clear pattern builds up over more than a decade.

El Niño–Southern Oscillation (ENSO)

The **El Niño–Southern Oscillation (ENSO)** is one of the world's most important climate cycles. El Niño episodes result from quasi periodical changes in

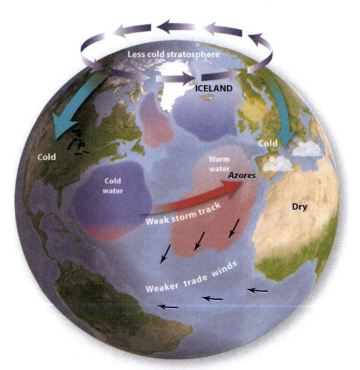

▲ **Figure 4.49:** When both Icelandic low pressure and the Azores high pressure weakens, the NAO is said to be in a negative phase, allowing the jet stream to follow a more southerly path. The light blue arrow illustrates how cold air spills from the Arctic to lower latitudes, and the red arrow, how warmer air in North Atlantic storms takes a southerly path towards the Mediterranean. The line of arrows at high latitudes is symbolic of a weaker polar jet stream.

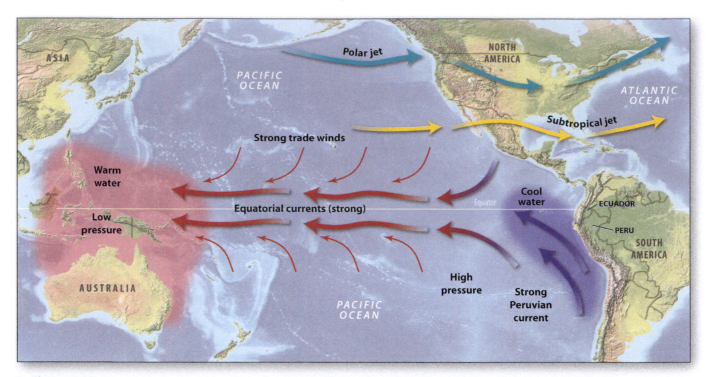

▲ **Figure 4.50:** Strong winds converging along the equator/ITCZ drive warm surface water across the Pacific Ocean. The wide yellow and light blue arrows over North America represent upper air winds. Ocean currents and ocean temperatures are shown as red=warm, blue=cold.

ocean temperature that affect the tropical Pacific. The *Southern Oscillation* part of this name refers to a reversal in surface atmospheric pressure over the Pacific that accompanies El Niño episodes. ENSO has a major impact on weather around the globe and plays a critical role in the redistribution of heat from the tropics toward the poles and in the amount of heat radiated from the atmosphere into space.

ENSO originates in the tropical Pacific, where the Coriolis effect diverts the trade winds so that they blow persistently from east to west across the ocean (**Figure 4.50**). Tropical surface water is entrained (moved) by this flow and accumulates over the western Pacific around Indonesia and northeast Australia to form a large deep pool of very warm water (>28°C [82.4°F]) that suppresses the thermocline by up to 200 meters (650 feet) and elevates sea level in this area by as much as 60 centimeters (2 feet) (**Figure 4.51**).

At the same time, on the opposite side of the Pacific, trade winds blow surface ocean water away from the South American coast, causing upwelling that draws the thermocline and cold water up to within 30 meters (100 feet) of the surface. This allows surface waters to mix with deep, nutrient-rich water, creating a diverse marine ecosystem that is of great biologic, economic, and social importance, especially to the fishing industry.

Air above the accumulating pool of warm water to the west of the Pacific is soon heated and charged with moisture, leading to the development of intense tropical storms that pump heat and moisture high into the troposphere. These storms feed air into the

meridional circulation of a Hadley cell and also set up an important zonal (east–west) circulation in the equatorial Pacific of air known as a **Walker cell** (**Figure 4.52**).

For reasons that are not completely clear, every 2 to 7 years, this well-established pattern of atmospheric circulation breaks for periods of up to 8 or 9 months. During this time, the trade winds weaken or reverse and allow some of the warm water that has accumulated over the western Pacific to spill back across the surface of the ocean toward the coast of Peru,

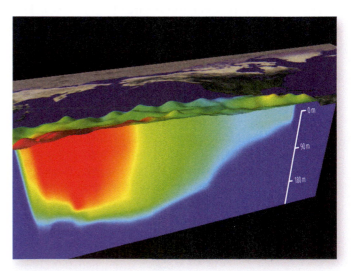

▲ **Figure 4.51:** Warm water accumulates in the western Pacific as a deep pool of exceptionally warm water that can depress the thermocline to a depth of over 200 meters (600 feet).

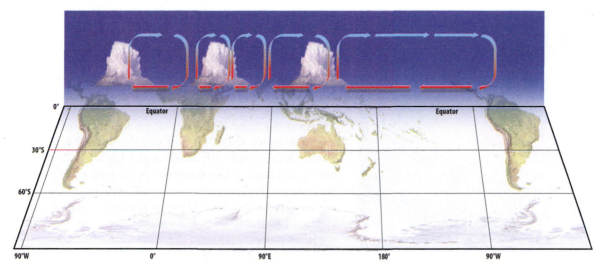

▲ **Figure 4.52:** The Walker circulation is driven by deep convection in the troposphere and is similar to Hadley cell circulation, except the movement of air is zonal (east–west) rather than meridional (north–south). This diagram illustrates the vertical section of the atmosphere over the equator during the typical winter (December, January, and February) mean pattern of circulation.

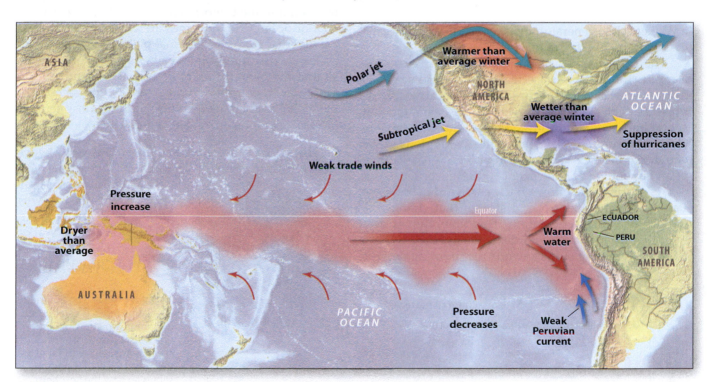

▲ **Figure 4.53:** During the warm phase of an ENSO event, the trade winds weaken or reverse, and water from the western Pacific warm pool spills across the Pacific from west to east. Weather conditions over North America are strongly affected by El Niño events via teleconnections. The wide yellow and light blue arrows over North America represent upper air winds. Ocean currents and ocean temperatures are shown as red=warm, blue=cold. The orange color of Australia indicates hot and dry weather.

deepening the level of the thermocline. As this water spreads out to cover a much larger surface area than before, it enables the transfer of sensible heat (heat you can feel), latent heat, and moisture to the air above, rapidly heating the troposphere. This is a warm ENSO event (**Figures 4.53** and **4.54**).

Changes in ocean temperature are accompanied by a modification of the pattern of Walker circulation.

This is measured using the Southern Oscillation Index, which compares the historical records of air pressure at Darwin on the north coast of Australia with Tahiti in the South Pacific (**Figure 4.55**).

Under neutral conditions (in the absence of ENSO), persistent low pressure is sustained over Darwin as air rises to feed tropical storms, while over Tahiti and further to the east, high pressure develops as air descends from

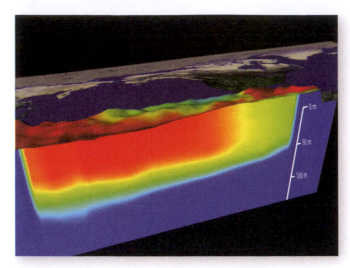

▲ **Figure 4.54:** As the Pacific warm pool spills to the east, the depth of the warm pool decreases to around 100 meters (300 feet), and energy that was progressively stored at depth in the western Pacific is released rapidly into the atmosphere.

a Walker cell to generate surface winds that enhance the overall impact of the trade winds (**Figure 4.56**).

The first signs of an impending ENSO event are a rapid rise in sea surface temperature and sea level in the central Pacific and a decrease in atmospheric pressure over Tahiti. These events have such important social and economic consequences that the data are closely monitored over four indicative regions of the equatorial Pacific (**Figure 4.57**) in order to give warning of an impending El Niño event.

As warm water flows back eastwards across the Pacific, it generates tropical storms that change the pattern of Walker circulation. This time, the air rises over the eastern Pacific and descends over the western Pacific, generating surface winds that blow against the prevailing direction of the trade winds. On the eastern side of the ocean, weaker trade winds and warmer surface water suppress the thermocline, cutting off nutrient-rich water from the surface. This has a devastating impact on both marine and related ecosystems and a catastrophic impact on the local fishing industry (**Figure 4.58**).

As an El Niño event develops, the focus of tropical storm activity shifts so that countries in the western Pacific tend to suffer from drought, high temperatures, and forest fires, while those in the eastern Pacific suffer from an increase in intense rainfall, flooding, and landslides. The impact of El Niño on the weather is not confined to the tropics. **Figures 4.58** and **4.60** show how weather patterns around the world are affected.

El Niño events release heat to the atmosphere that has accumulated in the western Pacific Ocean over many years. Water has a very high heat capacity. It is a good thermal insulator and a very effective medium for heat storage. When this vast amount of heat is rapidly released into the troposphere over such a large surface area of ocean, the outcomes are extreme. Jet streams are diverted from their normal paths, seasonal climate cycles are disrupted, and weather systems are affected so that the overall flux of both heat and moisture to the poles and upper troposphere is enhanced, and the global average temperature of the atmosphere can increase temporarily by as much as 0.6°C.

La Niña   **La Niña** episodes often, but not always, follow warm ENSO events. They develop as very strong trade winds blow across the tropical Pacific from east to west and move the warm water pool at the core of a Walker cell even further to the west than normal. Strong easterly

◀ **Figure 4.55:** The Southern Oscillation Index is established by comparing atmospheric pressure at Darwin on the north coast of Australia with Tahiti in the South Pacific. This graph shows (a) how the Southern Oscillation Index (measured in standard deviation from normal) is related to changes in sea surface temperature. (b) As El Niño develops, average atmospheric pressure over the central Pacific to the eastern Pacific falls as surface convection develops above warming water.

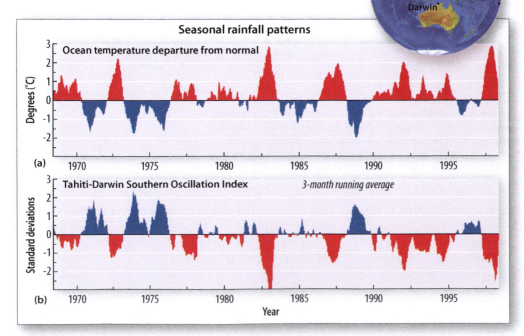

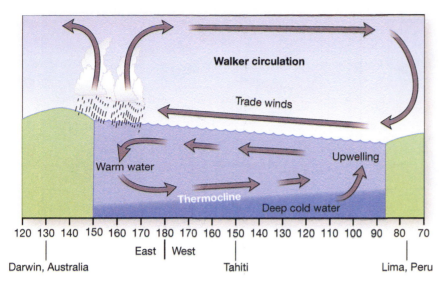

Walker circulation

Trade winds

Upwelling

Warm water

Thermocline

Deep cold water

120 130 140 150 160 170 180 170 160 150 140 130 120 110 100 90 80 70

East | West

Darwin, Australia   Tahiti   Lima, Peru

▲ **Figure 4.56:** Under neutral ENSO conditions, warm water pooled over the western Pacific sets up vigorous convection that establishes Walker cell circulation in the atmosphere and the normal pattern of the trade winds. The persistent flow of the trade winds creates a pool of warm water in the western Pacific and deepens the thermocline, while in the eastern Pacific, the movement of water offshore draws the thermocline toward the surface.

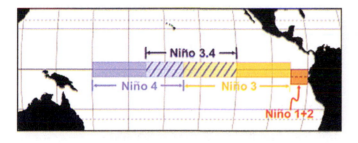

▲ **Figure 4.57:** The state of El Niño is closely monitored over specific regions of the west, central, and eastern Pacific: Niño 1+2, Niño 3, Niño 3.5, and Niño 4 are the names given to these regions.

(from the east) winds blow offshore from the coast of Peru and draw the thermocline close to the surface, while strong divergence across the tropics caused by Ekman transport draws a tongue of cold water across the central Pacific (**Figures 5.59** and **4.61**). These events tend to cool the global atmosphere as the surface air slowly warms the cold water rising to the surface. The net effect is to recharge the oceans with the heat they lost during the El Niño. Just like El Niño events, La Niña events have impacts on weather across the world via teleconnections. They can, for example, set up drought conditions, as happened in Texas in 2010 and 2011.

CHECKPOINT 4.25 ▶ In the short term, a La Niña episode cools Earth. What is the long-term impact of a La Niña episode on ocean temperature?

What Causes ENSO Events? There have been many recent advances in monitoring, modelling, and predicting ENSO events, but there is much still to be discovered. It is possible that they have occurred more frequently over the past 40 years than in the past, but a lot of research is required before we can link ENSOs to anthropogenic global warming.

ENSO episodes are certainly part of a natural cycle in which thermal energy is stored in the western Pacific and then periodically discharged across the central to eastern Pacific. As the warm pool grows deeper, and the ocean above more elevated, gravity tends to force some of this water to flow back across the Pacific, against the direction of the prevailing winds. This is made easier when the strength of the prevailing winds weakens due to the passage of a large-scale shorter-period

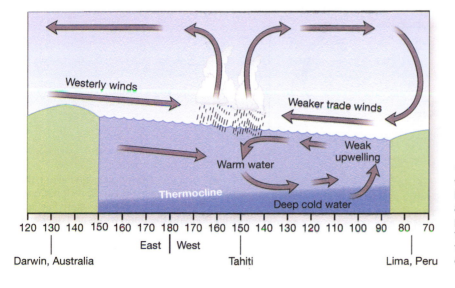

Westerly winds

Weaker trade winds

Warm water

Weak upwelling

Thermocline

Deep cold water

120 130 140 150 160 170 180 170 160 150 140 130 120 110 100 90 80 70

East | West

Darwin, Australia   Tahiti   Lima, Peru

◀ **Figure 4.58:** During an El Niño event, warm water spills out over the central Pacific. The core of active convection in the atmosphere also moves and changes the pattern of Walker circulation, weakening the trade winds. At the same time, the thermocline rises in the western Pacific and is depressed in the eastern Pacific.

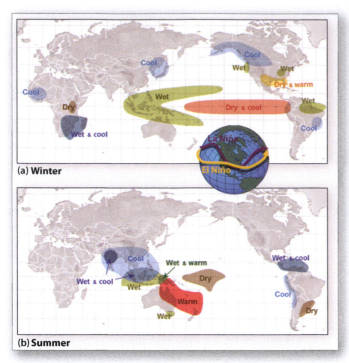

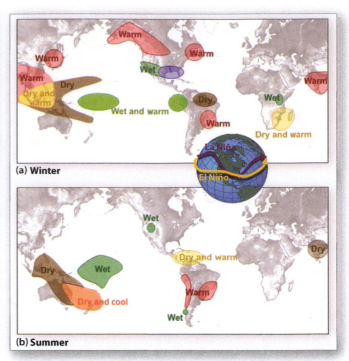

▲ **Figure 4.59:** La Niña: Global weather patterns are affected by the warm pool of water that collects in the far western Pacific during a La Niña episode. The upper panel (a) represents winter conditions, and the lower panel (b) shows conditions during the summer. Note the warm, wet conditions over the western Pacific in association with active convection and the influence on weather patterns of North America via teleconnection. The inserted globe indicates the impact of La Niña (purple) and El Niño (yellow) on the position of the polar jet stream.

▲ **Figure 4.60:** An El Niño episode changes weather patterns around the world by affecting the flow of air in the upper atmosphere. Note the change in the path of the polar jet stream. The panels represent (a) winter and (b) summer conditions. Note the decrease in precipitation over the western Pacific that can lead to drought and forest fires. Note also the increase in summer rain over parts of California and Peru that can lead to floods and landslides. Look beyond the Pacific to see the impact of El Niño elsewhere in the world. The inserted globe indicates the impact of La Niña (purple) and El Niño (yellow) on the position of the polar jet stream.

atmospheric-oceanic cycle known as the *Madden–Julian Oscillation (MJO)*.

The Madden–Julian Oscillation (MJO)

The **Madden–Julian Oscillation (MJO)** originates over the western Indian Ocean and is an eastward-propagat-ing atmospheric disturbance that moves into the Pacific, accompanied by a decrease in surface pressure and intense rainfall (**Figure 4.63**). MJO storms draw up air and moisture (latent heat) from the surrounding ocean and transfer this momentum and thermal energy to the atmosphere. A typical pattern moves at around 5 to 10 ms^{-1} and repeats every 30 to 60 days (**Figure 4.64**). MJO

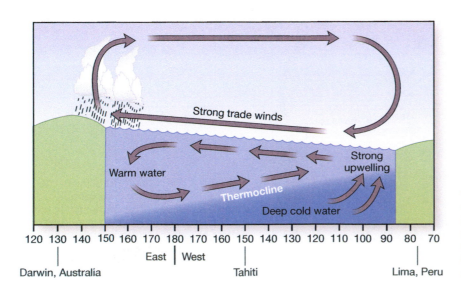

◀ **Figure 4.61:** During a La Niña event, the core of active convection and precipitation moves further west into the western Pacific, strengthening the trade winds and deepening the warm pool. Over in the eastern Pacific, the thermocline is drawn closer to the surface, cooling surface waters.

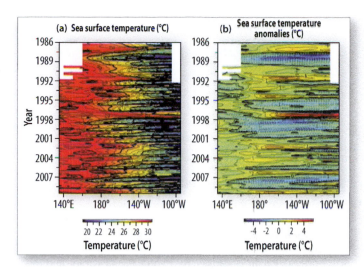

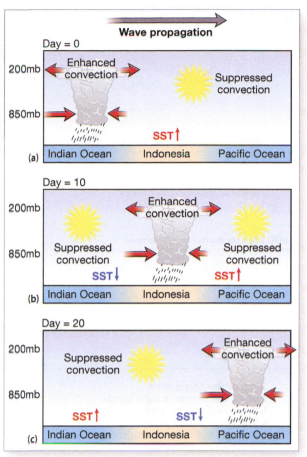

▲ **Figure 4.62:** This type of diagram is used commonly to illustrate both (a) sea surface temperatures and (b) temperature anomalies. The graph plots longitude along the x-axis against time along the y-axis. Temperatures are color coded as indicated on the diagram and are the average monthly mean within 2° north and south of the equator. Read these diagrams like they are a layer cake of time, showing how sea surface temperature changes from youngest (bottom) to oldest (top) from the West Pacific (left side) to Eastern Pacific (right side). Note the strong El Niño in 1998. Note the cyclicity over time related to the development of El Niño and La Niña events.

storms may not have the same global impact as ENSO events, but they can impact the weather in mid-latitudes and increase the frequency and intensity of hurricanes (Figure 4.65).

The MJO–ENSO Link   As they pass over the western pacific, MJO storms generate strong surface winds that are locally persistent and counter to the prevailing trade winds. The momentum that builds from these winds penetrates deep into the ocean and generates subsurface waves along the thermocline. These allow some of the warm water above the deep thermocline in

▲ **Figure 4.63:** (a) The Madden–Julian Oscillation is an eastward-propagating atmospheric anomaly marked by active convection (air rising) with precipitation at the core and areas of suppressed convection (air descending) both in front and behind. (b) Surface winds generated by these anomalies act to reinforce Walker circulation ahead and (c) counter Walker circulation after they have passed.

the western Pacific to surge across the Pacific as short pulses called Kelvin waves that have been observed to build into full El Niño episodes. The MJO is active every year, so it is not a unique trigger for an El Niño episode, but when the warm pool is well developed and only

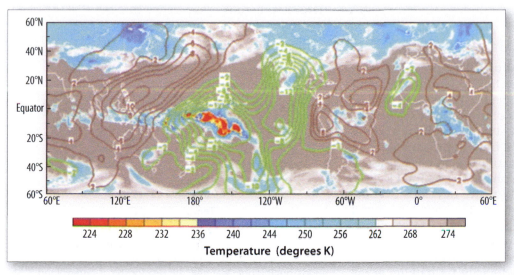

◀ **Figure 4.64:** This illustration maps upper-air anomalies indicating convergence and divergence associated with eastward-propagating convection as part of the Madden Julian Oscillation. Note the alternating patterns that indicate where air is rising (shown in green) and descending (in brown). The diagram also maps outgoing infrared radiation associated with the release of latent heat during precipitation (blue, yellow, and red colors).

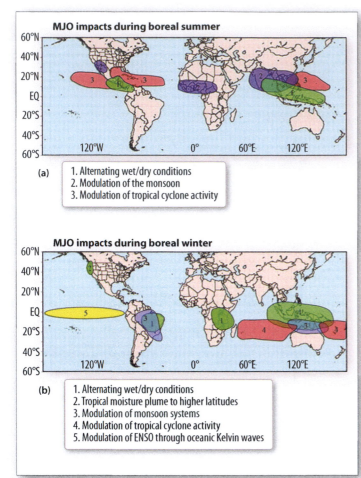

(a)
1. Alternating wet/dry conditions
2. Modulation of the monsoon
3. Modulation of tropical cyclone activity

(b)
1. Alternating wet/dry conditions
2. Tropical moisture plume to higher latitudes
3. Modulation of monsoon systems
4. Modulation of tropical cyclone activity
5. Modulation of ENSO through oceanic Kelvin waves

▲ **Figure 4.65:** The impact of MJO storms is greatest in tropical latitudes in the Indian and Pacific Oceans, but their impact on the circulation of upper air flow and their possible role as a catalyst in the generation of El Niño episodes gives them truly global impact. This diagram highlights some of the main effects associated with this phenomenon during (a) the summer and (b) the winter.

held in place by strong trade winds, the MJO may be one of the triggers to release an El Niño episode.

CHECKPOINT 4.26 ▶ How is it possible that the MJO could play a role in triggering the release of an El Niño event?

The Pacific Decadal Oscillation (PDO)

The **Pacific Decadal Oscillation (PDO)** is a major oscillation in sea surface temperature, pressure, and wind patterns that affects most of the Pacific Ocean. The PDO establishes different patterns of troposphere circulation, redirects the path of the jet streams, and has a strong impact on the transfer of heat from tropics to poles.

The origin of the PDO is complex, but its effects are far reaching. In the positive phase, the central Pacific is warmer; a deep pool of cold water develops over the northern Pacific, while cooler waters develop around the Southern Ocean. During the cold, negative phase, the central Pacific cools, but the western Pacific and seas around Antarctica tend to warm (**Figure 4.66**). The cycles can be identified easily by looking at the

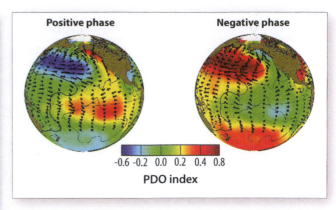

▲ **Figure 4.66:** During the positive phase of the PDO, the northern Pacific is cooler, the central Pacific is warmer, and the southern Pacific is cooler. During the negative phase, the northern Pacific is warmer, the central Pacific cooler, and the southern Pacific warmer. These changes have a major impact on global climate.

temperature of the ocean off the northwest of the United States. Southerly winds during the warm phase push warm Pacific waters toward the shore, while during the cold phase, northerly winds transport water offshore, allowing cool, deeper water to well up to the surface.

The overall spatial pattern of the PDO positive phase is similar to that of ENSO, except that the PDO operates over a much longer timescale, reversing only every 20 to 30 years, and it has greater impact in the northern Pacific (**Figure 4.67**). Changes in surface pressure and winds with the PDO also follow ENSO patterns.

The PDO index is determined from sea surface temperatures in the northern Pacific and exhibits strong cyclicity (**Figure 4.68**). Take a close look at this pattern and compare it to the rate at which the world has been warming over the past 100 years. There is a close correlation between the warm (ENSO-like) positive phase of the PDO and changes in global temperature. This is especially true of the cold phase between the 1940s and the late 1970s. The PDO clearly has a significant impact on the overall release of heat from the oceans to the atmosphere. It is important to note that there is no evidence that the PDO is in any

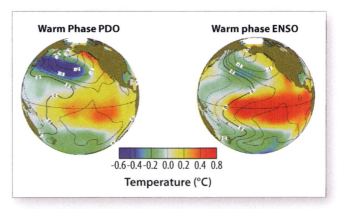

▲ **Figure 4.67:** The positive phases of the PDO and the ENSO have similar spatial patterns of temperature, pressure, and surface winds, but it is not clear how this relationship is determined.

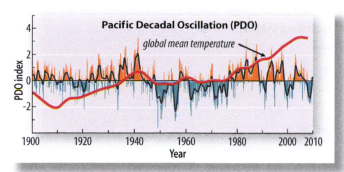

▲ **Figure 4.68:** The PDO index is determined from sea surface temperatures in the northern Pacific and exhibits strong cyclicity every 20–30 years, and evidence of further cyclicity periods as short as 10 years.

way driving underlying trend global warming, but like an El Niño episode, it does modulate an underlying warming trend by affecting the rate at which heat is released into the atmosphere.

CHECKPOINT 4.27 ▶ In what ways are El Niño and PDO events similar?

The Atlantic Multidecadal Oscillation (AMO)

The **Atlantic Multidecadal Oscillation (AMO)** creates periodic changes in sea surface temperature of around 0.6°C in the North Atlantic Ocean, with a dominant periodicity of 60 to 90 years. The warm phase of this cycle occurs when the North Atlantic warms relative to the South Atlantic and is associated with major weather events in North Africa, Europe, and the Americas, ranging from persistent droughts to more frequent tropical cyclones. These effects are to be expected as the warmer water pumps more moisture and energy into the atmosphere above the North Atlantic. The cool phase of the AMO is simply the reverse of the warm phase, with cooler water to the north and warmer water to the south (**Figure 4.69**).

A rise in temperature of less than 1°C may not sound like much, but the heat capacity of water is so high that even a small change in surface temperature represents a very large increase in the heat content of the ocean.

The origin in the AMO is complex, and there isn't a very good historical record, but there is some evidence that it has been present throughout the Holocene. The cycle may be due to natural changes in the rate of deep-ocean circulation associated with the thermohaline current. When the thermohaline current is strong, more warm water is diverted to high northern latitudes; when the thermohaline current is weak, the warmer waters stop circulating further south (**Figure 4.70**).

CHECKPOINT 4.28 ▶ What impact does the AMO have on the frequency of Atlantic hurricanes?

4.7 PAUSE FOR... THOUGHT | *What is natural, and what is anthropogenic?*

The interaction between the coupled atmospheric–oceanic cycles is complex and not fully understood. The idea that a disturbance in the atmosphere in one part of the world can have a large impact across the globe in known as *teleconnection* and was first established with certainty for El Niño events. One way to think of these is as if they were all cogs in a machine. Some are large, some are small, but they are all interconnected, and any change in one will be transmitted with varying effects around the globe. The overall impact is to create a great deal of natural variation in climate that results from the redistribution of existing energy in Earth's climate system and has nothing to do with the addition of any new energy through the enhanced greenhouse effect or any positive feedback.

Remember how we discussed that these natural changes can be thought of as ocean waves. Looking for a slow rise in global temperature against this background of natural variation is analogous to sitting on the beach trying to spot the incoming tide. Large and small waves lapping against the shore make it difficult for the casual observer to discern how the tide is moving. Only by observing over a longer period of time will the direction of the tide become clear. In the same way, a gradual increase in global temperature emerges from the background of natural climate variation only after years of careful observation. A casual observer will not notice the underlying trend in global temperature and will be confused by the ebb and flow of each succeeding wave of natural climate change.

▶ **Figure 4.69:** The AMO is measured by changes in sea surface temperature in the Atlantic and shows a clear cyclical pattern, with each major positive and negative phase of the cycle lasting between 20 and 40 years, with frequent short-period reversals.

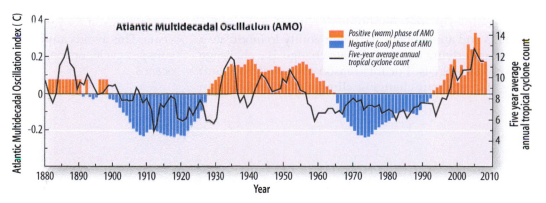

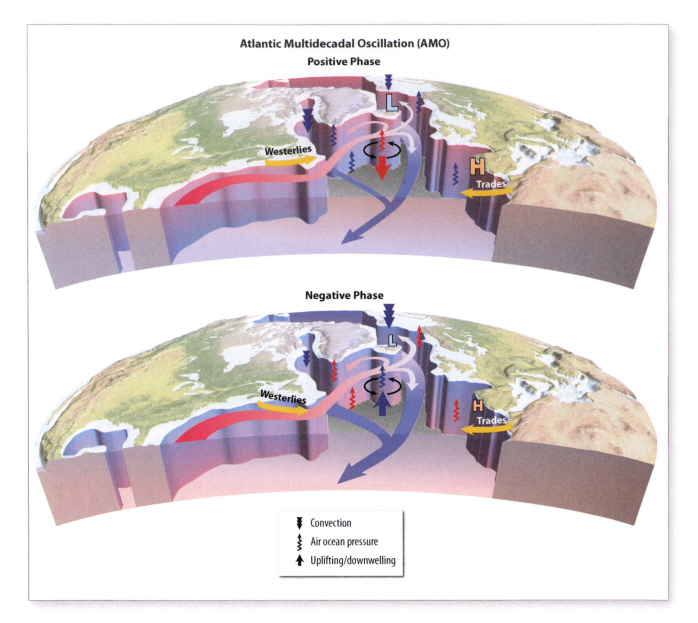

▲ **Figure 4.70:** The AMO may be driven by changes in the thermohaline current. (a) When the thermohaline current is strong, in a positive phase, more warm water (red color on the diagram) is diverted to high northern latitudes and both strong high and low pressure systems develop over the Azores and Iceland that affect the course of westerly winds. (b) When the AMO is weak, in a negative phase, warmer waters circulate further south with weaker pressure systems developing over the Azores and Iceland. As surface temperatures then fall in the Arctic region, the flow of North Atlantic Deep Water (NADW) is enhanced.

The complexity and amplitude of natural cycles in the oceans and atmosphere, and their ability to modulate a slow underlying trend of climate change, is a challenge for climatologists. For this reason they prefer to limit any claims about changing climate to data that span a minimum of 15 years and preferably more than 30 years. This avoids any confusion with the impact of the sunspot cycle, or major cycles in the atmosphere and oceans.

Summary

- Major climate cycles involve the movement of energy and mass within the atmosphere and oceans. Thermally driven convection and advection in the atmosphere dominates wind and weather patterns. Circulation in the oceans is driven by thermohaline circulation and the impact of surface winds. Some of the greatest complexity exists at the interface between atmosphere and oceans where the exchange of mass and energy across the boundary creates major weather events with global impact, such as the effect of an El Niño or the impact of Atlantic hurricanes.

- The weather we experience from day to day occurs within the troposphere. This is the part of the atmosphere that is in constant motion, and most important in the transfer of energy from low to high latitudes. In contrast, the stratosphere is relatively stable due to the presence of naturally occurring ozone that increases air temperature and puts a thermal lid on convection from below.

- Energy is transported through Earth's climate system in different forms. In the atmosphere water vapor plays a critical role because of the vast amounts of heat required to turn liquid water into water vapor. The energy released into the atmosphere by the condensation of water vapor is the fuel that drives the global climate engine. The heat capacity of water is a critical factor in the oceans, as water is able to absorb and retain a large amount of energy with only a small rise in temperature. Ocean currents then transport this water to higher latitudes, where this energy is transferred to warm the atmosphere.

- The polar fronts play a significant role in the latitudinal (North-South) transport of energy. Changes in the global jet streams at altitude along with the movement of air masses at the surface generate frontal systems that facilitate the transfer of energy. Embedded in these fronts are atmospheric conveyor belts that take warm, moist air from the surface high into the troposphere, where it crosses over the Polar front and transfers energy from the subtropics to the poles.

- Surface ocean currents also transport energy from the subtropics to the poles. The high heat capacity of water allows the North Atlantic Current to release a large amount of energy directly into the cold Arctic atmosphere. When covered in winter ice, this energy is retained in the water, but now that the number of ice free days in the Arctic region is increasing, additional energy is transferred from the ocean to the atmosphere, thus changing the dynamics of the Arctic climate and impacting the weather across many regions at high latitude, including North America and Europe.

- The Coriolis effect is like a force that acts on moving water and air. In the northern hemisphere the net effect is to deflect any flow, moving in any direction, to the right of the direction of flow. In the southern hemisphere the effect is reversed and any flow is deflected to the left of the direction of flow. The Coriolis effect is a major influence on the direction of mass movement in both the oceans and atmosphere and is responsible for much of the overall pattern of global circulation.

- The physical interaction between the atmosphere and oceans is complex, producing long-term oscillations in the rate of exchange of energy. These major ocean-atmosphere oscillations such as El Niño and the Pacific Decadal Oscillation take years to complete and are so large that they can mask any underlying trend of global climate change for more than a decade.

- The complexity of natural cycles in the oceans and atmosphere, and their ability to modulate an underlying trend of climate change, is a challenge to climatologists. For this reason they prefer to limit any claims about changing climate to data that span a minimum of 15 years and preferably more than 30 years. This avoids any confusion with the impact of the sunspot cycle, or major cycles in the atmosphere and oceans.

Why Should We Care?

As scientists increase their understanding of Earth's climate system and unravel the complexities of teleconnection, it is becoming possible to project future climate change with increasing accuracy. Over the past 10 years, both the precision (reproducibility) and probable accuracy (correctness) of global climate models have increased, and these models have become reliable at global and even regional levels. Their projections are not reassuring about the prospects of future climate change, but there is still an important missing element in these calculations. That essential element is our inability to decide what action to take to meet the challenge of global climate change. Projections that aim beyond 50 to 100 years in the future are weakened by our inability to decide what we are going to do about heat-trapping greenhouse gas emissions. All the models can do is present an array of possible futures that depend on which action we take today. It is then our collective decision which path to follow.

Looking Ahead....

To achieve a new international agreement that will reduce greenhouse gas emissions it is important to find out just how sensitive our climate is to small changes in global radiative balance. One way to determine climate sensitivity is to look back in time and see what the impact of changing solar radiation and greenhouse gas concentration has been in the past. Chapter 5 will look at the past 60 million years of Earth history to see if there are any lessons that can point to what is most likely to happen in the future.

Critical Thinking Questions

1. We talk about energy often—for example, how it can neither be created nor destroyed, how we experience its different forms, and how we quantify its effects. But what is energy? Where does it come from, and where does it go?

2. The Sun is the source of all the energy on Earth, apart from the slow release of energy from radioactive decay deep inside Earth and some residual energy from Earth's core. How would you determine how much energy is stored in the hydrosphere, atmosphere, and biosphere?

3. The atmosphere is in constant vertical motion, driven by differences in temperature at Earth's surface and at the top of the atmosphere. We observe that it gets colder with altitude, but if the top of the atmosphere is closer to the Sun, why isn't it warmer than the surface?

4. The temperature of the deep oceans is not much warmer today than it was during the depths of the last glaciation. Why is this?

5. For all the knowledge that scientists have accumulated about Earth's climate system, there is still much to learn. How certain should scientists be about the accuracy of their conclusions before they make a public call for political action? Explain your answer.

6. With so much natural variation in climate and so many different oscillations in the atmosphere and oceans that affect global temperature, how can we be sure that the rise in global temperature we have observed over the past 30 years is not just a natural phenomenon?

Key Terms

Please make sure you are familiar with the key terms introduced and highlighted in this chapter. A full glossary is available at the end of the book.

Abyssal plain, p. 99
Air mass, p. 91
Antarctic Bottom Water
    (AABW), p. 100
Antarctic Circumpolar
    Current, p. 107
Antarctic Coastal
    Current, p. 107
Anticyclonic systems, p. 90
Anticyclonic wind, p. 90
Atlantic Multidecadal
    Oscillation (AMO), p. 121
Bottom-ocean water, p. 98
Convergence, p. 98
Conveyor belt, p. 92
Coriolis effect, p. 91
Cyclonic weather
    system, p. 90
Deep-ocean water, p. 98
Divergence, p. 103
Doldrums, p. 90

Ekman transport direction,
    p. 103
Electrical energy, p. 86
El Niño–Southern Oscillation
    (ENSO), p. 113
Electromagnetic energy,
    p. 86
Ferrel Cell, p. 90
Frontal system, p. 91
Great Ocean Conveyor, p. 99
Gyre, p. 104
Hadley Cell, p. 87
Halocline, p. 98
Heat capacity, p. 99
Heat content, p. 95
Infrared radiation, p. 86
Intermediate water, p. 98
Intertropical convergence
    zone (ITCZ), p. 89
Jet stream, p. 91
Kinetic energy, p. 86

La Niña, p. 116
Latent heat, p. 86
Madden–Julian Oscillation
    (MJO), p. 118
Marine atmosphere boundary
    layer (MABL), p. 101
Meridional, p. 93
Momentum, p. 108
Monsoon, p. 110
Nonlinear system, p. 109
North Atlantic Deep Water
    (NADW), p. 99
North Atlantic Oscillation
    (NAO), p. 112
Ocean convection, p. 96
Ocean ridge, p. 98
Pacific Decadal Oscillation
    (PDO), p. 120
Polar front, p. 90
Potential energy, p. 86
Pycnodine, p. 97

Ridge axes, p. 99
Rossby wave, p. 91
Salinity, p. 96
Spatial and temporal scales,
    p. 109
Stochastic resonance, p. 109
Stochastic variability, p. 109
Subtropics, p. 90
Sverdrup, p. 100
Thermal capacitor, p. 108
Thermocline, p. 98
Thermohaline circulation, p. 99
Transform fault, p. 99
Transition zone, p. 97
Tropical cyclones, p. 109
Tropical disturbance, p. 109
Tropical divergence, p. 103
Turbulent action, p. 96
Walker cell, p. 114
Western boundary current,
    p. 105

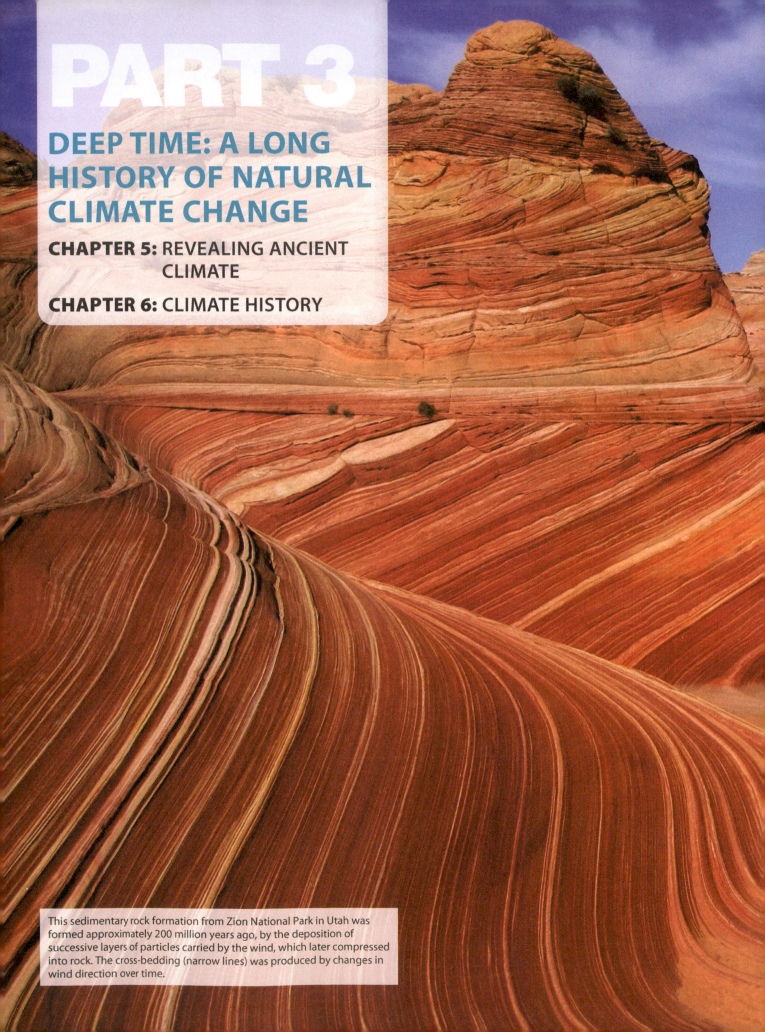

PART 3

DEEP TIME: A LONG HISTORY OF NATURAL CLIMATE CHANGE

CHAPTER 5: REVEALING ANCIENT CLIMATE

CHAPTER 6: CLIMATE HISTORY

This sedimentary rock formation from Zion National Park in Utah was formed approximately 200 million years ago, by the deposition of successive layers of particles carried by the wind, which later compressed into rock. The cross-bedding (narrow lines) was produced by changes in wind direction over time.

One of the best ways to model the future of any natural system is to observe how it has performed in the past. Geological evidence of ancient climate change gives us important clues about climate change today and in the future. Is the climate change we see today unusual? How rapidly has climate changed in the past? What is likely to happen if global warming continues in the future as projected by climate models? Chapter 5 introduces some of the major scientific tools used by geologists and climate scientists to uncover evidence of ancient climate change, and Chapter 6 explores examples of climate change from geological history.

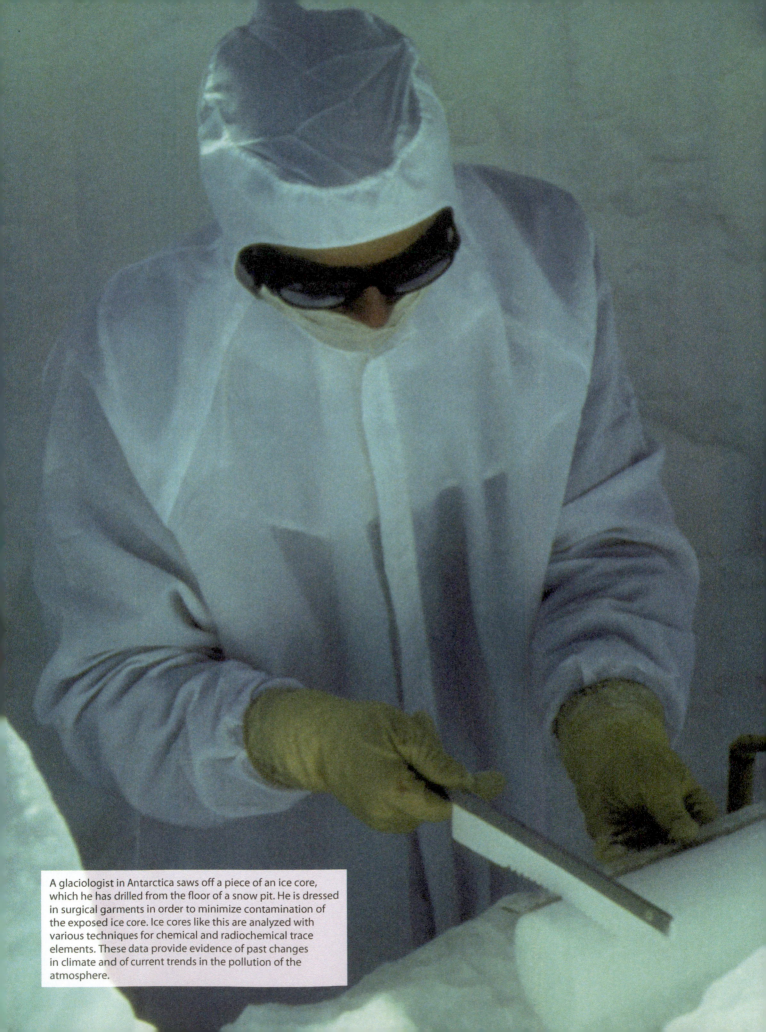

A glaciologist in Antarctica saws off a piece of an ice core, which he has drilled from the floor of a snow pit. He is dressed in surgical garments in order to minimize contamination of the exposed ice core. Ice cores like this are analyzed with various techniques for chemical and radiochemical trace elements. These data provide evidence of past changes in climate and of current trends in the pollution of the atmosphere.

CHAPTER 5

Revealing Ancient Climate

Introduction

Chapters 1 and 2 reviewed evidence for recent climate change and considered many of the different factors that may be responsible. Both natural and anthropogenic climate factors appear to be present, but most of the evidence points to human activity as the primary factor that is driving climate change today. Chapters 3 and 4 investigated how energy flows through the climate system, how sensitive Earth's climate is to external and internal radiative forcing, and how motion in the atmosphere and oceans maintains a balance between incoming energy from the Sun and outgoing energy from Earth. The same forces that control climate today have been in action throughout 4.5 billion years of geological time and have left abundant evidence of ancient climate change in rocks, ice, minerals, and fossils. Climate scientists use these data to project how Earth's climate will change in the future. The best way to gauge the future behavior of any system is to observe how it reacted under similar conditions in the past, and Earth's rich geological history is full of examples of climate change that cast light on the future of climate change today. By studying the ice, rocks, minerals, and fossils preserved in the geological record, climate scientists have established that global climate is very sensitive to even small changes in global temperature.

Learning Outcomes | *When you finish this chapter you should be able to:*

- Understand the many tools that can be used to interpret ancient climate
- Describe how, why, and when each of these tools should be used
- Explain how to determine basic information about climate from the lithology of sedimentary rocks and ice cores
- Assess the value of animal and plant fossils and their chemical remains as indicators of ancient climate
- Discuss how to use oxygen isotopes from a variety of sources to determine ocean and atmospheric temperature
- Analyze carbon isotope data in terms of ocean productivity and the origins of carbon in the atmosphere
- Describe how to use tree ring data as a proxy to compare ancient climate change with modern climate change

Decoding the Past

Geologists have known for some time that Earth's climate is not static. As long ago as 1837, Louis Agassiz, a Swiss-born American geologist, recognized that the deep glaciated valleys of Switzerland provided evidence of the existence of past ice ages (**Figure 5.1**). Fifty years later, when the remains of elephants, lions, hippos, and monkeys were exposed during shallow excavations in England, we learned that parts of the world were once much warmer than they are today.

CHECKPOINT 5.1 ▶ What evidence first helped geologists understand that climate can change significantly?

Climate change is even more evident in rocks and fossils that formed millions of years ago, during what geologists call "deep time." These rocks and fossils not only indicate that climate changes but that it can change frequently and dramatically. They show that much of the Triassic Period (about 200–250 million years ago) was hot and dry, while the Cretaceous Period (about 65–145 Ma) was generally warm and humid (**Figure 5.2**). Why did the climate change in this way? Did CO_2 play a role in these changes? Can we find out how rapidly the climate changed? What do these answers tell us about the possible future of climate change?

5.1 PAUSE FOR... THOUGHT | *What is "deep time"?*

James Hutton (1726–1797) was a Scottish naturalist who first understood the importance of "deep time" and played a major role in the development of modern geology. His field observations of geological unconformities (where young rocks lie immediately on top of much older rocks) led him to understand that Earth is much older than the 5,000 years first suggested by the biblical account of creation. In a 1788 paper, he concluded that ". . . we find no vestige of a beginning—no prospect of an end." For the first time it became clear that to understand Earth history we would need to look deeply into the geological past.

Many clues in the geological record inform us about the nature of ancient climate. Some are so obvious you can find them by examining a rock specimen with a hand lens. Other clues are more cryptic and require detailed study to unlock hidden information. In this chapter, we take a brief look at some of the techniques Earth scientists commonly use and explore how to use them to interpret ancient climate.

A Story in Stone

Rocks may appear to be hard and impenetrable, but we can recover a lot of information from them through field observation and detailed laboratory analysis.

▼ **Figure 5.1:** U-shaped valleys are common in areas of the world that were recently glaciated. They are formed by the grinding actions of thick glaciers that carve this characteristic shape from the walls and floor of the valley beneath.

Era	Period	Epoch	Duration in Millions of Years	Millions of Years Ago
CENOZOIC	Neogene	Holocene	0.0117	0.0117
		Pleistocene	2.6	2.588
		Pliocene	2.7	5.332
		Miocene	17.7	23.03
	Paleogene	Oligocene	10.9	33.9 ±0.1
		Eocene	21.9	55.8 ±0.2
		Paleocene	9.7	65.5 ±0.3
MESOZOIC	Cretaceous		80	145.5 ±4.0
	Jurassic		54.1	199.6 ±0.6
	Triassic		51.4	251.0 ±0.4
PALEOZOIC	Permian		48	299.0 ±0.8
	Carboniferous	Penn-sylvanian	19.1	318.1 ±1.3
		Missis-sippian	41.1	359.2 ±2.5
	Devonian		56.8	416.0 ±2.8
	Silurian		27.7	443.7 ±1.5
	Ordovician		44.6	488.3 ±1.7
	Cambrian		53.7	542.0 ±1.0
PRECAMBRIAN				

▲ **Figure 5.2:** The geological timescale; Geological history is divided into Eras, Periods and Epochs usually based on major periods of mountain building and significant periods of biological extinction.

The spatial distribution and physical characteristics (**lithology**) of sedimentary rocks can tell us a great deal about the surrounding environment when they were deposited. **Igneous rocks** provide evidence of ancient volcanic activity and the likely impact of volcanic emissions on the atmosphere. Ice deposits record information about the growth and decline of ice sheets and changing sea level. We can see even more at the microscopic, molecular, and atomic scales. **Microfossils**, organic remains, and common isotopes can reveal quantitative data on air and ocean temperature, the composition of the atmosphere, and the acidity of the oceans.

The Climatologist's Toolbox

Scientists draw many different diagnostic methods from their toolbox to study and interpret ancient climates and environments. These methods range from simple equipment like a hand lens up to expensive instruments like mass spectrometers and also include computers and algorithms for analyzing the data. As with any toolbox, it is important to use the right tool for the job, learn how to use it correctly, and ensure that the results are both accurate and precise enough to stand the test of time.

Interpreting Lithology

Sedimentary and igneous rocks provide important clues about ancient geography and climate. Geologists use this information to map out ancient rock formations in detail and use the physical properties of rocks to identify what the local environment was like when they were deposited. Without this information, none of the geographical reconstructions that allow us to study ancient climate change would be possible.

Useful Sedimentary Structures Sedimentary rocks often preserve fine and detailed internal structures that provide useful clues about their original environment of deposition. **Ripple marks** that form beneath flowing water can identify the position of ancient rivers and shorelines (**Figure 5.3**). **Desiccation cracks** that form when wet sediment dries out can indicate periods of drought (**Figure 5.4**). **Bioturbation** (sediment disturbed by the action of animals) may indicate biological activity and the amount of available oxygen. When considered together these and other sedimentary structures can build an overall picture of an ancient environment.

CHECKPOINT 5.2 ▶ Describe three sedimentary structures that help us understand the environment of deposition of a sedimentary rock and explain what these structures can tell us about ancient climate.

(a)

(b)

◀ **Figure 5.3:** Ripple marks form under flowing water or waves. Geologists are able to distinguish between ripples formed in rivers and in the ocean by looking at their detailed geometry. (a) A rock outcrop with abundant ripple marks preserved on the bedding surface. (b) This example of modern ripple marks illustrates how little the ancient examples have changed since they formed millions of years ago.

5.2 PAUSE FOR... THOUGHT | *How can fine sedimentary structures be preserved for such long periods of geological time?*

As a lifelong field scientist, I am often amazed at the minute details that can be preserved in rocks. We often see worm burrows, ripple marks, and even raindrop marks in sedimentary rocks that are millions of years old. Surface features such as mud cracks and ripple marks are often gently covered by younger sediment soon after they form, and this sediment preserves their shape like a mold. Small changes in sediment composition affect the lithification process and often preserve internal features such as invertebrate and crustacean burrows.

Interpreting Alluvial Fans and Fluvial Deposits Alluvial fans are thick fluvial (river) deposits that are often associated with rising mountain chains that interrupt regional airflow, increase precipitation, and enhance erosion (Figure 5.5). Erosion is very effective at removing CO_2 from the atmosphere because rainwater dissolves CO_2 from the atmosphere and forms a weak acid that reacts with rocks to form minerals that lock the CO_2 away. The recent rise of the Alps, Himalayas, Andes, and Rocky Mountains may have played a significant role in the development of our current ice age by drawing down CO_2 in this manner.

CHECKPOINT 5.3 ▶ How might the presence of abundant alluvial fans in the geological record be associated with decreasing levels of carbon dioxide in the atmosphere?

◀ **Figure 5.4:** Desiccation cracks most often form when soft, wet sediment is exposed at the surface. As it dries out, it shrinks and cracks in a characteristic polygonal geometric pattern. In ancient rocks, these cracks can be a sign of periodic drought and flooding and are often found in association with rocks formed in arid environments.

Coal and Carbon Sequestration   Known coal reserves contain as much as 3,000 Gigatonnes (Gt, or 10^9 metric tonnes) of carbon that have been removed from the atmosphere over the past 400 million years. Coal forms where dead vegetation from tropical and temperate forests accumulates in swampy areas. Thick deposits are preserved where regional subsidence, rapid burial, and rising temperatures bake the buried vegetation into coal and isolate organic carbon from the atmosphere for hundreds of millions of years (**Figures 5.6** and **5.7**). Geologists can interpret many details about the environment from the lithology of coal deposits, but they are also able to predict the likely direction of global climate change from the amount of CO_2 that was removed from the atmosphere.

CHECKPOINT 5.4 ▶ Why is the presence of abundant coal deposits in the geological record often associated with falling levels of atmospheric carbon dioxide?

Oil Shales and Carbon Sequestration   Oil shales are marine clay rocks that are particularly rich in organic material and may contain as much as 300 Gt of carbon sequestered from the atmosphere. They form from the remains of marine **plankton** that thrive in the

◀ **Figure 5.5:** Alluvial fans form at the base of mountains and are typical of arid environments, such as Death Valley in California. Their spatial geometry, lithology, and texture are very characteristic and easy to identify in the geological record.

▲ **Figure 5.6:** Shallow coal deposits like these in Wyoming can be mined directly from the surface. The organic matter in coal can form only when the climate is warm and wet enough to allow high biological productivity, as in the artist's creation of a forest from the Carboniferous Period in Figure 5.7.

nutrient-rich waters of the continental shelf (**Figure 5.8**). The presence of oil shales indicates both high surface biological productivity and the accumulation of particulate organic residues (detritus) on the sea floor under anoxic (oxygen-starved) conditions. They tend to form in areas isolated from open ocean circulation, such as in the northern North Sea during the late Jurassic Period, or in the open ocean following the breakdown of thermohaline circulation, as happened several times during the Cretaceous Period.

CHECKPOINT 5.5 ▶ What do organic rich marine shales indicate about ocean circulation?

▼ **Figure 5.7:** An artist's view of the luxuriant forests that grew during the late Carboniferous. These swamps were in many ways similar to the swamps we find in Florida today, and the organic matter they produce each year removes carbon from the atmosphere.

▲ **Figure 5.8:** High levels of surface productivity produce abundant organic matter that falls to the ocean floor, where it becomes buried and later forms rocks rich in organic material like these shales from the Jurassic Kimmeridge Clay formation in England.

5.3 PAUSE FOR... THOUGHT | *Fire in the oven!*

Some marine shale is so rich in organic matter that it burns like coal. I discovered this personally while drilling through the Upper Jurassic Kimmeridge Clay Formation in the North Sea. Just a few minutes after a colleague placed a specimen of rock cutting in the oven to dry it out, the oven caught fire due to the high organic content of the shale. (The Kimmeridge Clay Formation is the source of most North Sea oil and gas.)

Limestone and Carbon Sequestration   The many limestone formations around the world may contain as much as 66,000,000 to 100,000,000 Gt of CO_2 sequestered directly from the atmosphere (**Figure 5.9**). Limestones most often form in a marine environment and are largely composed of the minerals **aragonite** and **calcite** (both $CaCO_3$). Shallow-water deposits are often coarse grained and crowded with the shelly remains of marine fauna. Deep-water deposits form carbonate muds that are full of lithified **fecal pellets** from small invertebrates that lived in the sediment. All these limestone deposits form in warm tropical and subtropical waters where the temperature of the sea can sustain high biological productivity. Limestones are strong indicators of both warm climates and the active sequestration of CO_2 from the atmosphere by organisms whose remains are preserved in the limestones.

CHECKPOINT 5.6 ▶ The slow release of all the carbon locked in coal, oil, and gas reserves would have a profound impact on global climate, but life would probably survive. What would happen if all the carbon locked in limestone were released into the atmosphere? (Hint: 1 Gt of carbon is equivalent to 3.67 Gt of carbon dioxide. To increase the level of carbon dioxide in the atmosphere by 1 ppm by volume requires 2.13 Gt of carbon.)

Chalk and Carbon Sequestration   Chalk is a variety of limestone that is made up almost entirely from the remains of calcareous algal phytoplankton known as **coccolithophores** (**Figure 5.10**). Still present in the oceans today, the coccolithophores form seasonal blooms that cover hundreds of square kilometers (**Figure 5.11**). During the **Cretaceous Period,** when CO_2 levels in the atmosphere were higher than 1,000 ppm, these **phytoplankton**

▼ **Figure 5.9:**   Thick deposits of chalk, a variety of limestone, remove large amounts of carbon dioxide from the atmosphere as they form.

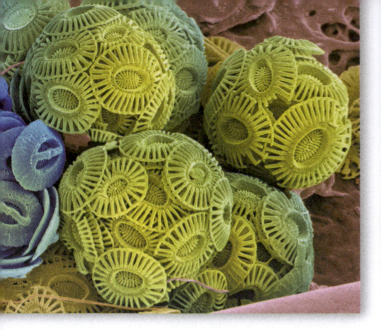

▲ **Figure 5.10:** The coccolithophores are single-celled protists that formed extensive algal blooms during the Cretaceous Period and deposited hundreds of feet of sediment on the sea floor. When this sediment is buried and turns into chalk it removes many gigatonnes of carbon from the atmosphere.

▲ **Figure 5.11:** A modern coccolithophore algal bloom turns the water a milky green and blue color off the coasts of England, Ireland, France, and Spain.

were so abundant that their tiny remains deposited hundreds of feet of chalk across the world and removed billions of tonnes of carbon from the atmosphere. Coccolithophores also release **dimethyl sulfide (DMS),** a natural aerosol that promotes cloud formation and increases the albedo of the atmosphere, while their light color increases the albedo of the oceans. These are two powerful forces that act to cool global climate.

CHECKPOINT 5.7 ▶ Why is the presence of abundant chalk in the geological record often associated with cooling global temperature and falling levels of atmospheric carbon dioxide?

5.4 **PAUSE FOR... THOUGHT** | *Why were coccolithophores so critical?*

Upper Cretaceous chalks are widely distributed around the world. These chalks formed in shallow continental seas when sea level was at least 100 meters (300 feet) above the current level. This warm, nutrient-rich water offered ideal conditions for the coccolithophores to bloom, and their thick deposits represent a very significant sequestration of CO_2 from the atmosphere.

Volcanoes and global warming Thick sequences of basaltic lava erupted at the surface strongly indicate both increased levels of CO_2 in the atmosphere and elevated global temperature. A number of times in the geological record, sustained, effusive eruptions of lava have occurred. These eruptions are often associated with extreme global temperatures and an enhanced rate of extinction. Such eruptions are either associated with the final rifting phase

of a continent splitting apart or the arrival of a deep-seated **plume** of hot mantle at the surface.

Volcanic Ash and global cooling Thick, widely dispersed volcanic ash deposits in the geological record indicate that violent (**Plinian**) eruptions have occurred. Plinian eruptions produce enormous eruption columns that reach high into the stratosphere and leave behind sulfate aerosols that stay for years and cool the atmosphere by increasing its albedo. This global cooling is rarely sustained for very long on a geological timescale, but can persist for decades and have a major impact on climate (**Figure 5.12**).

CHECKPOINT 5.8 ▶ Explain why hundreds of meters of lava erupting over a period of time of less than a million years, and covering an area of 10,000 square kilometers (3,860 square miles), would have had a major impact on climate at the time of eruption?

How to Identify Ancient Desert Rocks Rocks that form in arid (dry) environments provide an excellent record of the prevailing wind direction, rainfall patterns, and even monsoon activity in ancient deserts. Rocks that form in these environments are usually sandstones, siltstones, and mudstones that have a distinctive deep purple-red color due to the presence of iron oxide. Desert sandstones are composed of quartz grains that are round and frosted by the constant action of the wind,

▲ **Figure 5.12:** Large plumes of volcanic dust, like this example from the eruption of Eyjafjalljökull, Iceland, in 2010, can reach as high as the stratosphere and cool the planet. Volcanic release of carbon dioxide can also lead to longer-term global warming.

and they exhibit large, cross-cutting strata that formed as ancient dunes moved in the direction of the prevailing wind (**Figure 5.13** and **Figure 5.14**). Strata of fluvial and **ephemeral** lacustrine (lake) deposits often occur within large desert sandstone formations, and may be related to the changing activity of seasonal monsoons.

How to Identify Past Glacial Activity

Glacial drift (moraine or till) is formed by rocks, sand, silt, and clay left behind after ice sheets or glaciers have melted and retreated. The lithology of this rock (tillite) is so characteristic that it has been used to identify glacial deposits as far back as the Proterozoic Eon (**Figure 5.15**). More recent deposits often exhibit a variety of easily identifiable surface structures such as the sinuous **terminal moraines** that develop at the snout (end) of a glacier. These structures are usually soft and easily eroded and seldom survive subsequent periods of glacial advance.

Loess: Tracing the Path of Glacial Winds

Strong winds that blow over sediment discharged from glacial ice during

▼ **Figure 5.14:** Sandstones formed in a desert are often red in color and frequently indicate the presence of ancient sand dunes in the pattern of their internal structure. Formed by the constant movement of dunes this pattern, called cross stratification, is typical of rock formations exposed in Zion National Park in Utah.

▼ **Figure 5.13:** Under a microscope, you can often see how grains of normally clear quartz have been rounded and frosted by the constant action of wind.

▲ **Figure 5.15:** Ancient tillite deposits indicate that glaciations have occurred many times in the geological past. Note typical rock texture for tillite, with larger individual multicolored rocks embedded in a finer red-colored matrix. The geological hammer near the top of the photograph gives a sense of scale.

▲ **Figure 5.16:** Dropstones are isolated blocks of often-angular rock found embedded in much finer-grained marine or lacustrine (lake) sediments. They are clear signs of past glacial activity.

the summer pick up fine silt and carry it hundreds of miles. Eventually, the silt settles to form thick **loess** deposits that mantle the terrain in a soft blanket of yellow to yellow-brown silt. Loess deposits form only during major phases of glaciation, and their dispersal provides information about the strength and direction of prevailing winds.

Dropstones: Tracing the Path of Ancient Icebergs

Dropstones are isolated rocks embedded in finer-grained glacial sediments that almost always indicate the presence of ice in the geological record (**Figure 5.16**). They form when icebergs melt and release embedded rocks

that fall to the ocean or lake floor. A detailed examination of the sediment surrounding a dropstone may reveal smaller fragments of rock, also released by the ice, that provide important information about the provenance (geographical origin) of the icebergs that carried them.

Ikaite: A Sign of Cold Water Ikaite is a hydrated form of calcium carbonate (calcite) that forms crystals in sediment close to the ocean floor under very cold conditions. These crystals are not very stable and readily dehydrate to the mineral calcite, but they tend to keep the shape of the original crystal of ikaite intact. These so-called **pseudomorphs** of ikaite are known by different names around the world, but most commonly as Glendonite. They clearly indicate past glacial conditions (**Figure 5.17**).

CHECKPOINT 5.9 ▶ What features of the sedimentary record indicate the presence of abundant ice?

Feeling the Heat: Evaporites and Calcretes

Evaporites form when water evaporates and leaves behind dissolved salts as a mineral residue. These deposits are common in the geological record and indicate hot, dry, and subtropical conditions (**Figure 5.18**).

Calcretes (caliches) are soils that are typical of arid and semiarid regions and are commonly preserved in the geological record. Calcium carbonate is deposited as veins and concretions (lumps) within the soil profile, especially during dry times of the year (**Figure 5.19**). Scientists use the isotope geochemistry of calcite in calcretes to estimate the concentration of CO_2 in the atmosphere.

How to Identify Tropical Soils Laterites are deep red colored iron-rich soils that form only in tropical environments (**Figure 5.20**). The concentration of iron in these rocks is often so high that they have been mined around the world.

▲ **Figure 5.17:** The presence of Glendonite in shallow sediments on the ocean floor is a clear indication of very cold conditions on the ocean floor.

▲ **Figure 5.18:** As water evaporates, dissolved mineral salts start to precipitate out of solution and are left behind as a mineral residue. In this example from the the Dead Sea in Israel, the white salt deposits are clearly visible at and just below the water surface.

CHECKPOINT 5.10 ▶ What different conditions are necessary for the formation of laterites and evaporites?

▲ **Figure 5.19:** Calcium carbonate is deposited as veins and concretions (lumps) within the soil profile, especially during arid times of year. In this photograph of rock from the Permo-Triassic age, the fossil soil layer (calcrete) is clearly visible as white mineral embedded in a red-colored rock matrix.

5.5 **PAUSE FOR...** | *Reconstructing*
THOUGHT | *paleoenvironments.*

By gathering all the available information from rock outcrops, boreholes, body fossils, trace fossils, and microfossils, research geologists are able to reconstruct an accurate picture of how ancient environments and climate have changed over time. These clues reveal a long history of mountain building, volcanism, erosion, sediment transportation, deposition, burial, and changing sea level. Deep sedimentary basins around the world store an almost continuous record of hundreds of millions of years of climate change. Much of this geological information was used by the petroleum industry to explore for more oil and gas. Climate scientists now use the information to argue that we should use as little oil and gas as possible.

Using Chemistry to Interpret Ancient Climate

Chemical fingerprints of past climate remain permanently embedded in the rocks, fossils, ice, and minerals left behind in the geological record. Radioactive isotopes provide information about past temperature, ice volume, and even how rapidly mountains were weathered. Trace elements can help us determine temperature, and they also provide clues to the composition and origin of ancient ocean currents. Organic residues provide information about the ambient temperature and biological productivity of ancient seas.

139

▲ **Figure 5.20:** Laterites are deep iron- and aluminum-rich soils that form in tropical environments and are useful in identifying ancient tropical zones. In this photograph of a laterite soil profile in Western Australia, iron-rich laterite lies on top of an aluminum-rich mottled zone.

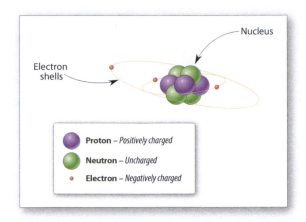

▲ **Figure 5.21:** The nucleus of each atom is composed of a fixed number of positively charged protons and a variable number of neutrons. Isotopes of an element can occur when there are a variable number of neutrons in the nucleus. When there are just a few extra neutrons the nucleus may remain stable, but the addition of too many extra neutrons makes the atom radioactively unstable and it decays at a constant rate into a different daughter atom with the release of elementary particles and energy.

Using Isotopes to Interpret Ancient Climate   Each element contains an exact number of **protons** and a variable number of **neutrons** in its nucleus. Protons are positively charged and give an element its unique physical identity. Neutrons are the same size and mass as protons, but they carry no charge, and they combine with protons in the nucleus to give each unique atom a characteristic range of physical properties (**Figure 5.21**).

The Difference Between Stable and Radioactive Isotopes
Isotopes are variants of an element with slightly different physical properties due to the presence of additional neutrons in the nucleus. These small differences in physical properties alter the way they are exchanged between major chemical reservoirs in the atmosphere, oceans, and lithosphere.

Radioactive isotopes have so many additional neutrons in their nuclei that they become unstable and break down by radioactive decay at a constant rate to produce smaller **daughter atoms** (**Figure 5.22**). The **half-life** of a radioactive isotope is the time taken for just half of the nuclei in a sample to decay, and because the half-life is a constant for each radioactive isotope, isotopes are like clocks that are used to date their host material. The most commonly used "atomic clock" in climate science is based on carbon (^{14}C, or carbon 14), a **cosmogenic isotope** formed in the atmosphere. ^{14}C slowly breaks down to its daughter nitrogen (^{14}N), with a half-life of 5,740 years. Other radioactive isotopes commonly used for dating are uranium (^{234}Ur), a heavy metal found in rocks that decays to its daughter, Thorium (^{230}Th), with a half-life of 245,500 years, and ^{238}Ur, which decays to its daughter ^{206}Pb (lead), with a half-life of 4.47 billion years.

Stable isotopes do not decay over geological time. The stable isotopes most commonly used in climate science are oxygen (^{18}O, or "oxygen 18," and ^{16}O, or "oxygen 16"), hydrogen (1H), deuterium (2H), and carbon (^{13}C, or "carbon 13," and ^{12}C, or "carbon

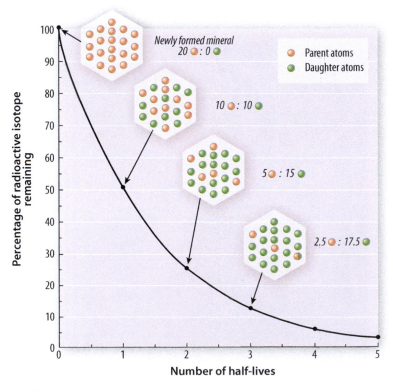

▲ **Figure 5.22:** The half-life of a radioactive isotope is the time it takes for exactly half of the initial parent atoms to decay into daughter atoms. Half-life is a constant, and it is *independent* of the parent/daughter ratio, temperature, pressure, and time.

12"). Small differences in the physical properties of each isotope affect how they are partitioned between different parts of the climate system, and these small differences help scientists measure environmental change. Oxygen and hydrogen isotopes, for example, help us measure temperature and ice volume. Carbon isotopes enable us to estimate past levels of CO_2 in the atmosphere.

CHECKPOINT 5.11 ▶ In your own words, explain the difference between stable and radioactive isotopes.

Using Oxygen Isotopes to Interpret Environmental Change
Nearly all naturally occurring oxygen atoms have eight protons and eight neutrons in their nucleus (^{16}O), but around 0.2% have two additional neutrons that create the heavier isotope, ^{18}O. The 11% difference in mass between these two isotopes has a significant impact on their physical properties.

Oxygen Isotope Fractionation Water containing both ^{16}O ($H_2^{16}O$ molecules) and ^{18}O ($H_2^{18}O$ molecules) evaporate easily at the surface of the ocean, but the less massive $H_2^{16}O$ molecule needs less energy to make the transition to water vapor. Given time, energy, and billions of atoms, this selective evaporation (called *fractionation*) leaves the water vapor relatively enriched in the $H_2^{16}O$ molecule and the oceans relatively enriched in the $H_2^{18}O$ molecule.

The $H_2^{18}O$ molecules that do manage to enter the atmosphere are further fractionated when it rains because their greater mass means that they are also more likely to condense and precipitate. Through repeated cycles of evaporation and precipitation, water vapor becomes progressively richer in $H_2^{16}O$ and depleted in $H_2^{18}O$ as it moves toward the poles. By the time water vapor finally condenses and falls near the poles (usually as snow), it is significantly enriched in $H_2^{16}O$, and the oceans are enriched with all the $H_2^{18}O$ that is left behind (**Figure 5.23**).

The degree of enrichment or depletion of ^{18}O relative to ^{16}O is determined by comparing the ($^{18}O/^{16}O$) ratio of a sample to the ($^{18}O/^{16}O$) ratio of an international standard. The results are multiplied by a factor of 1,000 to keep the units manageable and are expressed in the formula below as an isotope difference "per mille" between sample and standard, signified by the Greek symbol delta (δ).

$$\delta^{18}O\ ‰ = \frac{(^{18}O/^{16}O)_{sample} - (^{18}O/^{16}O)_{standard} \times 1{,}000}{(^{18}O/^{16}O)_{standard}}$$

The international standard used to measure these changes is calculated from synthetic samples of very pure distilled water controlled by the Atomic Energy Agency (Vienna Standard Mean Ocean Water (VSMOW; originally just SMOW).

Some of the most ^{18}O-depleted ice on Antarctica has $\delta^{18}O$ as low as −60‰ to −80‰ in the interior and −20‰

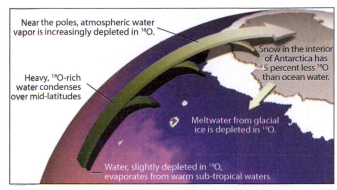

▲ **Figure 5.23:** Atmospheric water vapor is progressively enriched in ^{16}O as it moves toward the poles. This isotopic fractionation leaves polar ice relatively enriched in ^{16}O and the oceans relatively enriched in ^{18}O. By measuring changes in the ($^{16}O/^{18}O$) ratio of marine shells, it is possible to estimate how much ice is locked up at the poles.

to −30‰ at the coast. So much of the lighter $H_2^{16}O$ is preferentially locked up as ice at the poles during glacial periods that the composition of global seawater is significantly enriched in $H_2^{18}O$ (larger positive values of $\delta^{18}O$). When the ice caps melt during periods of warmer climate, they flood the oceans with lighter $H_2^{16}O$, and the global value of $\delta^{18}O$ slowly returns to preglacial levels with ocean circulation (**Figure 5.24**).

CHECKPOINT 5.12 ▶ What is the relationship between the level of ^{18}O in the oceans and the abundance of polar ice?

Using Oxygen Isotopes to Measure Air Temperature
The oxygen isotope ratio of polar ice is related to the temperature of the troposphere. As the climate cools, and the temperature of the troposphere falls, an increasing amount of ^{18}O is precipitated at lower latitudes (away from the poles). This leaves the water vapor that reaches the poles enriched in ^{16}O. When this water vapor precipitates as snow, the surface ice is also enriched in ^{16}O.

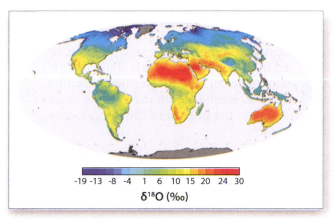

▲ **Figure 5.24:** The oxygen isotope values of global precipitation. Note the very low values of $\delta^{18}O$ in the Arctic. When polar ice melts it floods the oceans with low $\delta^{18}O$ water.

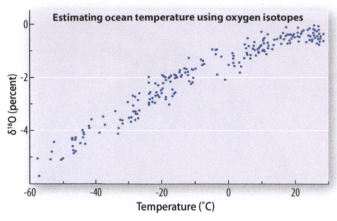

Figure 5.25: Decreasing $\delta^{18}O$ in ice cores indicates lower temperatures in the atmosphere.

There are many variables in this analysis, such as the effect of changing altitude, the isotopic composition of the oceans, and the distance travelled from the source of the water vapor, but there is a now strongly established **empirical** relationship between the temperature of the troposphere and the isotopic composition of ice. Even when accurate temperatures are hard to tie down, it is still possible to determine relative change. As a rule of thumb, when the level of $\delta^{18}O$ gets lower in an ice core, it indicates that the original snow fell from a cooling troposphere, and when the level of $\delta^{18}O$ gets higher, it indicates that the polar troposphere was warming (**Figure 5.25**).

CHECKPOINT 5.13 ▶ Oxygen isotopes are routinely extracted from ice cores and used to determine temperature. Explain what properties of oxygen isotopes allow them to be used in this way.

Using Oxygen Isotopes from Carbonate Shells

Marine invertebrates such as corals and foraminifera use carbon and oxygen from the surrounding water to construct their calcium carbonate ($CaCO_3$) shells. The isotopic composition of these shells, therefore, reflects the isotopic composition of the seawater they lived in, and this can be analyzed in the laboratory. These data provide a snapshot of ocean isotope geochemistry that can be used to estimate both the volume of ice locked away at the poles and the temperature of the oceans.

Using Oxygen Isotopes to Determine Ice Volume

Benthic (bottom-dwelling) shells taken from marine cores are used to study changes in polar ice volume. When ice accumulates at the poles, the oceans are relatively enriched in ^{18}O and depleted in ^{16}O, and this change is reflected in the isotopic composition of benthic shells that grow in this water (**Figure 5.26**). When ice melts, and the oceans are flushed with $\delta^{16}O$-rich meltwater, this is also reflected in the composition of the benthic shells. The changing ratio of ^{18}O to ^{16}O in these shells is therefore a measure of how global ice volume changes over time.

Using Oxygen Isotopes to Measure Ocean Temperature

The oxygen isotope composition of carbonate shells is also affected by ocean temperature. The isotope ^{18}O is preferentially partitioned into $CaCO_3$ *at lower temperatures*, so as long as the bulk isotope composition of the ocean remains constant (no changes in ice volume), the $\delta^{18}O$ value of any shell will increase as the temperature falls (**Figure 5.27**).

CHECKPOINT 5.14 ▶ If you observe that the shells of marine planktonic (surface) foraminifera become enriched in ^{18}O over time, while coeval marine benthic (bottom dwelling) foraminifera show no change, what could you deduce about the temperature of the atmosphere?

▼ Figure 5.26: These records compare $\delta^{18}O$ data recovered from the Byrd ice core in Antarctica with a bottom-dwelling foraminifera *Uvigerina hispida* recovered from a marine rock core sample. Please note the scales: $\delta^{18}O$ values on the core data (upper curve) increase downwards, while on the ice data (lower curve) they increase upwards. This is so that temperature always increases towards the top of the Y-axis. As the temperature of Antarctica falls the level of $\delta^{18}O$ in the Byrd ice core data decreases due to isotope fractionation in the atmosphere (note the larger negative number). At the same time, the level of $\delta^{18}O$ in marine shells increases—partly because of falling temperature but also because an increase in global ice volume enriches the oceans in $\delta^{18}O$. It can be difficult to distinguish between the relative influences of falling global temperature and changing ice volume when studying the isotopic composition of foraminifera, as both increase the relative amount of ^{18}O in their shells. (www.sciencedirect.com/science/article/pii/S0277379109001607)

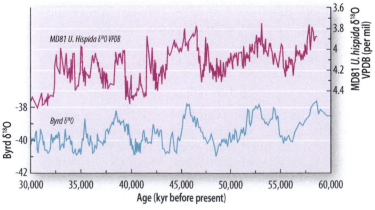

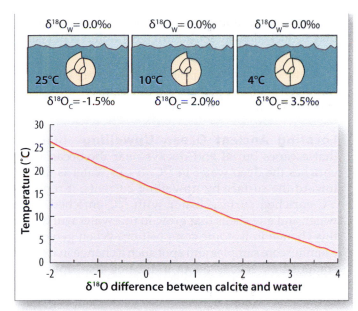

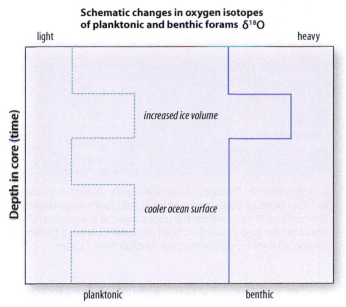

▲ **Figure 5.27:** As the water temperature falls, the level of $\delta^{18}O$ increases in marine shells. This is exactly the opposite of the relationship observed in ice cores, and it is a common cause of confusion. In this diagram, the isotope composition of the oceans is held constant to show how the $\delta^{18}O$ content of shells falls with rising ocean temperature.

Is This Very Confusing?   The fact that high $\delta^{18}O$ in ice indicates warm temperatures, while high $\delta^{18}O$ in marine shells indicates low temperatures, often causes confusion. It is also confusing that there are two factors that control the level of $\delta^{18}O$ in marine shells: Higher $\delta^{18}O$ in marine carbonate shells can mean that there is more ice on the poles, or that the water temperature is colder, or even both at the same time! To resolve this dilemma and get a reliable record of sea surface temperature, geologists study the isotope chemistry of benthic foraminiferal shells to establish that ice volumes were stable and then use the isotope chemistry of planktonic foraminiferal shells to measure sea surface temperature. They can do this because water temperature does not vary much in the deep ocean water where the benthic foraminifera live, even during warm interglacial periods. If geologists confirm that there is no change in ice volume by studying the isotope chemistry of benthic shells, then

▲ **Figure 5.28:** If benthic oxygen isotope ratios in shells remain unchanged, then changes in $\delta^{18}O$ in planktonic fauna must represent changes in sea surface temperature. In this vertical profile from an idealized sediment core, from top to bottom, an increase in ice volume first enriches the entire ocean- and thus the shells of both planktonic (surface) and benthic (bottom-dwelling) foraminifera- in $\delta^{18}O$ but this was preceded (below) by a decrease in ocean surface temperature that affected the $\delta^{18}O$ of planktonic foraminifera, leaving the deep ocean benthic foraminifera unaffected (see text for details).

any observed change in the oxygen isotope chemistry of planktonic foraminifera must be due to changes in sea surface temperature (**Figures 5.28** and **5.29**).

Using Carbon Isotopes to Interpret Environmental Change   Carbon is a very common element in the natural environment and occurs as two stable isotopes. Carbon 12 (^{12}C; 99%) has six protons and six neutrons in the nucleus, and carbon 13 (^{13}C; 1%) has one additional neutron, which gives it a slightly higher mass. A third isotope, Carbon 14 (^{14}C), is an unstable cosmogenic isotope that is present in very small concentrations, with a half-life of 5,730 ± 40 years.

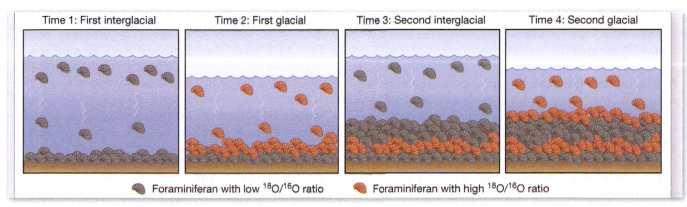

▲ **Figure 5.29:** The shells of foraminifera reflect the changing oxygen isotope composition of seawater as the polar ice caps expand and contract.

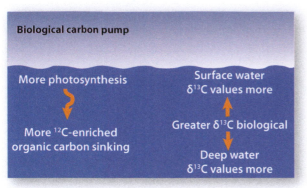

Biological carbon pump

More photosynthesis

More ¹²C-enriched organic carbon sinking

Surface water δ¹³C values more

Greater δ¹³C biological

Deep water δ¹³C values more

▲ **Figure 5.30:** The upper layer of the ocean is relatively enriched in ¹³C as biomass that is enriched in ¹²C falls to the ocean floor. The removal of carbon from the upper layers of the ocean and its burial in the deep ocean floor is an important part of a biological "pump" that cycles carbon through Earth's climate system.

The reference standard against which all changes in stable carbon isotopes are measured was first established from a fossil belemnite from the Pee Dee Formation in South Carolina (the **PDB standard**), but as the belemnites ran out, this was exchanged for an identical synthetic standard. As with oxygen isotopes, the units are multiplied by a factor of 1,000 simply to keep the numbers readable. The resulting units are expressed as an isotope difference per mille between the sample and the standard according to the equation:

$$\delta^{13}C\ ‰ = \frac{(^{13}C/^{12}C)_{sample} - (^{13}C/^{12}C)_{standard}}{(^{13}C/^{12}C)_{standard}} \times 1,000$$

The Fractionation of Carbon Isotopes The lighter ¹²C isotope is selectively partitioned into plant biomass by photosynthesis, leaving the surrounding environment (air and oceans) relatively enriched in ¹³C (**Figure 5.30**). In the marine environment, plants (phytoplankton) form the bottom of the food chain, so all living tissue becomes relatively enriched in ¹²C, but shells are enriched in ¹³C because they grow directly from seawater. The δ¹³C ratio of shells is thus a measure of how much photosynthesis is taking place in the oceans and how much organic material is removed from the surface to the deep ocean, as both processes enrich the water in ¹³C.

CHECKPOINT 5.15 ▶ Why are organic remains enriched in the ¹²C isotope?

A Biological Carbon Pump A constant stream of organic matter in the form of dead plankton, fecal pellets, and other organic remains leaves the surface of the ocean and falls towards the deep ocean, where it is either dissolved in the water or buried under the ocean floor. This burial or dissolution of organic matter is an important **biological pump** that removes CO_2 from the atmosphere and leaves the surface water enriched in ¹³C. Carbon that dissolves in the deep ocean can be isolated from the atmosphere for hundreds of years before upwelling finally returns this water to the surface. Carbon that is stored in ocean floor sediments may be locked away for hundreds of millions of years.

CHECKPOINT 5.16 ▶ Explain how high surface ocean biological productivity affects the concentration of ¹³C in the shells of marine invertebrates.

Locating Ancient Ocean Upwelling Biomass that escapes burial and decays near the ocean floor enriches the deep water in ¹²C. If this water is drawn toward the surface by upwelling currents, it replaces ¹³C-enriched surface water with ¹²C-enriched deep water, and any shells that grow in this water will reflect this change in isotopic composition. When geologists examine core samples and spot such a change in isotope composition, it may be an indication that upwelling once occurred at this site. They can use this information to trace the geological history of major ocean currents.

A Sign of Disaster? A sudden global increase in the ratio of ¹²C to ¹³C (decrease in δ¹³C) in plankton shells indicates a decrease in biological productivity or the release of ¹²C from a large geological reservoir. These events are commonly associated with times of extinction due to extreme events such as a meteorite impact, or the catastrophic release of methane gas from shallow marine reservoirs (**Figure 5.31**).

More Evidence for the Origin of Global Warming Scientists have recently observed a significant increase in the relative amount of ¹²C in both the atmosphere and the surface of the oceans. This is evidence that most of the CO_2 added to the atmosphere since the industrial revolution is derived from the burning of fossil fuels. All fossil fuels are naturally enriched in ¹²C because they are largely derived from plant materials. This evidence is supported by an analysis of the amount of cosmogenic ¹⁴C in the atmosphere. Fossils fuels are so old that the ¹⁴C they once contained has decayed, and the release of all this ¹⁴C-depleted gas into the air by combustion has reduced the concentration (not the mass) of ¹⁴C in the atmosphere. Upwelling ocean currents can also release CO_2 gas that is enriched in ¹²C into the atmosphere, but this carbon is not depleted in ¹⁴C, and this difference allows scientists to be confident in their claim that fossil fuels are largely to blame.

CHECKPOINT 5.17 ▶ You suspect that a large volume of organically derived carbon was suddenly released into the atmosphere. How would you prove your hypothesis?

Terrestrial Isotopes of Carbon The level of carbon dioxide in the atmosphere over geological time can be estimated by studying the carbon isotope ratios of fossil soils (caliches). The details of this method are complex because there are many variables, and past results have tended to exaggerate the level of CO_2. Researchers are addressing these errors because the abundance of caliches in many geological formations makes this an attractive and useful tool.

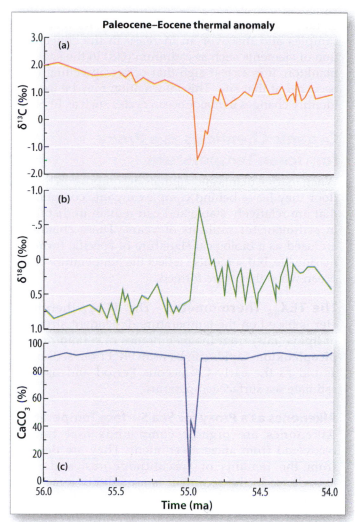

Figure 5.31: During the Paleocene–Eocene Thermal Anomaly, CO_2 levels in the atmosphere rose rapidly, along with average global temperature. This diagram shows that a major **isotope excursion** event is recorded in the marine sediment record from this time (top panel). (a) A strong negative anomaly (decrease) in $\delta^{13}C$ reflects an increase in the fraction of ^{12}C in surface waters. (b) The $\delta^{18}O$ record from foraminifera (middle panel) also indicates a concurrent sudden rise in temperature of up to 5°C (9°F). (c) The $CaCO_3$ content of sediments deposited over this interval (lower panel) reflects a relative scarcity of shell residue when temperatures reached a maximum. This was most likely due to the sudden release of a large volume of biogenic gas from destabilized **methane hydrates** or from coal beds that are baked by hot magma that eventually increased the acidity of the oceans.

Boron Isotopes as a Proxy for Carbon Dioxide

The addition of CO_2 to the atmosphere increases the level of dissolved CO_2 in the oceans and increases surface acidity. This in turn affects the distribution of two stable isotopes of the element Boron (^{10}B and ^{11}B) that occur in the shells of marine invertebrates. Because the value of $\delta^{11}B‰$ in these shells is directly related to ocean pH, and the pH of the ocean is directly related to the level of CO_2 in the atmosphere, the value of $\delta^{11}B‰$ reflects the concentration of CO_2 in the atmosphere when the shells formed.

Strontium Isotopes as a Proxy for Erosion

Strontium 87 (^{87}Sr) is a stable isotope that is produced by the radioactive decay of an isotope of Rubidium (^{87}Rb), an

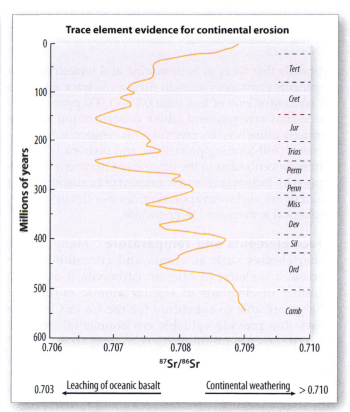

Figure 5.32: The isotope ^{87}Sr is concentrated in crustal rocks, and its increased concentration in marine shells can be an indication of enhanced continental weathering (erosion), commonly associated with mountain building. Note the peaks of higher ^{87}Sr at the end of the Silurian (Sil), Devonian (Dev) and Pennsylvanian (Penn) Periods and during the mid Triassic (Trias) and Tertiary (Tert) Periods, that indicate periods of enhanced continental erosion due to mountain building.

element that is abundant in crustal rocks. The concentration of ^{87}Sr in the oceans varies with the amount of weathering on land, and these changes are recorded in marine shells. This allows ^{87}Sr to be used as a potential proxy for mountain building (orogeny) and erosion, two factors that are strongly linked to the removal of CO_2 from the atmosphere (**Figure 5.32**).

CHECKPOINT 5.18 ▶ Why would an increase in the concentration of ^{87}Sr in the shells of marine invertebrates be associated with a slow reduction of carbon dioxide in the atmosphere?

5.6 PAUSE FOR... THOUGHT | *Can we trust isotopic and geochemical analyses?*

These analyses are very accurate if conducted at a recognized facility, but as climate research becomes ever more removed from direct field observation, it is increasingly probable that some scientists will make incorrect assumptions about data they did not collect themselves. Climate science is a field science, and anyone who interprets climate data must become familiar with the "ground truth" of these data. I tell my students that all data are just an academic source that must be assessed for quality and veracity.

Trace Elements as Environmental Markers

Minerals that form in both marine and terrestrial environments commonly contain diagnostic **trace elements** in concentrations of less than 0.1% (1,000 ppm). In the terrestrial environment, their concentration is most strongly influenced by precipitation, temperature, pedogenesis (soil-forming processes), and bedrock lithology. Their concentration in the marine environment is determined by factors such as the amount of freshwater input from rivers, surface evaporation, deep-water upwelling, biological activity, and temperature.

Trace Elements and Temperature   Many marine invertebrates such as corals and coccolithophores secrete a skeleton of calcium carbonate ($CaCO_3$). A number of elements of similar **atomic radius** and charge are able to substitute for the Ca in $CaCO_3$ in ways that provide valuable environmental information. Ratios of strontium (Sr/Ca), magnesium (Mg/Ca), and barium (Ba/Ca) to calcium, for example, are all closely related to temperature. These data are relatively easy to extract from marine core samples, and the trace element analysis of corals and foraminifera is certainly a much cheaper and easier way to determine temperature than using isotope analysis (**Figure 5.33**).

Trace Elements as a Proxy for Upwelling   When water masses circulate deep within the global thermohaline conveyor they acquire characteristic trace element signatures. When scientists analyze core samples and discover an increase in the concentration of elements such as cadmium (Cd) in the shells of plankton, it is a clear sign that ocean upwelling once occurred at that site. This information may be used to identify changes in major ocean cycles such as El Niño.

Organic Chemicals as a Proxy for Ocean Temperature

When marine plants and animals decompose on the sea floor they leave behind complex organic compounds that are relatively stable and can remain undisturbed in sediment for millions of years. These chemicals are used as a biological signature of specific forms of marine life, but some also reflect the temperature of the ocean when they were formed.

The TEX$_{86}$ Thermometer   The **TEX$_{86}$** thermometer is based on the temperature dependent structure of lipids (fats) from the membranes of certain marine **picoplankton**. Lipids can be recovered from sediments as old as the early Cretaceous Period and used to estimate sea surface temperature.

Alkenones as a Proxy for Sea Surface Temperature   **Alkenones** are organic compounds that can be recovered from ancient sediment. They are derived from the remains of coccolithophores, and their abundance in the sediment is a qualitative measure of past ocean productivity. Their detailed chemical

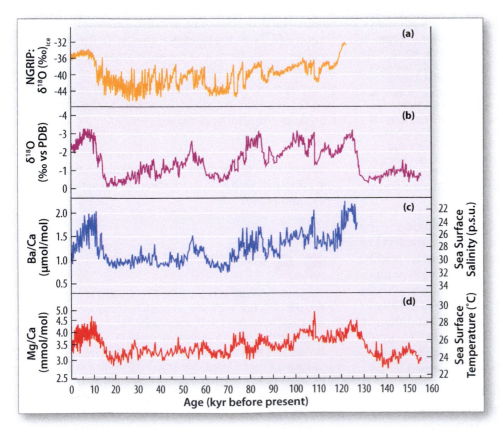

◀ **Figure 5.33:** This figure compares the temperature record over the past 160,000 years, based on (a) oxygen isotopes from ice cores, (b) oxygen isotopes from shallow marine foraminifera shells, (c) the Mg/Ca ratio from foraminifera shells, and (d) the Ba/Ca ratio from foraminifera shells. The detailed record of climate change recorded in these data is discussed in detail in Chapter 6.

structure depends on the ambient temperature when they formed, so they can be used as a proxy for sea surface temperature.

Fossils as Environmental Indicators

Fossils offer some of the most convincing evidence of ancient climate change. The analysis of fossil diversity and abundance can reveal amazing detail about an ancient ecosystem and be a great indicator of climate change. Although large fossils (**macrofossils**) are attractive and sometimes informative, the fossils that are of greatest use to climate scientists are the microfossils that include plant spores, pollen, and the shells of small phytoplankton and zooplankton. Unlike the macrofossils, these can be recovered in large numbers from rock and sediment cores on land and in the oceans.

Fossil Leaves as a Proxy for Precipitation and Temperature   Fossil leaves are among the most common fossils in the geological record. Many aspects of leaf geometry and physiology are related to their environment, and these can be used to track ancient climate change. Leaves that have large surface area, for example, are more likely to form in very wet environments, and leaves with more serrated edges are more likely to be found in cooler climates. These are not perfect relationships, but they can be used to delineate the geographical boundaries of ancient biomes (**Figures 5.34, 5.35,** and **5.36**).

Leaf Stomata as a Proxy for Carbon Dioxide Leaf structures regulate plant respiration and respond

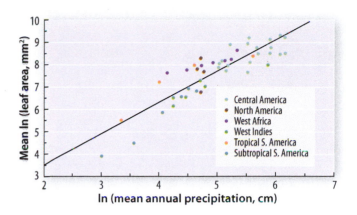

▲ **Figure 5.35:** Data collected from a range of modern environments indicate that leaves with large surface area are more likely to form in very wet environments than in dry ones. Individual samples from different regions are plotted on this graph that uses natural logarithms because of the large range in values.

to environmental change. Stomata are small pores in a leaf that take in CO_2 from the atmosphere and release water. There is a useful relationship between the number of **stomata** per unit area on a leaf (the stomatal density) and the concentration of CO_2 in the atmosphere. Plants respond to an increase in the concentration of CO_2 in the atmosphere by decreasing the number of stomata (**Figure 5.37**). This is desirable for a plant because fewer stomata limit water loss and reduce the possibility of infection. If the concentration of CO_2 falls, the plant responds by increasing the stomatal density in an effort to enhance photosynthesis (**Figure 5.38**). We can therefore use stomatal density to measure the concentration of CO_2 in the atmosphere, one of the most important factors that drive climate change.

CHECKPOINT 5.19 ▶ Plants reduce the density of leaf stomata when the level of carbon dioxide in the atmosphere is high. What benefit do they gain by doing this?

Smooth (entire) margin   Jagged (toothed) margin

▲ **Figure 5.34:** Despite their delicate nature, leaves are frequently preserved in the fossil record and can tell us a lot about the environment in which they grew. Leaves with more serrated edges like the example on the right hand side are more likely to be found in cooler climates than in warm ones.

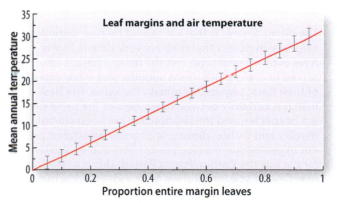

▲ **Figure 5.36:** Leaves with more serrated edges are more likely to be found in cooler climates than in warm ones. The vertical bars indicate the range of possible error in the data.

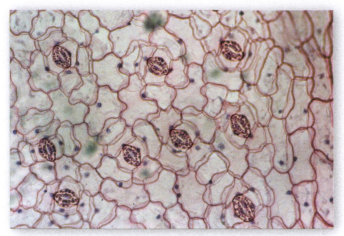

▲ **Figure 5.37:** Stomata are small pores in the surface of a leaf that control the exchange of water and CO_2 with the atmosphere.

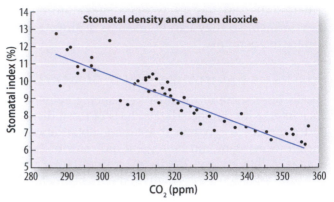

▲ **Figure 5.38:** If the concentration of CO_2 falls, a plant responds by increasing the stomatal density in an effort to feed photosynthesis. On this graph of experimental data, stomatal density is plotted against changing levels of CO_2. The relationship is not perfect, but the regression curve for the data (also plotted) can be used to estimate changing levels of CO^2 from geological samples.

Pollen and Spores as Environmental Indicators

Leaves are exciting fossils to collect and analyze, but most rocks studied by climate scientists are collected from cores, and the chance of finding a complete leaf in a core is very small. Fossil pollen and spores, however, are so small that even a small core sample may contain hundreds of grains for analysis. Plant spores and pollen have shapes that are species specific, and this makes it possible to monitor the changing composition of an ecosystem over time (Figure 5.39). Scientists must be careful to avoid some obvious pitfalls. For example, pollen and spores are widely dispersed by wind and may eventually settle far from their point of origin, but when properly collected and analyzed, they can be used as clear evidence of past vegetation and climate.

Foraminifera as Environmental Indicators

Foraminifera are tiny unicellular **protists** that are abundant in many of the world's oceans (Figure 5.40). They are mentioned many times in this book and are probably the most important fossil groups that climate scientists use to understand ancient climate change. Many species secrete an ornate calcareous skeleton that makes identification relatively easy. The foraminifera ("forams" for short) are so small and widely dispersed that many individuals can be recovered from a single core sample. Forams inhabit both epipelagic (as plankton) and benthic (bottom-dwelling) zones, and they are highly sensitive to the environment. They also evolve rapidly, making them excellent fossil markers (called *zone fossils*). As described earlier in this chapter, study of their trace element and isotope geochemistry can yield quantitative data on environmental factors such as salinity, deep and shallow ocean temperatures, ice volume, and possibly even atmospheric CO_2.

5.7 PAUSE FOR... THOUGHT

How we can be sure that plants that lived millions of years ago had the same physiology as plants living today?

The direct answer is that we cannot be sure, although evidence indicates that they are very similar. Plants have evolved over time, but the minute details preserved in many plant fossils suggest that many aspects of their basic physiology remain the same. The best match is certainly between members of the same genera or species, and the relationship between stomatal density and carbon dioxide appears to have been present in gymnosperms (such as conifers and cycads) as far back as the Carboniferous Period. This is a relatively new field of science, and we have a lot to learn, but it is clear that plant fossils retain a lot of information about climate embedded in their morphology, isotope chemistry, and physiology.

▲ **Figure 5.39:** Plant pollen and spores have characteristic shapes and sizes (20-90 micrometers), and a high preservation potential. For this reason they are used extensively by scientists to reconstruct how climate change affected ancient environments.

▲ **Figure 5.40:** Foraminifera are tiny unicellular protists that are abundant in many of the world's oceans. Foraminifera are usually less than 1 mm in size, but some are much larger, the largest species reaching up to 20 cm.

CHECKPOINT 5.20 ▶ Why are foraminifera commonly used to determine the age of marine rock formations?

▲ **Figure 5.41:** As trees grow, they add new wood beneath the bark each year, forming concentric growth rings.

The Recent Past

Samples from recent geological deposits offer more opportunities for the quantitative analysis of climate change. Unlike the analysis of ancient deposits, the analysis of tree rings, stalactites, ice cores, corals, and lake cores offer a more continuous, and sometimes annual, record of climate change.

Dendrochronology: Using Tree Rings to Date Climate Change

As trees grow, they add new wood beneath the bark each year, forming concentric **growth rings.** Each ring has two parts: an inner, thicker band of less dense wood that forms during the first part of the growing season when the tree grows rapidly, and an outer, thinner band of denser wood that forms late in the season, as the growth rate slows. Scientists sample trees by extracting small core samples for analysis in the laboratory (**Figures 5.41** and **5.42**).

In the tropics, where growth occurs year round, growth rings are absent or poorly defined. In temperate and polar regions, strong seasonal differences in growth rate leave a clear pattern of growth rings. Many different kinds of trees have been sampled, including spruce, larch, bristlecone pine, and Scotch pine (**Figure 5.43**).

Environmental factors such as temperature, precipitation, wind, slope, soil moisture, parasites, and shading are all critically important in determining the width and structure of tree rings. Age analysis (dendrochronology) is much more reliable when multiple trees from the same region are sampled and show similar growth patterns. Scientists can then eliminate from analysis any individual tree that has been adversely affected by local factors.

Once a regional pattern has been identified, a scientist can correlate the inner (oldest) rings from a living tree with the outer (youngest) rings of an older fallen tree and so on back in time, giving a continuous chronological record that is limited only by the availability of wood. In southern Germany, river oak trees provide a continuous tree ring chronology that reaches back as far as 10,000 years, and bristlecone pines in the southwestern United States provide a record 8,500 years long. The tree ring patterns from wood of unknown

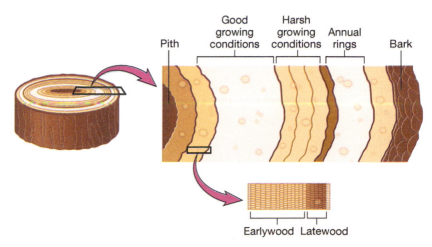

Pith | Good growing conditions | Harsh growing conditions | Annual rings | Bark

Earlywood Latewood

▲ **Figure 5.42:** Scientists sample trees by extracting small core samples for analysis in the laboratory.

Figure 5.43: Slow-growing trees like the bristlecone pine that live at the margins of their range are most affected by small changes in temperature or precipitation and provide the best record of climate change.

age can be matched with these records to provide an accurate estimate of when it grew (**Figure 5.44**).

Dendroclimatology: Tree Rings as a Proxy for Climate Change

In addition to providing data through dendrochronology, trees are useful indicators of past climate (dendroclimatology). The width of each tree ring and the density of the wood are very sensitive to the local environment. Scientists must select suitable trees from stable locations because local changes in any of the factors that impact growth could affect any one tree. The most suitable trees are usually old and growing at the limits of their range where they are especially sensitive to environmental variables.

CHECKPOINT 5.21 ▶ Why do climatologists study trees at the limits of their natural range?

Dendroclimatologists first collect tree ring data from living trees and correct these data for known statistical effects, such as a slight decrease in ring width that always occurs toward the margins of older trees. They then correlate the thickness of each ring and the density of wood with the local climate records. Scientists use these data to correlate historical and more ancient tree ring data with temperature across the same region. The reliability of tree ring records has been called into question, but these criticisms are unfounded. The record of climate change recovered from tree rings has been reproduced successfully and repeatedly by a variety of alternative temperature proxies.

CHECKPOINT 5.22 ▶ If most trees live for just a few hundred years, how can dendrochronologists use tree rings to date samples that are thousands of years old?

5.8 PAUSE FOR... THOUGHT | *Hiding the decline?*

The release of stolen emails from the University of East Anglia Climate Research Unit (CRU) in 2009 made headlines around the world. Articles stated that climate researchers were caught "fixing" the data to hide a decline in temperature in order to maintain an illusion of global warming. Climate skeptics hoped to use this information to derail sensitive climate talks that were taking place in Copenhagen that year. The "decline" at the center of this controversy is an unexpected change in the relationship between tree ring width and temperature that first appeared around 1960. This anomaly is well known among climate scientists and does not call earlier data into question. The scientists involved were completely exonerated by multiple inquiries on both sides of the Atlantic.

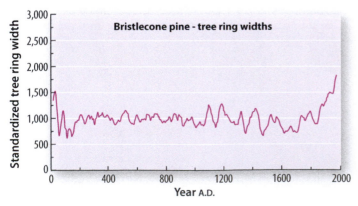

◀ Figure 5.44: When data on the width of tree rings can be directly correlated with the known temperature record over some interval, this relationship can be used to estimate temperatures from tree ring data older than the historical record. This diagram shows how tree ring data from bristlecone pines have been used to construct a continuous 2,000-year record of local climate change.

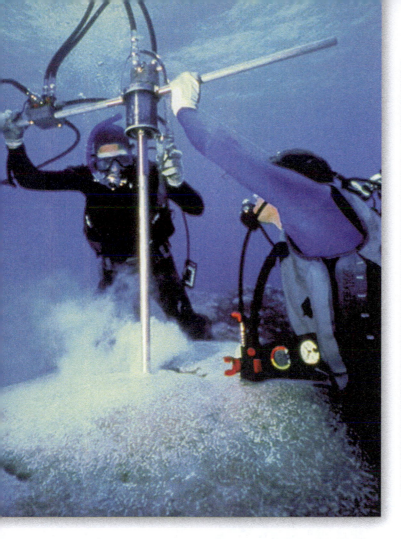

▲ **Figure 5.45:** Researchers extracting cores from corals. Small cores taken from scleractinia corals exhibit seasonal banding when X-rayed, with more dense growth over the winter, when the corals grow slowly, and less dense growth over the summer, when the corals grow faster.

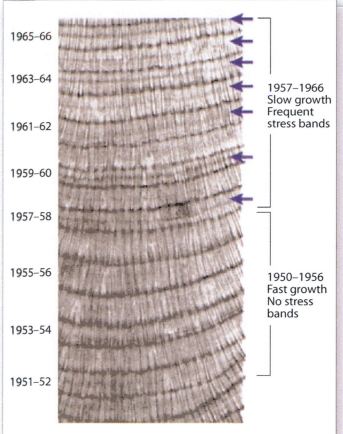

1965–66

1963–64

1961–62

1959–60

1957–58

1955–56

1953–54

1951–52

1957–1966
Slow growth
Frequent
stress bands

1950–1956
Fast growth
No stress
bands

▲ **Figure 5.46:** Like dendrochronology, coral banding can be used to provide a qualitative record of past climate. Detailed analysis of the trace element and isotope chemistry of these bands provides additional quantitative data on sea surface temperature and deep-sea upwelling. In this photograph, there is a difference in the patterns of bands that grew between 1950 and 1956 when conditions were good for coral growth and between 1957 and 1966 when conditions deteriorated. The bands that grew between 1957 and 1966 exhibit light gray-colored "stress bands" (indicated by arrows within the annual growth ring) that reflect conditions less suited to coral growth during this time.

Scleroclimatology: Corals as a Proxy for Climate Change

Reef-building corals (**scleractinia**) are widely distributed across the tropics and secrete calcareous skeletons that are often preserved in the geological record. Corals offer important insights into recent climate change (scleroclimatology), especially since few other climate proxies and records are available in the tropics. Scleractinia grow in shallow, warm (20°C [68°F]) water between 30° north and 30° south of the equator and are very sensitive to sea surface temperature. X-rays of small cores taken from these corals exhibit seasonal growth banding that develops from dense growth over the winter and less dense growth over the summer (**Figures 5.45** and **5.46**).

Like dendroclimatology, coral banding can be used to provide a qualitative record of past climate. Detailed analysis of the trace element and isotope chemistry of these bands can also provide quantitative data on sea surface temperature and deep-sea upwelling. Recent studies of coral **luminescence** suggest there may be a relationship between the amount of luminescence and freshwater runoff from adjacent land surfaces, a good proxy indicator of precipitation (**Figure 5.47**).

Coral samples that grew as long ago as the early 1600s have been analyzed so far, and the results from the luminescence data correlate well with known variation in sea surface temperature from El Niño events. Once the control experiments on modern corals are fully tested, scientists may be able to extend coral luminescence data well beyond the last interglacial period.

Stalactites and Stalagmites as Proxies of Climate Change

Sinkholes, deep caves, and subterranean rivers are all features of limestone **karst** terrain that have attracted the interest of climatologists. Most attractive are the stalactites (growing downward) and stalagmites (growing upward) that form over many thousands of years and preserve a record of the changing environment at the surface (**Figure 5.48**).

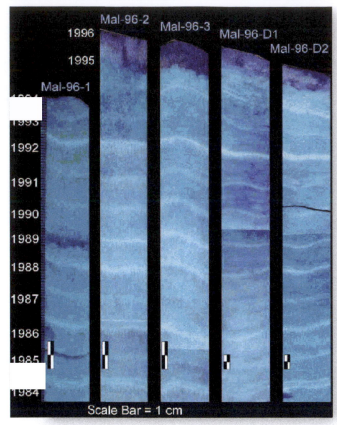

▲ **Figure 5.47:** Recent luminescence studies of cores extracted from the coral *Porites* in Kenya have shown a relationship between the intensity (brightness) of luminescence and the amount of freshwater runoff from adjacent land surfaces. In this photograph, these bands appear as bright, almost white-colored layers.

Rainwater is naturally acidic and dissolves calcium carbonate when it percolates though calcium-rich rocks such as limestone. Water dripping from the roof of underground caves slowly evaporates and leaves behind a thin residue of calcium carbonate. Drops that are suspended from the roof or that land on the floor of the cave can build up massive structures that can be as old as 350,000 years.

The interior structure of **stalactites** is complex. Evaporation leaves behind fine layers that often show many signs of interruption or erosion due to changes in climate. Their growth depends on many factors that make it difficult to extract a meaningful age from growth rings, but more accurate radiometric dates can be obtained from uranium isotopes. Trace elements and oxygen isotopes can add further quantitative data about changes in rainfall, temperature, and vegetation (**Figure 5.49**).

CHECKPOINT 5.23 ▶ What are some of the problems associated with analyzing the climate record based on stalactites?

The Record of Climate Change in Ice Cores

Ice cores have provided some of the most useful and reliable information about climate over the past 800,000

▲ **Figure 5.48:** Stalactites and stalagmites form over thousands of years, as mineral-rich water slowly percolates from the surface and evaporates underground, leaving a carbonate residue that preserves an isotope trace of changing climate, as in these examples from the Meramec Caverns that extend for 7.4 kilometers underneath the Ozark mountains in Missouri, USA.

years. To obtain cores, scientists drill boreholes as close as possible to the center of large ice sheets and high mountain glaciers, where the ice is least disturbed (**Figure 5.50**).

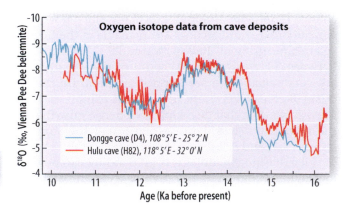

▲ **Figure 5.49:** Oxygen isotope data from the Dongge and Hulu caves in China record changes in the isotope composition of rainwater. This record correlates closely with known events from the ice and marine records. Cave deposit (Speleothem) data thus provide further corroborative evidence of past climate change.

◀ **Figure 5.50:** Recovering an unbroken and uncontaminated ice core is both a human and a technological challenge. When drilling for deep cores, the drilling rig must be totally enclosed to protect it from the harsh Antarctic weather.

Dating Ice Cores Near the top of an ice core, the annual layers of ice and snow are undeformed and make dating easy and reliable (**Figure 5.51**). Deeper under the surface, this layering is no longer visible because the ice flattens and stretches under the increasing weight of the ice above. Scientists can still identify seasonal layers in this older ice by studying small changes in the concentration of trace elements and other seasonal markers, such as dust, hydrogen peroxide, and oxygen isotope chemistry. Correlating depth and age between cores from different sites is difficult but is made possible by distinctive and frequent **marker horizons** of volcanic ash and atmospheric dust that fall over a wide geographical area and mark specific moments in time.

CHECKPOINT 5.24 ▶ As ice gets older, why is it increasingly difficult to see the annual layers in an ice core?

How to Measure Ice Accumulation Rates
Scientists use the width of annual layers in an ice core to determine how the amount of precipitation has changed over time. Thicker layers correlate with warmer atmospheric temperature because warm air can hold more moisture. The most useful cores come from sites where

▼ **Figure 5.51:** Seasonal layering is often clearly visible in the upper sections of an ice core but loses definition with depth. Gently sloping annual layers are clearly visible in this large ice wall exposure.

there is a good accumulation of snow each winter, with little subsequent melting over the summer.

Oxygen and Hydrogen Isotope Ratios as Proxies for Air Temperature

As described earlier in this chapter, scientists sample ice along the length of an ice core and use oxygen (and hydrogen) isotope ratios to calculate the temperature at which ice formed in the atmosphere prior to its burial. Isotope analyses provide critical information about climate change but can also help scientists resolve cryptic seasonal layering.

Collecting Ancient Atmospheric Gases

Firn is permeable, porous, uncompacted snow and ice found in the first 50 to 100 meters (164 to 328 feet) beneath the surface of an ice sheet. Accumulating ice and snow gradually compact the firn, but it can take hundreds of years for trapped air to become isolated from the surface. Therefore, the air that is finally trapped may be much younger than the ice that surrounds it. Scientists must correct this "age gap" before they can interpret any data. Because the concentrations of carbon dioxide, methane, and nitrous oxide in the trapped air do not change after burial and isolation, ice cores can preserve a continuous, unbroken record of how the atmosphere has changed over thousands of years (Figure 5.52).

The Application of Electrical Conductivity Measurements (ECM)

Ice is slightly acidic because rainwater dissolves CO_2 to form weak carbonic acid. During cold periods, when winds are often more intense, alkaline mineral particles from the ocean become trapped on the surface of the ice and reduce its natural acidity. As any change in acidity affects the electrical conductivity of ice, scientists are able to use ECM to establish marker horizons that help them to date ice cores.

Using Volcanic Ash to Date Ice Cores and Monitor Volcanicity

Large volcanoes eject dust, gas, and aerosols high into the stratosphere, where they circulate around Earth before falling back to the surface. Volcanic deposits in ice cores create important marker horizons that scientists use to correlate ice cores from different sites that may be up to thousands of kilometers apart. Because volcanic activity is a strong influence on climate, it is also important to mark the occurrence of all eruptions as a record of climate forcing.

CHECKPOINT 5.25 ▶ How do ice cores show a record of large volcanic eruptions?

5.9 PAUSE FOR... THOUGHT | *How can we be sure that ice core dating is reliable?*

There are many techniques that scientists use to try and make the dating of ice cores more accurate, but this task becomes increasingly difficult at depth because ice flows under pressure. Scientists can correlate data between adjacent ice cores using layering and conductivity as a major guide, but accurate correlation with more distant cores requires the use of dust, sulfate, chloride isotope, and volcanic ash markers.

Dust as a Proxy for Climate Change

Atmospheric dust is disseminated throughout most ice cores, but its concentration changes over time. Dust is picked up as wind blows over recently glaciated surfaces where there is little or no vegetation. Exactly how much dust gets into the atmosphere depends on rainfall, wind velocity, and the amount and composition of exposed terrain. The greatest concentration of dust in cores is found in ice deposited during the early stages of an interglacial period. This is because retreating ice leaves behind large outwash plains that are a ready source of dust.

Salt as a Proxy for Climate Change

The concentration of sodium and chlorine ions in ice depends on the amount of sea salt reaching the site of ice accumulation. Environmental factors such as the strength and direction of the wind, distance from the ocean, and frequency of coastal storms all affect the concentration of salt.

Ammonium as a Proxy for Burning Biomass

The concentration of ammonium in ice cores has been related to the burning of biomass. Peaks in the level of ammonium might be a strong indicator of early anthropogenic activity.

Sulfur-Related Chemicals as a Proxy for Volcanicity and Biological Activity

Sulfate and **methane-sulphonate (MSA)** are the two major sulfur species trapped in polar ice. Sudden spikes in sulfate are often due to volcanic eruptions and can be used to record distant eruptions in the absence of volcanic ash. Changes in MSA can be related to the emission of dimethyl sulfide (DMS) due to biological activity in the

▲ **Figure 5.52:** A thin section of an ice core, showing trapped bubbles of air. The photograph was taken under polarized light that adds colors to enhance the detail.

◀ **Figure 5.53:** Drill ships like the *JOIDES Resolution* contribute to the Integrated Ocean Drilling Program, which is a systematic survey of deep-ocean cores from many sites and is critical to our understanding of climate change.

surrounding ocean. Scientists can use sulfur isotopes to discriminate between the biogenic and abiogenic origins of some other sulfur compounds.

Beryllium 10 as a Proxy for Solar Activity

Beryllium 10 is a cosmogenic isotope in the atmosphere that is related to sunspot activity and acts as a proxy for both solar irradiance and the strength of Earth's magnetic field (see Chapter 3).

CHECKPOINT 5.26 ▶ You are the lead scientist on a project to extract ice cores from an ice field in the Himalayas, and you are on a limited budget. Which analyses would you prioritize, and why?

The Record of Climate Change from Sediment Cores

The sediment cores that geologists recover from lake and ocean beds provide a longer, although less detailed, record of climate change than ice cores. Ice core data lose precision and accuracy with depth, but sediment cores recover samples from millions of years ago that can still provide detailed information about past climate change (Figures 5.53 and 5.54).

Core Lithology as a Proxy for Climate Change

Changes in lithology are often related to changes in climate. This is especially true in sediments that

◀ **Figure 5.54:** Ocean cores can lack the fine temporal resolution of ice cores, but they still offer a wealth of information about changing climate. Changes in lithology, geochemistry, and microfossil content parallel the changes observed from ice cores but extend this record far back in time, to a period when Earth was ice free. Here a geologist is cutting the core into smaller sections for analysis.

▲ **Figure 5.55:** Varved sediments form in deep or undisturbed water where there is a significant change in the seasonal input of sediment. In this core section, the annual layers appear as thin dark bands of summer deposition with lighter colored bands of winter deposition.

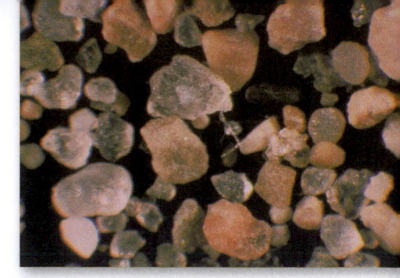

▲ **Figure 5.56:** Small rock and mineral fragments from sand like these are dropped from the base of melting icebergs, and their composition can point to their region of origin.

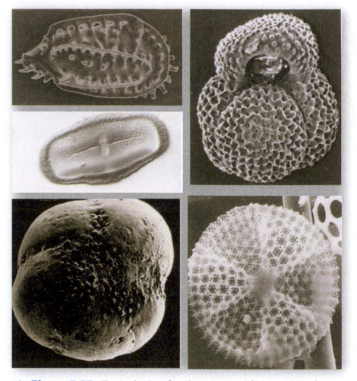

▲ **Figure 5.57:** Typical microfossils recovered from rock and sediment cores. Clockwise from top left: ostracod (a variety of crustacean), planktonic foram, diatom (a variety of phytoplankton), benthic foram, and pollen grain. One way scientists reconstruct past climate is by comparing the spatial distribution of these microfossils with the distribution of their relatives living in today's climate. Calcareous microfossils (foraminifera and ostracoda) carry additional climate information in their elemental ratios; for example, the ratio of magnesium to calcium varies with ocean-water temperature, and the ratios of different isotopes of oxygen vary with global ice volume and, therefore, sea level.

accumulate in deep undisturbed water, isolated from the impact of storms and burrowing invertebrates.

Glacial lake sediments often show clear annual layering in grain size and organic content due to seasonal variation in precipitation and algal growth. Under ideal conditions, an analysis of the pollen, microfauna, geochemistry, and isotope chemistry of these annual layers (called **varves**) can provide a detailed, undisturbed record of climate change over many thousands of years (**Figure 5.55**).

Deep-marine sediments provide the longest continuous record of climate change available. International cooperation programs such as the Deep Sea Drilling Project (1968–1983), the **Ocean Drilling Program** (1983–2007), and the current Integrated Ocean Drilling Program have accumulated a large database of data from deep-sea cores. One of the easiest ways to spot past glacial activity in cores is to look for small rock and mineral fragments dropped from the base of melting icebergs. These rock

and mineral fragments can be highly diagnostic of their region of origin (provenance) and can be easily identified using a simple binocular microscope (**Figure 5.56**).

As described earlier in the chapter, many species of marine microfauna and microflora, such as foraminifera, diatoms, and coccolithophores, are sensitive to changes in water temperature and salinity. Scientists often recover large numbers of their shells from sediment cores, and they can use the oxygen and carbon isotope composition of these shells to determine surface and bottom-ocean temperatures and the possible level of atmospheric CO_2. Scientists have also used biochemical markers preserved in sediment to estimate both temperature and atmospheric CO_2 (**Figure 5.57**).

CHECKPOINT 5.27 ▶ What are the advantages and disadvantages of using sediment cores and ice cores to determine climate change?

5.10 **PAUSE FOR... THOUGHT** | *Why do sediment cores lack the temporal (time) resolution of ice cores?*

Ice sheets grow each year due to the annual addition of winter snowfall. Most sediment cores lack this resolution for two reasons. The first reason is that the process of sedimentation is not regular. Major storms, such as hurricanes, can strip off the top layers of sediment and then dump more sediment in a day than was deposited in decades. The second reason is that the action of invertebrates and crustaceans mixes the upper layers of sediment in most shallow marine environments, in a process called bioturbation. Both of these processes make it almost impossible to recover annual records from sediment cores, but on the scale of centuries to millennia sediment cores can provide a continuous record of Earth's climate stretching back millions of years.

▼ **Figure 5.58:** An aerial view of research stations in the South Pole. Working in Antarctica is a technological and human challenge, especially for those who stay over the long dark polar winter. Most scientists leave when the darkness arrives at the end of a field season that is often punctuated by severe storms.

Summary

- The geological tools that allow Earth scientists and climatologists to study ancient climate are based on field-based observations and laboratory analysis. Clues to ancient climate can be observed in almost every sedimentary rock formation and ice core.

- When scientists study rocks and ice cores in the laboratory, they use trace element and isotope chemistry to extract quantitative details about temperature and the level of CO_2 in the atmosphere. Even the organic residue of animals dead for millions of years can reveal details about the environment in which they lived and the climate that surrounded them. Many of these tools are still under development, and the data they obtain need to be treated with caution, but they have the potential to reveal details about ancient climate change that were considered impossible just a few decades ago.

- Sedimentary rocks contain information about climate in their structure, fossil content, chemistry, and mineralogy. The resolution of the record is not as fine as in ice cores, but the record stretches back millions of years and can be recovered from sites all around the world. Sites for good ice cores are harder to find and limited to high altitude and latitude but offer the potential of a clear annual record of climate change.

- Terrestrial and marine plant and animal fossils provide direct evidence of past climate and ancient ecosystems. Terrestrial plant fossils provide quantitative evidence of past levels of atmospheric carbon dioxide through their stomatal density and qualitative data on rainfall and temperature. Marine plankton can also provide quantitative data about ocean temperature from the structure of their chemical remains.

- Oxygen isotopes are one of the most important tools available to climate scientists. They provide an estimate of air temperature from the water recovered from ice cores and stalactites, and water temperature and ice volume from the shells of marine shells.

- Carbon isotopes extracted from the carbonate shells of marine organisms reflect changes in marine bioproductivity and have been used as evidence that the recent rise in atmospheric carbon dioxide is from the burning of fossil fuels.

- The pattern of tree rings is used to estimate regional temperatures and climate over large areas of the world where trees survive at the margins of natural tolerance.

Why Should We Care?

The best way to predict the future of any system is to look at the past. By studying rocks and fossils deposited over time, it is possible to evaluate the impact of climate change on the environment. Geological observations and geochemical tools open windows into ancient worlds that allow us to construct a detailed picture of both climate and climate change. Many of these tools now offer enough accuracy and precision to determine both the rate of ancient climate change and the sensitivity of ancient climate to radiative forcing. As you will read in the following chapter, they indicate that Earth's climate is not always benevolent.

Looking Ahead . . .

Chapter 6 illustrates how scientists apply the different geological tools introduced in this chapter to study climate change over geological history. Earth's climate is in a constant, natural, and gradual state of change, but periods of rapid climate change are common, especially when driven by sudden changes in the composition of Earth's atmosphere that can upset the balance of radiative forcing.

Critical Thinking Questions

1. Geologists exploring for oil reserves off the coast of Virginia are examining a section of core from dark-colored shale. What geological tools should they use to indicate what the environment and climate were like when this shale was deposited?

2. Geologists are accustomed to working in time units of millions and even billions of years, and it is often difficult for students to understand the power of time and slow geological processes to create very significant change. If the African Plate started to drift south toward the equator at a rate typical of the drifting continents, how many years would it take before tropical vines started to grow around the ancient city of Jerusalem? Is this a long time in terms of the geological record?

3. While exploring a thick sequence of lava flows in Northern Ireland, you discover the remains of lake deposits between many of the flows. On closer examination, you find that the lake deposits are full of fossil leaves. You wonder if this volcanic activity could have increased the concentration of carbon dioxide in the atmosphere. What would you do to test your hypothesis?

4. The rate at which the cosmogenic and unstable isotope carbon 14 is produced in the upper atmosphere varies with the flux of high-energy neutrons from the Sun. Once it is created, the carbon 14 immediately starts to decay into nitrogen, at a half-life of 5,730 ± 40 years. Geologists measure the amount of carbon 14 remaining in geologically recent organic material (such as wood) to determine the age of a specimen. Highlight the weaknesses of this technique and explain your answer.

5. Data from Greenland's ice cores show that the Arctic climate can change very rapidly. Data from plant and animal remains show how the Arctic ecosystem also changed rapidly in response to these changes, with no signs of major extinction events. Why are we so worried about changes in the Arctic today?

Key Terms

Please make sure you are familiar with the key terms introduced and highlighted in this chapter. A full glossary is available at the end of the book.

Alkenone, p. 146
Alluvial fan, p. 132
Aragonite, p. 135
Atomic radius, p. 146
Benthic, p. 142
Biological pump, p. 144
Bioturbation, p. 131
Calcite, p. 135
Calcrete, p. 138
Caliche, p. 138
Coccolithophore, p. 135
Cosmogenic isotope, p. 140
Cretaceous Period, p. 135
Daughter atom, p. 140
Desiccation crack, p. 131
Dimethyl sulfide (DMS), p. 136

Dropstone, p. 138
Empirical, p. 142
Ephemeral, p. 137
Evaporite, p. 138
Fecal pellet, p. 135
Firn, p. 154
Growth ring, p. 149
Half-life, p. 140
Igneous rock, p. 131
Ikaite, p. 138
Isotope, p. 140
Isotope excursion, p. 145
Karst, p. 151
Laterite, p. 138
Lithology, p. 131
Loess, p. 138

Luminescence, p. 151
Macrofossil, p. 147
Marker horizon, p. 153
Methane hydrate, p. 145
Methanesulphonate (MSA), p. 154
Microfossil, p. 131
Neutron, p. 140
Ocean Drilling Program, p. 156
Oil shale, p. 133
PDB standard, p. 144
Phytoplankton, p. 146
Picoplankton, p. 135
Plankton, p. 133
Plinian, p. 136

Plume, p. 136
Protist, p. 148
Proton, p. 140
Pseudomorph, p. 138
Radioactive isotope, p. 140
Ripple mark, p. 131
Scleractinia, p. 151
Stable isotope, p. 140
Stalactite, p. 152
Stomata, p. 147
Terminal moraine, p. 137
TEX$_{86}$, p. 146
Trace element, p. 146
Varve, p. 156

The world today may seem as if it will never change, but fossil evidence reveals a very different story. During the late Cretaceous Period, Pteranodons commanded the skies in a hothouse world sustained by high levels of carbon dioxide in the atmosphere. If we continue to add greenhouse gases to the atmosphere, we may return to a world that would seem very familiar to the dinosaurs, but very different from our own.

Climate History

Introduction

Deep under the ground and entombed in the geological record is evidence of ancient climate change and the forces that shaped it. Data from rocks, ice, and minerals record the pattern of Earth's orbit, the trail of moving continents, the impact of volcanism, the growth and decay of ice sheets, the rise and fall in global temperatures, and variations in sea level.

Geologists have investigated patterns of ancient climate for decades, and this chapter will show how they been able to construct a detailed picture of climate change over the last 700 million years. There are a number of times when ice sheets advanced from the poles and almost covered the globe, but for most of geological history, Earth has been relatively ice free and much warmer than it is today due to higher levels of carbon dioxide in the atmosphere. The transition between the warm and humid global climate of the Cretaceous Period and the world we see today is a case study of the many factors that control climate. Embedded in this long geological record is evidence that global climate can change very rapidly due to the release of greenhouse gases. Whether due to volcanism or the release of organically derived methane, climate change forced by greenhouse gases has been the root cause of many major extinction events throughout geological history, and may be again.

Learning Outcomes | *When you finish this chapter you should be able to:*

- Describe how climate has changed throughout geological time
- Identify key factors that determine the direction of climate change
- Discuss the sensitivity of climate to changes in temperature
- Apply what you know about climate history to predict what is likely to happen in the near future
- Analyze recent changes in climate in the context of natural climate change throughout geological history
- Evaluate whether natural processes alone can explain recent climate change

Climate, Life and Geological Time

Earth's climate has changed frequently over the 4.5 billion years of geological history (**Figure 6.1**). For the first 500 million years the atmosphere and oceans were toxic, violent, and unsuitable for the evolution of life, but by 3.8 billion years ago, conditions had improved enough for the first basic forms of life (**prokaryotes**) to appear in the oceans. It took another 1.8 billion years for more complex cellular life (**eukaryotes**) to evolve, and a further billion years before the fossil record reveals the presence of more complex plant and animal life forms that depended on oxygen for respiration. It was only around 480 million year ago, fairly recently in geological time, that the first plants and animals moved from the oceans onto the land. In this rocky and largely uninhabited world, life was much more vulnerable to changes in climate than it had ever been in the oceans.

By using data from rock and ice cores, geologists can trace a continuous climate history over this time and contrast the geological record of natural climate change with our observations today.

6.1 **PAUSE FOR...** | *How do we determine the*
THOUGHT | *age of geological events?*

We determine the age of ancient sedimentary rock formations by analyzing radiogenic isotopes (produced by radioactive decay) from igneous (once molten) rocks that were deposited within, or cut across, the formation. We can also use fossil remains if their age has been determined elsewhere and if their age range is limited. We cannot use the sedimentary rocks themselves to derive an accurate age because they are made from the remains of other rocks and their isotopic composition is not directly related to their age.

What Controlled Ancient Climate?

The world looked very different in the deep geological past than it does today, but the same laws of physics applied then as now. You can still use the rules you learned in Chapter 3 to understand how the same forces have shaped climate change over the past 65 million years and more. For example:

- The Sun has always been the strongest single factor controlling Earth's climate.
- Milankovitch cycles have always acted to alter the total amount of solar radiation reaching Earth's surface and affected its seasonal intensity.
- The slowly changing position of the continents has always affected the distribution of energy by altering patterns of circulation in the atmosphere and oceans and the growth and decay of polar ice sheets.
- The impact of carbon dioxide and other natural greenhouse gases has varied naturally over time with changing biological productivity, volcanic activity, ocean temperature, and rate of chemical weathering.
- Changes in sea level and the growth and decay of ice sheets have affected changes in global albedo and rates of erosion by covering or exposing rocks to subaerial weathering.
- Changing air temperatures have varied the amount of water vapor, a very important greenhouse gas, in the atmosphere.
- Major ocean–atmosphere cycles operated in the past as they do today. Their location, intensity, and geometry may have changed, but their broad patterns are determined by physical principles that have not changed over time.

We can use these observations as guiding principles to help us understand why and how climate has changed in the past.

Fire and Ice: The Story of "Snowball" Earth

During the **Proterozoic Eon** (2,500 Ma to 542.0±1.0 Ma million years ago), the Sun was 30% weaker than it is today, and only the presence of much higher levels of CO_2 and other greenhouse in the atmosphere kept Earth from freezing permanently. Glacial sediments, **dropstones**, and **rock striations** from around 750 to 630 million years ago record three very significant periods of ice advance. During those three periods, global albedo exceeded a critical threshold that allowed ice to spill beyond the poles, past mid-latitudes, and as far as the equator. From space, Earth would have looked like a large snowball—although rather slushy in places and stained with ash from explosive volcanic activity. Before reading further, consider what impact this would this have had on Earth's climate system.

CHECKPOINT 6.1 ▶ What evidence led geologists to conclude that the world was more than once engulfed in global ice ages during the late Proterozoic?

The Power of Ice

The global ice sheet that covered snowball Earth put an effective lid on the exchange of mass and energy between the atmosphere and oceans:

- Much more of the Sun's energy would have been reflected directly back out to space than today, and surface temperatures would have plummeted.
- The strong temperature gradient that exists between the equator and poles today (Figure 3.27) would have been weaker, and atmospheric circulation must have been very different.
- The rate of surface evaporation would drop to zero once the oceans became entirely ice covered, and the atmosphere would gain only a small amount of moisture by the direct sublimation of ice.

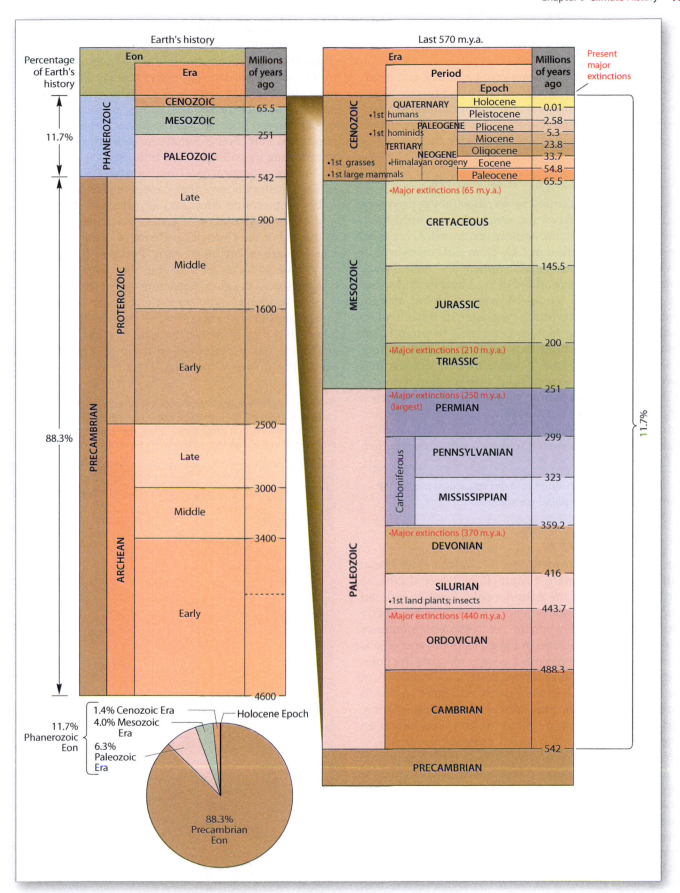

▲ **Figure 6.1:** Earth history is divided into geological time units that were stacked together to create this geological column. The different time units are often defined by major upheavals in Earth history such as periods of mountain building, extreme climate, and mass extinction. The largest time unit geologists use is the Eon, followed in descending order by Era, Period, Epoch and (not illustrated here) Age.

- Total global precipitation would have decreased as the atmosphere cooled and the air would have held less water vapor.
- Earth's atmosphere became cold and dry, and much of the land surface of the surface was a frozen desert.
- Under the oceans, vertical thermohaline density currents would have dominated global circulation as cold, dense brines formed beneath the sea ice.

The Power of Greenhouse Gases

When Earth was engulfed by ice, most of the processes that remove CO_2 from the atmosphere would have shut down, but volcanic activity continued. Over time, the slow accumulation of CO_2 from volcanoes would have started to warm the atmosphere and melt the ice. Volcanic ash may have added to this effect by darkening the surface of the ice and lowering global albedo. Geological evidence (discussed later in this chapter) suggests that these factors combined to push Earth past a critical threshold that led to the rapid melting of ice around the globe. Climate models suggest that Earth would have experienced extreme temperatures at this time. To overcome the cooling effect of such a large global albedo, the atmosphere must have reached temperatures as much as 6°C to 11°C (11°F to 20°F) higher than today's—even while ice still covered much of Earth's surface.

CHECKPOINT 6.2 ▶ What role did volcanism play in the termination of each global ice age?

From Ice to Oven

Once the ice started to melt, positive (warming) feedbacks within Earth's climate system forced global climate further away from the extreme cold. Freshly exposed land and ocean surfaces would have lowered albedo and warmed the atmosphere. As tropical temperatures started to rise, circulation in the atmosphere and oceans would have transferred more energy towards the poles and initiated melting at higher latitudes. Once areas of the oceans were free of ice, they would have started to remove CO_2 from the atmosphere by **aqueous solution**, and weathering on land would have proceeded at an accelerating rate under higher atmospheric temperatures. The warming oceans would have rapidly saturated with CO_2 near the surface and become very acidic. At the same time, rapid weathering of the freshly **exhumed landscape** must have fed the oceans with calcium and magnesium ions that combined with the $H^+CO_3^-$ from dissolved carbon dioxide to precipitate vast amounts of limestone and dolomite (magnesium-rich limestone). As the oceans continued to warm to depth, the sudden release of CH_4 from methane ice hydrates would have accelerated warming even further. Together, these factors caused Earth to become hot very quickly.

Geological evidence for these events comes from Namibia, where outcrops expose rocks that formed in the tropics at this time (Figure 6.2). These exposures contain dropstones in marine sediments that are clearly of glacial origin but are overlain across a sharp boundary by marine carbonate sediments that could only have been deposited in very warm water. This evidence suggests that the runaway icehouse of snowball Earth quickly turned into a global oven over a short period of geological time. Some climate models suggest that the temperature transition took as little as 2,000 years.

Reliable geological data from the next 600 million years of Earth history shows that continental drift continued to have a profound effect on the pattern of atmospheric and oceanic circulation. Further glaciations occurred at the end of the Ordovician Period and during the late **Carboniferous Period** to early **Permian Period**—and they continue today as well—but they have never been as extensive as the glaciations during the late Proterozoic.

The sudden climate transitions of the late Proterozoic are not likely to occur in the immediate geological future, but they illustrate the kind of climate extremes that Earth has experienced in the past. They demonstrate the power of albedo, greenhouse gases, and

▼ **Figure 6.2:** This geological outcrop in Namibia shows that carbonates formed in very warm water lying directly above glacial deposits, with evident dropstones (the large white boulders embedded in the rock face at the same level of the geologist).

◀ **Figure 6.3:** Moving ice sheets and glaciers carve deep grooves, called glacial striations, in the rocks at their base, as in this example from Westchester County, New York. Among the best-preserved striated and grooved pavements in the world are those of Permo-Carboniferous age (around 290 million years ago) in South Africa.

volcanic activity to flip global climate between temperature extremes. They also demonstrate the ability of early life to either adapt or retreat from such rapid and extreme environmental changes.

CHECKPOINT 6.3 ▶ What evidence do we have that the level of carbon dioxide in the atmosphere was very high when the global ice sheets melted?

Ice and Coal: The Great Permo-Carboniferous Glaciation

The Permo-Carboniferous glaciation occurred between 310 and 290 million years ago. Southern Hemisphere ice extended as far as 30° south, leaving geological evidence in the form of glacial till, rock striations (**Figure 6.3**), and exhumed **U-shaped valleys** in all the countries that were then gathered around the South Pole (South America, Africa, Madagascar, Arabia, India, Antarctica, and Australia). There is less evidence of glaciation from the Northern Hemisphere, but dropstones suggest that there was at least some ice present (**Figure 6.4 (a)**).

Why a Carboniferous Glaciation?

The Permo-Carboniferous glaciation, which lasted about 30 million years, was the coldest period on Earth during the past 500 million years. Some similarities between the Carboniferous Period and Earth today could explain our current glaciation—and could possibly even predict the long-term future of climate change.

The fact that continents surrounded the South Pole during the Carboniferous Period may be one of the reasons why there was a major period of glaciation at this time. Heat is lost to space much more rapidly from land than from the oceans, and when there is a lot of land

at high latitudes the polar regions cool quickly with the onset of winter. It is therefore possible that winter heat loss helped to initiate and sustain glaciation during the Carboniferous Period, as it did again when ice sheets first started to develop on Antarctica ca. 35 million years ago. Cooling of the southern oceans would have also resulted in greater absorption of CO_2 and further global cooling.

The lithology and distribution of rocks from the Carboniferous Period show that it was a time of significant mountain building (**orogenesis**) following the collision of major continental plates. This led to very active erosion, as large rivers transported sediment from the mountains onto wide coastal plains covered with great forests on swampy deltas that are now preserved as extensive deposits of sandstone and coal.

Both erosion and coal formation led to the removal of CO_2 from the atmosphere and further cooling (Figures 6.4 (a) and **6.4 (b)**). We can estimate the level of CO_2 in the atmosphere based on fossil plant and soil samples, and it seems likely that it fell from around 1,500 ppm to present-day values, at around 300 to 400 ppm (**Figure 6.5**).

The combination of high winter heat loss, cooling oceans, enhanced erosion, high biological productivity, snow and ice accumulation, and increased albedo was enough to push the climate into a prolonged ice age. This ice age ended only when the continents slowly moved away from the poles some 30 million years later, during the late Permian to early Triassic Periods.

CHECKPOINT 6.4 ▶ What combination of factors during the late Carboniferous Period led to global cooling and the formation of polar ice sheets?

These conditions present during the Permo-Carboniferous glaciation are also present (in different proportions) on Earth today. Our knowledge of that

▶ **Figure 6.4:** (a) The glaciation at the end of the Carboniferous Period (around 300 million years ago) coincided with a major phase of enhanced erosion and the appearance of a large continent over the South Pole. (b) During the Carboniferous Period, the continents were gathered into one large supercontinent that rested over the South Pole. Climate proxies indicate the presence of cold conditions over both poles, but the widespread growth of tropical and subtropical forests deltas that led to the deposition of many of the world's major coal deposits.

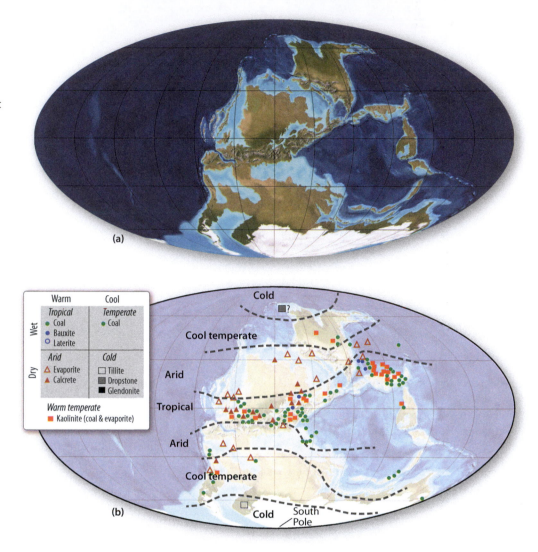

▼ **Figure 6.5:** This view of the Carboniferous is based on data from many proxies, especially the abundant plant fossils that are found in association with coal deposits.

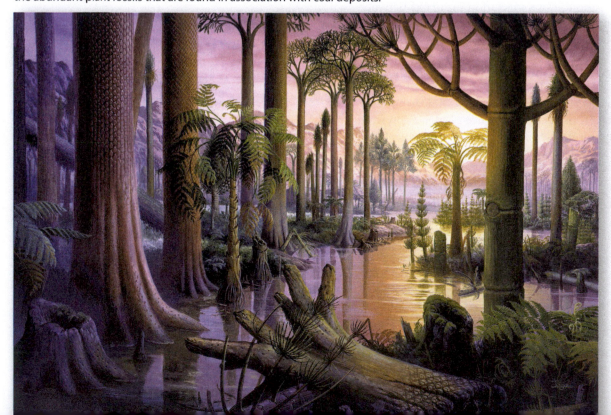

glaciation allows us to predict that the current cold phase of Earth's climate history has a long way to go before it comes to a natural end.

6.2 **PAUSE FOR...** **THOUGHT** | *What is the evidence of changing sea level during the late Carboniferous glaciation?*

A characteristic feature of Upper Carboniferous sedimentation is the appearance of cyclothems, repeating sequences of sedimentation driven by changes in sea level. A typical cyclothem starts with marine sediment deposited in at least 40 meters (130 feet) of water. This is progressively overlain by dark organic-rich shale, then sandstone, and finally coal—a sequence that represents the progressive shallowing of water and the advance of the coastline. A new cycle begins as sea level rises again and the shoreline retreats, covering the coal with marine sediment. Analysis of cyclothem periodicity suggests that changes in sea level were driven by Milankovitch cycles that controlled the growth and melting of ice sheets, just as we experience today during the current icehouse phase of Earth's climate history.

The Great Extinction: The Permo-Triassic Climate Crisis

The climate during the late Permian and early Triassic Periods pushed a fragile biosphere into a deep and almost fatal crisis. Only the most extreme anthropogenic global warming today might repeat those events, but that period nevertheless provides a sobering case study of the very real danger of extreme climate change. It is simply not true, as some still believe, that global climate can always adjust to maintain conditions essential for the preservation of life. Evidence from the Permo-Triassic makes it clear that global climate can create toxic environments that can have a disastrous impact on life.

The Permo-Triassic World

By the end of the Permian Period (~251 million years ago), the continents had moved off the South Pole and gathered into a large supercontinent called **Pangaea** (pronounced "Pan-GEE-ah") that straddled the equator and stretched almost from pole to pole (**Figure 6.6**).

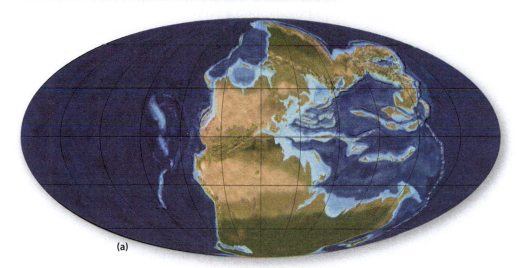

(a)

◀ **Figure 6.6:** (a) Much of the Permo-Triassic world (299 to 206 million years ago) was dominated by arid continental climates on land and a large ocean covering much of the planet. (b) Climate proxies for the Permo-Triassic world show that deserts were much more extensive than they are today and that the poles were free of permanent ice. Coal deposits left behind by temperate forests were not as extensive as those of the Carboniferous Period, but were mined in places such as Virginia in the United States.

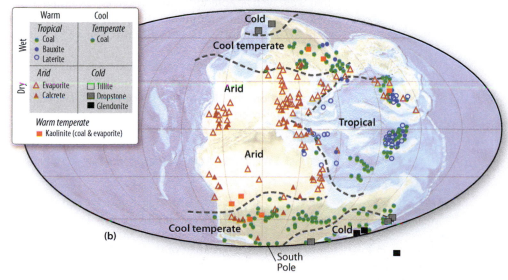

(b)

Extreme Warming   The widespread occurrence of rocks deposited in arid environments during the Permo-Triassic shows us that large areas of desert covered much of the supercontinental interior and restricted the range and diversity of the terrestrial fauna. The Siberian Traps, thick formations of volcanic rock in Siberia that formed at this time, are evidence of rapid, voluminous, and non-explosive (**effusive**) **volcanism.** This volcanism would have released vast quantities of CO_2 into the atmosphere and elevated global temperature significantly.

Fossil evidence shows that forests still existed around the continental margins where moisture was available, but further inland, even in the tropics, there were no sources of moisture to generate rainfall. Thin clays deposited in seasonal lakes, and occasional beds of fluvial sandstone indicate that summer monsoons were modulated by Milankovitch cycles and generated enough seasonal rainfall to sustain some life in parts of the continental interior. However, water was very seasonal (ephemeral), and life was hard (**Figure 6.7**). To make matters worse, carbon isotope evidence suggests that warming oceans destabilized deep-ocean methane hydrates, releasing gigatonnes of CH_4 into an atmosphere that was already warm from the impact of volcanism. Fossil, geochemical, and isotope data suggest that the combined impact of CO_2 and CH_4 drove the global temperature as much as 12°C (22°F) higher than it is today.

▼ **Figure 6.7:** Continental sedimentary rocks from the Permo-Triassic are typically colored red due to the formation of iron oxide cements that form soon after deposition in warm climates due to high rates of evaporation in hot arid to semiarid environments. This towering example of an outcrop is from Fisher Towers near Moab, Utah.

The Oceans Become Toxic   Following the end of the Permo-Carboniferous glaciation, around 260 million years ago, the oceans warmed quickly. Dark organic-rich marine rocks provide evidence that weakly circulating, oxygen-poor bottom waters replaced the vigorously circulating cold oxygen-rich bottom waters that are more typical of glacial conditions. Fossils and geochemical evidence show that by 252 million years ago, large **anoxic** (oxygen-deficient) and toxic zones soon developed on the deep-ocean floor, where **anaerobic bacteria,** which are well adapted to these conditions, fed on biomass falling from the upper ocean and released poisonous hydrogen sulfide (H_2S) into the surrounding ocean. Fossil evidence shows how these toxic zones built upward into shallower water over time and eventually spilled over onto the continental shelf, with devastating impact on marine life. Some studies suggest that the level of toxicity may have increased to a point where H_2S gas eventually escaped directly into the atmosphere, filling the air with droplets of sulfuric acid, and reducing the level of oxygen. This was a difficult time to be alive for many marine and terrestrial organisms.

The Great Extinction: An Oxygen Crisis?   Geological models and fossil evidence suggest that the rapid oxidation of both H_2S and CH_4 in the atmosphere, along with surface oxidation under dry desert heat, finally lowered the level of oxygen (O_2) in the atmosphere to around 15%. (The current oxygen level is 21%.) This had a profound impact on the creatures that had evolved during the Carboniferous Period, when the luxuriant growth of coal forests boosted oxygen levels to more than 35%. This low oxygen level led to the most critical extinction event to have occurred over the past 600 million years of geological history. Fossil evidence shows that more than 95% of all marine species and 70% of known continental vertebrates became extinct over a period as short as 10 million years (**Figure 6.8**).

This Permian climate crisis created a world that favored the evolution of the proto-dinosuars, animals that were well adapted for low-oxygen environments and whose descendents remained the dominant vertebrates until the end of the **Cretaceous Period** (**Figure 6.9**).

CHECKPOINT 6.5 ▶ What combination of factors during the late Permian Period created the conditions that led to the extinction of 95% of marine species and 70% of known continental invertebrates?

Could the Permo-Triassic Extinction Happen Today?

There is no supercontinent today, and we see no signs of impending effusive volcanism, but there are still lessons to be learned from the Permo-Triassic extinction. Perhaps the greatest of these is that global climate

▲ **Figure 6.8:** The Permian extinction event was the largest extinction event of the **Phanerozoic Eon**. More than 95% of all marine species and 70% of known continental vertebrates became extinct over a period as short as 10 million years.

is capable of extreme change, and the impact of that change on life can be catastrophic.

It takes a long time and a lot of energy to warm the deep oceans, so it is highly unlikely that anthropogenic warming today will reproduce the catastrophe of the Permo-Triassic event. The appearance of large dead zones in the deep ocean and on the continental shelf is not impossible, however, as this appears to be a typical feature in a hothouse world.

CHECKPOINT 6.6 ▶ Why are the deep oceans not as likely to develop dead zones today?

6.3 PAUSE FOR... THOUGHT | *Can climate change cause a mass extinction?*

Complex life nearly came to an end at the end of the Permian Period, a reminder that the oceans and atmosphere can become toxic to life. No one is suggesting that such an extinction is a likely outcome of global warming today, but it serves as a timely warning that Earth is not always able to adapt to change in ways that are well suited to the continuity of life. The future of life on Earth depends on many factors, and it's in our best interests not to add to the many uncertainties.

◀ **Figure 6.9:** The Permian extinction was a tragedy for more than 70% of land-dwelling vertebrates, but it left a world primed for the evolution of the early dinosaurs, such as these shown in an artist's rendering of life during the Triassic Period.

The Cretaceous Hothouse World

Evidence from rocks and fossils shows that during the Cretaceous Period (145.5 to 65.5 million years ago), the climate was hot and humid, with deciduous forests extending as far as 82° north. Sea level was up to 250 meters (820 feet) higher than it is today, and the ocean floors were periodically starved of oxygen. This was the hothouse world of the dinosaurs and giant marine reptiles, a time that saw the first emergence of flowering plants and grasses.

The great Pangaea supercontinent had started to tear apart during the Triassic and Jurassic Periods (ca. 200 million years ago), and deep rift valleys ripped open the continental interior. By the beginning of the **Jurassic Period** (199.6 to 145.5 million years ago), new mid-ocean ridges had displaced water onto the continental margins, flooding the rift valleys and increasing global sea level by as much as 100 meters (328 feet). The resulting climate was very different from the climate in Permian and Triassic times (**Figure 6.10**). Both the atmosphere and the oceans were much hotter than they are today. For most of the Cretaceous, the poles were ice free (some evidence suggests that there may have been temporary ice sheets during parts of the Cretaceous), and although there were

seasonal changes in temperature, the ocean surface temperatures stayed above 0°C (32°F).

The oceans that flooded the continental shelves were biologically productive, with large reefs of mollusks, referred to as **rudists,** and other carbonate rocks dominating tropical sedimentation. The old high mountain chains of the Carboniferous and Permian Periods were extensively eroded by this time, and the only new mountains that were forming occurred at destructive plate margins, similar to the Southeast Asian archipelago today (**Figure 6.11**).

Cretaceous Volcanism

Most proxy data suggest that intense and extensive volcanism during the Cretaceous Period produced levels of CO_2 in the atmosphere that were as much as four to eight times higher than today's. This high concentration of CO_2 would have been even higher if it were not for the extensive deposition of marine carbonate rocks that locked away some of the excess. Vast blooms of coccolithophorid algae alone were able to sequester billions of tonnes of CO_2, leaving hundreds of meters (feet) of chalk deposits on the continental shelf during the late Cretaceous and early Paleogene Periods (Figure 5.9).

▶ **Figure 6.10:** (a) The world of the Cretaceous Period (145.5 to 65.5 million years ago) was very different from the Carboniferous world. The supercontinent Pangaea had broken apart, and the world was starting to look more familiar, despite the fact that sea level was over 100 meters (328 feet) higher than it is today. Look at the position of India and Australia relative to Antarctica, and also note the position of Antarctica relative to the geographic South Pole. (b) Climate proxies for the Cretaceous Period show that cool temperate conditions extended as far as the poles. Some evidence (such as dropstones) suggests that there may have been temporary ice sheets during parts of the Cretaceous.

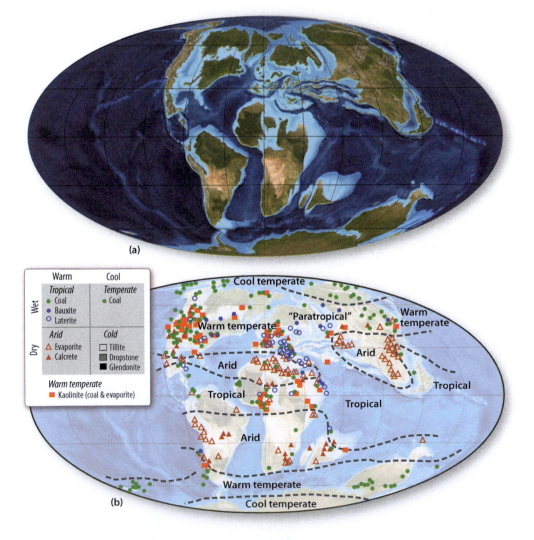

The story of global climate change in the Late Cretaceous

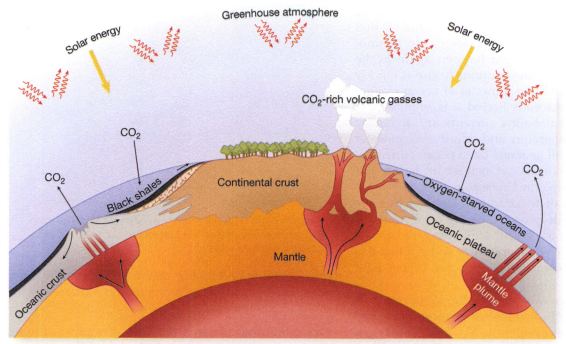

▲ **Figure 6.11:** The Cretaceous world was very different from the world today. There was extensive global volcanism, the level of CO_2 in the atmosphere was so high that global temperatures were 8°C to 10°C (14°F to 18°F) higher than today, and there was no permanent ice at the poles. Deep-water circulation in the oceans was interrupted at times, and black, lifeless, organic-rich clays were deposited in dead zones on the ocean floor.

CHECKPOINT 6.7 ▶ Why was the level of carbon dioxide in the atmosphere during the Cretaceous Period so much higher than it is today?

Cretaceous Atmospheric Temperatures

It is difficult to determine exactly how much warmer the atmosphere was during the Cretaceous Period than it is today. Fossil evidence suggests that global temperature may have been as much as 8°C (14°F) higher, with average Arctic surface temperatures at least 18°C to 20°C (65°F to 68°F) during the summer and dropping to –4°C to 0°C (25°F to 32°F) during the long, dark polar winter. During the warmest periods in the Arctic, the difference between seasonal maxima and minima appears to have been as little as 12°C (22°F)—much less than today.

Some recent research suggests that there may have been more climate variability during the Cretaceous than scientists first thought. Cooling of the atmosphere during the long, dark polar winters may have allowed some ice to form, even during the warmest periods, and there is some evidence for the existence of small Antarctic ice sheets occurring at least twice during the Cretaceous. Oxygen isotope analyses of some foraminifera show excursions to higher levels of $\delta^{18}O$ around 144 million years ago and again at 65 million years ago, suggesting that small polar ice sheets may have developed for as long as 200,000 years each time.

There is, however, no indication of any permanent or extensive polar ice during times when the concentration of CO_2 remained higher than 1,000 ppm.

CHECKPOINT 6.8 ▶ What evidence indicates that limited volumes of polar ice may have been present for at least some of the Cretaceous period?

Carbon Dioxide in the Cretaceous Atmosphere

There is a lot of uncertainty about the exact concentration of CO_2 in the Cretaceous atmosphere. Early estimates based on the study of isotopes from fossil soils suggested concentrations in excess of 2,000 ppm, but more recent studies from the stomata of plants and from isotopic studies of fossil **liverworts** and phytoplankton suggest a maximum as low as 1,000 ppm and a minimum not much higher than today's level.

The lower levels found in more recent studies are easier to reconcile with observed climate variability, but they are not reassuring. They suggest that global climate is much more sensitive to atmospheric CO_2 than many have assumed. If the higher global temperatures that were typical of the Cretaceous actually coincided with levels of CO_2 around 1,000 ppm, then we might expect much more warming over the next 200 to 300 years.

The Cretaceous Oceans

During the Cretaceous Period ocean temperatures rarely dropped below 0°C (32°F), so it is unlikely that the poles were a major source of deep, cold thermohaline circulation. This might happen again in the future if global temperatures continue to rise as quickly as they are today. To build a model of ocean circulation during the Cretaceous Period it is important to remember that thermohaline currents are driven by both changes in temperature and salinity. Warm water at the surface can still generate deep-ocean currents if strong surface evaporation can increase its density. But did this happen during the Cretaceous Period?

Isotope studies of foraminifera suggest that the temperature of the deep ocean during the Cretaceous Period was as high as 18°C (64°F), compared to just 2°C (36°F) today. This temperature is so high that it is unlikely the water originated at the poles. Where did it come from?

Equatorial surface water during the Cretaceous was as warm as 30°C (86°F), and intense evaporation would have produced brines as saline as 37 0/00. These brines were certainly dense enough to sink to the ocean floor and form deep, warm thermohaline currents of tropical origin. A similar process can be observed today in the Mediterranean Sea, where dense surface brines sink to the ocean floor and flow out into the Atlantic Ocean.

It is still possible that some high latitude deep-water currents were generated, especially during long dark polar winters when surface ice may have formed, but these currents must have been less dense than the

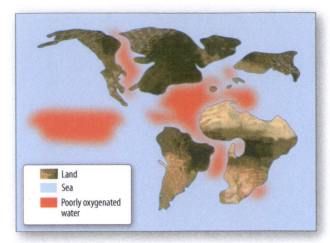

▲ **Figure 6.12:** This schematic map shows where the main dead zones developed on the Cretaceous ocean floor due to poorly oxygenated water.

tropical brines and flowed above their surface in order to explain the high water temperatures indicated by the benthic foraminifera.

Oceanic Anoxic Events (OAEs)   A number of times during the Cretaceous Period, the ocean floor appears to have been starved of oxygen, leading to the development of toxic dead zones that deposited dark, organic-rich shales with little sign of benthic (bottom-dwelling) life. The warm, deep waters of the Cretaceous held much less oxygen than today, and OAEs were probably triggered by changes in the strength of thermohaline circulation that limited the transport of oxygen from the surface to depth (**Figures 6.12** and **6.13**).

▼ **Figure 6.13:** This cliff section of deep-ocean sediment shows the transition between surrounding lighter-colored sediment that was deposited when there was enough oxygen at depth to support life and a thick, dark, organic-rich layer of lifeless rock that was deposited in a dead zone on the Cretaceous ocean floor.

CHECKPOINT 6.9 ▶ What evidence indicates that ocean circulation was sometimes limited during the Cretaceous Period?

Cooling the Cretaceous

Volcanism added so much CO_2 to the atmosphere during the Cretaceous that it pushed Earth into an intense hothouse. However, other factors offset some of this warming, removing large amounts of carbon from the atmosphere and oceans:

- **The loss of carbon at depth.** The organic-rich shales that formed during the extensive OAEs may have helped to cool the planet because they locked away many gigatonnes of organic carbon, preventing it from being recycled into the oceans and atmosphere.
- **The capture of carbon at the surface.** Coccolithophore blooms became so common toward the end of the Cretaceous Period that the tiny coccolithophore shells drew down many gigatonnes of CO_2 from the atmosphere. The high reflectivity of these extensive blooms also increased global albedo.

Working together, these very different forces acted to offset the impact of CO_2 from earlier effusive volcanism, and as volcanism waned toward the end of the Cretaceous Period, the world started to cool. But not for long . . .

Does the Cretaceous Provide a Vision of the Future?

The Cretaceous world was very different from both the hothouse world of the Permian and the icehouse world of today. The continents were dispersed, and high sea levels brought moisture to all but the deepest continental interiors. Australia, South America, and India were joined to Antarctica, and there was no Antarctic Circumpolar Current. North and South America were separated, and ocean currents were able to circulate freely between the Pacific and Atlantic Oceans. Periodic ventilation of the deep ocean by thermohaline currents prevented the runaway development of toxic dead zones as happened during the Permian Period. Both marine and terrestrial ecosystems were complex and diverse, and the first flowers and grasses appeared during this time. The dinosaurs were well adapted to this world, even to the dark, cold polar winters.

If we do nothing to reduce greenhouse gas emissions by the end of the 21st century, the concentration of CO_2 in our atmosphere could reach a level last seen during the Cretaceous Period. Everything is not the same today as it was during the Cretaceous Period, but if we rapidly returned to a Cretaceous-like climate, sea level and global temperatures would rise very much higher than they are today, and the impact on humanity and our familiar environment could be devastating.

The K–T Extinction

The Cretaceous Period came to an abrupt end 65 million years ago, as the dinosaurs and 60% of other species alive on Earth became extinct. Many scientists believe that this extinction was due to the impact of a large meteorite that struck Earth just north of the Yucatan Peninsula in northern Mexico. The boundary between rocks of the Cretaceous Period below and the Tertiary Period (often divided into Paleogene and Neogene Periods) above is known as the **K–T boundary** and has become an icon for mass extinction (**Figures 6.14, 6.15, and 6.16**).

◀ **Figure 6.14:** Every global catastrophe seems to have winners and losers, and who wins and loses often has more to do with luck than with any inherent evolutionary advantage. Last time, mammals won and the dinosaurs lost. What will happen next time? This artist's impression shows the Yucatán Peninsula as it may have looked shortly after the impact of the meteorite with an expanding shock wave in the atmosphere and rising plume of incandescent vaporized rock that would soon encompass the globe and devastate many ecosystems.

A large increase in the proportion of the lighter ^{12}C isotope in rocks above the K–T boundary may show that biological activity decreased significantly after this event, as photosynthesis normally depletes ^{12}C from surface marine environments. However, a lot of ^{12}C was almost certainly released into the atmosphere from sediments vaporized in the impact zone. Despite the enormity of this event on the biosphere, the K–T event was short-lived and is not well recorded in the climate record (Figure 6.16).

CHECKPOINT 6.10 ▶ Why is there not a clear record of climate change following the K–T extinction event?

▲ **Figure 6.15:** The impact site for the K–T event is thought to be at the location of a large buried crater of the correct age that lies offshore of Chicxulub in the Yucatán Peninsula in northern Mexico.

6.4 PAUSE FOR... THOUGHT | *Will a massive extinction happen again?*

For much of geological history, large destructive meteorites have bombarded Earth. It is almost inevitable that another impact the size of the one that hit the Yucatán Peninsula at the end of the Cretaceous will occur again. What would happen if such an impact occurred today? Because Earth is mostly covered in water, the most likely point of collision would be in the ocean, where a large impact would create a tsunami over 300 meters (1,000 feet) high—large enough to wipe out most coastal cities and fill the air with incandescent debris. Do you think the fabric of modern society and culture could withstand such a catastrophe?

The Last Days of "Summer"

The **Cenozoic Era** is our current geological era, and it is subdivided into three geological periods: the **Paleogene Period**, the Neogene Period, and the Quaternary Period (**Figure 6.17**). It has been a time of rapid environmental and ecological change, with the evolution of mammals, grasses, and other flowering plants that have transformed the landscape and many ecosystems.

The climate of the Cenozoic was initially warmer than during the late Cretaceous due to renewed volcanism, but temperatures gradually declined toward present levels as Earth moved from a Mesozoic hothouse to a Cenozoic icehouse (**Figure 6.18**). (We will keep referring to Figure 6.18 many times over the following section, so you might want to bookmark page 175.)

▼ **Figure 6.16:** The thin layer that marks the K–T boundary belies its fundamental importance for humanity. From this point on, mammals were in the ascendency, but at the price of 60% of all living species at the end of the Cretaceous Period.

Eon	Era	Period		Epoch	Start date (*mya*)
Phanerozoic	Cenozoic	Quaternary		Holocene	0.01
				Pleistocene	2.58
		Tertiary	Neogene	Pliocene	2.58
				Miocene	23.3
			Paleogene	Oligocene	35.4
				Eocene	56.5
				Paleocene	65

▲ **Figure 6.17:** The Cenozoic Era.

Too Hot for Comfort: The Paleocene Epoch

The **Paleocene Epoch** (65.5 to 55.8 million years ago) was a time when greenhouse gases were released naturally into the oceans and atmosphere, triggering rapid global climate change. By studying how Earth's climate system responded to greenhouse gases in the Paleocene, we may be able to project how global climate will respond to the release of anthropogenic greenhouse gases today. The critical boundary between the Cretaceous and Paleogene Periods is marked by the sudden faunal changes that occurred at the K–T extinction boundary and not by a sudden change in climate. Earth remained a hothouse world during the Paleocene Epoch due to sustained volcanism that continued to add CO_2 to the atmosphere. Toward the end of the Cretaceous Period, enormous volumes of basaltic lava had erupted in northern India,

and this was followed during the Paleocene by extensive volcanic activity as the North Atlantic Ocean opened between Greenland and Europe (**Figure 6.19**).

This sustained emission of CO_2 from volcanic activity was periodically enhanced by the further addition of CO_2 and CH_4 to the atmosphere by volcanic activity that baked carbon-rich coals and shales in sedimentary basins around the North Atlantic. Yet more CO_2 was probably released into the atmosphere from limestone rocks being buried and baked (metamorphosed) when the Indian Tectonic Plate started to slide beneath southern Asia (**Figure 6.20**).

CHECKPOINT 6.11 ▶ How did major tectonic events at the start of the Paleogene Period have a significant impact on the level of carbon dioxide in the atmosphere?

The Paleocene–Eocene Thermal Anomaly (PETA): A Vision of the Future?

Rising global temperatures reached a maximum 55.5 million years ago, at the boundary between the Paleocene and Eocene Epochs. This period of time is particularly interesting to climate science, as proxy evidence indicates that greenhouse gases were rapidly but naturally released into the atmosphere and had a catastrophic impact on global temperature (Figure 6.18 (a)). Some scientists believe that the **Paleocene–Eocene Thermal Anomaly (PETA)**, also known as the Paleocene-Eocene Thermal Maximum (PETM), is a close geological analogy to what is happening today with the release of anthropogenic greenhouse gases. If they are correct, the prognosis is not good.

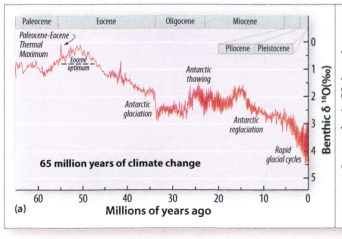

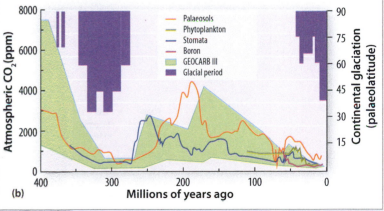

▲ **Figure 6.18:** (a) The climate at the start of the Cenozoic was initially warm, but by the middle Eocene had started to cool, interrupted by later periods of rapid climate change that can be tentatively related to changes in the position of the continents and major ocean currents. Temperature is represented here by measured changes in the oxygen isotope composition of benthic foraminifera (see Chapter 5). Note the position of the Paleocene-Eocene thermal maximum at 55.5 million years ago. (b) These data on the concentration of CO_2 in the atmosphere are compiled from a range of climate proxy sources, including observational data and a theoretical model (Geocarb III) that give various estimates of the levels of carbon dioxide in the atmosphere as far back as the Carboniferous Period. The blue bars descending from the top of the diagram mark periods of the Permo-Carboniferous glaciation and Quaternary (current) glacial advance. The scale on the right side indicates the maximum latitude of ice advance so that the further down the diagram the bars descend, the greater the degree of glacial advance at that time. Note the degree of uncertainty in the GEOCARB estimates indicated by the green shaded area.

▶ **Figure 6.19:** The famous basalt columns of the Giant's Causeway in Northern Ireland are part of the large North Atlantic Volcanic Province that developed during the Paleogene as Greenland broke away from Europe.

▶ **Figure 6.20:** (a) An artist's reconstruction of how the world might have looked around the time of the K–T event. Note the northward movement of India and the fall in global sea level when compared to Figure 6.10. (b) Climate proxy indicators for the Paleocene. Note the presence of coal deposits from temperate forests that reached almost as far as the North Pole.

(a)

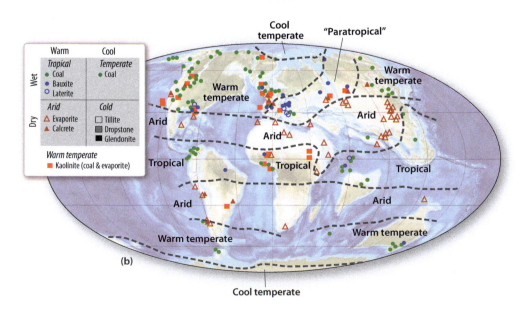

(b)

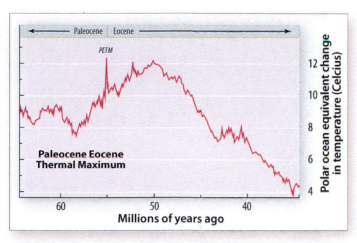

▲ **Figure 6.21:** A series of short, rapid changes in temperature during the late Paleocene culminated 55.5 million years ago with the PETA. The rate of temperature change associated with this thermal anomaly was high and appears as a spike in the climate record. This figure presents an estimate of polar ocean temperature based on the measured oxygen isotope ratios of foraminifera recovered from ocean cores.

Changes in Global Temperature During the PETA

Foraminifera recovered from marine cores that sample the short PETA time interval have $\delta^{18}O$ levels and Mg/Ca ratios that record a sudden and global rise in temperature of greater than 6°C (11°F) in less than 10,000 years. Ocean temperatures increased by as much as 4°C to 5°C (7°F to 9°F) in the tropics, 6°C to 8°C (11°F to 14°F) at mid-latitudes, 9°C (16°F) at high latitudes, and 4°C to 5°C (7°F in 9°F) in the deep oceans (**Figure 6.21**).

CHECKPOINT 6.12 ▶ What geological evidence points to a very significant warming event at the boundary between the Paleocene and Eocene Epochs?

The Impact of Ocean Acidification

Marine cores record a major change in sedimentation at the time of the PETA. The sediment both above and below the PETA is full of light-colored carbonate shells that are typical of marine sedimentation at this depth, but a layer of dark brown, organic-rich clay marks the actual boundary. The presence of clay and absence of shells here suggests that the oceans may have become too acidic for carbonate deposition and that any shells that fell from the surface were dissolved in the water before they reached the ocean floor (**Figure 6.22**).

CHECKPOINT 6.13 ▶ What geological evidence suggests that the oceans absorbed a lot of carbon dioxide at the time of the PETA?

A Record of Extinction Events

The sudden release of heat-trapping greenhouse gases during the PETA affected changes in atmospheric temperature and ocean chemistry that resulted in major changes to the biosphere. **Marine benthic extinction events (MBEs)**

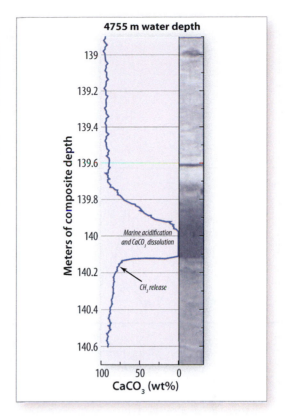

▲ **Figure 6.22:** Deep marine cores sampling the interval of the PETA show a change from light-colored, carbonate-rich sediment that formed beneath well-oxygenated water to dark brown, organic-rich clay sediment that formed beneath water that was starved of oxygen.

caused the widespread disappearance of many species of foraminifera; at the same time, changes in terrestrial ecosystems led to the rapid evolution of mammals. In some ways, we owe our current existence to the evolutionary events that took place at this time.

The Origin of the PETA

The combination of elevated temperature and increased ocean acidity is best explained by the rapid release of heat-trapping greenhouse gases into the atmosphere. The initial rise of CO_2 in the atmosphere was probably a result of sustained volcanism, but two short 1,000-year thermal pulses staged just 20,000 years apart are marked by large *negative* shifts of −2.5% in the $^{13}\delta C$ values recovered from fossil foraminifera. These data are best explained by the sudden influx of ^{12}C of organic origin into the biosphere (**Figure 6.23**).

There are two possible sources for the lighter ^{12}C isotope, and both involve methane. Hot magma injected into rich coal deposits along the North Atlantic continental margin could have baked the coal and released CH_4 (coal gas) into the atmosphere; the existence of vertical gas vents above coal measures intruded by hot magma supports this theory. It is also likely that large volumes of organically derived CH_4 were released by the destabilization of methane hydrates as the oceans warmed. Climate models estimate that methane hydrate

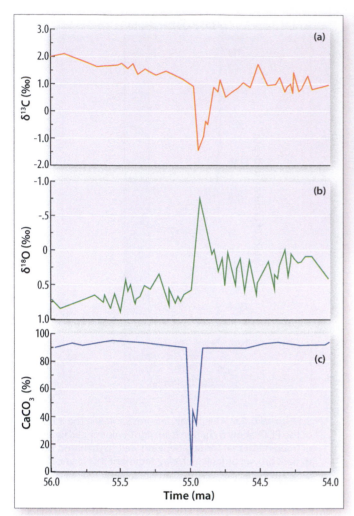

▲ Figure 6.23: During the Paleocene–Eocene Thermal Anomaly, CO_2 levels in the atmosphere rose rapidly, along with average global temperature. This diagram shows that a major isotope excursion event (rapid change) is recorded in the marine sediment record from this time. (a) A strong negative anomaly (decrease) in $\delta^{13}C$ reflects an increase in the fraction of ^{12}C in surface waters. (b) The $\delta^{18}O$ record from foraminifera also indicates a concurrent sudden rise in temperature of up to 5°C (9°F). (c) The $CaCO_3$ content of sediments deposited over this interval reflects a relative scarcity of shell residue when temperatures reached a maximum. This was most likely due to the sudden release of a large volume of biogenic gas from destabilized methane hydrates or from coal beds that are baked by hot magma that eventually increased the acidity of the oceans.

▲ Figure 6.24: These rising bubbles show where methane is escaping from hydrates beneath the sea floor. These small leaks may forecast a more catastrophic release of methane in the near future if the temperature of the Arctic Ocean continues to rise.

Climate Sensitivity During the Paleocene-Eocene Climate models suggest that our current estimates of climate sensitivity to CO_2 are too low to explain the changes in global temperature that took place during the PETA. Unrecognized sources of feedback in the Paleocene climate system must have been capable of multiplying the impact of smaller changes within Earth's climate system. *If this feedback is still active today, our estimates of future temperature may be much too low.* It is vitally important that scientists identify more of the factors involved in this process. As discussed in some detail in Chapter 3, there may be other sources of feedback in the climate system that could act over timescales longer than we have been able to observe. The PETA is a clear warning to the global community about what can happen when global climate is rapidly disturbed and a stark indicator of the importance of understanding the sensitivity of modern climate to increasing levels of CO_2.

Implications for Climate Change Today The PETA is a good example of how the geologically sudden release of greenhouse gases into the atmosphere can cause rapid climate change. It was the largest of a number of similar events that took place during the late Paleocene to early Eocene. All those events were associated with acute climate instability. The main temperature and climate excursion took just over 10,000 years to develop, but global climate did not return to an unperturbed state for another 200,000 years.

Methane hydrate instability appears to occur regularly in Earth history, and it may become a climate hazard again if ocean temperatures continue to rise. The PETA is a sobering warning about the long-term implications of modern climate change because the current rate of release of greenhouse gases into the atmosphere is faster than it was during PETA events.

was stable at depths greater than 920 meters (3018 feet) during the Paleocene but would have been destabilized by both rising temperature and falling sea level.

It is not easy to discriminate between these two potential sources of CH_4, but it is really unlikely that one source alone was responsible. Whatever the source of the carbon, each high-temperature event was accompanied by an increase in CO_2 by as much as 70% over background levels, with organic carbon isotope ratios reaching as low as –4% to –6%. These data require the release of at least 2,000 gigatonnes of carbon into the atmosphere (**Figure 6.24**).

6.5 PAUSE FOR... THOUGHT | *Rolling Nature's dice.*

Both the end-Cretaceous and Paleocene–Eocene extinction events created opportunities for new species to evolve. If the Cretaceous event had not occurred, mammals may have forever languished in the shadow of the dinosaurs. If the Paleocene-Eocene event had not occurred, the great grazing herds that migrated across new grasslands and fed our forefathers may not have existed. So is extinction always a bad thing for the natural world? Hominids have been on Earth for a few million years, while the dinosaurs ruled for tens of millions. We probably have some time left in our prime position, but our long-term prognosis is not good.

The Great Global Cooling

This section considers how natural processes drove global climate from a hothouse in the early Paleocene Epoch to an icehouse in the Oligocene. Over this time, large ice sheets developed over the Antarctic, and glaciers started to form on Greenland. High-resolution isotope studies of shells recovered from deep-sea cores show how Milankovitch cycles controlled the advance and retreat of these ice sheets over millennia and how deep-water circulation played a critical role in modulating global climate. These processes are still at work today.

Following the end of the PETA, continuing volcanism in the North Atlantic pushed global temperature to a maximum more than 6°C (11°F) warmer than the temperature today. This was a world of rapid evolution, a world still building on the unsettled foundations of the K–T ecological disaster. The first extensive grasslands started to transform the landscape, and mammals evolved to claim many of the ecological niches left vacant by the dinosaurs. But just as the biosphere had adapted to these new evolutionary challenges, the climate started to change once more.

What Turned Down the Thermostat?

Starting around 52 million years ago, factors that had sustained a hothouse world for so long changed direction, and Earth started to cool rapidly. Very different feedback mechanisms set Earth on a course of sustained cooling that continues to the present day.

As you have seen, many factors can bring about a dramatic change in global climate. Some are **external factors** in the climate system, such as changes in solar irradiance, the impact of orbital cycles, volcanic activity, and the movement of tectonic plates. Others are **internal factors** that result from natural feedback processes, such as changes in vegetation, rainfall, ocean circulation, and water vapor and other natural greenhouse gases.

Study the following list of factors that changed between the Paleogene and Neogene Periods and try to determine which factors are external and which represent internal feedback:

- The configuration of the continents changed as South America and Australia broke away from Antarctica. The Indian subcontinent started its protracted collision with southern Asia.
- A large continent was once more in position over the South Pole.
- Continents crowded around the North Pole, restricting the flow of warming ocean currents.
- New mountains formed—the Himalayas, the Rockies, the Alps, and the Andes—and rapid erosion associated with their uplift removed more CO_2 from the atmosphere.
- New pathways for deep-ocean currents opened, changing the way that energy was distributed from the tropics to the poles.
- The rate of seafloor spreading decreased, and sea level fell.
- Volcanism was less active and reduced the input of CO_2 into the atmosphere.
- Falling sea level exposed more land to erosion and replaced warm seas on the continental shelf with land that cooled quickly during the winter season.
- The higher albedo of land (~14% higher than the oceans) increased the amount of energy reflected back out to space, and as soon as ice started to form, the higher albedo of ice and snow reflected even more energy.
- As Earth cooled, more methane was trapped on the continental shelf and under newly frozen tundra.
- Organic carbon was also trapped under tundra and isolated from the atmosphere.
- The cooling atmosphere could hold less H_2O, a potent greenhouse gas.
- The cooling oceans could remove more CO_2 from the atmosphere.
- Less water vapor in the atmosphere meant that less latent heat was transferred from the tropics to the poles.

The first eight factors on this list are *primary drivers* of climate change resulting from major changes in the Earth system. The last six factors indicate *feedback* in the climate system resulting from the primary changes. An increase in albedo, for example, means that Earth reflects more energy from the Sun, making Earth colder. A colder Earth has more ice and snow, which increases albedo even more. This is a positive feedback loop that accelerates the rate of change. Similarly, the other factors at the end of this list generate positive feedback.

These changes knocked Earth's energy budget out of equilibrium. Even more heat was lost at the poles than was gained at the equator, and net global cooling started to take effect. The slow development of ice sheets, first around Antarctica and then later in the Northern Hemisphere, began the onset of the current icehouse phase of Earth history.

> **CHECKPOINT 6.15** ▶ What major factors were responsible for the development of the overall global cooling trend between the Paleogene and Neogene Periods?

The First Chills: The Eocene Epoch

Global temperatures first started to fall from their peak around 52 million years ago, following a peak in global temperature known as the Eocene Climate Optimum. **Continental rifting** in the North Atlantic slowed, and the amount of volcanism decreased, causing the concentration of CO_2 in the atmosphere to fall from around 1,000 ppm at the start of the Eocene to less than 600 ppm by the end. As the average rate of sea floor spreading also decreased, the mid-ocean ridges subsided, and sea level started to fall, much as the water level changes as you get out of the bath (**Figure 6.25**).

The **Eocene Epoch,** which occurred 55.8 to 33.9 million years ago, was also a time of rapid mountain building. The Himalayas, Andes, Rocky Mountains, and Alps all started to form around the same time, and extensional rifts started to separate the continents that surrounded Antarctica.

These new mountains grew into an atmosphere that was still much warmer and more humid than the atmosphere today, and high levels of rainfall and associated erosion started to reduce the amount of CO_2 in the atmosphere. A higher level of strontium (Sr) in marine shells around 40 million years ago is a clear sign of this increased rate of continental erosion (**Figure 6.26**).

By the end of the Eocene, the world had cooled considerably. There is even some geological evidence that small ephemeral ice sheets existed on Antarctica as early as 38 million years ago. Further rapid cooling of Antarctica occurred during the Oligocene, as the continent became progressively isolated at the southern pole and cut off from the warm inflow of tropical energy.

> **6.6 PAUSE FOR... THOUGHT** | *Why are some periods of Earth history more mountainous than others?*
>
> Mountains are most common in geological history when tectonic plates converge and continents collide, a process known as orogeny. Earth was mountainous prior to the great Permo–Carboniferous glaciation, when the continents converged to form the global supercontinent called Pangaea. Earth is again mountainous today, as India and Africa continue to collide with southern Asia and Europe and ocean crust is consumed beneath the western seaboard of the United States of America. The Alps, Himalayas, Andes, Rocky Mountains, Cascade Mountains, Sierra Nevada Mountains, and Colorado Plateau are all products of tectonic plate convergence. This process is likely to continue and intensify over millions of years as the continents slowly converge to form a new global supercontinent.

The First Ice Sheets: The Oligocene Epoch

The **Oligocene Epoch,** which occurred 33.9 to 23.03 million years ago, was marked by a rapid fall in the level of CO_2 in the atmosphere to as low as 300 ppm. Temperatures close to the North Pole fell as low as 0°C (32°F), and Antarctica became much colder as large ice sheets started to form, lowering sea level by more than 50 meters (164 feet). Oxygen isotope data show that the Oligocene oceans became richer in ^{18}O (with more ^{16}O locked up in ice), and sea surface temperatures were

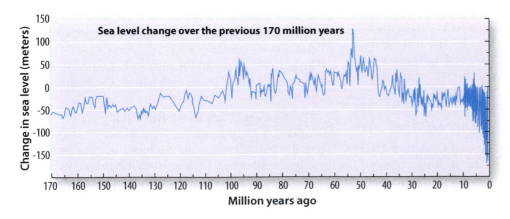

▲ **Figure 6.25:** Global sea level from 170 to 36 million years ago was controlled by the rise and fall of the mid-ocean ridges. Since then, the most important factor appears to have been the formation of major continental ice sheets.

◀ **Figure 6.26:** The Mont Blanc range near Chamonix, French Alps. High mountain ranges, such as the Alps and Himalayas, interrupt the flow of air in the atmosphere and increase the level of precipitation and erosion, helping to reduce the amount of CO_2 in the atmosphere. The high Himalayas also made it easier for orbital changes to enhance the monsoons of Asia.

3°C to 4°C (5°F to 7°F) cooler than they were during the Eocene. The Oligocene did not have the ephemeral sheets of the Eocene; rather, the large continental ice sheets of that time may have even exceeded the size of the ice sheets today. Shorter periods of sharply colder climate around Antarctica were probably driven by Milankovitch cycles. The first dropstones in North Atlantic sediments are from this time, suggesting that some ice had also taken hold on Greenland in the Northern Hemisphere.

The Final Breakup of the Southern Continent

The continents of South America and Australia had started to move away from Antarctica during the Eocene, forming large rift valleys around the continental margins. By the late Eocene and early Oligocene, the sea had started to flood these rifts, and by 30 million years ago, both South America and Australia had broken free from Antarctica, forming the Drake Passage and Tasman Straits. A proto-Antarctic Circumpolar Current soon formed as these rifts deepened and widened. As Antarctica became isolated from warmer tropical currents, the ocean temperature around the continent dropped by a further 10°C (18°F) (**Figures 6.27 (a)** and **6.27 (b)**).

The Antarctic Ice Sheet was not fully stable throughout the Oligocene. Sediment cores show that there were multiple transient periods of glacial advance and retreat related to Milankovitch cycles. Data from leaf stomata even suggest that, by the end of the Oligocene, the level of atmospheric CO_2 had increased again to as much as 400 ppm to 500 ppm due to sustained volcanic activity (possibly in Europe).

CHECKPOINT 6.16 ▶ When and why did ice sheets first start to form on Antarctica?

Changing Oligocene Ocean Circulation

From the Oligocene to the early Miocene, deep-water circulation around the globe was dominated by thermohaline flow from the Antarctic. The ocean floor around Antarctica warmed slightly at the end of the Oligocene, as the reorganization of deep-water currents allowed some warmer waters from the north to encroach southward. Sediment cores from the Atlantic seafloor record no evidence of bottom water from the North Atlantic or Arctic Basins at this time, as all the cold polar water was trapped in a fresh-to-brackish lake that was isolated from the rest of the Atlantic Ocean (**Figure 6.28**).

Antarctic Ice Sheet Instability

The Oligocene Epoch ended with a major advance of the Antarctic Ice Sheet that was probably caused by changes in Earth's orbital eccentricity and obliquity and a decrease in global volcanic activity. The stability of the Antarctic Ice Sheet during this time is of great interest. Global temperatures were as much as 4°C (7°F) higher than today, but they were within the temperature range predicted for the end of the 21st century. If the ice sheets became unstable at these temperatures during the Oligocene, it is possible that they would become unstable again in the near future, and sea level would then rise by as much as 6 to 72 meters (20 to 236 feet). Further research is required to determine how much and how quickly these ice sheets melted during interglacial periods in the Oligocene.

CHECKPOINT 6.17 ▶ Why is the stability of the Antarctic Ice Sheet during the Oligocene Epoch of special interest today?

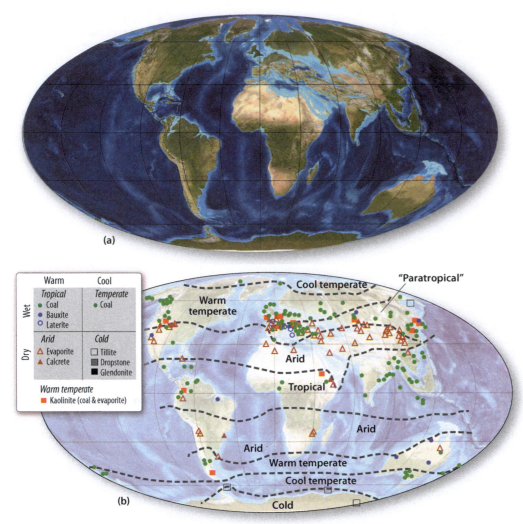

▲ **Figure 6.27:** (a) The world of the Oligocene. Note the positions of Australia, India, and South America. (b) Climate proxy indicators for the Oligocene. Note the cold climate indicated in Antarctica and the cool temperate conditions in the far north.

6.7 PAUSE FOR... THOUGHT | *Antarctica: A cold window on the world?*

The continent of Antarctica has been close to the South Pole for hundreds of millions of years. At the moment, all the surrounding continents are moving away, leaving Antarctica isolated in its current position over the pole. The Antarctic landmass loses energy to space much faster than does the open ocean during the long, cold polar winter. As long as Antarctica stays in its current position over the South Pole, the world will remain in a climatic icehouse that is interrupted by only short periods of limited warming at high latitudes, as the intensity of the Sun changes with variations in Earth's orbit.

▶ **Figure 6.28:** During the Oligocene, the Arctic Ocean was a large, deep, freshwater lake largely isolated from the North Atlantic Ocean by a narrow, shallow strait, the Barents Sea Gateway.

The Neogene Period

This section explores the **Neogene Period** and shows how different factors that affect global climate worked together to bring about change. The Neogene Period includes the Miocene and Pliocene Epochs. From the mid-Miocene until the present, the temperature of Earth has fallen rapidly, interrupted only by brief warm interglacial periods modulated by Milankovitch cycles.

Passing the Precipice: The Miocene Epoch

During the **Miocene Epoch,** which occurred 23.03 to 5.33 million years ago, CO_2 levels in the atmosphere remained as high as 450 ppm to 650 ppm. Sea level was around 25 meters (82 feet) higher, and global temperatures were as much as 3°C to 4°C (5°F to 7°F) higher than they were during the Oligocene. Although the Miocene was warmer than today, it was still an icehouse world, with permanent ice sheets present on Antarctica. The proto-Antarctic Circumpolar Current was well established, and geochemical and isotope data suggest that cold water

from the Antarctic ice shelf fed bottom water that flowed out along the floor of the Pacific, Atlantic, and Indian Oceans (**Figure 6.29 (a)**).

Glaciers probably existed on Greenland at this time, but there is no evidence of major ice sheet development. Fossil evidence in the North Atlantic shows that the Arctic Ocean was still a large, isolated freshwater lake prior to the opening of the Fram Strait between Greenland and Iceland (**Figure 6.30**). Deep-ocean cores demonstrate that free circulation between the Arctic and Atlantic did not start until 17.5 million years ago, during the mid-Miocene, and full exchange may not have developed until as late as 12 to 13 million years ago, when cold Arctic water finally began to spill over into the North Atlantic as proto-Atlantic Deep Water (Figure 6.30).

CHECKPOINT 6.18 ▶ Why did Arctic waters make little contribution to global circulation in the oceans from the Oligocene to the early Miocene?

Rapidly Falling Temperature During the Miocene

Around 14 million years ago, the temperature fell rapidly from a Mid-Miocene maximum (Figure 6.18). It is not yet clear why this happened. There are many

(a)

◀ **Figure 6.29:** (a) During the Miocene, the world underwent significant global cooling. India had collided with southern Asia. The proto-Antarctic Circumpolar Current was well established, and cold water from the Antarctic Ice Shelf fed Antarctic bottom water into the Pacific, Atlantic, and Indian Oceans. (b) Proxies for climate during the Miocene. Note that cold conditions are indicated at both poles.

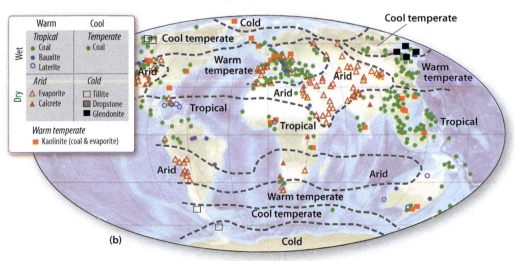

(b)

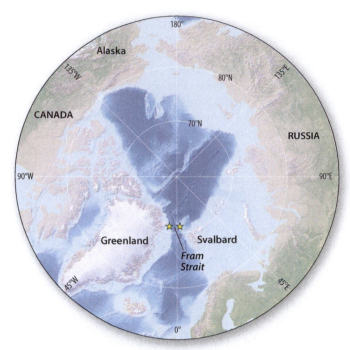

Figure 6.30: Once the Fram Strait opened as late as 12 to 13 million years ago, cold water from the Arctic Ocean was able to circulate freely in the Atlantic Ocean. Modern geographical names are added for reference.

competing factors to consider, all of which probably had some impact:

- Earth's orbital configuration was at a stage when both eccentricity and obliquity were low, a configuration that can favor the growth of ice sheets.
- The growth of the Alps, Andes, Himalayas, and Tibetan Plateau would have changed atmospheric circulation and continued to draw down the level of CO_2 in the atmosphere.
- Deep-ocean and bottom water began to circulate from both the Arctic and Antarctic. This represented a significant reorganization of oceanic circulation and would have changed the meridional (north–south) heat flux, as it displaced warmer intermediate waters from the vicinity of the current Mediterranean Sea.
- A large, positive ~1‰ excursion in $\delta^{13}C$ over this period suggests that a lot of $\delta^{12}C$ was buried in organic matter.
- It is possible that explosive volcanism in the United States and in central Europe caused transient cooling due to the ejection of sulphate aerosols high into the stratosphere, where they increased global albedo.

With so much cooling feedback in the climate system, it is unlikely that any one factor dominated temperature change during the Miocene.

The Appearance of Grasslands   The Miocene was a period of significant change on land. This is the first time in evolutionary history that grasslands started to replace more open woodland. Grasses are more tolerant than trees of lower levels of CO_2, and their widespread appearance indicates a climate that was cooler and drier, with less CO_2 in the atmosphere. The appearance of grasslands had a profound impact on the evolution of the mammals, as new species evolved to make use of this new and abundant source of food.

CHECKPOINT 6.19 ▶ What do botanists deduce about changing climate from the first widespread appearance of grasslands during the Miocene? What impact did the appearance of grasslands have on the evolution of mammals?

Back to the Future: The Pliocene Epoch

The **Pliocene Epoch,** which occurred 5.33 to 2.58 million years ago, is of special interest to climate scientists because the arrangement of the continents and oceans then looked very much as it does today. The Pliocene is thus an excellent model for how climate may change in the future if we continue to add greenhouse gases to the atmosphere. During the early Pliocene, global temperatures were 2°C to 3°C (4°F to 5°F) higher than they are today, and the level of CO_2 in the atmosphere had fallen to around 300 ppm to 400 ppm, which is similar to the level today (see Figure 6.18). The East Antarctic Ice Sheet was stable, but the West Antarctic Ice Sheet underwent periodic collapse during warmer periods. Sea level was still at least 15 to 25 meters (50 to 82 feet) higher than it is today, as there was no permanent ice sheet on Greenland, and the Arctic had only seasonal ice. During the summer, the North Pole was ice free, and in the absence of any ice, the average temperature in Greenland may have been as much as 12°C (22°F) higher than it is today. Higher temperatures around the world meant that there was much more moisture in the atmosphere than there is today; it also appears that there was much less continental aridity.

Turning on the North Atlantic Current   The Isthmus of Panama that separated North and South America finally closed during the early Pliocene. This blocked the flow of mass and energy between the Pacific and Atlantic Oceans, directing a flow of warm tropical water northward toward the Arctic. The warm western boundary currents that today form the Gulf Stream and North Atlantic Current were enhanced by this flow, and together they started to transfer more heat and moisture to the Arctic regions. The geography of the world as we recognize it today had finally arrived (Figure 6.31).

CHECKPOINT 6.20 ▶ How would the emergence of the North Atlantic Current have affected the Arctic climate?

A Permanent El Niño?   Oxygen isotope data from the eastern Pacific show that there was little upwelling during the Pliocene and that warm surface water kept the thermocline at depth. A "permanent El Niño" may have existed, and it seems that there was an enlarged pool of warm surface water over the eastern Pacific.

Growth of the Northern Ice Sheets   Some evidence suggests that an enhanced flow of warm water in the North Atlantic generated more snow over the uplands

Chapter 6 Climate History 185

▲ **Figure 6.31:** The Isthmus of Panama closed during the early Pliocene. This blocked the flow of mass and energy between the Pacific and Atlantic Oceans and directed the flow of tropical water in the Atlantic toward the Arctic.

of Greenland, Scotland, and Scandinavia during the winter. This snow accumulated to form the northern ice sheets. Climate models have difficulty reproducing this effect, so there may be important factors missing in the analysis. It is possible that enhanced precipitation slightly decreased surface salinity in the Arctic Ocean, allowing the oceans to freeze more easily. Ice would have isolated the atmosphere from the warming effect of the ocean and increased albedo, causing more cooling, the reverse of the trend we observe today.

By the end of the Pliocene, an ice sheet had developed over Greenland, and dropstones in both the Arctic and the North Pacific suggest that ice shelves had also developed. Oxygen isotope data show a significant increase in marine $\delta^{18}O$, as more of the lighter $\delta^{16}O$ was locked up in continental ice around both poles.

The Emergence of Mankind In Africa, the area covered by forest continued to be replaced by more grassland and savannah. Deforestation in East Africa was an important factor in the evolution of our own species. The first human-like fossils date back to the late Miocene and early Pliocene. These changes were partly due to changes in global climate as the gradual cooling of the Indian Ocean reduced precipitation, while uplift of the East African Rift created a rain shadow that decimated the forests and forced our distant ancestors out onto grasslands that were already well developed by the Miocene. This transition from forest to grassland would have favored the upright stance of early hominids, and the abundant herds of grazing animals provided a reliable source of protein for brain development (**Figures 6.32** and **6.33**). The climates in these areas would also have varied over a 20-thousand-year and longer period as the monsoons varied due to Milankovitch orbital forcing.

CHECKPOINT 6.21 ▶ How did all the changes in climate and vegetation during the Pliocene affect the evolution of humans?

A World Like Our Own After the early Pliocene, the configuration of the continents and patterns of atmospheric and oceanic circulation were much the same as they are today. All the major climate cycles that emerge from interaction between the oceans and the atmosphere were also very similar. The level of CO_2 in the Pliocene atmosphere was close to the level we now predict for the end of the 21st century, and global temperatures were well within the range that climate models predict. The Pliocene can thus help us predict the impact of future climate change.

If we do not limit the current production of greenhouse gases, the transition from an ice-free Arctic at the start of the Pliocene to an ice-bound Arctic by the end of

▼ **Figure 6.32:** The Miocene landscape changed rapidly as the climate became generally colder and more arid. Grassland replaced forest in many areas, and this had a profound impact on the evolution of mammals, including humans. This scene recreates the Miocene landscape of North America.

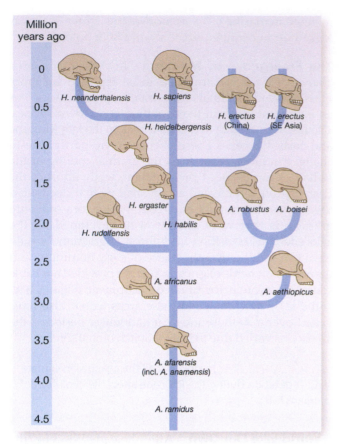

Million years ago

0	H. neanderthalensis H. sapiens
0.5	H. heidelbergensis H. erectus (China) H. erectus (SE Asia)
1.0	
1.5	H. ergaster A. robustus A. boisei
2.0	H. rudolfensis H. habilis
2.5	A. africanus
3.0	A. aethiopicus
3.5	A. afarensis (incl. A. anamensis)
4.0	
4.5	A. ramidus

▲ **Figure 6.33:** Around 4 million years ago, during the Pliocene, the first human-like species developed. Rapid climate change and habitat change would have favored the natural selection of intelligence. First letters are the abbreviations of different genera. Oldest is Ardipithecus *ramidus*; in all other cases, A is the abbreviation of "Australopithecus" and H the abbreviation of "Homo".

the Pliocene will be reversed. This is not reassuring. The Greenland Ice Sheet seems almost certain to melt over time, raising sea level by 6 meters (20 feet), and although the Antarctic ice sheets are likely to remain stable (though there is some question about the West Antarctic Ice Sheet), the total loss of ice would be enough to raise sea level by 25 meters (82 feet) although the full process would probably take well over a thousand years.

Early climate models treated ice sheets as if they were large static blocks of ice that would melt only slowly. More realistic computer models show how instabilities within ice sheets can lead to their catastrophic collapse over a much shorter time scale than initialy thought. The Pliocene world was certainly not the climate disaster some environmental catastrophists predict for our future, but it was a very different world that favored the evolution of our ancestors. Another such change in climate might force further social and economic evolution of our own society.

CHECKPOINT 6.22 ▶ The Pliocene world was warmer than the world today, and it does not appear to have been a time of unusual environmental stress. Why should we be afraid of similar global temperatures in the near future?

6.8 **PAUSE FOR... THOUGHT** | *Ice cubes or sand castles?*

Early climate models assumed that ice sheets would melt like a block of ice sitting on a table. More recent research suggests that ice sheets are more like sand castles, where the addition of water (meltwater) can lead to rapid structural collapse.

The Quaternary Period

The **Quaternary Period** (including the Pleistocene and Holocene Epochs) began 2.58 million years ago and continues up to the present day. It is characterized by as many as 50 periods of high-latitude cooling, punctuated by shorter periods when global temperatures were as high as or even higher than the temperatures of today. These cycles are set to continue into the foreseeable future, as the ice sheets advance and retreat with the incessant rhythms of Milankovitch orbital cycles.

The Quaternary is our recent past, present, and future, and its many important and immediate lessons can guide us as we plan for the future. Not so long ago, during the last glacial maximum New York was covered by over 1,000 meters (3,281 feet) of ice, Chesapeake Bay was a river valley that extended far out onto the continental shelf, and in Europe, Neanderthals struggled to survive on the margins of great ice sheets. Given the momentous changes in climate that have taken place over the past 25,000 years, it is naïve to think that the climate we enjoy today is here to stay.

In the future we face a natural and inevitable return to the icehouse world of the Pleistocene unless controlled global warming (quite possibly an oxymoron) is able to maintain global climate as it is today. Can we learn to limit and control emissions at a level that can stabilize climate and prevent either of these extremes from developing? Greenhouse gases may turn out to be the cheapest and most readily available agent of climate change that future generations can use to prevent the next ice sheets from pulverizing Central Park (once again).

A Story of Ice: The Pleistocene Epoch

The geological history of the **Pleistocene Epoch**, which occurred 2.58 million to 12,000 years ago, has unfolded in more detail than any other period of Earth history because of the remarkable annual record of climate change recorded in ice sheets and glaciers around the world. We used to think of the Pleistocene as a time when both hemispheres lay in the stable grip of ice and cold, punctuated by brief (~10,000-year) interglacial periods when Earth's climate became at least as warm as it is today. It turns out that the climate was anything but stable during the cold periods of the Pleistocene and that rapid swings in climate occurred regularly every few thousand years. The

◀ **Figure 6.34:** During periods when the ice sheets advanced, the Pleistocene world was cold and inhospitable in the Northern Hemisphere, with extensive ice sheets covering much of Europe, Canada, and the northern United States. Ice sheets were at their maximum extent as shown here for around 10% of the time, but the climate warmed and cooled at other times in rhythm with Milankovitch orbital cycles.

detailed 800,000-year record of climate recovered so far is a warning to climate scientists and policymakers alike. It shows clearly that small changes in climate forcing can initiate rapid climate change due to feedback from the cryosphere. As large areas of the globe are still covered by ice sheets and sea ice today, there is a significant risk that small changes in radiative forcing from greenhouse gases could become amplified in the cryosphere and cause rapid and damaging climate change.

The beginning of the Pleistocene Epoch marked the onset of major glaciation in the Northern Hemisphere. At their maximum extent, ice sheets in both hemispheres expanded to cover over 30% of Earth's land surface. New ice sheets, developed in North America, Europe, and Siberia, joined the existing Greenland Ice Sheet that had started to form as far back as the Pliocene. Large, thick ice sheets developed over both eastern and western Antarctica, and at sea, wide ice shelves formed and buttressed the flow of ice from the continental interior (**Figure 6.34**).

Falling Pleistocene Temperatures The Pleistocene was a chilly time. When the ice sheets were at their maximum extent, the average global temperature was at least 5°C (9°F) lower than it is today. Surface temperatures in the Arctic were 9°C (16°F) colder than today, and the Antarctic was as much as 21°C (38°F) colder than today. Even in the tropics, sea surface temperatures were 2°C to 3°C (4°F to 5°F) lower than today (**Figure 6.35**).

CHECKPOINT 6.23 ▶ Using your knowledge of the geologist's toolbox introduced in Chapter 5, suggest what analyses and tools scientists would use to reach the conclusions presented in this section about global and polar temperatures.

Greenhouse Gases in the Pleistocene The level of greenhouse gases in the atmosphere fell rapidly during the Pleistocene but varied as global temperature continued to fluctuate due to Milankovitch cyclicity.

◀ **Figure 6.35:** Sea surface temperature during the last glacial maximum (between 26,500 and 19,000–20,000 years ago) was much lower in the North Atlantic, in the North Pacific, and around the Antarctic than they are today, but they may have been slightly warmer in some tropical areas. Dark gray areas indicate insufficient data.

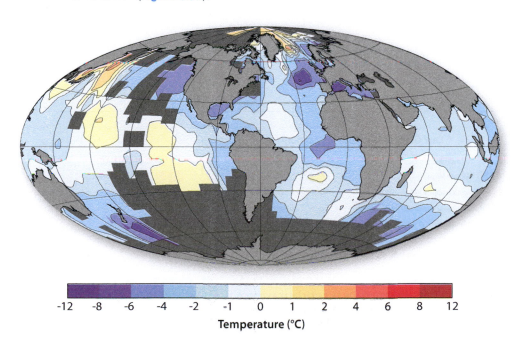

-12 -8 -6 -4 -2 -1 0 1 2 4 6 8 12
Temperature (°C)

The level of CO_2 fell to as low as 180 ppm during the last glacial maximum, and CH_4 fell to as low as 320 ppb. The cold, biologically productive southern oceans appear to have been very successful at pumping CO_2 out of the atmosphere by fixing carbon in marine biomass. The colder oceans were also able to hold more CO_2 in solution. On land, ice and subzero temperatures reduced the amount of CO_2 and CH_4 produced by biological processes, especially the breakdown of organic remains by bacteria.

CHECKPOINT 6.24 ▶ Scientists estimate that carbon dioxide levels were much lower during the Pleistocene than they are today. What analyses would they use to reach this conclusion?

Impact on the Ecosystem in the Pleistocene

During the Pleistocene, changing climate had a dramatic impact on ecosystems around the world. Pollen records indicate that the boreal tree line retreated for hundreds of kilometers during periods

when the ice sheets advanced. Desert regions expanded around the world, and winds intensified. In fact, the winds of the time created massive dunes in the Sahara that are over 100 meters (328 feet) high and survive to this day. Winds also picked up a lot of fine glacial dust and deposited extensive **loess** formations around the world. Tropical forests, including the Amazon Basin, were reduced in size and replaced by grassland habitats that were able to thrive with less CO_2 in the atmosphere (**Figures 6.36** and **6.37**).

CHECKPOINT 6.25 ▶ Climatologists have shown that the ecosystem underwent major changes as the ice sheets advanced during the Pleistocene. What analyses would they have used to reach this conclusion?

Human Evolution During the Pleistocene

Modern hominin species appeared during this time (Figure 6.33). Many earlier species, such as the *Australopithecines*, *Homo rudolfensis*, *Homo habilis*, *Homo ergastus*, and *Homo erectus*, had disappeared, but stocky, barrel-chested

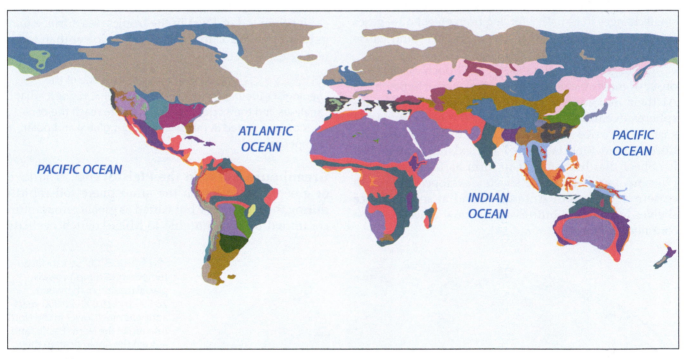

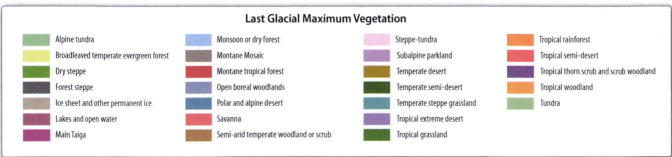

Last Glacial Maximum Vegetation

Alpine tundra	Monsoon or dry forest	Steppe-tundra	Tropical rainforest
Broadleaved temperate evergreen forest	Montane Mosaic	Subalpine parkland	Tropical semi-desert
Dry steppe	Montane tropical forest	Temperate desert	Tropical thorn scrub and scrub woodland
Forest steppe	Open boreal woodlands	Temperate semi-desert	Tropical woodland
Ice sheet and other permanent ice	Polar and alpine desert	Temperate steppe grassland	Tundra
Lakes and open water	Savanna	Tropical extreme desert	
Main Taiga	Semi-arid temperate woodland or scrub	Tropical grassland	

▲ **Figure 6.36:** The distribution of terrestrial ecosystems during the last glacial maximum of the Pleistocene Epoch, 19,000–20,000 years ago. Compare with Figure 2.1 and note the expanded range of desert and the broad expanses of grassland and tundra.

▲ **Figure 6.37:** Mammals of the Pleistocene era. Artist's impression of wildlife believed to have existed in the Northern Iberian Peninsula during the Upper Pleistocene era (125,000 to 10,000 years before present). These include horses (Equus caballus, left), woolly mammoths (Mammuthus primigenius, center) and a woolly rhinoceros (Coelodonta antiquitatis, right). At center right, cave lions (Panthera leo) are eating a reindeer (Rangifer tarandus).

Homo neanderthalensis was particularly suited to harsh glacial environments and even managed to survive the late appearance of our own species, *Homo sapiens*. Neanderthals then became extinct around 25,000 years ago, during the last glacial advance (**Figure 6.38**).

6.9 PAUSE FOR... THOUGHT | *The forced evolution of the forest apes.*

There is little doubt that our species is the product of climate change. As Earth slowly cooled, a reduction in the area covered by tropical forests forced our distant ancestors to adapt to open savannah. Much later, as the ice gradually retreated from high latitudes towards the end of the last glacial maximum, modern humans were well adapted to compete with and displace the more sturdy but less adaptable Neanderthals.

Living in a Land of Ice and Snow

As ice slowly accumulated at the poles over thousands of years, the ice sheets during the last glacial maximum grew to be as much as 3,000 meters (9,843 feet) thick. This polar freezing, together with a smaller contribution from subpolar, temperate, and tropical glaciers, removed so much water from the oceans that sea level fell more than 100 meters (328 feet). In addition, all this freezing raised $\delta^{18}O$ by as much as 2.5‰ in benthic foraminifera, and more ^{16}O was locked away in fresh water at the poles (**Figure 6.39**).

Ice Sheet Growth and Decay   Ice starts to accumulate in upland regions when the air temperature cools enough that winter snow can survive the heat

▲ **Figure 6.38:** *Homo neanderthalensis*, as shown in this artist's rendering, was particularly suited to harsh glacial environments and even survived the late appearance of *Homo sapiens*. Neanderthals became extinct around 25,000 years ago, during the depths of the last glacial maximum.

▶ **Figure 6.39:** The δ^{18}O data from marine foraminifera over the last 6 million years reflects the periodic growth and decay of ice sheets as progressively more δ^{16}O was locked up in the ice sheets of both Antarctica and Greenland.

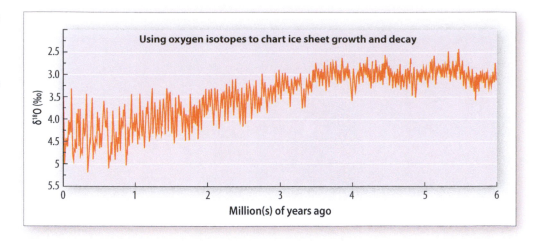

of the following summer. As ice thickens and starts to flow, it forms glaciers that carve out great U-shaped valleys from the surrounding rock and slowly crush and pulverize anything that stands in their way. These glaciers eventually coalesce as the climate cools further. With more cooling, they overflow the valleys that confine them to form large ice sheets that can grow into thick ice domes over 3,000 meters (9,843 feet) deep (Figure 6.40).

As ice sheets grow thicker, they tend to freeze against the solid rock at their base and stop moving. If you try scraping ice off the smooth surface of a car window, you will feel just how strong this surface ice bonding can be. Internal flow can still take place within the ice sheet, but this is a very slow solid-state deformation process that resists the force of gravity.

As time passes, and the ice grows thicker, pressure starts to build at the base of the ice sheet. At the same time, the temperature at the base starts to increase due to the constant flow of **geothermal heat** from below.

Ice is a good insulator (think igloos and ice caves), and much of this heat is trapped close to the bottom of the ice sheet.

The combination of high pressure and rising temperatures causes the basal ice to melt and turn to water. This water is a lubricant that starts the ice moving again in a process known as **basal sliding.** Once the ice starts to move, friction at the base of the ice keeps the temperature high enough to prevent the water from freezing, and the ice starts to flow increasingly rapidly downslope toward the sea (Figure 6.41).

CHECKPOINT 6.26 ▶ In your own words, describe how a permanent ice sheet develops.

Ice Streams   Recent studies have shown that many large ice sheets contain rapidly moving **ice streams** that transport ice from the frozen continental interior to the oceans. These streams are often located over parts of the ice sheet where the base rock is softer and

◀ **Figure 6.40:** In this aerial photo of southeastern Greenland, glaciers almost fill the valleys that confine them and have started to coalesce at their margins.

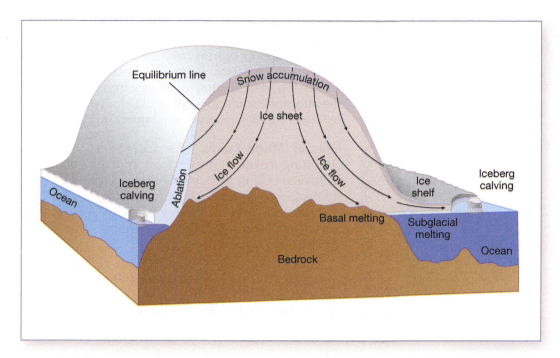

▲ **Figure 6.41:** Model of an ice sheet, illustrating some of the main concepts discussed in the text.

more easily deformed, allowing the ice to flow more freely (**Figure 6.42**).

Binge and Purge: The Periodic Growth and Collapse of Ice Sheets   Evidence indicates that Pleistocene ice sheets grew slowly when the ice stayed firmly frozen to the rock beneath but purged rapidly, collapsing into the surrounding ocean, as soon as the basal ice started to melt. Following collapse, they slowly recovered over a period of 5,000 to 10,000 years before repeating the cycle. These **binge and purge** cycles had a major impact on the climate of the North Atlantic.

The Role of Ice Shelves   Ice shelves appear to play an important role in maintaining the stability of continental ice sheets by buttressing thick continental ice and slowing the rate at which it flows into the sea. During periods of prolonged warming, winds and ocean currents break apart and melt these ice shelves, allowing the thick continental ice to collapse into the ocean more rapidly under the force of gravity. As ice streams and glaciers reach the open ocean, they quickly break apart and float away, forming flotillas of massive icebergs that melt very slowly, dumping tonnes of detritus all over the ocean floor and lowering

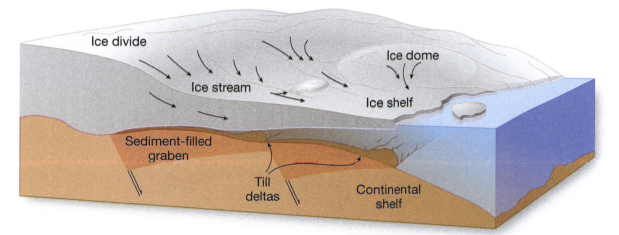

▲ **Figure 6.42:** Studies in Antarctica have shown that ice flows much more freely over softer, unconsolidated sediments than over bare rock. Recent studies have shown that within many ice sheets, are clearly defined streams that move rapidly from the frozen continental interior to the oceans. On this diagram of a typical ice sheet note how the ice streams help to drain ice from the continental interior. Beneath the ice, soft sediment helps ice to move faster and where the ice meets the ocean and melts the sediment (glacial till) once locked in ice is deposited on the sea floor to produce a till delta.

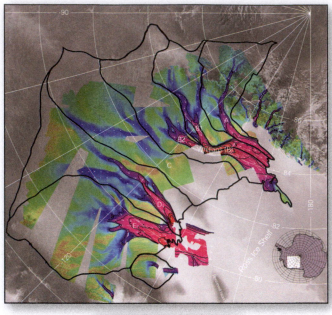

Flow of ice streams (m/yr)

0   800

▲ **Figure 6.43:** Ice streams like these in western Antarctica allow ice sheets that once surrounded ice shelves to rapidly melt and break up. The fast-moving central ice streams are shown in purple to red. Slower tributaries feeding the ice streams are shown in blue. Green areas depict slow-moving, stable areas. Thick black lines depict the areas that collect snowfall to feed their respective ice streams.

surface water salinity (**Figure 6.44**). This layer of cold, brackish water acts like a lid on the ocean and can prevent the flow of heat from ocean currents into the atmosphere. Brackish water also freezes earlier in the winter, blocking the flow of energy from the oceans to the atmosphere and increasing albedo; both (positive) feedback effects have a cooling impact on climate.

Glacial Mega-Lakes   As ice sheets melt, the retreating ice often traps extremely large volumes of meltwater behind ice dams that periodically break and discharge enormous flows of fresh water into the surrounding ocean. This can have an important impact on ocean circulation and climate (see Figure 6.63 of Glacial (Mega) Lake Aggasiz).

> **CHECKPOINT 6.27** ▶ In your own words, describe how ice streams, binge and purge, and the disintegration of ice shelves can contribute to the collapse of an ice sheet.

Climate Stability: Is It Always Cold During an Ice Age?

The Pleistocene climate was not permanently cold. Rather, it oscillated back and forth between long colder glacial stages and shorter warm interglacial stages.

Brief warm periods called interstadials and sharp cold snaps called stadials interrupt the major oscillation between colder glacial and warmer interglacial

▼ **Figure 6.44:** Melting sea ice along the shoreline of Sakhalin Island in Russia. Melting ice lowers surface salinity and can create a freshwater cap that prevents the flow of energy into the atmosphere from the water beneath.

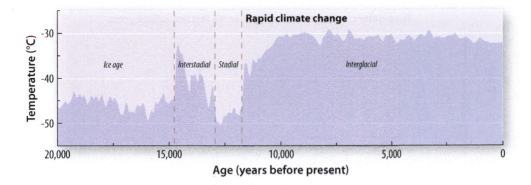

◀ **Figure 6.45:** The large climate swings that mark major glacial and interglacial stages are interrupted by brief warm periods called interstadials and sharp cold snaps called stadials.

stages (**Figure 6.45**). Look at the rate at which the climate changes in this figure. How do you think this compares to the rate at which climate is changing today? We will come back to answer this question later in the chapter.

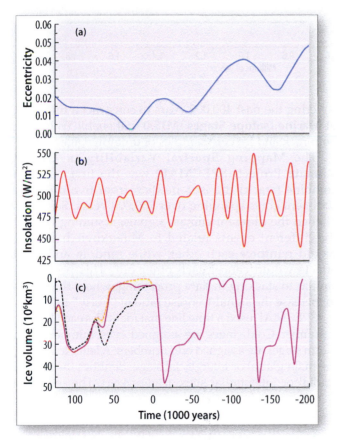

▲ **Figure 6.46:** (a) An overview of how long-term variations of eccentricity, (b) June insolation at 65° north, and (c) computer simulated Northern Hemisphere ice volume have changed over the previous 200,000 years and are projected to change up to 130,000 years in the future. The impact of Earth's obliquity (tilt) on summer insolation in the northern hemisphere is shown in the middle panel, where you can see a clear ca. 41,000 year obliquity signal in the amount of insolation reaching 65°N. Note that time on the x-axis is negative in the past and positive in the future. For the future, three CO_2 scenarios were used: last glacial–interglacial values (purple line), a human-induced concentration of 750 ppm (orange line), and a constant concentration of 210 ppm (black line).

Changing Polar Dominance During the Pliocene and Miocene, when there was little or no permanent Northern Hemisphere ice, climate change was driven by seasonal changes in the Southern Hemisphere. With the development of ice sheets in the Northern Hemisphere, however, positive feedback in the climate system upset this balance, and the Northern Hemisphere became the major driver of global climate change.

The advance and retreat of ice sheets and glaciers in the Northern Hemisphere is controlled by a delicate seasonal balance between the amount of ice that accumulates during the winter and the amount that melts over the following summer. This balance is a function of the intensity of the summer Sun that reaches the Northern Hemisphere at a latitude around 65° north of the equator. Changes in Earth's obliquity (tilt) appear to be most important in determining this amount of Sun. As the degree of obliquity increases, higher latitudes receive more intense sunlight in the summer and less during the winter. These are competing influences, but if summer melt is dominant, the ice sheets will retreat. As the angle of obliquity decreases (as it is doing today), northern summers become cooler, and snow and ice have a greater chance of surviving to the following winter, a pattern that allows scientists to project how Earth's climate is likely to change in the future depending on what we do to limit emissions today (**Figure 6.46**).

The Mystery of the Mid-Pleistocene Transition The **Mid-Pleistocene Transition (MPT)** was a global event that started 1.2 million years ago and was complete by 0.7 million years ago (**Figure 6.47**). Over this time, there was a gradual change in the dominant periodicity of the major climate cycles from 41,000 years per cycle to 100,000 years per cycle. It looks as if a long-established pattern of climate change driven by the obliquity cycle was succeeded in dominance by the much weaker eccentricity cycle. How could this be? It is not clear why this occurred, but many imaginative and creative suggestions have been proposed. For example, the rise of the Alps, Andes, Caucasus, Himalayas, and Rockies at this time may have affected changes in atmospheric circulation that reinforced the weaker eccentricity cycle and made it dominant.

▶ **Figure 6.47:** The Mid Pleistocene Transition (MPT) was a global event that started 1.2 million years ago and was complete by 0.7 million years ago. Over this time, there was a gradual change in the dominant periodicity of the major climate cycles from 41,000 years per cycle to 100,000 years per cycle. This change is reflected in (a) the isotope record from benthic (bottom dwelling) foraminifera; data that are used (b) to estimate ice volume and thus changing sea level.

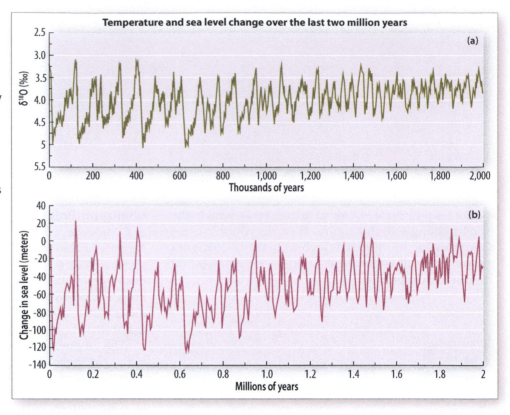

Colder and Colder After the MPT

The amplitude of climate oscillations increased dramatically after the MPT (see Figure 6.47), with much larger swings between warmer interglacial and colder glacial periods. The temperature of each interglacial stage stayed within the same range, but the glacial stages became colder and the ice sheets extended up to 10 degrees further south. Strong positive albedo and other natural feedback between the ice sheets and the climate system may have been responsible for this temperature shift.

The Rock Core Story: Unlocking the Milankovitch Code

The combined impact of the different Milankovitch cycles on radiative forcing over time is predictable and produces a climate signal that can be correlated across both ice and sediment cores. Scientists have therefore been able to use a detailed oxygen isotope record of marine foraminifera from the Pleistocene covering the past 400,000 years to construct a timescale of **Marine Isotope Stages (MISs)** that enable us to date both rock and ice core samples of unknown age.

The **Mapping Spectral Variability in Global Climate Project (SPECMAP)** was the first effort to combine (**stack**) the isotope data from five different sediment cores spanning over 750,000 years. Scientists aligned the different climate signals, assuming that all the different events related to Milankovitch forcing would reinforce each other, while random and local effects would cancel each other out. This process is similar to stacking in data processing, where the aim is to increase the signal-to-noise ratio of "noisy" raw data.

SPECMAP used a time line that identifies each MIS by a number. Cold stages are assigned even numbers, and warm stages are assigned odd numbers. A letter following the main glacial or interglacial stage identifies the related stadial or interstadial stage (**Figure 6.48**). The sequence

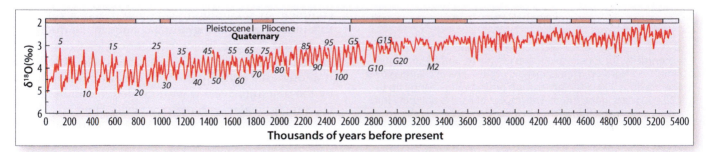

▲ **Figure 6.48:** Numbered marine isotope stages recognized from the combined record of 57 core sites. The data record 5.4 million years of temperature change recorded in the shells of foraminifera recovered from these cores. The scale at the top of the diagram is a record of reversals in Earth's geomagnetic field for reference.

starts with MIS-1 as the current interglacial stage and MIS-5e as the previous (Eemian) interglacial phase. More recent research has added the LR04 core stack, a complete stack of 57 cores that extend our detailed knowledge about the climate record back as far as the Pliocene 5.3 million years ago (Figure 6.48).

6.10 PAUSE FOR... THOUGHT | *Have Milankovitch cycles influenced climate throughout geological history?*

There is evidence of an orbital influence on both marine and terrestrial sedimentation stretching far back into geological time, to rocks deposited well over 400 million years ago.

The Ice Core Story: An 800,000-Year Record of Our Recent Climate History

Ice cores from Antarctica record the longest detailed record of climate change since the MPT (Figure 6.49). On the East Antarctic Ice Shelf, the ice core recovered by the **European Project for Ice Coring in Antarctica (EPICA)** still represents the longest available, detailed climate record, with more than 800,000 years of ice accumulation represented in a core over 3,300 meters (10,827 feet) long. The Russian Vostok ice core is the next longest record, providing evidence of more than 400,000 years of climate history. The West Antarctic Ice Sheet is less well sampled, but the U.S. Byrd ice core, one of the first cores to be drilled on the continent, in 1968, fully penetrated that ice sheet, recording more than 90,000 years of climate history.

In the Northern Hemisphere, the **Greenland Ice Sheet Project (GISP)**, the **Greenland Ice Sheet Project 2 (GISP2)**, the **Greenland Ice Core Project (GRIP)**, and the **Northern Greenland Ice Core Project (NGRIP)** have penetrated both central and southern ice fields, delivering a record of more than 120,000 years of climate data (Figure 6.50).

Ice Core Data   Ice core records from Antarctica and Greenland show the same overall pattern of climate change as the sediment record, but they are more detailed and contain much more information. Eight glacial cycles have been identified in the ice cores recovered by the EPICA team (Figure 6.51). Each cycle is characterized by a relatively sudden increase in atmospheric temperature, followed by a much slower cooling of climate toward the next glacial maximum.

Not all glacial phases share identical characteristics. Look at the overall shapes of the glacial cycles before and after about 440,000 years ago. More recent cycles are strongly asymmetrical, with a pronounced sawtooth pattern. Each major cooling cycle is interrupted by a number of small-amplitude interstadial and stadial stages, and a number of the warm interglacial periods have been interrupted by stadial episodes with more significant cooling.

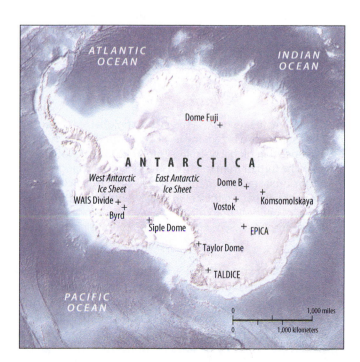

▲ **Figure 6.49:** The locations of major ice cores recovered from Antarctica.

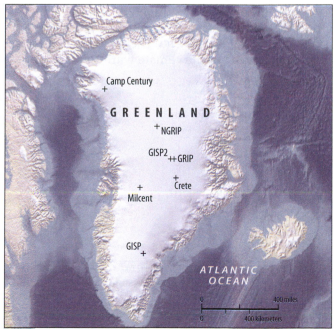

▲ **Figure 6.50:** The locations of major ice cores recovered from Greenland.

▶ **Figure 6.51:** The **European Project for Ice Coring in Antarctica (EPICA)** recovered ice as old as 800,000 years, covering eight complete glacial and interglacial cycles. Graphs show; (a) how solar insolation varied naturally over this period of time. (b) Changing levels of CO_2 recovered from ice core samples, (c) changing temperature determined from oxygen isotope data, (d) dust content from ice core samples, (e) changes in sea level over this period of time as determined from changing ice volume and other geological data.

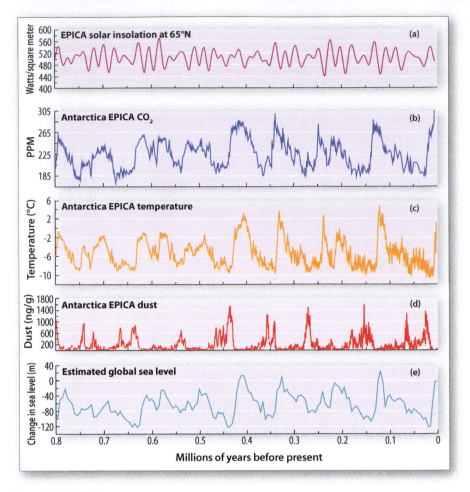

CHECKPOINT 6.29 ▶ With ice sheets as large as those covering Greenland and Antarctica, why do scientists recover ice cores from only a few select localities?

Temperature and CO_2   An increase in global temperature that is driven by Milankovitch cycles normally precedes the increase in the level of carbon dioxide observed towards the beginning of each interglacial stage during the Pleistocene. The Milankovitch climate forcing is small, but the climate warms enough for the oceans to start releasing more carbon dioxide into the atmosphere. Strong positive feedback then leads to a more rapid rise in global temperature and the release of yet more carbon dioxide (**Figure 6.52**).

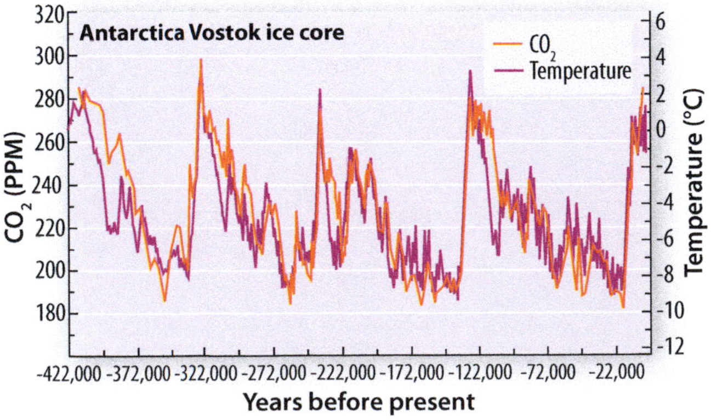

CHECKPOINT 6.30 ▶ In your own words, explain why we expect the temperature of the atmosphere to increase before the level of carbon dioxide increases.

6.11 PAUSE FOR... THOUGHT | *Which came first?*

Some climate skeptics use the fact that global temperatures start to rise before the level of carbon dioxide rises as evidence that global warming is entirely natural. Nothing could be further from the truth. It is clear that changes in Earth's orbit start the warming process, but these changes are so small that, on their own, they would have little impact. Feedback from the oceans as they release CO_2 into the atmosphere keeps global temperatures rising.

A Glimpse of the Future   The interglacial period that started around 400,000 years ago is particularly interesting because the orbital configuration at that time was similar to what we see today. Low orbital eccentricity both then and now means that climate is mainly affected by changes in Earth's obliquity (tilt)—a configuration that tends to produce unusually long stable interglacial periods lasting more than 40,000 years. It

▲ **Figure 6.52:** Changes in temperature precede the rise of carbon dioxide. Rising temperature is initiated by changes in orbital (Milankovitch) cycles, but once it becomes warmer, the oceans begin to release CO_2 into the atmosphere; a positive feedback for warming that reinforces and accelerates rising temperature.

appears that there is a low probability that Earth will enter a major new phase of glacial advance within the next 30,000 years.

The Eemian Interglacial Stage

The **Eemian interglacial stage** was the last time the world was at least as warm as it is today. Starting approximately 130,000 years ago, the Eemian stage came to an end 114,000 years ago, as the climate once more cooled and ice sheets advanced. Global temperatures were perhaps 1°C (1.8°F) warmer than they are today, due to differences in the eccentricity of Earth's orbit. However, the level of CO_2 in the atmosphere remained close to preindustrial levels, at around 280 ppm, and the climate was fairly stable.

Not everything was the same during the Eemian as today. During the Eemian, much more solar radiation arrived at 65° north of the equator due a higher angle of obliquity, resulting in average Arctic temperatures as much as 4°C to 5°C (7°F to 9°F) warmer than today, as well as much more seasonal variation at both poles.

Higher Arctic temperatures during the Eemian melted more ice in the Northern Hemisphere than we see today, and sea level was between 4 and 6 meters (13 and 20 feet) higher then than it is now. The southern Greenland Ice Sheet appears to have been the source of much of this water, but the world was generally more ice free. The level of insolation in the Northern Hemisphere finally began to decrease around 110,000 years ago, and the world started an intermittent cooling trend toward the depths of the last glacial maximum.

> **CHECKPOINT 6.31** ▶ Explain why the Eemian interglacial stage was warmer than today, when there were no anthropogenic greenhouse gas emissions.

Rapid Climate Change

One of the most important questions policymakers face today is how fast global climate is likely to change in the near future. The ice core record shows that rapid climate change is a normal part of climate evolution. More than 10% of the land surface is still covered by ice, and large ice sheets expand and contract with the seasons. Rapid climate change during the Pleistocene was driven by small changes in radiative forcing that were amplified by large changes in the cryosphere. Many of these factors are relevant today: Changes in albedo, the exchange of energy between the oceans and atmosphere, and the release of greenhouse gases feed energy into the climate system. The following section illustrates how rapid change has dominated climate over the past 800,000 years of glacial history.

Dansgaard–Oeschger Cycles

There is no complete ice core record from Greenland that covers the entire Eemian interglacial stage, but the NGRIP ice core does sample the end of the Eemian interglacial through to the present day. Together, the Greenland records reveal 24 minor cycles of rapid climate change over the glacial period between 110,000 and 10,000 years ago. During each cycle, the temperature in Greenland increased by as much as 8°C to 10°C (14°F to 18°F) in as little as 20 years. Each transient warm period lasted between 500 and 2,000 years, until the temperature fell slowly at first and then declined just as rapidly as it had risen (Figure 6.53). These recurring interstadial events, called **Dansgaard–Oeschger (D–O) cycles**, represent a cyclical redistribution of heat within the climate system.

The more extensive Antarctic ice records do sample the entire Eemian interglacial period through to the present day. It shows the same pattern of gradual overall cooling toward the last glacial maximum, also modulated by weak oscillations that occur at the same time but are strangely out of phase with the D–O cycles in Greenland. When it got warm in Greenland, it got cold in Antarctica. How could this be? Greenhouse gases do not seem to be responsible. The level of CO_2 in the atmosphere changed by just 25 ppm during each cycle, and although CH_4 levels increased by as much as 50%, this appears to be a consequence, and not the cause, of climate change at that time.

The Origin of the D–O Cycles

It is possible that the D–O cycles are due to changes in the pattern of thermohaline circulation in the North Atlantic. Ocean currents are not always perfectly stable, and they can periodically flip between two or more alternative patterns of circulation. As sea ice started to cover large areas of the Arctic and North Atlantic, the warm tropical waters of the North Atlantic Current could no longer flow as far north as they do today and stabilized just south of the

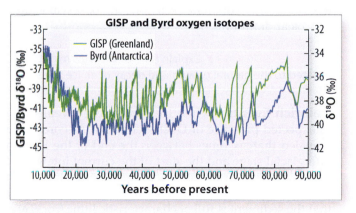

▲ **Figure 6.53:** The history of climate change in both the Northern Hemisphere and Southern Hemisphere is well preserved in ice cores from Greenland and Antarctica. This diagram plots data sets from both the West Antarctic Ice Sheet and the Greenland Ice Sheet. The Greenland data show that there were regular changes in climate every 1,500 years or so, when the temperature of the ice sheet warmed by as much as 10°C (18°F) in as little as 20 years. Data from Antarctica also show this cyclicity, although the change in temperature in the interior of this ice sheet was less noticeable than in the Greenland cores.

Icelandic sill. Along this shifting ice margin, surface water became cold and dense, and it sank to the sea floor, driving global thermohaline circulation.

This theory suggests that rapid warming at the start of each D–O cycle occurred when the pattern of thermohaline circulation "flipped" from a stable glacial mode with water sinking to the south of the Icelandic sill to a less stable, or just weakly unstable, mode with water sinking much farther to the north. This less stable pattern of ocean circulation drew tropical energy into the core of the Arctic and persisted for hundreds of years before falling global temperatures reached a critical point that flipped the pattern of thermohaline circulation back to its original state. When this happened, the temperature in Greenland suddenly plummeted, as observed. This process must have involved complex coupling between ice sheets, ice shelves, oceans, and some form of regular internal or external forcing, and the details are not yet clear (**Figure 6.54**).

This theory certainly explains the observable data. Such a flip in the pattern of thermohaline circulation could produce the observed temperature profile of each D–O event, but it is not clear why this should occur. Marine cores that cover these events show a slight increase in the amount of ice-rafted debris during the cold phase, suggesting that the ice sheet margins may have been unstable.

In addition, some data suggest that these events were triggered by a 1,470-year cycle of uncertain origin. The idea of this elusive cycle is widespread in popular science literature, and although there is some evidence of its existence as a weak source of climate forcing, it is not at all clear how it develops. There is no known solar cycle of that length, and stochastic (random) changes within an excitable climate system are unlikely to be so regular. Climate scientists have many theories to explain this apparent cycle, but the observable data lack clear definition.

> **CHECKPOINT 6.32 ▶** Explain how D–O cycles might be related to the natural growth and decay of an ice sheet.

Heinrich Events While ice cores have revealed the unexpected presence of D–O cycles, marine cores have revealed that six times between the end of the Eemian interglacial stage and the start of the current interglacial stage, immense armadas of icebergs flooded the North Atlantic. Marine cores that sample this interval show that the sediment is full of characteristic lithic (rock) and mineral fragments that could only have come from ice-rafted debris from the Laurentide Ice Sheet. The deposits occur around 50° north in a 3,000-kilometer (1,864-mile) band that is thickest close to the Hudson Strait and thins progressively toward northwestern Europe (**Figure 6.55**).

Heinrich events, which are numbered H1 to H6, appear to have developed as the **Laurentide Ice Sheet** became unstable and started to collapse into the ocean,

flooding the surface with cold, fresh water and ice debris. The impact on global ocean circulation was immediate. Freshwater incursions from melting ice lasted between 250 and 1,000 years, with average flow rates around 0.3 Sverdrups, or 300,000 cubic meters (80 million gallons) per second, and peak flow rates that were much larger. This put an effective lid on the generation of the cold saline currents that drives thermohaline circulation (**Figure 6.56**).

Heinrich Events and D–O cycles share some common features, but there are also many differences. During the coldest phase of many D–O cycles, ice-rafted debris came from sources scattered all across the

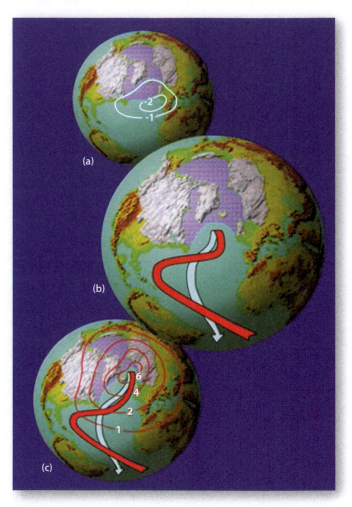

▲ **Figure 6.54:** (a) The upper globe illustrates the pattern of ocean circulation that developed during a Heinrich event, when the Atlantic conveyor collapsed and a cold anomaly spread over the mid-north Atlantic. Contours show the surface air temperature difference relative to the stable state. (b) The large globe in the center illustrates the pattern of ocean circulation that possibly existed in the Atlantic during cold (stadial) climate periods that had greatest climatic stability and prevailed during most of the Pleistocene. (c) The smaller globe at bottom illustrates the situation during one of the warm D–O events, during which the Atlantic conveyor belt temporarily advanced into the Nordic Seas, and a strong warm anomaly developed (note the contours). Ocean circulation is shown schematically, with warm surface currents shown in red and deep cold currents in light blue. Purple shading indicates the extent of continental ice sheets.

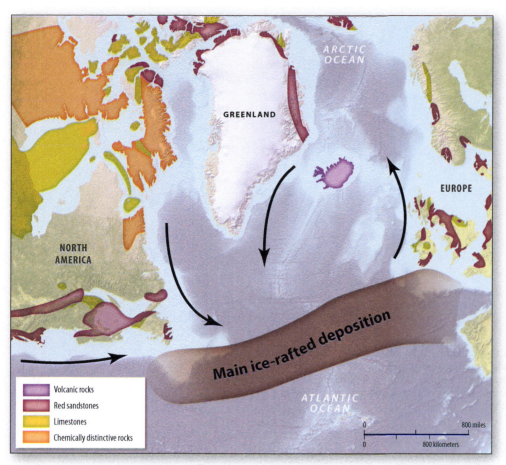

◀ **Figure 6.55:** Ice-rafted debris from the Laurentide Ice Sheet occurs in a 3,000-kilometer (1,863-mile) band across the North Atlantic (shown in brown) that is thickest close to the Hudson Strait and thins progressively toward northwestern Europe. The other shaded colors indicate the geographical sources (provenance) of different rock types recovered from the ice rafted debris.

North Atlantic. The much larger Heinrich events originated within the Laurentide Ice Sheet, and ice-rafted debris appears to have flowed directly into the Atlantic through the Hudson Strait. The timing of both events is also different. Heinrich events followed a 7,000-year to 10,000-year cycle, while D–O cycles were tuned to an enigmatic 1,500-year cycle. It is clear that both cycles are related to ice sheet instability, and both had an impact on thermohaline circulation, but it is not clear that they share a common cause.

The most likely cause of Heinrich events is the binge-and-purge process discussed earlier. As the Laurentide Ice Sheet grew, geothermal heating at its base allowed basal melting to occur, and the ice sheet became unstable and started to collapse into the ocean. Soft sediment on the floor of the Hudson Strait allowed the ice to slip even faster. Once this process started, it was not going to stop until it was over (**Figure 6.57**).

The North American Laurentide Ice Sheet would have started to build again once the volume of ice purged allowed the basal temperature to reach freezing point, once more locking the ice securely to the rock beneath. Ice sheet models suggest that it would have taken at least 7,000 to 10,000 years for the Laurentide Ice Sheet to become large enough to trigger another collapse.

The impact of the Heinrich events was global. As with the D–O events, the level of CO_2 in the atmosphere did not increase very much, but all the ice that melted increased sea level by as much as 15 meters (50 feet) over 250 to 750 years.

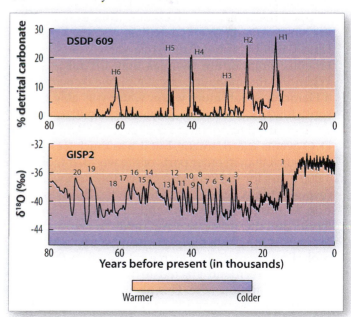

▲ **Figure 6.56:** The $\delta^{18}O$ record (top) from the GISP2 ice core in Greenland, showing 20 of the 25 observed D–O events during the last glacial period (top), and the rock record (bottom) of ice-rafted carbonate material that was deposited during Heinrich events in the North Atlantic and recovered from a deep-sea core (DSDP 609).

▲ Figure 6.57: The Barnes Ice Cap on Canada's Baffin Island is a remnant of the Laurentide Ice Sheet that covered much of North America during the Pleistocene epoch. As the Laurentide Ice Sheet collapsed, it released an armada of icebergs into the North Atlantic. These would melt and shut off thermohaline circulation, leaving evidence of their passing in mineral and rock fragments in sediment on the sea floor.

CHECKPOINT 6.33 ▶ Explain how the binge-purge model of ice sheet growth and decay can be used to explain Heinrich events.

A Bipolar "Seesaw" If Heinrich and D–O events stopped thermohaline circulation from developing in the North Atlantic, it could explain why the ice records from Greenland and Antarctica show almost simultaneous temperature changes that are out of phase. When the thermohaline circulation weakened during a Heinrich event, more warm water was able to flow south toward Antarctica and warm the Southern Ocean.

The rate of temperature change in the Arctic during D–O events was dramatic, but it was much less so near Antarctica than in other parts of the world. It took up to 1,000 years for the Southern Ocean to warm weakly in response to strong cooling in the Arctic. This was partly due to the immense heat capacity of the Southern Ocean and the strength of Antarctic circumpolar circulation that acted as a barrier to heat transfer. Once thermohaline circulation was reestablished in the North Atlantic, Greenland started to warm again, and the Antarctic started to cool. This phenomenon is known as the **bipolar seesaw**, and is clear evidence of the strong links that exist between the ice sheets in the Northern and Southern Hemispheres.

CHECKPOINT 6.34 ▶ Explain why warming in the Arctic is often associated with cooling around Antarctica.

6.12 PAUSE FOR... THOUGHT | *Are you confused by all these glacial cycles?*

The information in the past several pages may seem complicated and confusing. If you cannot recall the details you should remember the evidence that there is a strong interconnection between all the different parts of Earth's climate system. Even a small change in one part reverberates throughout the entire system, enacting change that may appear well beyond the spatial and temporal scale of the initial disturbance. Climate scientists have much to learn about these processes, and progress will require experts from many different fields of science to work together to make sense of this recent phase of Earth history.

The Last Glacial Maximum (LGM): The End of an Ice Age

The last glaciation was at its most intense between 26,500 and 19,500 years ago, in response to climate forcing from a decrease in northern summer insolation. Up to 30% of the global land surface was covered in ice, and ice sheets as thick as 3,000 meters (9,843 feet) covered much of North America and Europe. This was called the **last glacial maximum (LGM).**

Greenhouse Gases During the LGM During the LGM, the level of CO_2 in the atmosphere fell to as low as 180 ppm due to a combination of factors, including the slow rate of decay of vegetation trapped in permafrost,

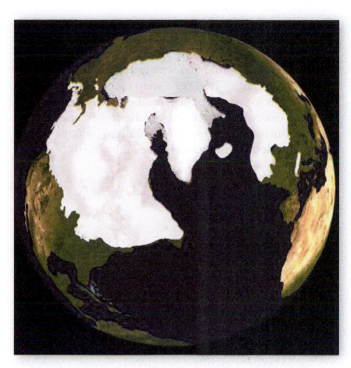

▲ **Figure 6.58:** During the LGM, ice sheets in the Northern Hemisphere reached as far south as 36°, as illustrated on this graphic simulation of the Arctic region, and sea ice around Antarctica (not shown) extended almost as far as South America.

more efficient absorption of CO_2 by the cold oceans, and a more effective biological pump in the cold, nutrient-rich Southern Ocean. More expansive ice sheets, extensive sea ice, and atmospheric dust also increased Earth's albedo and reflected more of the Sun's energy (Figure 6.58).

Global Land and Sea Temperatures During the LGM

In the tropics during the LGM, ocean surface temperature cooled by up to 3°C (5°F) and land surface temperatures by as much as 5°C (9°F). Central America and northern South America cooled by 5°C to 6°C (9°F to 11°F). It was 9°C (16°F) colder in the Antarctic and 21°C (38°F) colder in parts of Greenland. Both deserts and tropical glaciers expanded, and thick loess deposits formed around the globe due to the rapid erosion of freshly exposed bare rock and soil around the margins of the ice sheets.

Sea Level During the LGM

The expansion of polar ice caps during the LGM caused sea level to fall by as much as 120 meters (400 feet), once more exposing wide areas of the continental shelf to erosion. Many land bridges reappeared between previously isolated areas, joining Alaska with Russia, the United Kingdom with continental Europe, and Australia with New Guinea. Much of the Southeast Asian archipelago eventually formed one large landmass.

The Biosphere During the LGM

The overall impact of the LGM on the biosphere was large. Once-continuous tropical rain forests became highly fragmented. Boreal forests expanded southward, replacing temperate forests, but lost ground to the north, where they were replaced by steppe and tundra (Figure 6.36). Most deserts expanded, and many parts of the world were more arid, as the colder air was not able to hold as much water vapor. An exception was the Great Basin of the United States, where changes in the pattern of the jet stream led to an increase in precipitation, creating large freshwater lakes where only salty remnants now remain.

The Bølling–Allerød Interstadial Stage

The LGM came to an end with the advent of the **Bølling–Allerød interstadial stage**, around 14,700 to 12,700 years ago. This stage marks an important climate tipping point, where Milankovitch orbital cycles began to deliver more insolation to the Northern Hemisphere, and Earth started to warm. Temperature in the North Atlantic increased by up to 7°C (13°F), but at a rate still 10 times slower than recent warming (Figure 6.59).

It took time for the additional heat from the Northern Hemisphere to penetrate deeply into the Southern Hemisphere, and Antarctica took much longer to warm than the Arctic. Parts of the West Antarctic Ice Sheet seem to have melted during this time, resulting in a rapid rise in sea level known as Meltwater Pulse 1A. Sea level rose by 20 meters (66 feet) in less than 500 years, at rates exceeding 40 millimeters (1.6 inches) a year, compared to just 2 millimeters (0.08 inches) a year today (Figure 6.59). This was the dawn of the present interglacial period, but the return to warmer conditions was soon interrupted.

The Younger Dryas Event

The rapid warming of the Bølling–Allerød interstadial stage was terminated suddenly between 12,800 and 11,500 years ago, as temperatures once more plummeted over the North Atlantic. This intense but transient cooling event, known as the **Younger Dryas event,** after a small Alpine flower, lasted just 1,400 years before rapid warming was reestablished. The temperature over the entire North Atlantic may have dropped by as much as 7°C (13°F) in 20 years, and

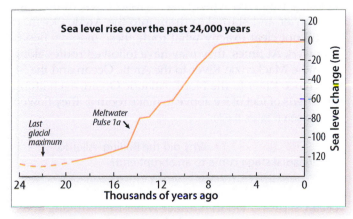

▲ **Figure 6.59:** Most of the early and rapid rise in sea level from this time, known as Meltwater Pulse 1A, is thought to have been due to melting of parts of the West Antarctic Ice Sheet.

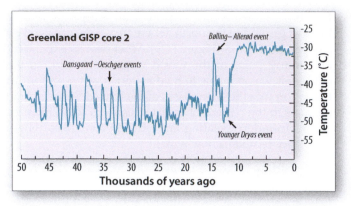

▲ **Figure 6.60:** The climate record from the GISP2 ice core shows that the warming of the Bølling–Allerød interstadial stage marked the end of the LGM and was a period of very rapid climate change. Note also the rapid changes in temperature associated with the Dansgaard-Oeschger events and rapid initial cooling associated with the Younger Dryas event.

in Greenland it may have dropped by as much as 10°C to 15°C (18°F to 27°F) in just 10 years (see **Figure 6.60**).

The Younger Dryas was a global event. Although most apparent in the Northern Hemisphere, it is also found in the Antarctic ice record. There is also evidence of drying in the tropics, suggesting that both the Intertropical Convergence Zone and Asian Monsoon were affected. Tree lines that had started to move north during the warmer Bølling–Allerød interstadial stage were forced to retreat again when faced with the advancing cold.

The brief intense time of cooling at the beginning of the YD was due to the sudden breakdown of thermohaline circulation in the North Atlantic. Warming during the Bølling–Allerød interstadial stage melted a lot of ice around the Arctic. The massive Laurentide Ice Sheet and smaller ice sheets in Europe and Siberia had all started to decay, and immense freshwater lakes had started to form around their margins. The largest of these was glacial **Lake Agassiz,** which covered much of Manitoba, western Ontario, northern Minnesota, eastern North Dakota, and Saskatchewan—an area as large as 440,000 square kilometers (170,000 square miles). The Younger Dryas event was possibly triggered when Lake Agassiz burst through an ice barrier and started to drain into the Arctic Ocean. It is still not clear exactly what route these massive floods followed. At times, they may have followed routes along both the Mackenzie River to the Arctic Ocean and the St. Lawrence River to the North Atlantic Ocean, even lifting hundreds of feet of ice above to make room as they flowed out to sea (**Figure 6.61**).

CHECKPOINT 6.35 ▶ Why did the Bølling–Allerød interstadial stage come to an abrupt end?

As this fresh water spread over the surface of the ocean, it prevented the surface water from sinking to the ocean floor and weakened the deep thermohaline circulation. This in turn reduced the surface flow of warm water into the still-icy North Atlantic, effectively starving

the ocean of subtropical energy. Surface temperatures in the North Atlantic started to cool rapidly, and for a while the glacial conditions returned and the Scandanavian ice sheet expanded as winter sea ice advanced into the North Atlantic. This cold snap came to a rapid end, however, as ice and freezing temperatures in the North Atlantic once more kick-started thermohaline circulation and orbital changes due to Milankovitch cyclicity increased global temperature (see Figure 6.60).

CHECKPOINT 6.36 ▶ If the Younger Dryas event was generated by events in the Arctic, why did it have a global impact?

The Holocene

The **Holocene Epoch** began at the end of the Younger Dryas event, some 12,000 years ago, marking the beginning of the most recent interglacial stage. The only real difference between the Holocene and the Pleistocene is the rise of human civilization, and as we are now poised to become architects of global climate, some have even proposed that we change the name from Holocene to **Anthropocene.** In fact, there is no sound *geological* reason to mark any major stratigraphic boundary at this point. As far as global climate is concerned, we are still in the Pleistocene Epoch, locked in the grip of an icehouse, with little prospect of immediate natural release. With the slow movement of the continents, it would normally take millions of years for nature alone to return Earth to a permanently warmer climate, but it is becoming clear

▲ **Figure 6.61:** Lake Agassiz formed to the south of the Laurentide Ice Sheet. As melting progressed, the lake spilled out into the North Atlantic ca. 12,800 years ago, shutting down thermohaline circulation and plunging Earth back into glacial conditions for 1,400 years.

that greenhouse gas emissions might now achieve this change in less than 100 years.

The Holocene is a time of global climate change driven by Milankovitch cycles, volcanic activity, and changes in solar intensity. Regional climate change has occurred due to naturally shifting patterns of circulation in the oceans and atmosphere, but global climate has been remarkably stable during the Holocene when compared to the Pleistocene.

> **CHECKPOINT 6.37** ▶ Discuss why some geologists and almost all climate scientists believe that the Holocene does not really exist as a separate and clearly defined geological epoch.

New Tools Enhance the Holocene Climate Record

The climate record from the Holocene has been enhanced by the addition of data from tree rings, corals, boreholes, and cave deposits (stalactites) that are not always available for older records. In more recent times, oral, written, and thermometer records provide an even more detailed record of recent climate change.

Global Recovery from the Younger Dryas Event

Global average temperature increased rapidly following the end of the Younger Dryas event (Figure 6.62). Greenland warmed most rapidly, possibly by as much as 15°C (27°F) in two short steps of 7°C to 8°C (13°F to 14°F) each. In as little as 10 to 20 years thermohaline circulation was reestablished and sea surface temperature warmed rapidly. Northeastern America and northwestern Europe were most affected by the loss of thermohaline circulation during the Younger Dryas event, but they recovered quickly, and forests expanded as the ice retreated. Sea surface temperatures also recovered in the tropical Pacific and Indian Oceans,

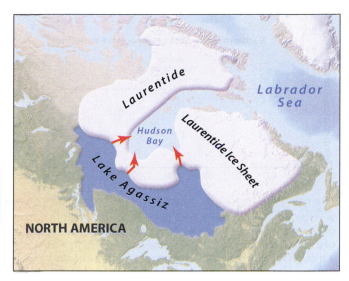

▲ **Figure 6.63:** Lake Agassiz is thought to have spilled out into the Arctic and North Atlantic Oceans via Hudson Bay and the Labrador Sea.

as deep-ocean circulation invigorated by renewed thermohaline circulation spread the heat around the globe.

> **CHECKPOINT 6.38** ▶ Why did the world warm rapidly following the end of the Younger Dryas event?

The 8,200-Year Cooling Event

The massive Laurentide Ice Sheet continued to melt and recharge Lake Agassiz during the early part of the Holocene. Around 8,200 years ago the lake once more broke through the weak barriers of ice and glacial sediment that held the water in place and spilled over into the Arctic and North Atlantic, flushing the ocean surface with fresh water (Figure 6.63). As had happened before, this resulted in a rapid breakdown in thermohaline circulation, and evidence from ice cores indicates that the temperature in Greenland fell by 2–3°C for around 150 years, and lake and ocean sediment cores indicate that temperatures in Europe fell by as much as 2°C. Climate records from other parts of the world are more sparse and make it difficult to assess the impact of this event beyond the North Atlantic (Figure 6.62).

The impact of the 8,200-year flood from Lake Agassiz was much less severe than the Younger Dryas event, because the world was much warmer. During the 8,200-year event, the level of greenhouse gases was higher, the albedo from ice was much reduced, and heat-absorbing forests had replaced reflective bare rock and soil. The average discharge from Lake Agassiz was also lower than during the Younger Dryas. The flood rate of the 8,200-year event is estimated to have been around 0.1 Sverdrup, although it may have reached as high as 5 to 10 Sverdrups over very short periods of time early in the flood. The volume of water released was enough to raise global sea level by 11 to 14 centimeters (4 to 6 inches).

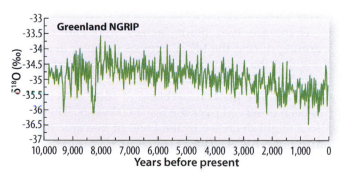

▲ **Figure 6.62:** The rapid cooling event 8,200 years ago indicated in this isotope record from the Greenland NGRIP core was related to the release of a vast reservoir of fresh meltwater from the Laurentide Ice Sheet. Notice how natural climate cycles create high frequency changes in oxygen isotopes that are superimposed on the long term trend determined by Milankovitch cyclicity.

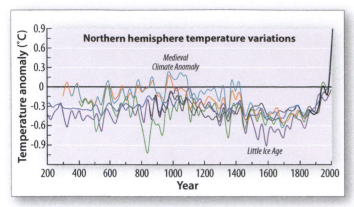

▲ **Figure 6.64:** Evidence from the combined data from many different proxy measurements does not support the existence of a global warming in medieval times (ca. AD 950 to 1250), although many regions did experience periodic warming.

The Holocene Climate Maximum

The Holocene climate maximum occurred between 9,000 and 5,000 years ago, during the early Holocene, when insolation at 65° north reached a maximum as much as 7% higher than it is today due to changes in Earth's precession. During that time, the temperature in the Arctic was 4°C (7.2°F) higher than it is today, and there is evidence that parts of North America and northwestern Europe were also warmer than they now are. More energy from the Sun intensified the African Monsoon, and parts of the Sahara had lakes, rivers, and abundant plant and animal life, much like parts of East Africa today.

Many climate skeptics take the Holocene climate maximum as evidence that temperatures have been higher in the past without any anthropogenic effects. This is true but irrelevant. First of all, solar forcing was higher then than it is today due to the impact of Milankovitch cycles. In addition, there is evidence for some cooling in the Southern Hemisphere at that time. Warming today is global, and it is taking place at a time when solar intensity is decreasing, not increasing. This is a very different situation from that found during the early Holocene.

CHECKPOINT 6.39 ▶ In the absence of human influence, why was the temperature during the Holocene climate maximum higher than the temperature today?

More Holocene Cooling

Between 5,000 and 4,000 years ago, the Northern Hemisphere cooled again (Figure 6.62). Archeological evidence gathered from North America to the Middle East shows that the impact of this cooling on early human society was severe, and many communities declined in population and distribution.

The Medieval Climate Anomaly

The **Medieval Climate Anomaly (MCA)** was a regional, periodic, and transient warming between A.D. ~950 and ~1400 that affected many parts of the Northern Hemisphere and at least some parts of the Southern Hemisphere. Climate skeptics have used the MCA to argue that recent climate change is natural, but there is very little data to support the idea that the MCA was a truly global event that can compare with the warming we observe today. Available proxy measurements from tree rings, corals, stalagmites, and boreholes do not indicate any significant or systematic change in *global temperature* over this period; some estimates even suggest that global temperatures were actually 0.1°C to 0.2°C (0.2°F to 0.4°F) cooler than the 1961–1990 average. Tree ring evidence from Tasmania, New Zealand, and the southern Andes suggest that some parts of the Southern Hemisphere were warmer around 1,000 to 700 years ago than they are today, a possible hint at the existence of teleconnection between hemispheres, but this kind of regional warming is not unusual and it is not clear that there is any direct link to similar warming in the Northern Hemisphere (**Figure 6.64**).

CHECKPOINT 6.40 ▶ What evidence suggests that the MCA was not a global event?

6.13 **PAUSE FOR... THOUGHT** | *A good time for English wine?*

Few events have stirred more controversy in climate science than the MCA, once called the Medieval Warming Period. For climate skeptics, the MCA is clear evidence that the warming we experience today has occurred in the past and that we have no need to worry about future climate change. The crucial point to remember is that warming over that period was *regional*, occurred at different times in different places, and was driven by natural factors. Those natural factors are missing from the world today. Unlike the relatively benign MCA, today we face a future of rapid and damaging *global* climate change.

There are two main factors that can explain warming during the MCA. Proxies for solar insolation show that the Sun was more active for a short period around the time of the MCA than it is today. This was counter to the general decrease in solar activity that has been observed over the last 1,000 years. There was also a slight decrease in the level of volcanic activity during the MCA. The absence of volcanic aerosols would reduce global albedo and account for some of the warming.

It is possible that there were periods during the MCA when parts of the Northern Hemisphere were as warm or warmer than today due to a combination of natural factors, such as the influence of the North Atlantic Oscillation, the Pacific Decadal Oscillation, the

Atlantic Multidecadal Oscillation, La Niña/El Niño, or the intensification of thermohaline circulation, but there is no evidence of the globally simultaneous and rapid warming we observe today. If small changes in radiative forcing were responsible for the MCA, it supports the view that regional climate is very sensitive to radiative forcing, a fact that should raise heightened concern for our immediate future.

A minority of climate scientists have suggested that the MCA may be linked to a source of weak radiative forcing that drives a mysterious and unproven 1,500-year climate cycle. This is intriguing, as there are no solar cycles that fit this periodicity, and it is difficult to see what could drive these changes. Most climate scientists do not agree with this idea.

CHECKPOINT 6.41 ▶ What evidence exists for an enigmatic 1,500-year climate cycle?

The Little Ice Age

The **Little Ice Age (LIA)** refers to a period of time between the 15th and 19th centuries when global temperatures were 0.3°C to 0.4°C (0.5°F to 0.7°F) colder than at present and when regional climate feedback produced temperatures as much as 1°C to 2°C (2°F to 4°F) lower than those of today (Figure 1.3c). Public and private records from Europe, North America, and the Southern Hemisphere tell of brutal years dominated by famine, disease, war, and hardship when glaciers were more advanced than today in almost all mountainous areas of the world.

A number of factors contributed to the development of the LIA:

- Earth had been cooling for at least a millennium, and less sunlight reached northern latitudes due to changing obliquity
- A period of intense volcanic activity accelerated cooling by ejecting highly reflective sulfate aerosols into the stratosphere, where they increased global albedo
- The area covered by polar ice expanded following the eruption of volcanic aerosols, further increasing global albedo and prolonging the period of cooling
- The thermohaline current in the North Atlantic may have slowed following an influx of meltwater produced during the MCA
- Solar activity was low. There were two major sunspot minima during the LIA

Some scientists have suggested that the world would have continued to cool along this path if it were not for the early slow release of heat-trapping greenhouse gases from industry and agriculture.

CHECKPOINT 6.42 ▶ What led to global cooling during the LIA, and what brought this cooling to an end?

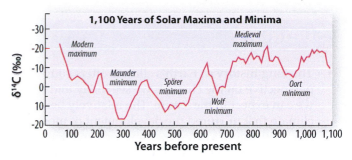

▲ **Figure 6.65:** Solar minima are related to colder periods in climate history. The Maunder Minimum, around 1645–1715, was associated with the advance of glaciers and times of great hardship. Solar activity is estimated on the basis of the amount of the cosmogenic isotope ^{14}C recovered from ice cores.

The Maunder Minimum

The Sun was much weaker at times during the LIA than it is today. During the **Maunder Minimum**, which occurred from 1645 to 1715, there were as few as 50 sunspots over a 30-year period compared to the 40,000 to 50,000 that would typically occur over 30 years. Global surface temperatures were at their minimum from the mid-1600s to the early 1700s. Proxy evidence from tree rings and temperature records suggests that winter temperatures may have been as much as 1°C to 2°C (2°F to 4°F) colder than they are today (**Figure 6.65**).

CHECKPOINT 6.43 ▶ What impact did the Maunder Minimum have on global climate, and how does this relate to climate change today?

Looking Forward to Future Climate Change

The relatively rapid rise in global temperature that marked the end of the LIA was due to a reduction in the level of global volcanism and an increase in solar activity. Computer climate models can reproduce this warming accurately, but after solar activity faltered in the middle of the 20th century they were no longer able to do this without including a significant contribution from heat-trapping greenhouse gases (Figure 1.7). This anthropogenic climate forcing has been augmented by strong positive feedback from the cryosphere, atmosphere, and oceans, and there are few, if any, indications of negative feedback to counter this effect. Evidence gained by studying ancient climate change suggests that Earth's climate may be even more sensitive to small changes in radiative forcing than currently accepted. If so, the social, environmental, and economic impact of adding more heat-trapping greenhouse gases to the atmosphere will be much greater than anticipated by most observers. A lesson we need to learn from the past is that climate change can be both rapid and extreme, and that Earth's climate is highly sensitive to small changes in radiative forcing due to strong feedbacks within the climate system.

Summary

- The long history of Earth's climate offers some perspective on the changes we see today and what we can expect in the future. The relatively benign climate that nurtured the development of modern civilization is not something we can take for granted. During the late Proterozoic (ca. 700 million years ago) and the Permo-Triassic (ca. 252 million years ago) extreme climate threatened the very existence of metazoan life on Earth. The Paleocene–Eocene thermal event shows us how hidden triggers in the climate system can cause rapid and destructive changes that are devastating for the biosphere.

- The major controls on climate are the power of the Sun, shifting continents, volcanoes, and orbital cycles, but these are strongly amplified by feedback from some greenhouse gases, albedo, ocean circulation, atmospheric circulation, and the movement and storage of carbon in the biosphere and lithosphere.

- Observations from the Miocene and Pliocene show that the climate may be more sensitive to changes in the concentration of greenhouse gases than currently thought and that ice sheet stability may be more fragile than currently modeled. There is little prospect of the East Antarctic Ice Sheet collapsing in the near future, but both the Greenland and West Antarctic Ice Sheets have melted in the past when global temperatures have reached the same level we may experience by the end of the 21st century.

- The past is very much the key to the future. Our knowledge of the forces that shape climate allows us to project that general path of climate change for millions of years into the future. The slow movement of the continents and low levels of highly explosive volcanicity leave Milankovitch cycles as the dominant natural force in climate change, and in the absence of human intervention we can expect to remain in an icehouse world for a long time to come.

- The ability of computer models to reproduce many of the features of past climate builds confidence in their ability to predict the future. These models make it clear that natural cycles alone are not able to explain the warming we have experienced since the middle of the 20th century, and their prognosis of future climate change is not good if we do nothing to cut our current level of greenhouse gas emissions.

Why Should We Care?

Climate change may be natural and normal, but it is not always favorable to life. We should embark along a path that might change global climate with timidity and great concern for the consequences of our actions. The recent appearance of humans as an agent of climate change is an untested factor with no geological precedent. All we know for certain is that the rapid addition of greenhouse gases to the atmosphere in the past has led to extreme climate change and extinction. This may be your future—caveat emptor (buyer beware!).

Looking Ahead . . .

So far you have considered the evidence for both recent and ancient climate change and reviewed the physical processes that determine how and when climate change will occur. In Chapter 7 you will investigate the environmental and human impact of climate change around the world today and consider what further changes are likely to occur in the near future.

Critical Thinking Questions

1. Why is it unlikely that the world will ever experience another period of global glaciation when ice sheets reach at the equator?
2. Compare the climate during the late Carboniferous Period with the climate today and highlight how global geography is an important factor in each case.
3. Discuss why it is unlikely that large areas of the sea floor beneath the Atlantic Ocean will become oxygen depleted and toxic in the near future.
4. Compare global climate during the late Cretaceous Period with global climate today. Highlight how global geography is an important factor in each case.
5. Compare and contrast rapid climate change during the time of the Paleocene–Eocene Thermal Anomaly with climate change today.
6. Describe the geological evidence that would indicate when ice sheets first started to grow on Antarctica.
7. Why is it not likely that the sudden cooling Earth experienced during the Younger Dryas event will happen again in the near future?
8. Explain why the global cooling Earth experienced during the Little Ice Age could happen again in the near future.
9. Compare and contrast global climate change during the Medieval Climate Anomaly with climate change today.
10. How does the study of ancient climate inform us about the sensitivity of climate to small changes in radiative forcing today?

Key Terms

Please make sure you are familiar with the key terms introduced and highlighted in this chapter. A full glossary is available at the end of the book.

Anaerobic bacteria, p. 168
Anoxic, p. 168
Anthropocene, p. 202
Aqueous solution, p. 164
Basal sliding, p. 190
Binge and purge, p. 191
Bipolar seesaw, p. 200
Bølling–Allerød interstadial stage, p. 201
Carboniferous Period, p. 164
Cenozoic Era, p. 174
Continental rifting, p. 180
Cretaceous Period, p. 168
Dansgaard–Oeschger (D–O) cycles, p. 197
Dropstone, p. 162
Eemian Interglacial Stage, p. 197
Effusive volcanism, p. 168
Eocene Epoch, p. 180
Eukaryotes, p. 162

European Project for Ice Coring in Antarctica (EPICA), p. 195
Exhumed landscape, p. 164
External factors, p. 179
Geothermal heat, p. 190
Greenland Ice Core Project (GRIP), p. 195
Greenland Ice Sheet Project (GISP), p. 195
Greenland Ice Sheet Project 2 (GISP2), p. 195
Heinrich event, p. 198
Holocene Epoch, p. 202
Ice stream, p. 190
Internal factors, p. 179
Jurassic Period, p. 170
K–T boundary, p. 173
Lake Agassiz, p. 202
Last glacial maximum (LGM), p. 200

Laurentide Ice Sheet, p. 198
Little Ice Age (LIA), p. 205
Liverwort, p. 171
Loess, p. 188
Mapping Spectral Variability in Global Climate Project (SPECMAP), p. 194
Marine benthic extinction event (MBE), p. 177
Marine Isotope Stages (MIS), p. 194
Maunder Minimum, p. 205
Medieval Climate Anomaly (MCA), p. 204
Mid-Pleistocene Transition (MPT), p. 193
Miocene Epoch, p. 183
Neogene Period, p. 183
Northern Greenland Ice Core Project (NGRIP), p. 195
Oligocene Epoch, p. 180

Orogenesis, p. 165
Paleocene Epoch, p. 175
Paleocene–Eocene Thermal Anomaly (PETA), p. 175
Paleogene Period, p. 174
Pangaea, p. 167
Permian Period, p. 164
Phanerozoic Eon, p. 169
Pleistocene Epoch, p. 186
Pliocene Epoch, p. 184
Prokaryotes, p. 162
Proterozoic Eon, p. 162
Quaternary Period, p. 186
Rock striations, p. 162
Rudist, p. 170
Stack, p. 194
U-shaped valley, p. 165
Younger Dryas event, p. 201

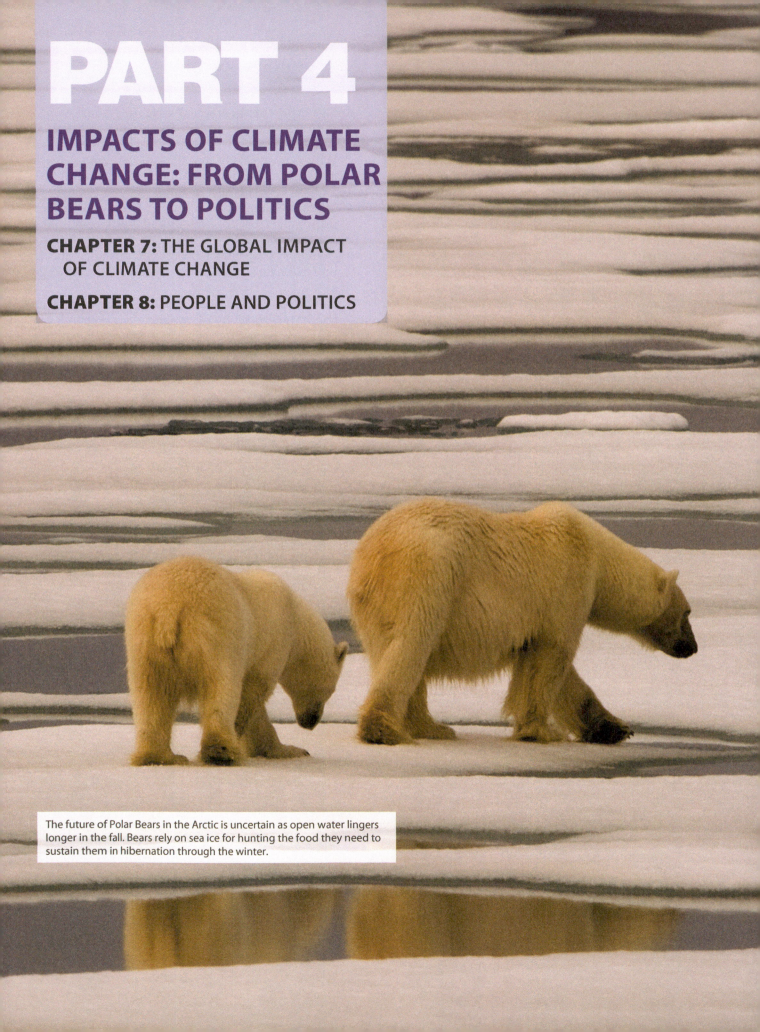

PART 4

IMPACTS OF CLIMATE CHANGE: FROM POLAR BEARS TO POLITICS

CHAPTER 7: THE GLOBAL IMPACT OF CLIMATE CHANGE

CHAPTER 8: PEOPLE AND POLITICS

The future of Polar Bears in the Arctic is uncertain as open water lingers longer in the fall. Bears rely on sea ice for hunting the food they need to sustain them in hibernation through the winter.

Climate change is a global problem
that requires a global solution. Chapter 7
illustrates how observations from around the
world, on land and in the oceans, show us that
climate change is a real, present, and deepening
threat to society. Many ecosystems now struggle
to cope with the combined impact of human
development and climate change. Chapter 8
considers how policymakers are using scientific
and economic data to search for a solution, but
political, social, and economic factors continue
to impede progress. In the near future, many
communities, unable to cope or adapt to the
impact of changing climate, will be forced to
move as climate refugees, creating new social
problems and enhancing the risk of conflict. We
must improve the quality of communication and
political debate if we hope to reach an agreed
upon solution to the risks of climate change.

Sandbags used as a form of coastal defense to prevent the erosion of an island in the Maldives. As global temperature rises and more ice melts, the level of the oceans will rise and small island nations will be among the first to be affected.

CHAPTER 7

The Global Impact of Climate Change

Introduction

The signs of climate change are visible around the world, and many people and ecosystems are under threat. The impact of rising temperature is greatest in Arctic regions, where rapid warming is melting ice sheets, glaciers, tundra, and sea ice at an alarming rate. In more temperate regions, the length of the seasons and weather patterns are changing, bringing the threat of severe drought to some areas and more intense rainfall to others. In many of the world's great mountain chains, the deep melting of snow pack increases the threat that major rivers will no longer carry enough water to support agriculture over the summer. In the tropics, the Amazon forests have become vulnerable to drought and the impact of deforestation. Many of the world's great coral reefs are in danger from the triple threat of rising temperature, ocean acidification and human development. Across the world, rising sea level threatens to drown deltas, small islands, and coastal cities. Future climate change is a threat to us all, but especially to impoverished communities that lack the resources to cope and adapt. The forced migration of climate refugees will exacerbate problems of crime and urban crowding in cities and increase the likelihood of conflict as refugees try to cross international borders.

There is still some uncertainty about the rate at which these changes will take place, and any actions taken to mitigate their impact involve a level of risk. . . . but failure to act when there is still time to stop these events from happening is a conscious act of neglect for which coming generations will hold us all accountable.

Learning Outcomes | *When you finish this chapter you should be able to:*

- Understand the differences between coping, adaptation, and mitigation
- Project how climate change will impact the environment over the next century
- Describe how human activity makes it more difficult for natural systems to adapt to climate change
- Discuss why climate change is a growing threat to national security and political stability
- Explain why any solution to the problem of climate change must include measures to address issues of poverty, equality, education, and environmental justice
- Use case studies to illustrate the impact of climate change on the natural environment and human development

A Global Problem

Satellite and ground-based data show that Earth's climate is changing rapidly as mean global temperature rises. This effect is most noticeable at high northern latitudes, but significant change is apparent almost everywhere around the world (**Figure 7.1**).

Projections of Climate Change

There is rising concern that our failure to limit heat-trapping greenhouse gas emissions will lead to rapid and irreversible changes in climate that will have damaging impacts on society and the environment. Computer models (see Chapter 1) project an array of possible future climates, but how climate will actually change depends on the actions we take to control the emission of heat-trapping greenhouse gases today. Most models project the widespread emergence of damaging climate change in the near future if nothing is done to abate emissions. Disturbingly, some of these projected changes are starting to appear earlier than anticipated.

Coping with, Adapting to, and Mitigating Climate Change

Many communities will be able to cope with the early stages of climate change by making simple changes to their existing lifestyle. These **coping strategies** require little structural change, and the costs are manageable. Examples include community education to inform the population about emerging risks and hazards, water conservation programs to manage resources, the planting of drought-resistant crops to maintain levels of food production, and deepening of wells to keep pace with falling water tables. Climate change response plans should consider how to help communities that are most at risk. Disaster mitigation and response plans must prepare for the timely evacuation, housing, and eventual return of communities most at risk from flooding and the impact of tropical storms. The case study of Bangladesh, later in this chapter, provides an example of how a developing nation has responded very effectively to the threat of tropical storms and saved thousands of lives in the process.

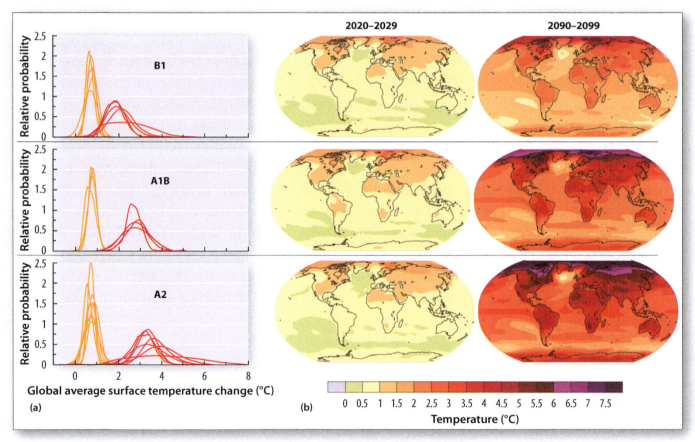

▲ **Figure 7.1:** Projected surface temperature changes for the early and late 21st century relative to the period 1980–1999. The central and right panels of globes show the multi-model average projections for the B1 (top), A1B (middle) and A2 (bottom) SRES scenarios averaged over the decades 2020–2029 (center) and 2090–2099 (right). The left panels show corresponding uncertainties as the relative probabilities of estimated global average warming from several different model studies for the same periods. Some studies present results only for a subset of the SRES scenarios, or for various model versions. Therefore the difference in the number of curves shown in the left-hand panels is due only to differences in the availability of results.

As the impact of climate change intensifies, communities most at risk will be forced to make substantial changes. Examples include building larger storm drains to cope with increased rates of precipitation, building reinforced coastal flood walls and flood gates to cope with stronger storm surges, digging new wells to reach deeper aquifers, and building desalination plants to provide fresh water. Many poor communities will not have the social, financial, or technical resources—what is called the **adaptive capacity**—necessary to respond in this way.

A community faces a **crisis point** when the impacts of climate change exceed its capacity to adapt. Entire regions will experience forced migration when people have no remaining choice but to leave their homes and move to nearby regions, cities, or countries as climate refugees.

The only way to prevent this from happening is to **mitigate** (diminish) the impact of climate change by taking early positive action to reduce the emission of heat-trapping greenhouse gases. This is still possible and is the subject of Chapters 8, 9, and 10 of this book.

CHECKPOINT 7.1 ▶ Describe what is meant by coping, adaptation, and mitigation.

Examples of the Impact of Climate Change on Society

The human impact of climate change will be devastating in many of the poorest nations of the world, where societies lack the social, technological, and economic capacity to adapt. Famine, disease, and conflict are already on the rise in regions such as the **Sahel** in Sub-Saharan Africa, and competition for diminishing water resources is expected to intensify across many regions.

Conflict and Forced Migration Many of the people who are forced to move away from their homes by the impact of climate change will not be welcomed at their destination. The countries affected will almost certainly resist the unwanted tide of refugees by reinforcing border restrictions and tightening controls on immigration. Major cities will become severely overcrowded and suffer from rising levels of unemployment, disease, and violent crime. As nations compete for access to dwindling water and food supplies, the social tensions created by forced migration and overcrowding will increase the chances of international conflict.

CHECKPOINT 7.2 ▶ Why will the probability of international conflict increase with climate change driven by global warming?

Floods and Drought Historical patterns of temperature and precipitation are already changing. Rising global temperatures appear to have increased the frequency and intensity of drought in some areas and the intensity of precipitation and floods in others. Each

change to a regional pattern of precipitation is a threat to the welfare of local communities that must cope or adapt to the change or move away.

Marginal deserts, such as the Sahel region of Africa, the southern Mediterranean, parts of Australia, and large parts of the southwestern United States, will experience more intense and longer periods of drought and increasing competition for dwindling water supplies. By 2020, between 75 and 250 million people in Africa are projected to be exposed to increased **water stress**, and yields from agriculture may be reduced by as much as 50% in some regions.

In contrast, areas such as the northeastern United States and parts of northwestern Europe will experience an increase in the frequency of intense periods of precipitation and damaging floods that exceed the coping capacity of existing infrastructure. Upgrading and replacing storm water systems in major cities alone will cost billions of dollars.

CHECKPOINT 7.3 ▶ What change in the character and pattern of regional precipitation in the United States presents the greatest challenge to existing storm water systems?

Snowpack and Water Resources Winter precipitation in the form of snow and ice at high elevations is a vital source of water for many areas of the world. Ice sheets and glaciers in mountainous regions such as the Alps, Andes, Himalayas, and Sierra Nevada provide meltwater that sustains agriculture and human consumption over hot, dry summers. Climate models project that the volume of **snowpack** that accumulates over the winter and survives into the following summer will decrease. Rising air temperature will deliver more early winter precipitation as rain, and the faster melting of snowpack in the spring will increase the risk from damaging floods. The Himalayas face the additional threat of meltwater being trapped behind fragile barriers of glacial sediment that will eventually break and release catastrophic floods into densely populated valleys.

CHECKPOINT 7.4 ▶ What combination of factors makes it likely that water resources in California will be adversely affected by climate change?

Rising Sea Level Rising sea level threatens low-lying coastal regions, and especially isolated communities on small Pacific islands, barrier islands, and delta plains. People living close to the coast are often among the poorest in the world. As sea level rises, coastal residents find it increasingly difficult to protect themselves from storm surges related to intense tropical depressions and hurricanes. **Coastal aquifers** and agricultural land are quickly contaminated by salt water driven inland by such storms and can take years to recover. **Storm surges** also threaten large coastal cities, and the eventual cost of keeping the rising water at bay may become a crippling economic burden on taxpayers, as happened in New York and New Jersey in 2012 after Hurricane Sandy.

CHECKPOINT 7.5 ▶ Why will rising sea level affect agricultural production long before coastal flooding becomes permanent?

Examples of the Impact of Climate Change on the Environment

Natural scientists closely monitor the environment for the impact of climate change. Many terrestrial and marine ecosystems already show signs of environmental stress, made worse by the impact of human activity. Under this combined pressure of climate change and human development the global species extinction rate is accelerating alarmingly.

Oceans and Rivers Ocean acidification and increasing sea surface temperature threaten the stability of many marine ecosystems. Important species such as reef-building corals are already suffering, and destructive species such as poisonous jellyfish and toxic phytoplankton are proliferating. Fish stocks are also declining in some rivers as rising water temperatures affect spawning. **Dead zones** are spreading at depth in rivers, in estuaries, and in low-lying areas on the continental shelf where rising temperatures act to reduce the level of vital oxygen in the water.

CHECKPOINT 7.6 ▶ What are dead zones, and why are they an important indication of rising temperature?

Tropical and Temperate Forests Tropical forests are under threat from deforestation and prolonged drought. Fires are reaching deep into untouched forests that have not evolved to cope with periodic burning. Temperate forests are under threat from insect pests such as the devastating pine beetles that are moving inexorably northward with climate change and rapidly devastating ancient evergreen forests across North America, Europe, Russia, and even Australia. The poisonous brown recluse spider in the United States is also gradually extending northward into cities where it has never been found before, and both alien and invasive species are proliferating in more favorable climes. In Virginia, people are starting to wonder when the first alligators will appear in the James River.

CHECKPOINT 7.7 ▶ How are temperate and tropical forests under threat from climate change?

Arctic Melting The Arctic tundra is changing more rapidly than any other ecosystem. Permafrost is melting to record depths, and long-frozen vegetation is starting to rot and release the heat-trapping gases methane and carbon dioxide into the atmosphere. The southern limit of permafrost has retreated by 39 kilometers (24 miles) on average and by as much as 200 kilometers (124 miles) in some parts of Arctic. Surface vegetation is also changing, as more evergreen shrubs and stunted tress replace the peatland grasses, sedges, mosses, and lichens that are more typical of the tundra. In some places, the boreal forest (**taiga**) tree line has already started to shift to the north, and some models project that by 2100, the tree line will have advanced by as much as 500 kilometers (311 miles), resulting in a loss of 51% of tundra habitat.

CHECKPOINT 7.8 ▶ Why is the latitude of the tree line a good indicator of climate change?

Dealing with Risk and Uncertainty

Computer models based on IPCC projections suggest that global temperature will rise a further 1°C by 2020–2029. Beyond 2030 the models project very different scenarios that depend on the future rate of greenhouse gas emissions (**Figures** 7.1 and **7.2**).

Under the socially ambitious and politically challenging IPCC B1-emissions scenario (where the level of emissions falls rapidly), the models project an increase in global temperature by 2100 of around 1.8°C relative to the 1980–1999 average. This is a tolerable level that would allow most countries to cope and adapt to the impact of climate change. The temperature projection rises to 2.8°C under the more realistic A1B-emissions scenario and to as much as 3.2°C for the "worst-case" (and increasingly likely) A2- emissions scenario. This is at least twice the magnitude of change in temperature that has been observed over the past 100 years and enough to generate

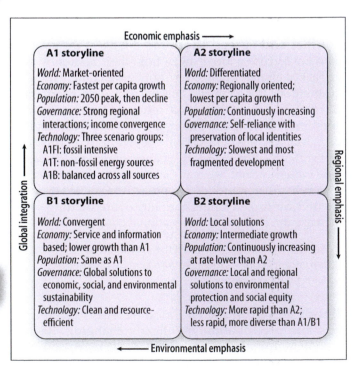

▲ **Figure 7.2:** The main IPCC (AR4) economic scenario storylines (see Chapter 1).

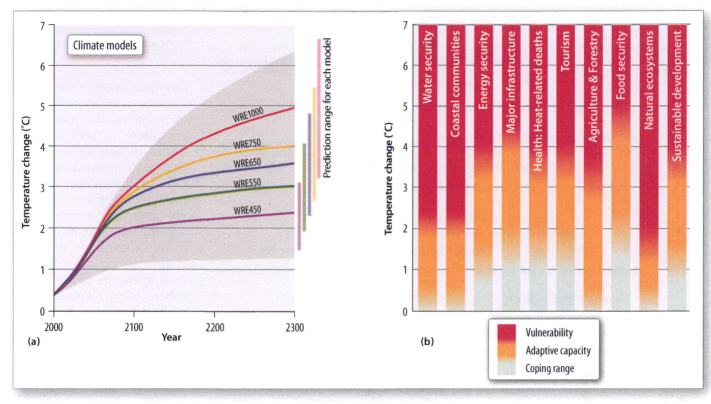

▲ **Figure 7.3:** IPCC prediction of global impact of temperature change. (a) The gray area on the graph encloses the range of possible temperature change according to the IPCC, if greenhouse gas emissions are stabilized at specific target values in parts per million by 2050 (450 to 1,000 ppm). (b) The chart estimates the impact of changing global temperature on 10 major areas of economic and environmental interest. Note how many are affected by a rise in temperature between 2°C and 3°C as projected for the end of the Century by many computer models.

significant hazards to both society and the environment. Each of these scenarios is still possible. The global temperature at the end of the 21st century depends entirely on what actions we take today to limit emissions.

The Increasing Risk of Climate Change To illustrate the risks associated with rising global temperature, the IPCC has identified how climate change will impact 10 areas of economic and environmental importance (**Figure 7.3b**). These data make it clear that the minimum 2.8°C of warming that scientists believe is already locked into Earth's climate system will adversely affect water scarcity, coastal communities, and natural ecosystems, and they believe it will require significant adaptation in other areas. If global temperatures increase by 3.2°C (5.8°F), as seems increasingly likely, all 10 key areas identified by the IPCC will reach a point of vulnerability where coping and adaptation are no longer options.

Beyond 2100 What will happen after the end of the 21st century? Climate models give us an accurate view of the *near* future, but intrinsic **uncertainties** in economic projections make it difficult to say exactly how much change will occur beyond the next 100 years.

If nothing is done to limit emissions for 100 years, the temperature could rise by more than 6°C (10.8°F).

Such a change would be truly catastrophic to both the natural environment and human cultures.

CHECKPOINT 7.9 ▶ Why do you think is it so difficult to project the future of economic development?

The Precautionary Principle Future climate change will be global, progressive, and increasingly destructive. The data from computer models and observations from ancient climate change are conclusive and incontrovertible (**Figures 7.4** and **7.5**). The international scientific community and all major government-funded scientific organizations, both in the United States and overseas, agree that climate change poses a risk of significant damage to economic development and global security. The only uncertainties that remain are climate sensitivity and the future level of greenhouse gas emissions (Chapter 3). According to the **precautionary principle**, any action or policy that might cause harm to the public or environment, such as the continued burning of fossil fuels, should be proven not to be harmful before it is allowed to continue unabated. Should the burden remain on climate scientists to prove that they are correct about climate change beyond any reasonable doubt before we take any action to limit greenhouse gas emissions? Or is it the responsibility of the fossil fuel–driven industries to

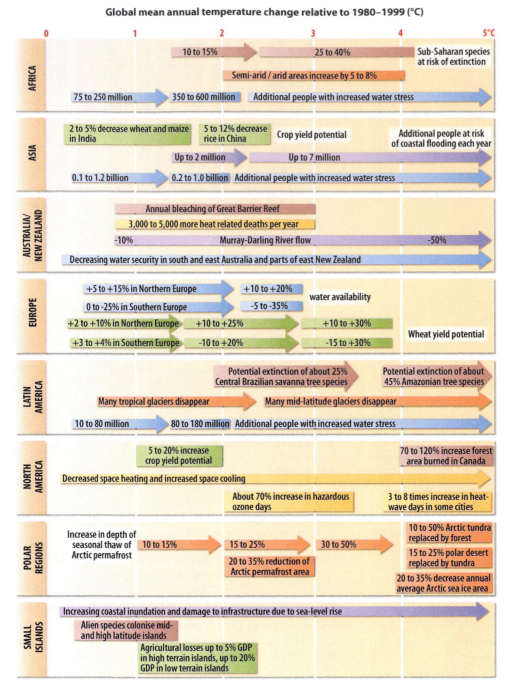

Global mean annual temperature change relative to 1980–1999 (°C)

◀ **Figure 7.4:** The progressive impact of rising temperature on eight major regions of the world.

change in one part may have a totally unexpected impact on another. The terrestrial biosphere is very closely linked to climate, and any change in climate is quickly reflected by changes in the biosphere. The deep-marine environment is relatively isolated from climate change in the atmosphere, but any change in the composition or temperature of the atmosphere is soon reflected in the surface layers of the ocean and eventually transmitted to the ocean floor. Changing global temperature affects the pattern of atmospheric and ocean circulation and the supply of nutrients to ocean ecosystems.

Terrestrial Biomes and Ecosystems

A **biome** is an area that covers a large region of the world where similar plants, animals, and other living things have adapted to regional climate and other physical conditions (**Figure 7.6**). Each biome is composed of a number of similar yet distinct **ecosystems**, where living organisms that are adapted to their physical environment compete for resources (**Figure 7.7**). Each ecosystem forms an evolving complex web of connectivity between species. Embedded in that web are certain **keystone species** that have a disproportionately large impact on the ecosystem and play a critical role in maintaining the overall structure of ecological communities. Some large **charismatic megafauna**, such as polar bears, arctic fox, fur seals, and wolves, attract great popular interest, but much smaller creatures, such as insects, tiny metazoans, and bacteria in the soil, may be equally important for the duration and stability of an ecosystem.

prove that emissions are harmless? The answer is partly political, and this is a theme that Chapter 8 will consider in some detail.

CHECKPOINT 7.10 ▶ In your own words, describe the idea of the precautionary principle.

Adaptation to Climate Change in the Natural World

The **biosphere** can be defined as the sum total of all living organisms on Earth. We have only partly explored the biosphere, and we understand it incompletely. There are many interconnected and interdependent parts, and

CHECKPOINT 7.11 ▶ What role do keystone species play in maintaining the integrity of an ecosystem?

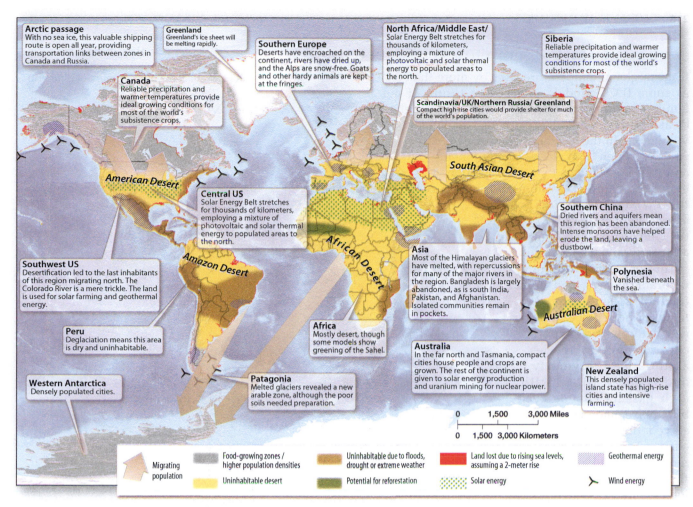

Arctic passage
With no sea ice, this valuable shipping route is open all year, providing transportation links between zones in Canada and Russia.

Greenland
Greenland's ice sheet will be melting rapidly.

Southern Europe
Deserts have encroached on the continent, rivers have dried up, and the Alps are snow-free. Goats and other hardy animals are kept at the fringes.

North Africa/Middle East/
Solar Energy Belt stretches for thousands of kilometers, employing a mixture of photovoltaic and solar thermal energy to populated areas to the north.

Siberia
Reliable precipitation and warmer temperatures provide ideal growing conditions for most of the world's subsistence crops.

Canada
Reliable precipitation and warmer temperatures provide ideal growing conditions for most of the world's subsistence crops.

Scandinavia/UK/Northern Russia/ Greenland
Compact high-rise cities would provide shelter for much of the world's population.

Central US
Solar Energy Belt stretches for thousands of kilometers, employing a mixture of photovoltaic and solar thermal energy to populated areas to the north.

Southern China
Dried rivers and aquifers mean this region has been abandoned. Intense monsoons have helped erode the land, leaving a dustbowl.

Southwest US
Desertification led to the last inhabitants of this region migrating north. The Colorado River is a mere trickle. The land is used for solar farming and geothermal energy.

Asia
Most of the Himalayan glaciers have melted, with repercussions for many of the major rivers in the region. Bangladesh is largely abandoned, as is south India, Pakistan, and Afghanistan. Isolated communities remain in pockets.

Polynesia
Vanished beneath the sea.

Peru
Deglaciation means this area is dry and uninhabitable.

Africa
Mostly desert, though some models show greening of the Sahel.

Australia
In the far north and Tasmania, compact cities house people and crops are grown. The rest of the continent is given to solar energy production and uranium mining for nuclear power.

New Zealand
This densely populated island state has high-rise cities and intensive farming.

Western Antarctica
Densely populated cities.

Patagonia
Melted glaciers revealed a new arable zone, although the poor soils needed preparation.

American Desert · Amazon Desert · African Desert · South Asian Desert · Australian Desert

Legend:
- Migrating population
- Food-growing zones / higher population densities
- Uninhabitable desert
- Uninhabitable due to floods, drought or extreme weather
- Potential for reforestation
- Land lost due to rising sea levels, assuming a 2-meter rise
- Solar energy
- Geothermal energy
- Wind energy

0   1,500   3,000 Miles
0   1,500   3,000 Kilometers

▲ **Figure 7.5:** A 4°C rise in global temperature is not expected, but it is not impossible by the end of this century if nothing is done to limit greenhouse gas emissions. This map illustrates the terrible impact a rise in global temperature of this magnitude would have across the world, but also shows that large parts of the world are suitable for the generation of alternative energy that can be used to reduce global greenhouse gas emissions and avoid the most damaging effects of climate change.

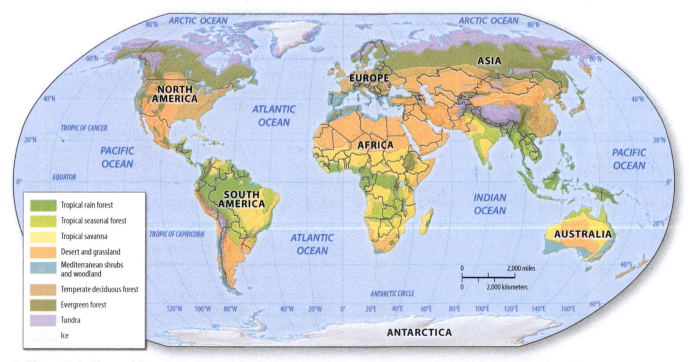

Legend:
- Tropical rain forest
- Tropical seasonal forest
- Tropical savanna
- Desert and grassland
- Mediterranean shrubs and woodland
- Temperate deciduous forest
- Evergreen forest
- Tundra
- Ice

0   2,000 miles
0   2,000 kilometers

▲ **Figure 7.6:** The world's major terrestrial biomes are not equally vulnerable to the impact of climate change, and it is even possible that some regions will experience positive changes in the near future. As the effects of rising global temperature intensify, however, the ability of many biomes to respond to climate change will be limited and the effects will be accentuated by the additional impact of urban development, agriculture, and deforestation. Which of these major biomes do you think is most vulnerable?

Environmental Vulnerability

If mean global temperature rises by more than 3.2°C, as some IPCC models project, the future of our familiar natural environment looks increasingly bleak. However, not all ecosystems are equally vulnerable. In the Northern Hemisphere, the study of pollen from glacial lake sediments and peat bogs shows that the transition between major ecosystems during the **Pleistocene Period** followed a consistent pattern. Every time the climate changed, less well-adapted species died out locally or migrated to more favorable climes, and more tolerant species moved in to take their place. Many ecosystems did experience changes in species abundance, diversity, and composition with each passing climate cycle, but their major characteristics stayed the same. This is because changes took place slowly enough for most plants and animals to keep pace with the rate of climate transition. There was an increase in the background rate of species extinction as climate swings became more extreme, but the fossil record does not show any evidence of widespread extinction each time the climate changed.

> **CHECKPOINT 7.12 ▶** What does the pollen record from lakes and bogs tell us about changes to ecosystems in the Northern Hemisphere during the past 4 million years?

In contrast to ecosystems at high latitude, ecosystems that evolved to cope with less extreme natural variation in climate, such as the forests of the Amazon Basin and the Great Barrier Reef, may be more vulnerable to climate change. Many species, such as the famous Amazonian tree frogs, have become so specialized that they are unable to cope with even small changes in their external environment. **Adaptation**, competition, and **specialization** may have been necessary for the survival of these species in the past, but they have not prepared them for the challenge they face today.

Human Impact on the Biosphere

Ecosystems may have been able to change in response to climate change in the past, but that does not mean they will be able to do so again in the future. Human activity has altered the physical and biological landscape, and decreased the capacity of many ecosystems to respond effectively to rapid climate change. Many large ecosystems around the world are

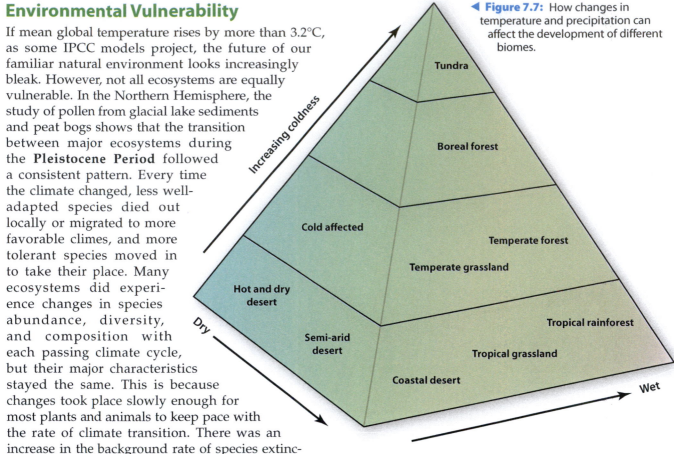

◀ **Figure 7.7:** How changes in temperature and precipitation can affect the development of different biomes.

already affected by human activity. Roads, cities, farms, coastal development, and deforestation have together fragmented a once-continuous landscape into smaller areas, leaving communities of plants and animals in barely sustainable pockets that lack connectivity. Cut off from surrounding communities, threatened by aggressive **invasive species**, and isolated from natural **migration corridors**, these communities have nowhere to go as the climate changes. Given time, some species may jump these human barriers, but time is something they may not have.

> **CHECKPOINT 7.13 ▶** How has the development of modern society made it more difficult for many animals and plants to adapt to climate change?

How Will Ecosystems Adapt to Modern Climate Change?

The impact of climate change today will be greater than at any other time in the recent geological past. Rapid climate change today is superimposed on a landscape that is often transformed by human activity, and the impact will prevent many species from adapting in ways that worked successfully before. Climate models project how terrestrial ecosystems will change by 2100, according to worst-case (A2) and best-case (B1) emission scenarios (see Chapter 1). Compare the two maps in **Figure 7.8** and consider where the changes are greatest and why that might be the case. In what parts

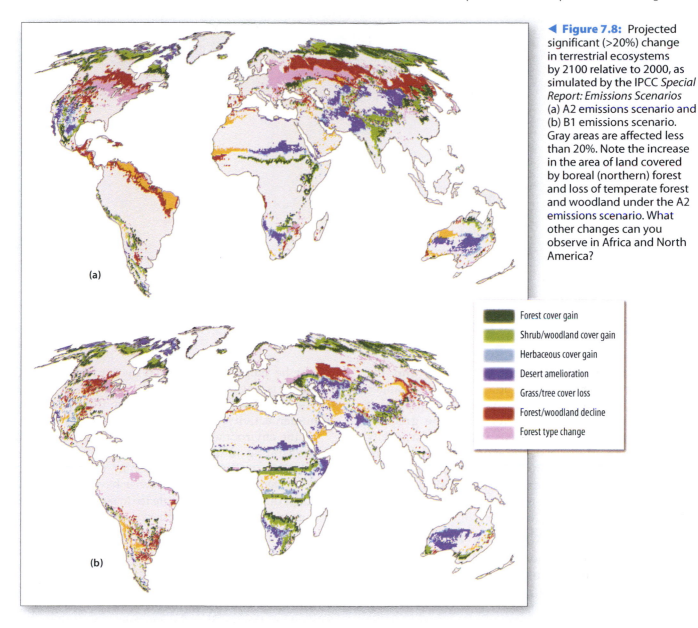

◀ **Figure 7.8:** Projected significant (>20%) change in terrestrial ecosystems by 2100 relative to 2000, as simulated by the IPCC *Special Report: Emissions Scenarios* (a) A2 emissions scenario and (b) B1 emissions scenario. Gray areas are affected less than 20%. Note the increase in the area of land covered by boreal (northern) forest and loss of temperate forest and woodland under the A2 emissions scenario. What other changes can you observe in Africa and North America?

Legend:

- Forest cover gain
- Shrub/woodland cover gain
- Herbaceous cover gain
- Desert amelioration
- Grass/tree cover loss
- Forest/woodland decline
- Forest type change

of the world will the human impact on the landscape have the greatest effect (**Figures 7.9** and **7.10**)?

CHECKPOINT 7.14 ▶ Explain why it is important to understand the rate at which climate is changing.

It is important to take time to reflect on the overwhelming impact humanity has had on the natural environment and how our activities make it increasingly difficult for ecosystems to adapt quickly in the face of climate change.

Case Studies of the Global Impact of Climate Change

The following case studies illustrate how Earth's climate system has already responded to a rise in global mean annual temperature in the atmosphere and oceans of just 0.63°C. Even this level of change

▼ **Figure 7.9:** The impact of urban and rural development on both animal and plant communities reduces their ability to adapt to rapid climate change.

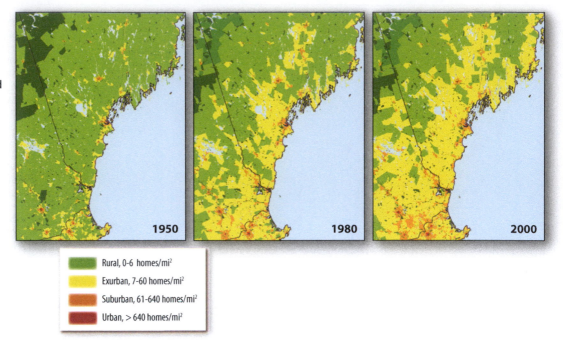

▶ **Figure 7.10:** Natural habitats around the world, like these in New England (in green) have been rapidly fragmented by the impact of rural and urban development (in yellow).

1950   1980   2000

Rural, 0-6 homes/mi²
Exurban, 7-60 homes/mi²
Suburban, 61-640 homes/mi²
Urban, > 640 homes/mi²

presents a significant threat to society and the environment. For society to adapt to these changes and mitigate their worst impact, it is important to understand where the effects will be greatest. Unfortunately, many of the areas affected most acutely are among the poorest in the world. In those areas, adaptation and mitigation are not always options due to the combined effects of poverty, political conflict, social injustice, tradition, and corruption.

How to Approach the Case Studies

Six main factors determine the social and environmental impacts of climate change:

1. *Natural variation* of climate due to solar and other natural forcing.
2. How much *human activity* will continue to release greenhouse gas emissions into the atmosphere over the next century.
3. How much global temperature will rise due to the radiative forcing of *greenhouse gases alone*. The best estimate for doubling the level of greenhouse gas in the atmosphere is 1.2°C (2.2°F).
4. How much *Earth's climate system* will then magnify the direct greenhouse gas radiative forcing. Many feedbacks in the Earth system can both intensify and attenuate the impact of rising temperature, but the scientific consensus is that positive (warming) feedback will predominate. By the end of the 21st century, an increase in global average temperature of between 2.0°C (3.6°F) to 4.5°C (8.1°F) is most likely.
5. The *rate of climate change*. Over the past 40 years, climate has been changing faster than at any other time in recent geological history. While most ecosystems have evolved to adapt to climate change, some

may be simply overwhelmed by the rate at which change is now taking place.
6. The additional pressure on the environment from human activity. We have had a profound impact on the environment, reshaping the landscape and devastating natural ecosystems through deforestation, pollution, and urban development. Ecosystems that might otherwise have been able to adapt may no longer be able to do so because of the human impact on the environment.

In each of the following case studies, consider how these six factors work together to impact different regions around the world and reflect on the following questions:

- How sensitive is this environment to climate change?
- How fast is this change occurring?
- What are the social and economic consequences of this change?
- How much has human activity on the ground made the problem worse?
- What steps could we take to *mitigate* the worst impact of climate change?
- What steps could we take to *adapt* to the worst impact of climate change?
- Is the best solution simply to adapt and mitigate when possible, in the hope that economic growth and technology will find an answer?

Case Study 1: Freshwater Resources

Nothing is more important to the health of an ecosystem and to the welfare of society than the availability of fresh water. The impact of climate change on the availability of water is expected to vary around the world. Some places will see more rainfall and others more drought. The worst impact will almost certainly be on countries already under significant water stress.

Water Supply

There are two main sources of fresh water. Groundwater accounts for 90% of fresh water resources, with over 1.5 billion people depending on it for survival. Some groundwater comes from shallow aquifers that can be tapped with a simple well and recharged annually by rainfall. Groundwater is also recovered from deep and sometimes ancient aquifers that hold substantial reserves but are recharged much more slowly—or not at all.

Rivers are the second major source of fresh water, and the supply depends on annual precipitation. Many rivers simply run dry or flow at much lower levels during the summer season, and in more arid regions, they are particularly vulnerable to changes in seasonal precipitation. Rivers fed by mountain snowpack are dependent on the depth of frozen winter precipitation and the timing of the spring melt, and they can sustain flow only as long as snow and ice at higher elevations survive through the summer.

Water Stress

The United Nations Environment Program (UNEP) estimates that two out of three people around the world will suffer from water stress by 2025. (Water stress is defined as per capita availability of less than 1,700 m³ [449 gallons] per year.) Even today, 20% of the world's population does not have access to safe drinking water, and more than 2 million people, many of them children, die annually due to infection related to water contamination. Of the fresh water in use today, 75% is used for agriculture, 20% for industry, and only 5% for direct consumption, but increasing population pressure is creating unsustainable demand, and many sources of water are depleting faster than they are being replenished.

Many areas of the world are already suffering from signs of water stress. The availability of water is determined by comparing the rate of demand with the rates of river flow and groundwater recharge (Figures 7.11 and 7.12). Climate models project that Sub-Saharan Africa (especially the Sahel region), North Africa, the Middle East, the southwestern United States, and southern Europe are all likely to suffer increased water stress as the climate changes. Reduced rainfall and higher-than-average temperatures increase the risk of drought, the rate of surface evaporation, and the buildup of damaging salt deposits in otherwise fertile soils (Figure 7.13).

CHECKPOINT 7.15 ▶ In your own words, describe water stress and why it is a rising problem across the world.

Water Stress in the United States

Water stress is a growing problem in certain regions of the United States. The rate of population growth, rising demands on agriculture, and the location of major metropolitan centers have all increased the demand for

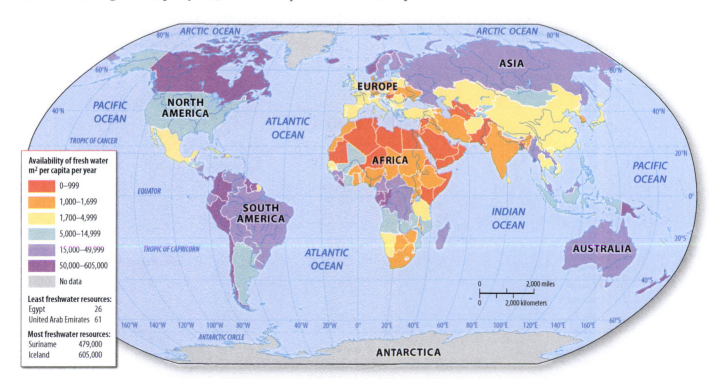

▲ **Figure 7.11:** The availability of fresh water from average river flows and groundwater recharge, in cubic meters per capita per year. Notice the low level of water availability in parts of Africa, China, India, and Europe.

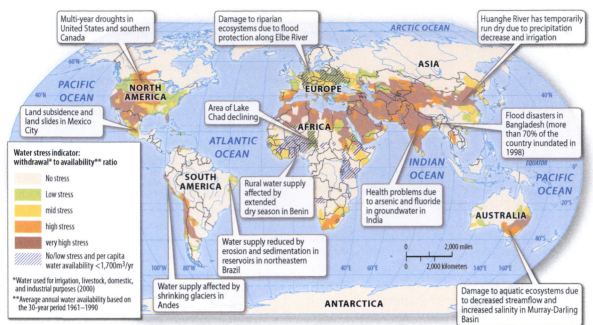

▲ **Figure 7.12:** A water stress map shows key areas of vulnerability. Many of the countries with limited or scarce supply of water are politically unstable and have been involved recently with some form of ethnic conflict. Any further reduction in the availability of water will increase the threat of future conflict. Water stress is also related to poverty. Notice that a number of richer countries with low water availability (Figure 7.11) have no or low level of stress, while some poorer countries with similar water availability have high stress levels (for example compare Europe with India). Why is this so?

water, and in many places demand now exceeds the sustainable capacity of water reserves.

California   The melting of snow that has accumulated over the winter (snowpack) is a vital source of water for many rivers in the United States. This water is needed to feed human consumption, support agriculture, and supply industry.

In the United States, climate models project a 25% reduction in the volume of the Sierra Nevada snowpack by 2050, as more winter precipitation falls as rain. The California Climate Change Center then projects that rising spring temperatures will melt this diminished snowpack much more quickly than before. This will lead to spring floods and prolonged summer droughts that will increase pressure on already stressed **aquifers**.

▲ **Figure 7.13:** Climate change's projected contribution (in 2050) to declining water availability, as a percentage change from the 1961-1990 average. Note the reduction of water availability in the Amazon Basin, parts of Africa and Southern Europe. Note also the general increase in water availability in parts of Asia. White areas indicate where there are no data.

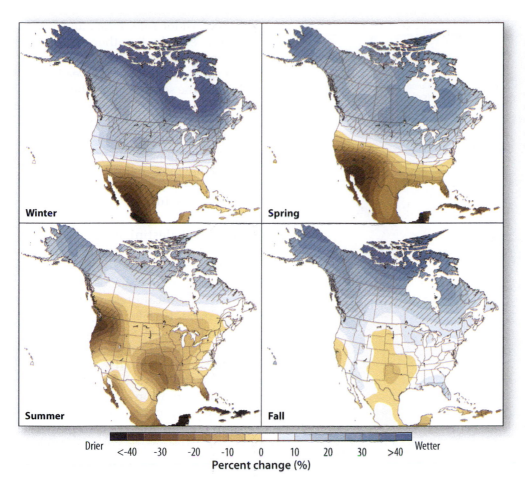

◀ Figure 7.14: Model projections of the expected seasonal impact of climate change on precipitation in the United States late in the current century. Note the general pattern of change, with northern areas getting wetter and southern areas drier. What does this imply for California?

Winter

Spring

Summer

Fall

Drier <-40 -30 -20 -10 0 10 20 30 >40 Wetter
Percent change (%)

The greatest impact will be in the San Joaquin, Sacramento, and Trinity **drainage basins**, where drought is expected to impact agricultural productivity and increase the frequency and intensity of natural fires (Figure 7.14).

CHECKPOINT 7.16 ▶ How might changes in snowpack affect the water supply in California?

The Southwestern States The American Southwest is a region prone to drought. Many times over the past 2,000 years, conditions have been worse than they are today, but in 2012 the current drought is the longest in over 100 years, and conditions are not expected to improve as the atmosphere warms. Regional temperatures are expected to increase by as much as 2°C to 5°C (3.6°F to 9.0°F) above historical baselines by the end of the century, depending on the future of greenhouse gas emissions. Higher temperatures are expected to shift the path of winter and spring storm tracks to the north, reducing the level of spring precipitation by between 20% and 40%. Some of this precipitation may be recovered from intensification of the summer monsoon, but the overall impact will be to intensify the impact of natural drought, reduce flow in major rivers such as the Colorado, and increase the area burned by natural fires each year. As major cities such as Phoenix, Albuquerque, and Las Vegas continue

to grow and increase the level of demand for dwindling water resources, more communities will become protective of their vital water supply, and the potential for local political conflict and even violence will increase.

Water Stress in Europe

The tourism industry and agriculture in southern Europe have benefited from a sunny climate moderated by proximity to the Mediterranean Sea and reliable seasonal precipitation. As global temperature changes, the climate of the region is projected to change and become much more like North Africa and the Middle East, where the availability of water is already a critical problem.

Southern Europe Climate models project that Southern Europe and parts of Eastern Europe will experience increasingly frequent and severe summer droughts over the next century. Ecosystems currently found in North Africa are expected to extend northward into Europe. If the models are correct, parts of Southern Europe bordering the Mediterranean Sea will soon experience an increase in average temperature of more than 4°C and a decrease in precipitation of as much as 40%. This would have a devastating impact on water resources, agriculture, and the ability of the region to support tourism.

▲ Figure 7.15: Major sources of water in Israel and other parts of the Middle East. Notice how the West Bank is situated on an aquifer sourced by rainfall in the mountains nearby, while the Gaza Strip relies on a narrow and vulnerable coastal aquifer.

The Middle East: Water Stress and Political Conflict

Climate models project that higher temperatures and reduced precipitation in Israel, Jordan, and the Palestinian territories will lead to more desertification and place increased stress on already scarce water resources. While Israel has the resources to obtain fresh water from deep aquifers and from the sea using desalination plants, Jordan and the Palestinian territories lack these resources and are increasingly at risk from severe water shortages as regional water reserves continue to fall (**Figure 7.15**).

The Dead Sea and Sea of Galilee The Dead Sea and Sea of Galilee add humidity to the air and create distinctive microclimates that are important for regional settlement and agriculture. Over the past decade, water resources in the region have been under increasing threat from the diversion of rivers to support both settlement and agriculture and from an increase in surface evaporation due to rising temperature. The level of water in the Dead Sea has been falling rapidly as Israel and Jordan use water from the Yarmouk and Jordan Rivers, and the level of water in the Sea of Galilee has reached 5 meters (15 feet) below the level that Israeli

authorities consider safe. Israel and Jordan are working closely together to address the problem; they have even considered creating a canal connecting the Dead Sea to the Red Sea to the south, but solutions of this magnitude could take decades to complete, and the demand for water may exceed supply before then. Continued political instability in the region makes it increasingly likely that competition over dwindling supplies of water will become a flashpoint for conflict and violence in the future.

Water Power Palestinians living on the West Bank of the Jordan River and Gaza Strip have been in conflict with Israel for decades. The Israelis have used access to water to exert political pressure for a negotiated settlement. People living on the West Bank are relatively well supplied by surface water from the lower Jordan River and an aquifer that is continually recharged in mountains to the west. In the Gaza strip, however, the situation is very different: Israel has tightly controlled the surface water supply following years of perpetual conflict, and Palestinians have been forced to depend on a coastal aquifer that is badly depleted due to over-pumping and that is contaminated by human and agricultural pollutants. Rising sea level increases the possibility that salt water will break through into this aquifer and cause irreversible damage. The destruction of the coastal aquifer would render the Gaza Strip effectively uninhabitable by the Palestinians because they do not have the economic resources necessary to obtain water by desalination.

CHECKPOINT 7.17 ▶ Describe how Israel uses water to exert political pressure on Palestinians.

Water Wars The example of one nation (Israel) using water to apply political pressure on another (Palestine) will be much more common in the future. Israel is a responsible member of the international community and fully answerable for its actions under international law, but some believe that Israel has already gone too far. Other nations that are in a position to wield power over water are not so responsible or accountable. The use of water to exert political pressure and control will become common in the future, and competition for limited resources is likely to lead to armed conflict in places where water is particularly scarce.

CHECKPOINT 7.18 ▶ Where and why are the first water wars most likely to occur?

7.1 PAUSE FOR... THOUGHT | *How do aquifers work?*

An aquifer is a porous reservoir of sandstone or limestone that is filled, like a sponge, with a large volume of fresh water. Many aquifers are slowly recharged at the surface with seasonal rain and melting snow, and they can supply a sustainable flow of water almost indefinitely. Others, like the giant North African Nubian Sandstone Aquifer System, contain "fossil" water that may be hundreds of thousands of years old and are no longer recharged. This fresh water beneath Sudan, Chad, Libya, and most of Egypt may cover more than 150,000 km^3 (58,000 square miles). It is an amazing resource that could bring life to the desert. But who owns this limited resource? Who decides how quickly the reservoir can be depleted and who will regulate the process? The Nubian reservoir is like a large underground tank, and once it is empty, it will be gone. Salt water will slowly take the place of fresh water, until the climate changes and once more brings sustained seasonal rains to North Africa. The water reserve in the Nubian Reservoir is a unique chance to do a lot of good for an area of the world in great need of water resources and under threat from climate change, but there is also a high risk of conflict if the use of this precious water is abused or overly restricted.

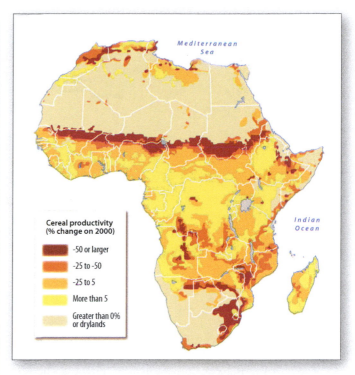

▲ **Figure 7.16:** How cereal production in Africa will be affected by climate change in 2050, according to the IPCC A2 emissions scenario. Note the large percentage decrease projected for cereal production in the Sahel region of Sub-Saharan Africa.

Sub-Saharan Africa: Drought and Migration

Some climate models suggest that the Sahel area of Sub-Saharan Africa will suffer an increase in drought that will make agriculture unsustainable. If the climate deteriorates as projected, the people living in this area will have no choice but to move as climate refugees. Such a forced migration would generate a major global humanitarian and security crisis.

The climate in this region depends on the position of the intertropical conversion zone (ITCZ) and the northern extent of the **West African monsoon** (Chapter 4). The IPCC A2 scenario projects the future of Africa's cereal crop production (**Figure 7.16**). The map in Figure 7.16 shows where the negative impacts of climate change are most likely to occur. The Sahel region to the south of the Sahara is projected to be one of the worst affected areas, with a loss of up to 50% of production capacity by 2080.

The example of the Sahel highlights how difficult it can be to project the impact of climate change in marginal areas that are affected by small shifts in major climate systems. Most models project an increase in drought and a fall in cereal production, but some models project slightly more precipitation due to an increase in the strength of the West African monsoon.

CHECKPOINT 7.19 ▶ What impact is the shortage of water likely to have on the mobility of people?

Case Study 2: The Amazon Forests

The Amazon ecosystem has remained fundamentally intact over the past 1 to 2 million years, surviving repeating glacial cycles. However, according to recent climate models, changes in ocean temperature and atmospheric circulation may contribute to natural fires and sustained drought, which could lead to the collapse of the world's greatest forest ecosystem (**Figure 7.17**).

A Treasure of Biodiversity

The Amazon covers an area of 5.5 million km^2 (21.2 million square miles) and is one of the most ancient, resilient, and complex ecosystems on the planet, with unparalleled **biodiversity**. For over 55 million years, this broad-leafed forest has at least partially covered the Amazon Basin. It contains 20% of all known birds, possibly 130,000 species of invertebrates, more than 150,000 types of higher (vascular) plants, and over 1 million kinds of insects. More than 438,000 species in the Amazon have been deemed as having economic and social interest, and an inestimable number of genetic and pharmaceutical resources are still locked away in the forest. The loss of this rich resource would be a disaster (**Figure 7.18**).

CHECKPOINT 7.20 ▶ Why would the loss of biodiversity in the Amazon have a global economic impact?

▲ Figure 7.17: The Amazon River Basin is so large and the Amazonian forests so extensive, that reduction in the area covered by forest would impact global climate significantly.

A Wealth of Human Knowledge

The Amazon is also home to more than 30 million people. Most live in a few large urban centers, but there are more than 350 indigenous communities, a few of which have little or no contact with the outside world. All these people are directly or indirectly dependent on the Amazon for food, shelter, and work. The indigenous people hold an immense store of knowledge about the plants and animals of the forest and of the healing properties of the plants. The continued rapid destruction and future contraction of the forest would almost certainly lead to the loss of most of this knowledge.

Carbon Storage in the Amazon

The loss of the Amazon would not only leave millions under threat of starvation but also lead to major changes in global climate and the acceleration of global warming. The thin Amazon soils are easily eroded once the trees are lost and the forest is replaced by unproductive scrublands. Billions of tonnes of carbon locked in wood and forest soils will be released into the atmosphere if the forest burns and decays. Some estimates suggest that as

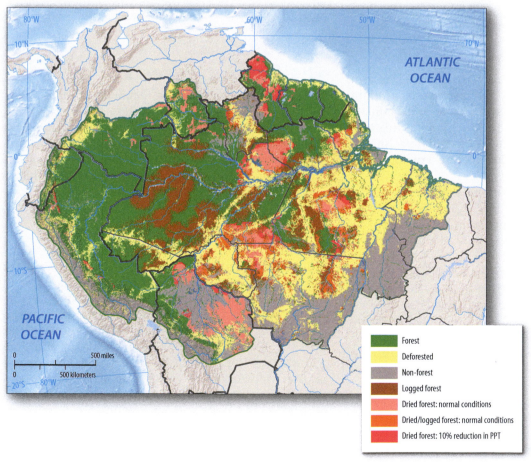

Forest
Deforested
Non-forest
Logged forest
Dried forest: normal conditions
Dried/logged forest: normal conditions
Dried forest: 10% reduction in PPT

▲ Figure 7.18: The projected impact of climate change and deforestation in the Amazon forests by 2050. Note that the loss of forest cover is due to a combination of deforestation and the impact of climate change, and that deforestation is one of the main sources of anthropogenic carbon dioxide and global warming.

little as 2°C (3.6°F) of global warming could result in the loss of 20% of the forest by 2100. The most devastating drought in 100 years occurred in 2005, with forest fires penetrating deeply into the heart of the untouched forest. These fires released as much as 5 billion tonnes of CO_2 into the atmosphere—more than the combined annual greenhouse gas emissions of Japan and Europe. The loss of the Amazon would be an ecological, economic, social, and global environmental disaster (**Figure 7.19**).

CHECKPOINT 7.21 ▶ Why is the loss of the Amazon forests likely to accelerate the rate of global climate change?

7.2 PAUSE FOR... THOUGHT | *Is it true that the Amazon makes its own weather?*

Tropical forests are unique ecosystems. The Amazon experiences both dry and wet seasons associated with the annual migration of the ITCZ. During the dry season, NASA satellite photographs show that the forests are blanketed with myriad small clouds that form above the forest canopy. Up to 70% of the water that falls on the forests is lost from the canopy as transpiration that keeps the air humid, cools the forest during the day, warms the forest at night, and leads to the formation of clouds that can increase albedo. As much as 25% to 30% of the rain that falls in the forest is generated from water vapor that actually came from within the forest. Each tree in the canopy can release about 760 liters (200 gallons) of water into the atmosphere each year, leading some to refer to the Amazon as the "green ocean."

▲ **Figure 7.19:** The current distribution of different ecosystem types in the Amazon Basin. The Amazonian forests area is so large and so important that even their partial loss would be an ecological disaster.

Case Study 3: Reef Systems

Barrier reef systems form part of a complex interconnected web of ecosystems that link near-shore with deep-water environments. Marine biologists believe that global warming could destroy many of these reefs by the end of the 21st century, affecting thousands of coastal communities and millions of people.

The Great Barrier Reef

The Great Barrier Reef system stretches for over 2,300 kilometers (1,429 square miles) off the northeast coast of Australia; it is the only biological structure that can be seen from space. The reef system is the largest of the maritime World Heritage areas and is composed of over 3,000 reefs and 940 islands. Together with the surrounding waters, the reef encompasses a diverse and complex range of communities. The barrier reef itself protects the Australian coastline, forming sheltered lagoons with abundant patch reefs and broad areas of seagrass, salt marsh, and mangrove swamp communities (**Figure 7.20**).

The Great Barrier Reef, like the Amazon forest, has evolved to cope with the natural pace of change associated with the waxing and waning of continental ice sheets. Borehole samples recovered from the base of the reef suggest that it is at least 600,000 years old, but other estimates extend the reef's age as far back as 18 to 20 million years.

This ancient reef has had time to evolve into a rich and complex ecosystem that includes more than 400 species of corals, 500 species of marine algae, 1,500 species of fish, more than 5,000 species of mollusks, and more than 40 species of whales, dolphins, porpoises, turtles, and (rare) dugongs. All these populations are interconnected, and any imbalance between species can have a dramatic impact on the overall health of the entire reef system.

CHECKPOINT 7.22 ▶ In what ways are reef systems similar to tropical forests?

Coral Bleaching

Scleractinian (stony, reef-building) corals live in a symbiotic relationship with **zooxanthellae**, a variety of dinoflagellate algae that provide nutrition to the coral through photosynthesis. Many of these algae are particularly sensitive to heat stress, suffer cell damage, and are expelled by the coral when the temperature rises above a critical threshold. It is the algae that give corals their unique and varied color, so when the algae are expelled, the coral appears white and **bleached** because all that remains is the natural white color of the calcium carbonate that forms the substance of the reef (**Figure 7.21**).

CHECKPOINT 7.23 ▶ In your own words, describe how and why coral bleaching occurs.

▲ **Figure 7.20:** The Great Barrier Reef off the coast of Australia appears to be thriving, but it is under threat from a combination of climate change, pollution, and exploitation. In this satellite photograph, the coral reef appears as light green patches in the blue ocean to the top right-hand side. The coast of Australia to the bottom left is dotted with small white cumulus clouds.

Coral bleaching occurs when the reef is exposed to a sudden large rise in temperature or more prolonged exposure at slightly elevated temperatures. NOAA defines the sensitivity of corals to bleaching in terms of "degree heating weeks" (DHW). One DHW is equivalent to one week of sea surface temperatures $1°C$ ($1.8°F$) greater than the expected summertime maximum. Two DHWs are equivalent to two weeks at $1°C$ ($1.8°F$) above the expected summertime maximum *or* one week of $2°C$ ($3.6°F$) above the expected summertime maximum.

▼ **Figure 7.21:** Symbiotic zooxanthellae add all the color we associate with healthy corals. Corals like the example in this photograph turn white and cannot live for long after rising temperature forces the coral to expel the colored zooxanthellae.

Bleaching is likely to occur when DHW >4, and severe bleaching when DHW > 8 (**Figure 7.22**).

CHECKPOINT 7.24 ▶ What are degree-heating weeks, and how do scientists use them to monitor the health of reef systems?

Lessons from El Niño

Severe bleaching occurred in the Great Barrier Reef in 1998 and 2002 and in the Caribbean in 2005. These events were closely associated with natural El Niño events. However, as ocean temperatures gradually warm around the world, more extensive bleaching will take place globally. There is evidence that some corals can adapt to higher water temperatures by hosting a different variety of symbiotic **dinoflagellate** known as "clade D," but scientists do not know whether this can offset the impact of annual to biannual bleaching events that are likely to occur within the next 30 to 50 years (**Figure 7.23**).

CHECKPOINT 7.25 ▶ What impact does El Niño have on coral bleaching?

An Eye in the Sky

In the United States, NOAA established a satellite tracking system for sea surface temperature (SST) data and launched a new coral reef watch that generates automatic alerts for coral bleaching. NOAA also set up a special Coral Reef Task Force and a new coral reef conservation program for the United States. The aim of these programs is to provide effective

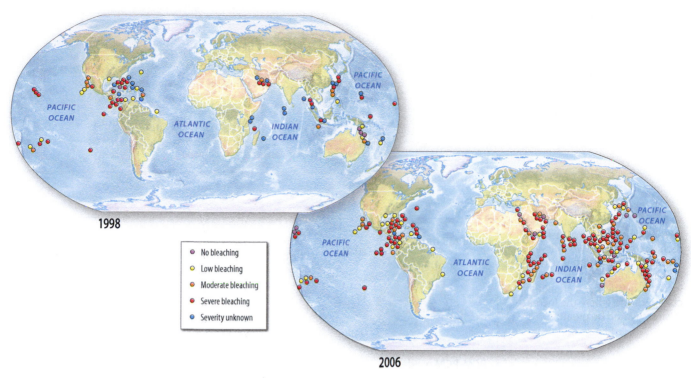

Legend:
- No bleaching
- Low bleaching
- Moderate bleaching
- Severe bleaching
- Severity unknown

1998

2006

▲ **Figure 7.22:** The number of bleaching events around the globe in 1998 and 2006. Note an increase in the frequency of bleaching events across the world, especially in the western Pacific and Indian Ocean.

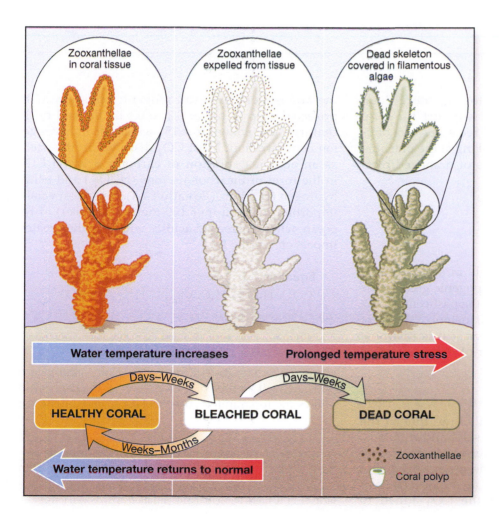

◀ **Figure 7.23:** The bleaching and recovery processes. The loss of Zooxanthellae from corals happens suddenly, and unless the temperature of the water falls dramatically, many corals are unable to recover and die.

Zooxanthellae in coral tissue

Zooxanthellae expelled from tissue

Dead skeleton covered in filamentous algae

Water temperature increases — Prolonged temperature stress

Days–Weeks

Days–Weeks

HEALTHY CORAL

BLEACHED CORAL

DEAD CORAL

Weeks–Months

Water temperature returns to normal

Zooxanthellae

Coral polyp

▶ **Figure 7.24:** The NOAA coral bleaching alert system gives advanced warning of bleaching events, based on satellite observations of ocean temperature that are plotted on maps such as this one showing global data shown for September 20, 2012. Where is the greatest threat of bleaching?

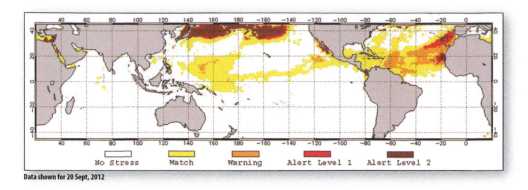

Data shown for 20 Sept, 2012

▶ **Figure 7.25:** The projected increase in the thermal stress of reefs according to two different climate models, using the IPCC A2a-emissions scenario. Circles of increasing radius and changing color indicate where the increase in stress would be greatest.

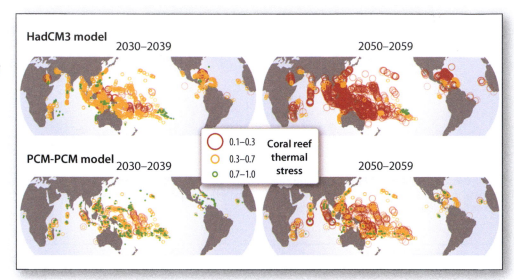

satellite coverage, onsite monitoring, more data, better modeling, and continued support for coral reef conservation. Because coral reefs are economically important, NOAA is also looking at the potential social and economic implications of damaged reefs (**Figures 7.24** and **7.25**).

Case Study 4: Ocean Acidification

Carbon dioxide is slightly soluble in seawater and increases ocean acidity. As anthropogenic greenhouse gas emissions have increased over the past century, the concentration of CO_2 in the top 100 meters (328 feet) of the ocean has also increased, and **ocean acidification** now poses a major threat to a number of delicate marine ecosystems (**Figure 7.26**).

Understanding Ocean Chemistry

Seawater is normally slightly alkaline due to the high concentration of dissolved salts and minerals. The historical **pH**[1] of ocean water has been around 8.2, but it has decreased (become more acidic) significantly over the past century, as the oceans have absorbed as much as 50% of the CO_2 added to

the atmosphere by anthropogenic emissions. The projected drop in pH of 0.3 to 0.5 by the end of the century is equivalent to a 150% increase in hydrogen ion concentration. Most CO_2 dissolves in the ocean as an aqueous solution of carbonic acid (H_2CO_3) in the form of bicarbonate ions (HCO_{3-}) and acidic hydrogen (H^+) ions (**Equation 7.1**). This is especially important in the upper 100 meters (328 feet) of the ocean, where increased acidity will have the greatest impact on marine life.

Equation 7.1:
$$CO_2 + H_2O \leftrightarrows H_2CO_{3\,(aq)} \leftrightarrows H^+_{\,(aq)} + HCO3^-_{\,(aq)}$$

(aq) = in aqueous solution

Making and Dissolving Shells

Marine invertebrates such as corals, clams, and gastropods, as well as tiny plankton such as the coccolithophores and foraminifera, use this HCO_3^- and the calcium

---

[1]pH is a logarithmic measure of the acidity of a solution and is defined by the concentration of highly reactive hydrogen ions (H^+) in that solution. Distilled water has a pH of 7.0, seawater a pH around 8.2, and vinegar a pH of around 3.0. Because pH is calculated on a logarithmic scale, a change of just 1 unit of pH is equivalent to an increase of 10 times the actual concentration of hydrogen ions.

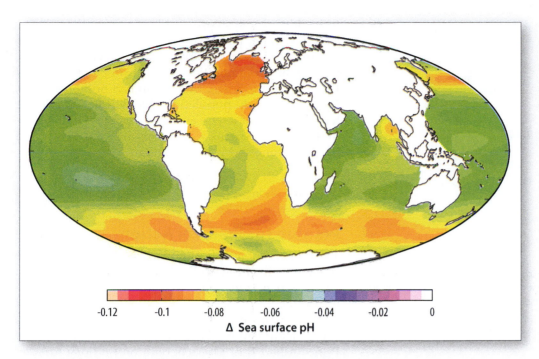

◀ **Figure 7.26:** Estimated change in annual mean sea surface pH between the preindustrial period (1700s) and the end of the 20th century. The more negative the change, the greater the relative increase in acidity, as shown on this map by increasing color intensity from green through yellow to orange and red.

ion Ca^{2+} in seawater to produce the calcium carbonate they need to build their shells (**Equations 7.2 and 7.3**).

Equation 7.2: $2HCO_{3-(aq)} \leftrightarrows 2H^+_{(aq)} + 2CO_3^{2-}_{(aq)}$

Equation 7.3: $2H^+_{(aq)} + 2CO_3^{2-}_{(aq)} + Ca^{2+}_{(aq)} \leftarrows$
$CaCO_3 + H_2O + CO_{2\,(aq)}$

The addition of H^+ to the oceans upsets the balance of Equation 7.3 and as acidity increases, the ocean adjusts to regain balance (equilibrium) by "mopping up" excess H^+, using carbonate ions (CO_3^{2-}) to form more HCO_3- (**Equation 7.4**).

Equation 7.4: $2H^+_{(aq)} + 2CO_3^{2-}_{(aq)} \leftrightarrows 2HCO3^-_{(aq)}$

As the acidity of the oceans increases, the solubility of $CaCO_3$ also increases (Equation 7.3 proceeds from right to left to produce more CO_3^{2-}), and the shells of living creatures become thinner and have difficulty growing. This process has a catastrophic impact on marine invertebrates. Today the pH of the oceans is changing faster than it has at any time over the past 100,000 years.

Rapidly Changing Ocean Chemistry

The rate of change of ocean chemistry is critically important for species that might otherwise survive if they had time to adapt. The level of CO_2 in the atmosphere has been much higher in the past (Chapter 6), and marine ecosystems have survived intact, but the oceans had time to adjust by removing much of the additional gas from the surface and mixing it into deep-ocean currents. The problem is greatest for common species that build their shells from **aragonite**, a common soluble form of calcium carbonate. Some studies suggest that parts of the oceans could become so acidic by 2100 that some marine invertebrates will no longer be able to make shells of aragonite.

Endangering Shallow Marine Ecosystems

Marine biologists have studied how shallow marine ecosystems are affected by ocean acidity by looking at places around the world where CO_2 from volcanic activity seeps naturally to the surface. Where this happens, seagrass communities flourish, but corals and coralline red algae, echinoderms (sea urchins), calcareous foraminifera, coccolithophores, oysters, and other mollusks are rare or absent. In essence, the high abundance and diversity of fauna that is typical of shallow marine carbonate ecosystems is replaced by a much lower-diversity ecosystem that is better adapted to high ocean acidity (**Figure 7.27**).

Struggling Reef Systems

Coral reefs have been carefully monitored to assess the impact of increasing ocean acidity. Studies have shown that the growth of some corals has slowed by as much as 13% since 1990, and corals could cease to grow and even start to dissolve if atmospheric CO_2 reaches 560 ppm. The disappearance of keystone species could push many major reefs toward a critical tipping point (**Figure 7.28**).

The broader reef ecosystem is also at risk. Many fish that are economically important spend part of their life in and around coral reefs. In addition, reef structures protect coast ecosystems from the impact of major storms. Reefs weakened by increasing ocean acidity may not withstand the increasing intensity of tropical storms.

CHECKPOINT 7.26 ▶ What impact will ocean acidification have on reef systems?

▲ **Figure 7.27:** The possible impact of ocean acidification on some marine ecosystems. These corals have died and have been replaced by algae and sea grasses that thrive under these new conditions.

The Social and Human Impacts of Ocean Acidification

Ocean acidification is not just a problem for marine life but also an economic problem. In Queensland, Australia, home of the Great Barrier Reef, fishing is a $360 million-per-year business that employs more than 2,000 people. The area also attracts 1.6 million tourists, who bring important additional income to the economy. Local indigenous communities are also affected by ocean acidification, as more than 70 aboriginal and Torres Strait Islander clans depend on a healthy reef for their livelihood. In the United States, coastal fisheries bring in over $30 billion a year, and many of the most important species depend on a healthy reef system as part of their life cycle. A global decline in the number

of healthy reef systems could cost more than 70,000 jobs in fishing, processing, and transport. However, the greatest impact would be in the developing world, where many communities are totally dependent on reef fisheries and tourism. More than 655 million people, 10% of the world population, live within 100 kilometers (62 miles) of a reef, and many of these people live in developing nations where reefs are already stressed by overfishing and water pollution.

CHECKPOINT 7.27 ▶ Discuss why reefs are so economically important.

Waiting for Political Action

The prospect of acidification has generated worldwide concern. In the **Monaco Declaration** in 2009, more than 155 scientists from 26 countries called on policymakers to stabilize CO_2 emissions "at a safe level to avoid not only dangerous climate change, but also dangerous ocean acidification." The declaration, supported by Prince Albert II of Monaco, built on findings from an earlier international summit.

The sea will eventually recover from this increase in acidity as the excess carbon is slowly absorbed into the vast reservoir of the deep oceans, but it can take hundreds of years for the surface waters to mix with deeper waters. By the time that happens, it may be too late to avoid widespread extinction in the shallow oceans. Some estimates put the likely extinction rate as high as 32%.

Case Study 5: The Roof of the World

The Himalayas form an active and growing mountain belt that separates the Indian subcontinent from the Tibetan Plateau (**Figure 7.29**). Tall, glaciated peaks

▶ **Figure 7.28:** Ocean acidification is a complex chemical process that threatens a broad range of invertebrate species, including coral reefs. As the oceans become more acidic the number of shelled animals living in the sediment decreases. In deep water, even the shells of plankton that fall from the surface dissolve before they reach the ocean floor.

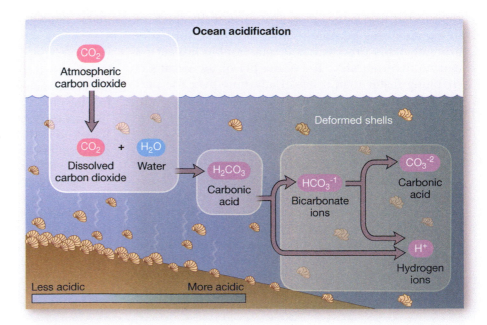

◀ **Figure 7.29:** The Himalayan Mountain belt is clearly visible in this NASA satellite image. These mountains are so high that they divert the jet streams, while air rising from the Tibetan Plateau behind the mountains has a major impact on global climate. Regional warming is melting many (but not all) Himalayan glaciers.

extend for 2,400 kilometers (1,491 miles) along a long mountain front that stretches from China as far west as the remote Hindu Kush Mountains in northeastern Afghanistan. The high Tibetan Plateau to the north of the Himalayas is a barren area of steppe, rock, and marginal desert where winter temperatures fall as low as −40°C (−40°F).

The Himalayas as a Reservoir of Water

The action of ice and rock over millennia has carved a rugged and majestic landscape where ice fields, glaciers, and glacial lakes hold a supply of surface fresh water that is second only to the polar icecaps. Many famous rivers flow from these mountains, including the Indus, Ganges, Brahmaputra, Irawaddi, Yangtse, and Huang He (Yellow) Rivers. Melting snow from the Himalayas sustains all these rivers. This region supports more than 1.3 billion people and some of the highest population densities on Earth. In countries such as Afghanistan, Pakistan, India, Nepal, Bhutan, and China, the impact of climate change on the seasonal flow of these rivers could be catastrophic.

CHECKPOINT 7.28 ▶ Why do so many countries depend on the Himalayas for their water supply?

Warming the Himalayas

The Himalayas are warming most rapidly at elevations above 4,000 meters (13,000 feet), at a rate of between 0.4°C per decade and 0.9°C per decade (0.72°F and 1.62°F). Climate models project that the temperature in Nepal could rise by a further 1.2°C (2.16°F) by 2025 and as much as 3°C (5.4°F) by 2100. As more snow melts, lakes form on the glaciers and ice fields, and more bare rock is exposed. More of the Sun's energy is then absorbed rather than reflected, heating the atmosphere above. Nepal saw a 6% loss in glaciated area between 1970 and 2000, and as much as 25% of glacial cover could be lost in northwestern China by 2050. Many glaciers that were retreating naturally due to warming during the early part of the 21st century are now retreating at unprecedented rates of up to 68 meters (223 feet) per year. Although the majority of glaciers are retreating, there are still some that continue to advance due to changes in the pattern of regional precipitation during the winter (**Figure 7.30**).

CHECKPOINT 7.29 ▶ Why is the loss of winter snowpack so important in the Himalayas?

The Danger of Glacial Outburst Floods As the climate changes, communities in the mountains are under increased threat from landslides and spring floods that can wash away entire villages. There is also a growing threat from glacial lakes that are blocked by fragile dams of ice and **glacial moraine**. These lakes fill rapidly with water from melting ice, and when the dams burst, they release glacial lake outburst floods (GLOFs; **Figure 7.31**). When the Dig Tsho glacial lake in Nepal burst its banks in 1985, it washed away 12 bridges and a hydroelectric plant, along with the homes and farms of many communities. The International Centre for Integrated Mountain Development (ICIMOD)

▲ **Figure 7.30:** In many places, glaciers flowing from the Himalayas are retreating rapidly, threatening water resources across the region. This photo is from the Khumbu Glacier, located in northeastern Nepal between Mount Everest and the Lhotse-Nuptse ridge.

estimates that 15,000 glaciers, 9,000 glacial lakes, and at least 20 other lakes are in imminent danger of GLOF in Nepal alone.

Scientists know how to control the GLOF threat, but the task is difficult and expensive. As an example, the government of the Netherlands spent over $3 million in 1998 to try to control the outflow of the Tsho Rolpa glacial dam, one of Nepal's most dangerous lakes. Engineers cut a notch in the moraine that holds back the lake and built a dam and sluice gate to allow the controlled release of water.

A regional solution requires making an up-to-date inventory of all glacial lakes that might pose a threat,

establishing remote-monitoring and early-warning systems, carefully mapping the area, analyzing the vulnerability of communities to flooding, and the active involvement of local communities. The glaciers and lakes hold deep religious significance for many remote communities, and the people are suspicious of outside intervention. Their full participation is vital for the success of any program that hopes to adapt and mitigate the impact of climate change in the high mountains.

CHECKPOINT 7.30 ▶ Describe how we can control glacial outburst floods.

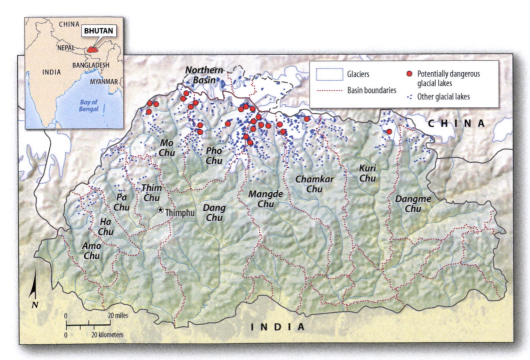

▲ **Figure 7.31:** Potentially dangerous glacial lakes in Bhutan. The risk of glacial lake outburst floods is real and growing across the region.

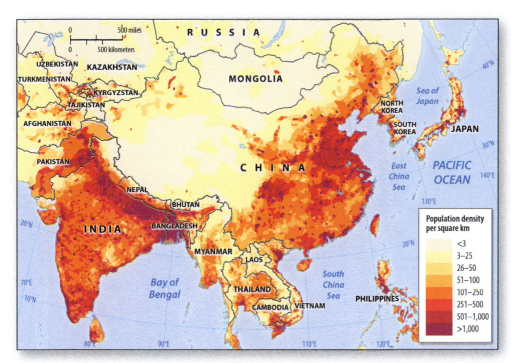

◀ **Figure 7.32** The population density of Asia.

Forest Fires and Flash Floods The high population density of the Himalayas (**Figure 7.32**) is supported by very effective yet simple farming practices that depend on terraced slopes to trap water and prevent soil erosion. In Nepal, 80% of the population still depends on the forests for wood for fuel and heat and on rivers for water and irrigation.

Climate models project that regional warming in Asia will increase the intensity of seasonal rainfall and the severity of summer drought. Intense rainfall in Nepal will damage terracing and increase the risk from flash flooding, soil erosion, and landslides. Intense summer heat will accentuate the impact of slash-and-burn farming practices and increase the frequency and intensity of forest fires. There is a very real risk that some important keystone species, such as the rhododendron (the native flower of Nepal) and alder trees, could be affected, which would seriously damage this fragile ecosystem.

A Future with More Famine Beyond the foothills of the Himalayas, 14 major watersheds in Asia contain some of the most productive agricultural land in the world. These rich soils and fertile floodplains sustain over 1.3 billion people, with many still living in poverty and dependent on the rivers for both food and water (**Figure 7.33**). The flow rate in these rivers varies throughout the year. It is highest when winter snow starts to melt in the spring and is sustained over the summer by the melting of snow higher in the mountains. The IPCC projects that these watersheds will experience a short-term increase in summer flow rate as more ice and snow start to melt with higher

air temperatures. Beyond 2050, there is an increased chance of sustained summer drought, as most of the available snow and ice will have melted.

In Afghanistan, Pakistan, India, Nepal, Bhutan, Myanmar, and China, as much as 35% of crop production depends on irrigation that needs constant recharging from rivers sourced deep in the Himalayas. This land accounts for 25% of the world's cereal production and as much of 56% of Asia's cereal production. Any decrease in seasonal flow would degrade regional agricultural yields by as much as 30%, which is equivalent to 1.7% to 5% of world cereal production.[3] This problem will be exacerbated as the region continues to develop; the demand for water is expected to rapidly increase by as much as 70% (**Figure 7.34**).

CHECKPOINT 7.31 ▶ What effect will climate change have on regional agriculture and food production in Asia? How many people are likely to be affected?

These factors have the potential to affect agriculture and food production in Asia very seriously and are major causes for concern. The forced migration of so many people looking for food would have serious political consequences.

There may be some hope in the future if we use more drought-tolerant species or genetically modified cereals. More efficient irrigation, better education and technical

---

[3]Increased rainfall may lead to an increase in cereal yield of as much as 20% in parts of East and Southeast Asia. This increase may partially offset the loss due to seasonal flow.

▶ **Figure 7.33** The Himalayas stretch from Pakistan in the west to China in the east and are the source of most of the major river systems that sustain the growing population of Asia.

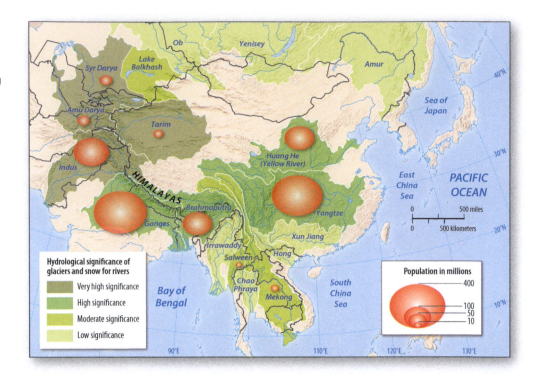

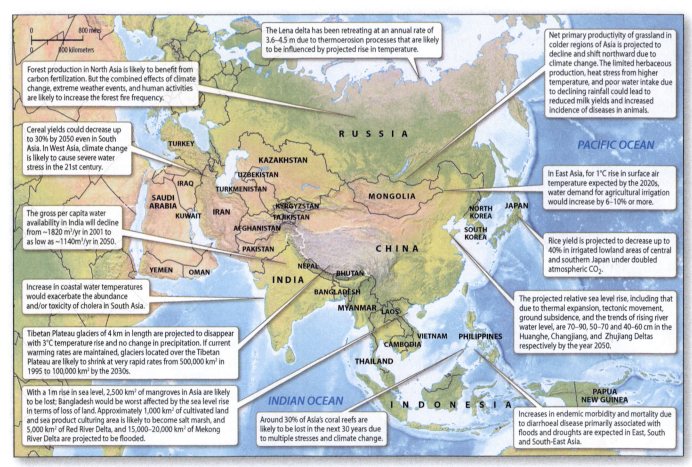

The Lena delta has been retreating at an annual rate of 3.6–4.5 m due to thermoerosion processes that are likely to be influenced by projected rise in temperature.

Net primary productivity of grassland in colder regions of Asia is projected to decline and shift northward due to climate change. The limited herbaceous production, heat stress from higher temperature, and poor water intake due to declining rainfall could lead to reduced milk yields and increased incidence of diseases in animals.

Forest production in North Asia is likely to benefit from carbon fertilization. But the combined effects of climate change, extreme weather events, and human activities are likely to increase the forest fire frequency.

Cereal yields could decrease up to 30% by 2050 even in South Asia. In West Asia, climate change is likely to cause severe water stress in the 21st century.

In East Asia, for 1°C rise in surface air temperature expected by the 2020s, water demand for agricultural irrigation would increase by 6–10% or more.

The gross per capita water availability in India will decline from ~1820 m³/yr in 2001 to as low as ~1140m³/yr in 2050.

Rice yield is projected to decrease up to 40% in irrigated lowland areas of central and southern Japan under doubled atmospheric CO_2.

Increase in coastal water temperatures would exacerbate the abundance and/or toxicity of cholera in South Asia.

The projected relative sea level rise, including that due to thermal expansion, tectonic movement, ground subsidence, and the trends of rising river water level, are 70–90, 50–70 and 40–60 cm in the Huanghe, Changjiang, and Zhujiang Deltas respectively by the year 2050.

Tibetan Plateau glaciers of 4 km in length are projected to disappear with 3°C temperature rise and no change in precipitation. If current warming rates are maintained, glaciers located over the Tibetan Plateau are likely to shrink at very rapid rates from 500,000 km² in 1995 to 100,000 km² by the 2030s.

With a 1m rise in sea level, 2,500 km² of mangroves in Asia are likely to be lost; Bangladesh would be worst affected by the sea level rise in terms of loss of land. Approximately 1,000 km² of cultivated land and sea product culturing area is likely to become salt marsh, and 5,000 km² of Red River Delta, and 15,000–20,000 km² of Mekong River Delta are projected to be flooded.

Around 30% of Asia's coral reefs are likely to be lost in the next 30 years due to multiple stresses and climate change.

Increases in endemic morbidity and mortality due to diarrhoeal disease primarily associated with floods and droughts are expected in East, South and South-East Asia.

▲ **Figure 7.34:** Possible impacts of climate change in Asia by ca. 2050.

skills, and the use of treated wastewater could also help. Still, the threat is very real. Much is known, but this is a vast and critical area, and more research is required to understand the large uncertainties in the data.

CHECKPOINT 7.32 ▶ In what ways could Himalayan communities adapt to climate change and minimize its impact?

7.3 PAUSE FOR... THOUGHT | *No ice left in the Himalayas?*

There are around 54,000 glaciers in the Himalayas, and only a few have been studied in any detail. The IPCC was embarrassed in 2009 when its claim that all glaciers in the Himalayas could melt by 2035 was revealed as inaccurate and not based on sound research. Recent studies have found that there has indeed been a significant reduction in snow cover over the past decade, and many glaciers are shrinking, but it is unlikely that the Himalayas will be glacier-free in the near future.

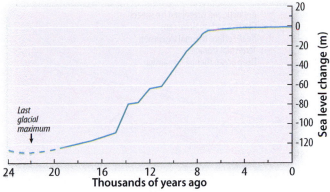

▲ **Figure 7.35:** Sea level rise since the end of the last glaciation 20,000 years ago.

Case Study 6: Rising Global Sea Level

While many of the early effects of global warming will be felt at regional levels, the impact of rising sea level will have truly global effects. We tend to think of sea level as uniform around the world, but the surface of the ocean can vary naturally by as much as 2 meters (6.5 feet) due to the effects of storms, temperature, salinity, winds, tides, and small changes in gravity around the globe. The volume of water in the oceans is also in a state of continual flux. Water falling as snow in Antarctica and Greenland each year would reduce sea level by as much as 8 millimeters (0.31 inches) per year if it were not for meltwater that returns to the oceans as runoff in the spring and summer. This became evident in 2011, when sea level fell by as much as 5 millimeters (0.2 inches) following a sharp increase in global precipitation the previous year. The oceans soon started to recover in 2012, as water drained from the continents.

All these variables can act over periods from a few hours to many decades and make it difficult to determine exactly how global sea level is changing. Add to this the further complication that in many places around the world, the land is sinking or rising relative to global sea level due to subsidence, **glacial rebound**, or tectonic activity, and it becomes clear how difficult it can be to get an accurate value. Some places will suffer a higher-than-average rise in sea level, and others will even see sea level fall. It is the global picture that is important.

The Last Great Flood

Sea level increased rapidly following the end of the last glacial maximum (**Figure 7.35**). This increase was as fast as 30 millimeters (1.2 inches) per year between 14,000 and 12,000 years ago, but it slowed to around 10 millimeters (0.4 inches) per year between 10,000 years and 8,000 years ago. Since then, sea level has been relatively stable, until it started to increase again slowly toward the end of the 19th century (**Figure 7.36**). Estimates range from an average of 1.8 millimeters (0.07 inches) per year between 1960 and 2003 to as high as 3.1 millimeters (0.12 inches) per year between 1993 and 2003. Recent estimates have become more accurate since satellites have started to make accurate measurements at the millimeter level (**Figure 7.37**).

CHECKPOINT 7.33 ▶ How fast did sea level rise following the end of the last glacial maximum?

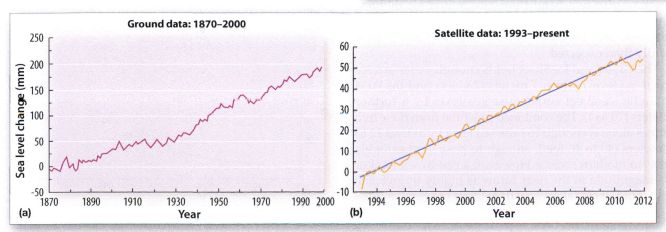

▲ **Figure 7.36:** (a) Sea level rise from 1880 to the present, measured on Earth and by satellite. Note the difference in scale interval used on both diagrams. (b) Sea level change between 1994 and 2004 as measured by satellite. Use the right-hand figure to calculate the rate of sea level change between 1994 and 2004.

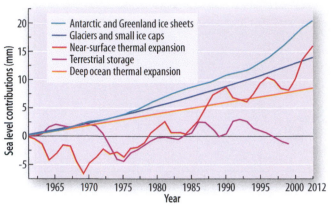

▲ **Figure 7.37:** Different components of sea level rise have contributed to the total increase in sea level since the 1960s.

Expanding the Ocean

Water expands as it warms, and this thermal expansion has been responsible for much of the observed increase in sea level over the past century. The oceans have absorbed 20 times as much heat as the atmosphere since 1960, and warming is slowly reaching into the deeper ocean from the upper 300 to 700 meters (980 to 2,300 feet), where the ocean is well mixed by waves and storm action. As long as thermohaline circulation carries this surface-derived heat energy into the deep oceans, thermal expansion will continue to drive rising sea level.

CHECKPOINT 7.34 ▶ What is thermal expansion? How much has it contributed to rising sea level (see Figure 7.37).

Melting the Last Great Ice Sheets

The last time global temperatures were as high as they are today was 125,000 years ago, during the last major interglacial period. At that time, sea level stood some 4 meters (13 feet) higher than today, probably due to more advanced melting of both the Greenland and West Antarctic Ice Sheets. Global mean temperature may now rise to a height unprecedented for millions of years, and both the Greenland and West Antarctic Ice Sheets may become unstable and melt much more rapidly than expected.

The Greenland Ice Sheet holds enough water to raise global sea level by over 7 meters (23 feet), and the West Antarctic ice sheet could raise sea level by a further 6 meters (20 feet). The combined potential from these two sources alone is enough to raise sea level by at least 12 to 14 meters (40 to 46 feet), enough to pose a catastrophic threat to modern society. However, a rise in sea level of this magnitude in the near future is highly improbable and would likely take many thousands of years.

CHECKPOINT 7.35 ▶ Explain why sea level was 4 meters (13 feet) higher during the last interglacial stage than it is today.

Understanding Ice Sheet Dynamics

Early climate models tended to underestimate the rate at which ice sheets would melt. They assumed that any rise in the surface temperature of an ice sheet would take a long time to reach its base. More detailed studies have since shown that large crevasses act as conduits that allow surface meltwater to penetrate deeply into the base of an ice sheet and act as a lubricant that speeds up the flow of ice from the interior. Once this ice meets the sea it breaks apart, forming flotillas of icebergs that disperse into the surrounding ocean.

There is even some geological evidence that this process can accelerate and lead to the catastrophic collapse of an ice sheet over a period of centuries or even decades rather than millennia. This debate is open and requires more research, but it is certain that observed processes are enough to cause sea level to rise as fast as 2 to 3 millimeters (0.12 inches) per year over the next 100 years—and faster if some of the recent research on rapid ice sheet collapse is confirmed (**Figure 7.38**).

Some research has shown that melting at the margins of ice sheets may be partially offset by increased snowfall in the interior. This is expected because warm air holds more moisture, and the ice core record clearly shows that snowfall on Greenland and parts of Antarctica increased steadily as the air warmed following the last glacial maximum, when the bitterly cold air was exceptionally dry.

CHECKPOINT 7.36 ▶ Discuss how current theories about the melting of ice sheets differ from early models.

The Impact of Rising Sea Level

With a realistic prospect of sea level rising from anywhere from 30 centimeters (11.8 inches) to over

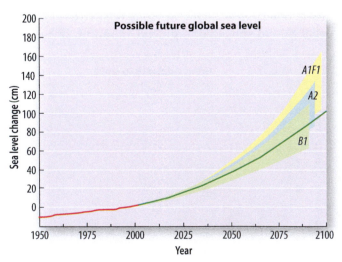

▲ **Figure 7.38:** The possible future of global sea level. Conservative estimates like this from the IPCC do not take account of more recent models that predict up to a meter of rise by the end of the century.

1 meter (3 feet) by 2100, society faces a significant challenge. Three major sectors are at immediate risk: coastal urban communities, communities living on deltas, and small island communities.

The Risk to Coastal Urban Communities The United Nations estimates that by 2080, the number of people living in coastal communities threatened by rising sea level could exceed 5.2 billion. Many of these people will be living in poverty in crowded cities. Long before these communities are actually flooded by the oceans, rising sea level will have devastating effects, as shown by the impact of Hurricane Sandy on New York and New Jersey in 2012.

An increase in the frequency of seasonal flooding will be the first sign of trouble. As sea level rises, even moderate storm surges will break through existing coastal defenses. Repeated flooding like this will allow salt water to intrude into shallow aquifers, and water for irrigation may become saline and toxic. Under such conditions, potable water may be contaminated as untreated sewage overflows into the water supply; this may, in turn, lead to an increase in the incidence of cholera, typhoid, and dysentery. As sea level continues to rise, increasing coastal erosion of beaches and dunes will discourage waterfront development, and tourism and agriculture will be damaged irreversibly.

Developed nations such as the United Kingdom, United States, and the Netherlands have the resources to mitigate the effects of moderate increases in sea level by building flood and sea wall defenses. Cities such as New York, London, Tokyo, Shanghai, Hamburg, Venice, and Rotterdam all face serious threat from flooding (**Figure 7.39**) but are at low to moderate risk due to their capacity to adapt. Even with their resources, however, cities like these will be at increased risk from low-probability extreme events that have the capacity to topple their technological defenses. An increase in sea level of just 1 meter (3 feet) would have a major impact along the east coast of the United States, as exemplified by the destructive storm surges associated with Hurricanes Katrina and Sandy. The effect of a larger increase would be truly devastating (**Figures 7.40** and **7.41**).

The Drowning of Delta Communities Deltas are attractive places for people to settle. The land is flat, productive, and easy to farm. Soils are rich in nutrients and replenished regularly when rivers flood their banks. The United Nations estimates that more than 300 million people live on deltas and depend on them for their livelihood. Of these, more than 15 million would be affected by as little as a 1-meter (3-foot) rise in sea level. Delta communities on the Nile, Ganges-Brahmaputra, Mekong, Yangtze, and other smaller deltas need to prepare for an increase in the incidence of floods and storm surges by the end of this century (**Figure 7.42**). However, most of these people live in poverty and do not have the capacity to adapt to rising sea level.

Many deltas around the world have already experienced increased flooding during storms, with the accompanying erosion of barrier islands and the loss of productive agricultural land. A further increase in global sea level will put these communities at even greater risk and guarantee that delta communities will be among the first to suffer the serious consequences of climate change.

CHECKPOINT 7.37 ▶ Discuss how many people may be under threat from rising sea level in 2080.

In much of Africa, southern Asia, and East Asia, deltas are important because of their high population

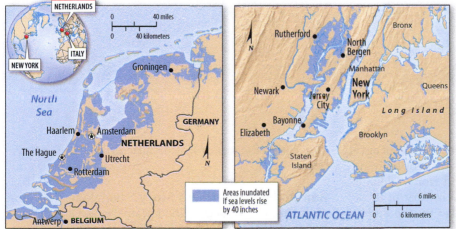

▲ **Figure 7.39:** Some examples of the expected effects of sea level rise in the Netherlands, the United States, and Italy.

▶ **Figure 7.40:** Changes in the coastline of the southeastern United States, Mexico, Cuba, Haiti, and the Dominican Republic after a sea level rise of 1 meter (3 feet), the maximum probable rise by 2100. Flooded areas are shown in red.

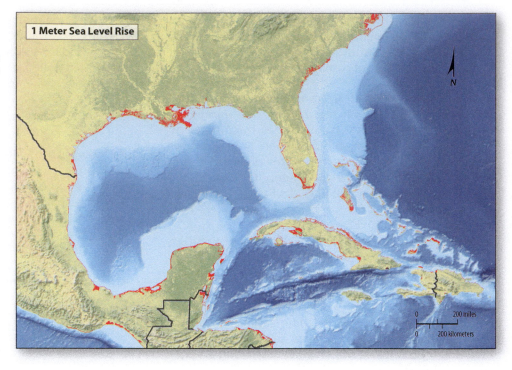

▶ **Figure 7.41:** Changes in the coastline of the United States, Mexico, Cuba, Haiti, and the Dominican Republic after a rise in sea level of 6 meters (20 feet)—the approximate rise scientists expect if the West Antarctic Ice Sheet melts completely. Flooded areas are shown in red.

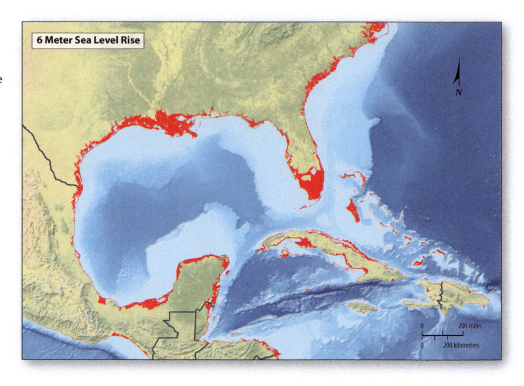

density. These deltas also serve as major sources of agricultural production to feed millions of people living in less fertile regions. The loss of land from a delta is not just measured by the immediate misery of those under threat from rising seas but also by the added threat of starvation to others who depend on the crops they produce. On the Nile delta, a 1-meter (3-foot) rise in sea level will impact an estimated 6 million people and lead to the loss of 15% of agricultural land in Egypt (**Figure 7.43**).

Bangladesh: A Success Story? Bangladesh has one of the highest delta and flood plain–bound populations in the world. More than 150 million people live closely together, with population densities as high as 1,000 people per square kilometer. More than 70% of the land

2000

2100

▲ **Figure 7.42** Compare the coastline of the Mississippi Delta in 2000 with the projected coastline in 2100 if sea level rise and regional subsidence continue at their present rate. In these false color satellite images the oceans are blue, forest and agricultural lands are green, and urban areas are red-purple.

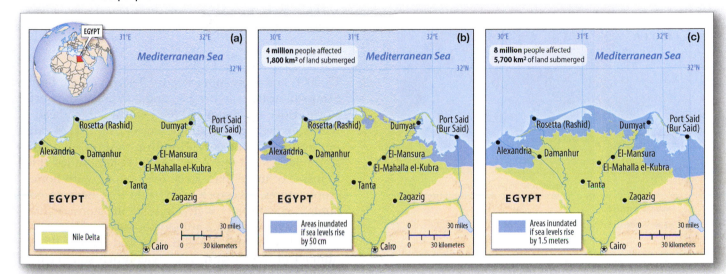

▲ **Figure 7.43:** The possible impact of sea level rise on the Nile Delta. The Nile Delta coastline (a) today, (b) after a sea level rise of 50 cm (20 inches), (c) after sea level rise of 1.5 meters (5 feet).

area is less than 1 meter (3 feet) above sea level and is at high risk of seasonal flooding and sea level rise. Flooding is expected to impact more than 13 million people and may destroy 16% of the national rice crop (**Figures 7.44** and **7.45**).

Two recent examples illustrate how the combined effects of atmospheric warming and sea level rise can devastate a nation. In 1998, the combination of heavy rain from tropical storms, higher-than-normal sea level, and unusually heavy runoff from the spring/summer snowmelt produced devastating floods. Meager levees, dams, and embankments made from mud and straw were no match for raging floodwaters that broke through and drowned hundreds of small farming communities.

This flooding covered two-thirds of the land area with water, drove more than 30 million people from

Ganges Delta

◀ **Figure 7.44:** A view of the Bay of Bengal and Ganges Delta from space.

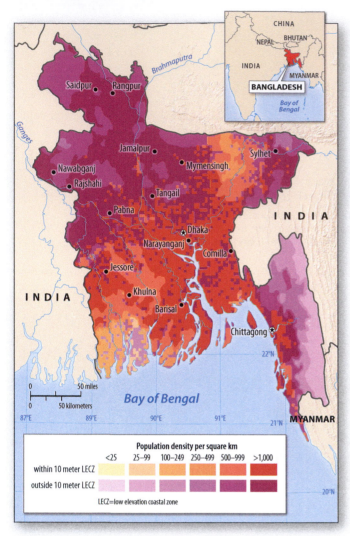

▲ **Figure 7.45:** A large part of the population of Bangladesh lives within 10 meters (33 feet) of sea level, putting them at high risk from storm surges.

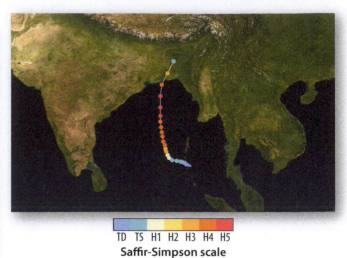

Saffir-Simpson scale

▲ **Figure 7.46:** The track of Cyclone Sidr from the Bay of Bengal to the Ganges Delta in 2007.

15, 2007, and grew to a category 5 storm on the **Saffir-Simpson scale** (**Figures 7.46** and **7.47**). This powerful storm drove a 3-meter (10 feet) storm surge ahead of winds exceeding 200 kilometers per hour (124 miles per hour). The massive flooding that followed this storm devastated low-lying coastal communities. It led to losses of more than 3,400 lives, along with homes, crops, stores, livestock, and soils that became contaminated with salty water. In total, the storm severely affected as many as 3.2 million people.

The loss of any life is tragic, and thankfully, the number of lives lost in 2007 was much smaller than might be expected. In 1970, more than 300,000 to 500,000 died after a cyclone made landfall in Bangladesh. In 1991, 150,000 to 200,000 more died as a large cyclone drove storm surges across the delta. That so few lives were lost in comparison in 2007 is testimony to the power of good disaster preparedness. In the case of Cyclone Sidr, more than 600,000 people successfully evacuated before the storm hit and found housing, food, and clean water in temporary shelters. This type of disaster preparedness will work for some time to come, but people are becoming tired of the constant threat of death, disease, and starvation, and many are moving to the cities.

The future for Bangladesh will be challenging. Increased urbanization in places such as the capital, Dhaka, will bring the usual pressures of overpopulation. As time passes, more people are bound to try to move across the border as economic refugees into India. By the middle of the 21st century, India's rapid economic growth will make it a land of great opportunity for many displaced people, and India has already responded to this perceived threat by strengthening border security. The potential for future conflict will grow, with increasing pressure from as many as 15 million climate refugees by 2050.

their homes, and killed more than 1,070. The floodwaters covered over 6,680 km² (2,579 square miles) and destroyed over 50% of the crops in these flooded areas. Vital rice stocks were also destroyed, leading to a 20% decrease in national rice production over the next year. This event may not be directly related to global warming and sea level rise, but climate change will make events like this much more likely to happen in the future. More intensive storms, heavier rainfall, and the more rapid melting of snowpack in the spring are all expected from global warming. Together with increasing deforestation in the foothills of the Himalayas, this has the potential to make devastating events so common in Bangladesh that people will no longer be able to live in areas of higher risk.

Seasonal flooding is not the only hazard that threatens Bangladesh. Increasing sea surface temperatures are likely to increase the intensity of tropical cyclones in the Bay of Bengal. An example is Cyclone Sidr, which formed in the Bay of Bengal on November

◀ **Figure 7.47:** The devastating aftermath of Cyclone Sidr in Bangladesh in 2007.

April 15, 2008   May 5, 2008

▲ **Figure 7.48:** United States Geological Survey (USGS) images of the Irrawaddy Delta before and after Cyclone Nargis struck in 2008. On this satellite image the ocean and flood waters are blue, agricultural fields are brown, and on the May 5, 2008, image the light green color in the Gulf of Martaban is sediment-rich water. Note the amount of floodwater and the increase in visible sediment load in the water.

Myanmar: A Tragedy of Contrasts When Cyclone Nargis struck (Figure 7.48) the Irrawaddy Delta on May 2, 2008, with sustained winds exceeding 215 kilometers per hour (135 miles per hour), it was preceded by a devastating storm surge that flooded much of the lower delta, wreaking havoc by destroying houses, villages, crops, and stores. The storm killed at least 146,000 fishermen and farmers (Figure 7.49). The military government in Myanmar (Burma), suspicious of any help offered by the outside world, turned a natural disaster into a human tragedy by rejecting early offers of aid (Figure 7.50). This stands in stark contrast to the situation in Bangladesh, where the government saved so many lives when Cyclone Sadr struck the year before precisely because it had responded openly to earlier cyclones with education, training, and the development of effective information and evacuation plans.

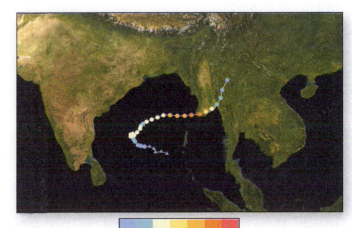

TD TS H1 H2 H3 H4 H5
Saffir-Simpson scale

▲ **Figure 7.49:** The track of Cyclone Nargis as it struck the Irrawaddy Delta in 2008.

243

▲ **Figure 7.50:** In the aftermath of Cyclone Nargis in Mayanmar, Burma, in May 2008, the indomitable spirit of the people was the only resource available to many.

The examples of Bangladesh and Myanmar (Burma) show how mitigation, adaptation, and effective planning over the next 50 years could save hundreds of thousands of lives around the world. It also serves as a warning. As great as the threat of global warming appears to be, political ineptitude, poverty, fear, ignorance, and violence can still pose an even greater threat to life and property. The world can show how effective adaptation and mitigation can limit the worst impacts of global warming, but if governments refuse to listen, or if they remain focused on conflict and division, the imminent future of their people will be marred by tragedy.

CHECKPOINT 7.38 ▶ Why would you expect coastal communities in Bangladesh to suffer much less than people in Myanmar (Burma) when the next tropical cyclone strikes?

Case Study 7: Small Pacific Island Communities

Few places feel the impact of global warming more intensely than communities living on small Pacific islands. These communities have what the IPCC calls high vulnerability and low adaptive capacity. The threat comes from many directions and is often exacerbated by a history of poor environmental management and building control (**Figure 7.51**).

Water lies at the root of many of these problems. The availability of fresh water is essential for the survival of any island community. Sea level is rising at around 3 millimeters (0.12 inches) per year, and when combined with storm surges, natural variation in sea level, and inevitable **tectonic subsidence**, this leads to increasing erosion and repeated flooding of coastal developments.

The probability of increasing storm intensity combined with the possibility of increased storm frequency will have a truly devastating impact on remote communities. Storm surges of 3 to 10 meters (10 to 33 feet) will easily override most coastal defenses, especially

▼ **Figure 7.51:** Many small island communities around the world live within 1 meter (3 feet) of sea level and are highly vulnerable to sea level rise. In this photograph, islanders construct a sea wall made of clam shells to halt coastal erosion on Han Island, Carteret Atoll, Papua New Guinea.

when natural barriers such as **mangrove swamps** and salt flats have been removed or drained for tourist development. Coral reefs can offer some protection, but as discussed earlier in this chapter, reefs themselves are under threat from a combination of bleaching, acidification, and overfishing.

Once flooded by salt water, normally fertile soils take years to recover, and agriculture is impacted severely. Beneath the surface, saltwater intrusion into shallow freshwater aquifers can easily turn potable water undrinkable, a problem exacerbated by overconsumption from aquifers in order to meet the needs of both tourism and agriculture. Poor sanitation combined with an increased threat from insect-borne disease creates a growing health problem that will impact both islanders and visiting tourists.

CHECKPOINT 7.39 ▶ How quickly do agricultural soils recover following flooding due to storm surge?

Tourism infrastructure depends on clean fresh water, pristine beaches, and sewage not leaking onto the streets and into the rivers with every moderate storm. Agriculture for local consumption and for important export revenue also needs a reliable supply of fresh water. Although rainfall may increase in intensity, some estimates project an overall reduction in Pacific rainfall of at least 10% by 2050. As the islanders demand more, nature may deliver less.

Many small island nations have come together to lobby for help and a place at the international table. The **Alliance of Small Island Nations** brings together 43 nations in the hope that, in combination, small voices may yet be heard above the quarrelling of the major nations. The future is not bright, as the threat of climate change is compounded by the way society continues to destroy natural habitats that are often the first and only line of defense.

7.4 PAUSE FOR... THOUGHT | *The fate of iconic Tuvalu.*

The small island of **Tuvalu** located in the southwest Pacific Ocean is only 4.6 meters (15 feet) above sea level and threatened by rising sea level. When Tuvalu's plight first became known, the country became an icon for global environmentalists, who saw sea level rising as fast as 5 millimeters per year (0.2 inches) as clear evidence of global warming. Not surprisingly, climate skeptics were soon suggesting that sea level was not rising at all, but actually falling. So who was correct? Recent studies have shown that the underlying level of the ocean is rising but that Tuvalu is strongly effected by ENSO and other natural changes in sea level that have major local effects and can at times mask a slow underlying upward global trend. El Niño and La Niña result in sea level fluctuations up to 20–30 centimeters in this part of the Pacific, which is around 40–60 times larger than the long-term annual increase (5 millimeters) found at Tuvalu.

The end may be inevitable for some very low islands. The highest point on the small island of Tuvalu, near the Solomon Islands in the South Pacific, stands only 4.6 meters (15 feet) or so above current sea level, and it is already suffering from rising sea level due to both natural and anthropogenic factors. These island nations need an international effort to help them mitigate and adapt to the effects of anthropogenic climate change. Some priorities are very clear: Water conservation, reef conservation, agriculture and forest management, natural coastal defense management, use of renewable energy, energy conservation, and education and health must be at the top of the list—but this is only the start. The end result will be the inevitable evacuation of thousands of climate refugees, but that does not need to happen immediately if we take the right precautions. The world needs to preserve and protect the rich natural and cultural heritage in these islands and consider who will take responsibility for the relocation of tens of thousands of islanders if no solution is found.

7.5 PAUSE FOR... THOUGHT | *Hasn't sea level been falling recently?*

Rising sea level, like rising temperature, is not uniform around the globe. The total volume of the oceans is increasing due to thermal expansion and melting ice, but some areas of the world may actually experience a slight decrease in sea level. We did not anticipate some recent changes. The overall trend in sea level remains positive and is even accelerating, but climate skeptics used a fall in sea level in 2011 to attack the projections of the scientific community. Paradoxically, the temporary fall in sea level was due to an increase in global precipitation—most likely because warming air has a greater carrying capacity for water vapor. The same skeptics argue that our experience of changing sea level this century will be no worse than we have already experienced over the past century. The problem with this argument is that 50 years ago, there was much less coastal development, and the population living close to the oceans was lower than it is today; in addition, the actual increase in sea level is likely to be much greater than the conservative IPCC models anticipate. Changes in sea level this century will quickly highlight the difference between the adaptive capacity of some nations and the vulnerability of others; even some developed countries will find it difficult to adapt.

Case Study 8: Sickness and Diseases

The relationship between climate change and disease is uncertain and complex. Each disease has its own unique method of transmission (vector) and associated epidemiology, which depend on factors that are often not well understood. Many diseases depend on physical conditions that are directly related to climate, such as water

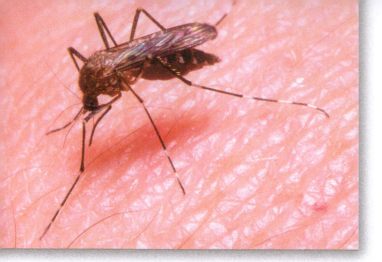

▲ **Figure 7.52:** There are approximately 3,500 species of mosquitoes, grouped into 41 genera. Only females of the genus *Anopheles* transmit human malaria. Of the approximately 430 species of *Anopheles*, only 30 to 40 transmit malaria (that is, are vectors) in nature. An increase in the incidence of malaria is an expected consequence of rising temperatures in some places, but it is also related to poor education, limited resources for protection, and growing parasite resistance to medication.

temperature, air temperature range, humidity, and precipitation, but many human factors are also involved.

The Deterioration of Human Health

Basic nutrition, sanitation, sewage management, waste disposal, access to medical care, health education, social welfare, building control, forest clearance, and storm-water management have major effects on disease. Maybe greatest of all is the impact of continuing social conflict, violence, and corruption that prevent social and medical intervention that could stop the spread of disease.

The Threat from Malaria

There is no doubt that global warming has the potential to increase the range of some diseases, such as malaria, as increasing temperatures allow mosquitoes and other disease vectors to extend their natural range (**Figure 7.52**).

Malaria is certainly a massive and chronic global health problem. Every year, somewhere between 300 million and 500 million people become infected with malaria, and more than 1.5 million of those infected die, most of them children in Africa and Asia (**Figure 7.53**).

There is, however, a lot of discussion about malaria in the media that is both misinformed and misleading. Rising concern is based on an idea that global warming will extend the range of malaria above 2,000 meters (6,562 feet) into the East Africa Highlands—and then farther afield into more temperate regions. But malaria is not a new threat to communities in the East Africa Highlands or even beyond the tropics. High tropical cities such as Nairobi have suffered from malaria many times before, as the mosquito *Anopheles* and the malarial parasite it carries can survive beyond 2,000 to 3,000 meters (6,562 to 9,841 feet). Less than 60 years ago, malaria was also present in the United States and Europe (including England), and it was common in many places during the 16th to 18th centuries (**Figure 7.54**).

> **CHECKPOINT 7.40** ▶ Why is malaria, a truly horrendous disease, as much a human problem as a climate problem?

Human behavior and **human ecology** (humans' interaction with the environment) remain the primary driving forces for disease. Developed countries have been able to eradicate many diseases that used to be common, such as malaria, typhoid, cholera, and polio, not because of any change in climate but because of political stability and economic development. We should focus our attention on these factors when we look at the developing world; any distraction based in uncertain science is a disservice to millions in crisis. Climate change is important and a real factor in spreading human disease, but it is only one small part of a more massive problem.

> **CHECKPOINT 7.41** ▶ Which is a greater threat to human health: climate or poverty? Explain your answer.

▶ **Figure 7.53:** States of the world where the population is at risk of malarial infection. It is unlikely that the threat of malaria ends so neatly at international borders as shown on the map, but this is a reminder that the incidence of malaria can be as closely related to politics as it is to climate.

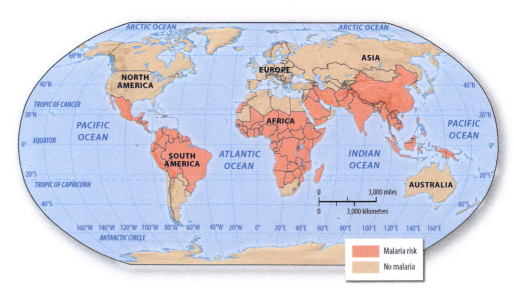

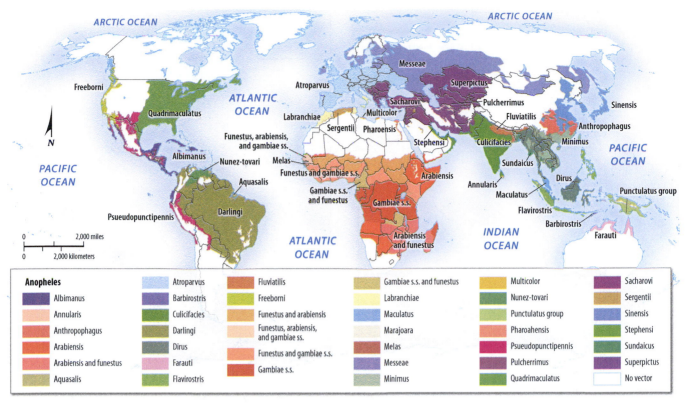

▲ Figure 7.54: Areas of the world where mosquito vectors have been responsible for malaria in the past and have the potential for future transmission with global warming. The colors represent the most prevalent species of Anopheles. The successful eradication of malaria from the United States and Europe was a great success for public health and education, and this must be the answer to the rising threat of malaria in other parts of the world due to global warming.

Case Study 9: The Melting of Arctic Tundra

Flying on my way to a meeting in Japan, I sat by the window and looked down on a barren and windswept landscape. This was Siberia in the early fall, before the winter snows blanketed the ground, and its vastness astonished me. The same scene unfolded hour after hour, as we crossed winding rivers, barren hillsides, and a myriad of small lakes and streams. We take the Arctic for granted; it seems remote and timeless, and so we give little thought for its future or the hidden threats that might lie beneath an unfathomable expanse of tundra and forested taiga. But all is not what it seems.

More than 4 million people from Canada, Finland, Greenland, Iceland, Norway, Russia, and the United States share this remote and hostile habitat. More than 10% of these are indigenous people whose cultures and traditions depend on cold, icy winters that freeze the oceans and open up hunting grounds on slowly shifting ice flows. Across the region, roads, buildings, pylons, and pipelines rest on thick, strong permafrost that thaws to form a thin active layer during the summer, freezing again at the onset of winter.

Climate scientists are very concerned about the impact of global warming on the entire Arctic region, from the margins of the great boreal forests to the

south, through the barren landscape of icy tundra, to the Arctic shoreline and great ice sheets that blanket the polar seas. The Arctic is warming—and it is warming faster than any other place on Earth.

Disappearing Tundra and Permafrost

North of the great evergreen taiga forests of Siberia and North America, a vast expanse of Arctic tundra extends to the icy borders of the Arctic Ocean. Climate is changing faster here than anywhere else on the planet. Harsh winters and a short growing season of only 50 to 60 days limit the growth of vegetation, and life is made even more difficult by the presence of permanently frozen ground, or *permafrost*, that forms as deep soil and bedrock remain frozen for more than two years in a row. The permafrost extends beyond the shoreline to the shallow continental shelf that was exposed during the last glaciation.

Arctic permafrost covers 23 million km² (8,900 square miles), occupying 24% of the land area of the Northern Hemisphere (**Figure 7.55**). With only 15 to 20 centimeters (6 to 8 inches) of rainfall a year, and with winter temperatures that average –28°C (–18°F) and dip as low as –50°C (58°F), this is a challenging ecosystem. But there is more diversity than the austere landscape suggests. The Arctic ecosystem is dominated by low shrubs and sedges, mosses, liverworts, and grasses. A variety of herbivores graze on this growth, including

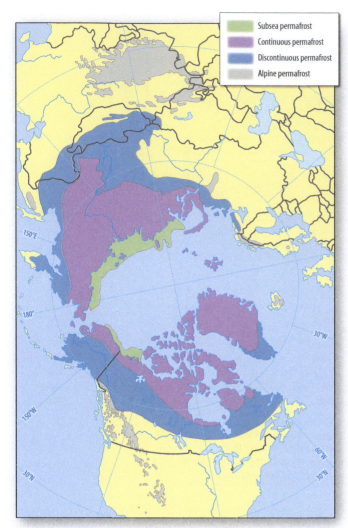

Subsea permafrost
Continuous permafrost
Discontinuous permafrost
Alpine permafrost

▲ **Figure 7.55:** The distribution of permafrost around the Arctic today still covers a vast area.

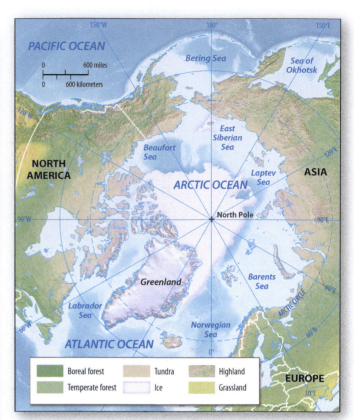

▲ **Figure 7.56:** A polar view showing the current vegetation and land cover of the Arctic.

Boreal forest   Tundra   Highland
Temperate forest   Ice   Grassland

herds of caribou and musk ox together with arctic hare, arctic fox, snowy owl, lemmings, and (in the far north) polar bears (**Figure 7.56**). Insects, especially beetles, are abundant, and swarms of ravenous mosquitoes plague indigenous residents and visitors alike.

The permafrost layer naturally melts at the surface for some time each summer, forming an unstable **active layer** (see Chapter 2). The depth and persistence of this active layer has been increasing over recent years, and any natural or human-made edifice not grounded on solid permafrost beneath is likely to collapse. Even trees are affected, and "drunken trees" often tilt and fall over due to the melting of near-surface permafrost (**Figure 7.57**).

As summer progresses, the surface layer becomes waterlogged, and thousands of small lakes and marshes form on the solid ice beneath. More recently, as

▶ **Figure 7.57:** Melting of the permafrost leaves surface structures such as roads, buildings, and pipelines unstable. Here "drunken trees" in Fairbanks Alaska lean in all directions, as their roots can no longer support their weight.

the climate has changed and the summers have become warmer, the deeper permafrost has started to melt, and the lakes are starting to disappear as the surface water drains away.

CHECKPOINT 7.42 ▶ What is the active layer? Why is it a problem for engineers and builders in the Arctic?

The Carbon Bomb: A New Threat from Methane

Permafrost soils are rich in accumulated plant material that has been frozen *in place* and preserved over time. Known as **yedoma**, this organic-rich permafrost contains anaerobic bacteria that thrive in oxygen-poor environments and start to break down organic material as the permafrost warms and starts to melt. These bacteria release methane gas as a by-product of cellular respiration and add more greenhouse gas to the atmosphere. During the winter months, some of this methane is trapped beneath lake ice. It is possible to drill a hole and set fire to an explosive stream of methane as it escapes from beneath the surface (see Chapter 2).

No one is sure how much methane is trapped in the permafrost or how much will be released by the renewed decomposition of long-dead vegetation as the Arctic continues to warm. Some estimates put the value as high as 950 gigatonnes worldwide, but it is the rate of release that is of more immediate concern than the size of the overall reservoir. Until recently, levels of methane in the atmosphere were stable after a period of earlier growth, but they are now starting to rise again. One of the most likely sources is rapid warming of the Arctic. The higher, "business as usual" estimates from climate models that assume we do little or nothing to limit greenhouse gas

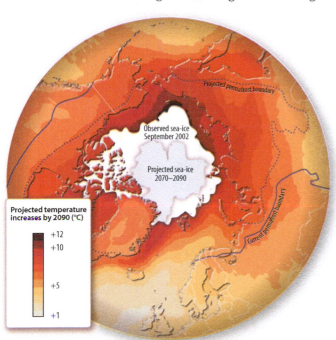

▲ **Figure 7.58:** Model projections of extreme Arctic warming by 2090, using the extreme, yet possible, A2 emissions scenario.

Projected *permafrost boundary*

Observed sea-ice September 2002

Projected sea-ice 2070–2090

Current permafrost boundary

Projected temperature increases by 2090 (°C)

+12
+10
+5
+1

▲ **Figure 7.59:** Methane gas escaping from ice-like methane hydrates burns easily

emissions suggest that some areas may warm by as much as 8°C to 10°C (14°F to 18°F); (**Figure 7.58**).

An Even Larger Source of Methane Deep under the surface of the tundra, where the pressure is much higher, and especially under the adjacent Arctic Ocean floor, methane and ice crystallize together to form **gas hydrates**, where molecules of methane are trapped in small holes or pockets within an ice-like crystal lattice. Methane hydrates look and feel like ice, but when warmed, they effervesce violently as the methane gas released expands (**Figure 7.59**).

Hydrates are most abundant under the ocean on the continental shelf, where they are estimated to have locked away as much as 2,000 to 5,000 gigatonnes of carbon, about the same amount as is left behind in all the coal reserves of the world. There is some concern that these methane deposits could decompose catastrophically as Arctic temperatures increase. This is physically possible, especially for deposits exposed or near the surface, but it is unlikely to be widespread due to the high depths and pressures at which most deposits exist. Warming sea surface temperatures will take a long time to reach down to deeply buried methane hydrates, and as the hydrostatic pressure increases due to rising sea level, it is likely that most will remain stable for some time to come.

7.6 PAUSE FOR... THOUGHT | *When is an ecological disaster not really a disaster?*

We often see the Arctic as a cold wasteland covering a vast area of around 4.5 million km² (1.7 million square miles) of the Northern Hemisphere. But on closer inspection, we see that the Arctic biome is actually quite rich and varied. Environmentalists are concerned with preserving the Arctic, especially from the ravages of oil and mineral production. When the population of Earth reaches 9 billion, however, and the vast area of Arctic land can only support 4 million, would it really be a disaster if climate change made the land more productive, more accessible, and more habitable? As far as Earth history is concerned, a cold and inhospitable Arctic is abnormal. As the Arctic cooled during the Neogene and tundra replaced temperate forests, was that not also a great ecological disaster? Or do we simply call any change in the environment from the way we knew it as a child a "disaster"?

Are There Any Positive Impacts of Climate Change?

Many aspects of climate change today are clearly threatening, but there are some limited advantages to a small increase in global temperature. Milder winters and a longer growing season in the subarctic region will enhance biological productivity and encourage greater biodiversity. The advance of taiga forests into areas once dominated by tundra is already evident in Siberia, and warmer coastal waters are now more favorable to herring, pollock, cod, and commercial salmon fisheries.

There are many other possible benefits to climate change in the Arctic. Migratory birds will benefit from increased vegetative cover, warmer spring temperatures, and more food to feed their chicks. In parts of the United States, Canada, and Russia, the range, growing season, and productivity of cereal crops will increase, and this is good news for a world that needs more food.

Fewer people will die from cold and related diseases in the wintertime. (Today many more die from cold in the winter than from heat in the summer.) Trade routes can open up across the Northwest Passage and across the coast of Siberia, cutting the costs of transporting goods to consumers (**Figure 7.60**).

New offshore resources of natural gas, a preferable alternative to coal power, could become available, reducing reliance on foreign imports (**Figure 7.61**). This has led to intensifying rivalry among all the nations that have legal claims on Arctic resources; the Russians even planted a flag at the North Pole (**Figure 7.62**). There is certainly potential for future conflict over increasingly rare and valuable resources.

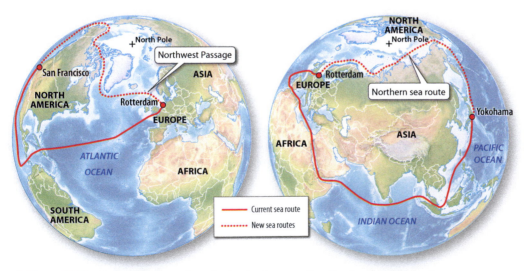

▲ **Figure 7.60:** New sea routes will make the Arctic very attractive for development.

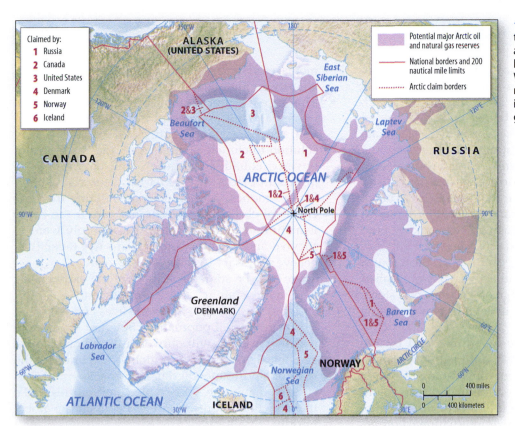

◄ Figure 7.61: The geology of the Arctic is under intense scrutiny as resources once locked away may become available for exploitation. With so many nations claiming mineral and transport rights in the Arctic, a potential future geopolitical crisis looms.

▼ Figure 7.62: When the Russians used two mini submarines to plant their flag two and a half miles beneath the North Pole in August 2007, it was a strong statement of "possession" to other nations competing for Arctic resources.

Summary

- The ability of a community to cope, adapt to, mitigate, and escape from the emerging hazards of climate change varies according to its social, financial, and technological capacity. The developed world can avoid some of the worst effects of climate change, but many developing countries lack the necessary financial and human capital resources.

- The emission of greenhouse gases has raised average global temperature and decreased the pH of the oceans. As a result, changing weather patterns, rising sea level, and increasing ocean acidity pose significant threats to global ecosystems.

- The fragmentation and pollution of the natural environment due to human activity has multiplied the impact of climate change on natural ecosystems and increased the risk of irreversible damage by the end of the century.

- Many millions of people are threatened by the impact of drought, disease, and rising sea level. Communities suffering from diminishing resources are at an increased risk from natural disasters and will inevitably migrate to find a better life. As the flow of climate refugees across international borders increases it will become a challenge to national security and political stability.

- The threat from climate change is real and immediate, but extreme poverty, political corruption, and violent conflict are parallel problems. Any solution to climate change must address issues of poverty, equality, education, and environmental justice. In places such as Africa, Central Asia, and South America, people who are suffering from the effects of climate change do not care if the cause is natural or anthropogenic. They simply want to survive and prosper, and will do whatever they believe necessary to achieve these goals.

- There may be some benefits from the early effects of climate change. Mild winters and a longer growing season may save some lives and help feed a growing global population. But very soon, more negative impacts will outweigh the advantages. A geologically rapid rise in global temperature, the risks of environmental damage, and the impact of climate change on vulnerable communities make the reduction of greenhouse gas emissions a global priority.

Why Should We Care?

We have only a short time left to enact real change by reducing the level of greenhouse gas emissions. Climate change is happening here and now, and it is threatening millions of people who lack the resources to either adapt to it or mitigate its effects. They want answers, they need help, and they are looking to developed nations to lead the way. Where do we concentrate our effort, who will pay, and where do we start?

To counter the threat of global climate change, we are faced with many difficult choices. Capitalism has a checkered environmental record, but despite some jarring failures in the fields of conservation and environmental justice, the free market working along side government has proven to be more robust than any alternative economic or political model at fighting poverty, disease, and inequality in the world. We have to find a way to make capitalism work for us to avoid the worst impacts of climate change.

Some environmentalists believe that any change in the environment is unacceptable, but they miss the geological perspective of climate change and fail to offer an alternative that is acceptable to most people in the developing world. On the other hand, some social conservatives, who believe that free-market forces alone can meet these environmental and social challenges, appear to have forgotten the burning rivers and chocking smog of the mid-20th century. It may be distasteful to some, but we need to look for a regulated capitalist solution that can tackle both poverty and climate change. These questions are addressed in Chapters 8, 9, and 10.

Looking Ahead . . .

In Chapter 8 you will explore the political, social and economic issues that surround climate change, consider the ability of communities in the developed and developing nations to respond effectively to these changes, and reflect on how different social groups perceive and communicate the risks involved.

Critical Thinking Questions

1. You are manager of a marine wetland sanctuary that is bounded to the east by the open ocean and to the west by a seawall and a major highway. To the west of the highway is a 2-kilometer (1.2-mile)

stretch of flat grassland leading to a major housing development. Sea level is rising fast, and you have to write a strategic plan for the next 50 years of development. What would you recommend as the best answer to combat rising sea level?

2. What do you believe to be the biggest threat to the Amazon forests? A local environmentalist believes that the answer is deforestation and not climate change. He thinks one is a clear and proven threat, while the other is still poorly understood. If you had a chance to talk to him, what would you say, and where would you recommend that he concentrate his limited local resources to have the greatest impact?

3. The Australian government is spending millions of dollars to protect and monitor the Great Barrier Reef. Your friend thinks this is a waste of money and resents the hike in beer tax to pay for it all. How do you explain to him that it really is important to protect the reefs?

4. Ocean acidification is a complex problem. When the boy next door refuses to get into the sea to swim because you told him the water is getting acidic, his parents suggest that you talk to the students at the local elementary school. What do you say to inform them correctly but get them back to some fun in the water?

5. You are the head of the government of a small tropical island nation. You know that the island is slowly sinking due to natural causes, but you also realize that it will not be submerged for hundreds of years. Recently, you noticed that sea level appears to be rising faster than normal, and scientists tell you that it is due to global warming. You can expect the island to disappear under the waves in just over a century. Some important tourist developers are already starting to ask difficult questions, and too many wealthy homeowners are putting their beachfront properties up for sale, threatening to damage the housing market. What policies will you recommend to the island government?

6. A large aquifer lies underneath the border of two competing nations. The geologists from each nation calculate that most of the aquifer lies within their territory (of course). The United Nations looks over the same data and decides that the water should be shared equally between the two countries. How would you monitor and enforce the fair use of water?

7. You are an aid worker in Bangladesh. After a major cyclone, you arrive at a village to help the few survivors you can find. A village elder tells you that this is the worst storm surge he has seen in 50 years and asks you if it is time to leave the farms behind and move the families from this village to the city or try to cross the border into India. What would you advise?

8. You are an Arctic ecologist. You understand that some polar bear populations are affected by changes in Arctic climate and that their future is uncertain. Bears have survived previous changes in climate, but this time, they face competition from humans. You are worried that bears will start to forage in the local towns and villages where there is rapid development due to changes in climate that allow drilling for oil and gas, and that many will be shot. Can you devise a strategy to deal with this problem? What can you do if there is little ice to send the bears back to?

Key Terms

Please make sure you are familiar with the key terms introduced and highlighted in this chapter. A full glossary is available at the end of the book.

Active layer, p. 248
Adaptation, p. 218
Adaptive capacity, p. 213
Alliance of Small Island
  Nations, p. 245
Anopheles, p. 246
Aquifer, p. 222
Aragonite, p. 231
Biodiversity, p. 225
Biome, p. 216
Biosphere, p. 216
Bleached, p. 227
Charismatic megafauna, p. 216

Coastal aquifer, p. 213
Coping strategy, p. 212
Crisis point, p. 213
Dead zone, p. 214
Dinoflagellate, p. 228
Drainage basin, p. 223
Ecosystem, p. 216
Gas hydrate, p. 249
Glacial moraine, p. 233
Glacial rebound, p. 237
Human ecology, p. 246
Invasive species, p. 218
Keystone species, p. 216

Mangrove swamp, p. 245
Marginal desert, p. 213
Migration corridor, p. 218
Mitigate, p. 213
Monaco Declaration, p. 232
Ocean acidification, p. 214
Ocean acidity, p. 230
pH, p. 230
Pleistocene Period, p. 218
Precautionary principle,
  p. 215
Saffir-Simpson scale, p. 242
Sahel, p. 213

Scleractinian, p. 227
Snowpack, p. 213
Specialization, p. 218
Storm surge, p. 213
Taiga, p. 214
Tectonic subsidence, p. 244
Tuvalu, p. 245
Uncertainty, p. 215
Water stress, p. 213
West African monsoon,
  p. 225
Yedoma, p. 249
Zooxanthellae, p. 227

The environmental movement in the United States has been successful in raising public awareness about the dangers of climate change. And yet despite active lobbying, there has been little progress in Congress towards limiting carbon emissions because of strong political oppositions from both Republicans and Democrats.

People and Politics

Introduction

The scientific evidence makes it clear that climate change is a serious threat to society and the environment and suggests that immediate political action is necessary to limit greenhouse gas emissions. From a political perspective, however, there are many additional social, economic, and strategic issues to consider before most nations will agree to limit the emission of greenhouse gases by reducing their dependence on fossil fuels for energy. The developing nations are concerned that these changes will slow economic development. They have to balance the need for growth today against the risk of climate change in the future. The developed nations are concerned about employment, economic competitiveness, and the growth of the richer developing nations. Early progress at the United Nations raised hopes that the international community could reach agreement on ways to reduce emissions and avoid the impact of climate change, but an accelerating demand for cheap energy, a global recession, and increasing international competitiveness interrupted this process, and negotiations are effectively stalled. The public debate about climate science in the United States has become increasingly polarized and contentious to a point where the basic scientific evidence is refuted and politicized by those opposed to any action to reduce greenhouse gas emissions. Scientists need new ways to communicate the risks of climate change to the people they live and work with if they hope to make progress towards an effective political solution.

Learning Outcomes | When you finish this chapter you should be able to:

- Explain what *stakeholders* are and describe how they help to shape climate policy
- Debate the appropriate role for science in guiding climate policy
- Summarize current international efforts to respond to climate change
- Compare the roles, responsibilities, and representation of developed and developing nations
- Understand the geopolitical environment that shapes climate negotiations
- Propose strategies to improve the quality of communication about the risks of climate change
- Understand the meaning of risk, how this is perceived, and how it can be amplified or attenuated within different social networks

Climate Policy

Concern about climate change is not new. Scientists have known for decades that rising greenhouse gas emissions present a growing threat to society and the environment. Despite this knowledge, there was no substantial political progress toward reducing emissions until a growing environmental movement finally succeeded in mobilizing international concern in the late 1980s. Early negotiations hosted by the United Nations toward an international climate treaty made rapid progress. Many people and nations recognized global warming as a problem that required an international solution, and the political pathway towards such a solution was much easier following the end of the Cold War. Since those first days of UN negotiations, however, strong corporate, industrial, and energy interests have tried to derail negotiations and counter the influence of the environmental movement. Together with their political allies, they are concerned that the cost of reducing emissions will limit economic growth at a time of increasing competition from China and India. They helped to form a strong anti-environmental countermovement that has succeeded in blocking most climate legislation in the United States for the past 30 years (Figure 8.1).

CHECKPOINT 8.1 ▶ How long have climate scientists known that greenhouse gases are a threat to the environment?

▼ **Figure 8.1:** Seen from space, Earth's atmosphere, a thin veneer of blue against the darkness of space, seems fragile and vulnerable. But for many people, the atmosphere appears to be so large that human activity could never have a major impact on it.

The Role of Stakeholders

The stakes in the climate change debate are high, and many different organizations, groups, and individuals are involved. These **stakeholders** come from all sides of the climate debate, and each has a vested interest in the outcome. They spend millions of dollars to influence climate negotiations, and they lobby extensively to make their views known at the highest levels of government (Figure 8.2). Many corporate, industrial, and energy stakeholders are concerned that government regulations to control and limit emissions will harm their business and the economy. They posit that trillions of dollars are at risk on world markets and that thousands of jobs will be lost if any legislation undermines economic competitiveness. Not surprisingly, they aim to block the progress of climate legislation. Lined up on this side of the debate in the United States are formidable organizations such as the American Petroleum Institute, the coal industry, manufacturing giants, utility giants, the Heartland Institute, the American Enterprise Institute, the George C. Marshall Institute, and Koch Industries.

Environmental stakeholders believe that corporate and industrial interests fail to consider the economic cost of *inaction*. If global temperatures continue to rise, they argue, future climate change will have such a devastating impact on the global economy that any short-term losses are worth the sacrifice. They reason that job growth in a new and growing low-carbon economy will offset any losses that result from climate legislation. These stakeholders include many well-established and relatively well-funded organizations,

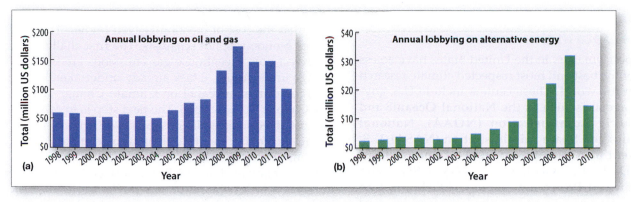

▲ **Figure 8.2:** Lobbying U.S. Congress. So much is at stake in the climate debate that many stakeholders, such as the fossil fuel and alternative energy industries, have invested millions of dollars to ensure that their viewpoint is well represented in government. (a) The amount of money spent on lobbying by the oil and gas industries increased dramatically between 2003 and 2008 because of rising concern about the impact of the Kyoto Protocol. Since then, activity has decreased, but the investment in lobbying is still many times higher than that (b) from the alternative energy industry.

such as Greenpeace, Friends of the Earth, the Union of Concerned Scientists, the World Wildlife Fund, and the Pew Research Center.

Both factions acknowledge that there is uncertainty involved with climate change. However, they propose to manage the *risks* in different ways, and these differences lie at the core of the political debate. Corporate stakeholders believe they safeguard the economy and lower the risks by delaying political action on climate change until there is more scientific certainty. Environmental stakeholders believe there is enough existing evidence to justify immediate intervention to combat climate change. This division has become highly contentious, and it has been a barrier to political progress. There are now so many opposing and contradicting arguments that it is difficult for politicians to decide which action to support. One main barrier is that most politicians are not trained as scientists and find the scientific arguments difficult to understand. As a result, scientists and policymakers, despite good intentions, often talk past each other and fail to establish important common ground (**Figure 8.3**).

CHECKPOINT 8.2 ▶ In your own words, explain the concept of *stakeholders* and consider the different ways that you are a stakeholder in the climate debate.

8.1   PAUSE FOR... THOUGHT   *Who are climate stakeholders?*

Make a list of all stakeholders who have some financial, political, or ideological interest in the climate change debate. What do you think drives the interest of each stakeholder? If you had to communicate an important message about climate change to each of these stakeholders, would you always frame the conversation in the same way?

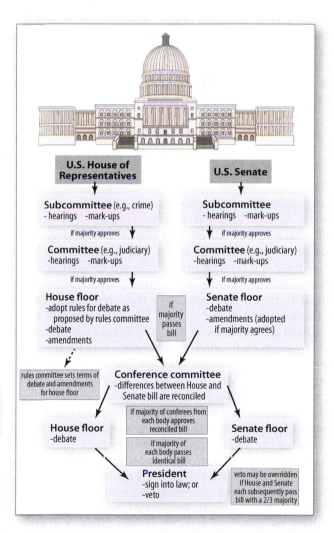

▲ **Figure 8.3:** The process of passing a bill through Congress is long and difficult, with many opportunities for lobbyists and political opponents to block or delay progress. When a bill has cross-party support, its progress through this legislative labyrinth can be surprisingly rapid. If a bill lacks cross-party support, however, clever political leaders may use innumerable methods to hamper its progress.

Informing Legislation: Sources of Climate Data in the United States

Climate policymakers in the United States have access to some of the best and most respected climate research centers in the world. These centers include U.S. government agencies such as the **National Oceanic and Atmospheric Administration (NOAA), National Aeronautics and Space Administration (NASA), U.S. Department of Agriculture (USDA), Department of Energy (DOE), U.S. Geological Survey (USGS),** and **Environmental Protection Agency (EPA).** Together with overseas agencies such as the UK Meteorological Office, the United Nations, and major universities around the world, the U.S. agencies have a long history of research excellence, and for years their scientists have published their data and opinions in the scientific literature. Political leaders regularly organize hearings to address key scientific questions, and there have been scores of such hearings about climate change over the past decade. The scientific data are now freely available, widely publicized, and summarized in a style accessible to nonspecialists. Despite this, the evidence that climate change is a growing risk is frequently ignored or bluntly denied.

Weighing the Costs and Benefits in the Climate Debate

Stakeholders on all sides of the climate debate try to make sure that politicians are acutely aware of the issues. As the climate debate became more polarized, politicians found themselves trapped in an argument that many recognize as being nuanced and complex. They hear the science but also listen to their constituents, and are acutely aware that climate legislation might harm the communities they represent. Much of the momentum for climate legislation comes from so-called green states such as California and Vermont that are pushing for alternative energy. Most of the resistance is coming from energy-producing and energy-manufacturing states in the Midwest and Great Plains, as well as smaller states such as West Virginia, which either produce coal or depend on cheap energy to stay competitive with the developing world. Consider the pressures on representatives from energy-producing **swing states** such as Colorado and Ohio, where consumer electricity prices are highly dependent on coal for generating power. California produces only 20.7% of its electricity from coal, but in Ohio, coal is used to generate 86% of power. This is a major dilemma for politicians who must win elections every few years and care deeply about the immediate needs of the people who vote for them.

The economic argument against climate legislation is so strong in some states that the only way to win the argument in favor of climate legislation is to make it clear that the cost of inaction is even higher. This is a job for economists, not scientists. The first challenge is to project how greenhouse gas emissions are likely to change in the future, a task already undertaken by the Intergovernmental Panel on Climate Change (IPCC) (see Chapter 1) (Figure 8.4). The next step is to compare the cost of reducing emissions today with the economic, social, and environmental cost of future climate change. It is important that people understand that the impact of future climate change will far outweigh the cost of mitigation today. Communities put at economic risk by legislation to reduce emissions need to believe that they will benefit from a green economy in the long term and receive adequate financial support to manage the transition from a high-carbon to a low-carbon economy in the short term.

> **CHECKPOINT 8.3 ▶** Is it reasonable to expect politicians to work in the interests of future generations if doing so is against the immediate interests of their electorate? Explain.

The Economic Analysis of Climate Change

To estimate the cost of future climate change, economists combine emissions data from the IPCC with climate models to project how factors that have great economic impact—such as heat waves, drought, forest fires, crop failures, storms, floods, and ocean acidification—are likely to change in the future. These data have been combined with other economic projections based on social and technological factors such as population growth, population movement, regional conflict, energy efficiency, energy conservation, fossil fuel consumption, and the impact of new technologies to arrive at an overall projection of the costs of climate change under a range of emissions scenarios, including some that take into consideration the possible impact of future climate legislation.

These scenarios allow policymakers to make informed decisions about the kinds of legislation necessary to balance the cost of reducing emissions today against the future costs of damaging climate change. In theory, this knowledge should allow them to propose economic solutions that balance sustained economic growth and emissions reduction.

The problem is that there are so many variables involved that it is impossible to arrive at a unique solution. Many factors are well known, but others are less well defined, and the accuracy of economic projections depends on getting all the factors right. One important example is that model solutions depend on knowledge of the sensitivity of climate to heat-trapping greenhouse

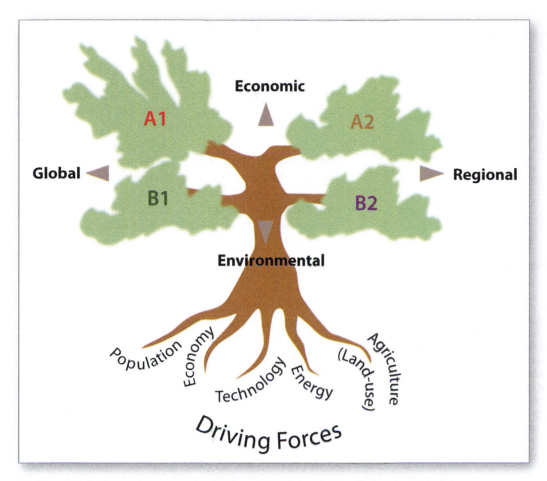

▲ Figure 8.4: The major economic criteria used to determine the four major emissions storylines outlined in the IPCC's *Special Report: Emissions Scenarios* (Figure 1.25).

gas emissions, but this is one of the few areas where it is widely accepted that climate data are incomplete. The broad consensus among climate scientists is that global temperatures will increase by as much as 4°C (7°F) by the end of the century if nothing is done to limit heat-trapping greenhouse gas emissions. However, there is still a small possibility that negative (cooling) feedback in Earth's climate system could limit this increase to between 1°C and 2°C (2°F and 3.6°F).

This critical uncertainty lies at the heart of the political debate. If the answer is 4°C (7°F), immediate action to reduce emissions is warranted because the economic cost of climate change is so high that even the swing states would have to change to ensure future prosperity. If the answer is 2°C (3.6°F), it is arguably better for the swing states to wait for the future, adapt to climate change, and look for technologies that offer a less expensive solution to the problem than exists today (Figure 8.5).

CHECKPOINT 8.4 ▶ What are the main driving forces that will determine the future level of greenhouse gas emissions?

The Stern Review: Immediate Action Is the Cheapest Option One economic analysis has influenced global climate policy more than any other. Lord Stern of Brentford, former chief economist and senior vice president of the World Bank, was asked to calculate the future cost of climate change for the UK government. His 2006 report warned that global warming is a very real threat and presents serious economic risks:

Our actions over the coming few decades could create risks of major disruption to economic and social activity, later in this century and in the next, on a scale similar to those associated with the great wars, and the economic depression of the first half of the 20th century.

Lord Stern concluded that the benefits of early action to counter global warming considerably outweigh the costs, estimating that global warming could cost up to 20% of global GDP if left unchecked. In comparison, he said, the cost of intervention could be as "little" as 1% of GDP. Lord Stern's conclusions are controversial, and not all economists agree with them. However, it is noteworthy that one of Lord Stern's most vocal

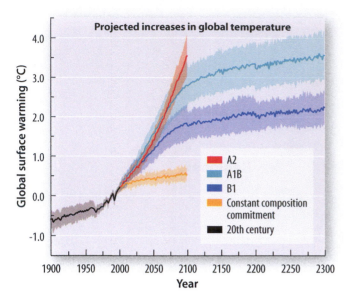

▲ **Figure 8.5:** The rise in global temperatures predicted by climate models depends on economic forecasts that are themselves dependent on many other variables (Figure 1.25). This figure shows how temperature will change according to the economic forecasts included in the IPCC's *Special Report: Emissions Scenarios*. Lines show the multiple-model means, and shading denotes the ±1 standard deviation range. The "constant composition commitment" scenario corresponds to a greenhouse gas concentration that remains constant throughout the 21st century.

critics, William Nordhaus at Yale University, still agrees that some immediate political and economic action is required to limit the impact of climate change. In March 2012, Nordhaus stated:

> *The cost of waiting fifty years to begin reducing CO_2 emissions is $2.3 trillion in 2005 prices. If we bring that number to today's economy and prices, the loss from waiting is $4.1 trillion. . . . The claim that cap-and-trade*

legislation or carbon taxes would be ruinous or disastrous to our societies does not stand up to serious economic analysis.

CHECKPOINT 8.5 ▶ What conclusion did Lord Stern reach in his report on the economics of climate change?

The World Beyond 2100 Unfortunately for policymakers, long-term economic computer models, like climate computer models, become less predictable with the progression of time, and no models are able to agree in any detail beyond the end of the 21st century. Inherently complex systems such as the climate and the economy are very sensitive to initial conditions, and small changes "upstream" can have profound impacts on the outcomes "downstream."

In many ways, the economic models are even more complex than the climate models. Consequently, any economic planning that extends beyond the end of the 21st century is highly speculative. Most of the uncertainty lies with the social factors. There are just too many unknowns. For example, how will advances in alternative energy, bioengineering, and nanotechnology affect the economy? What is the potential during this time for international conflict, pandemics, and other major natural disasters that could profoundly affect the global economy? Think of all the changes in technology and society that took place during the 20th century. What impact will future technology and population growth have on our ability to combat climate change (**Figure 8.6**)?

CHECKPOINT 8.6 ▶ Why is it difficult for computer models to make realistic predictions about the state of the economy beyond the end of the 21st century?

▶ **Figure 8.6:** It is difficult to predict economic growth and the demand for energy at the end of the century. This graph from the United Nations shows that we are still uncertain about such fundamental factors that drive the economy as world population growth. One factor that has been important historically is economic growth. More developed nations with higher average per capita income have a lower (or even negative) population growth rate. The high, medium and low projections of population growth shown here are based on models used by the United Nations.

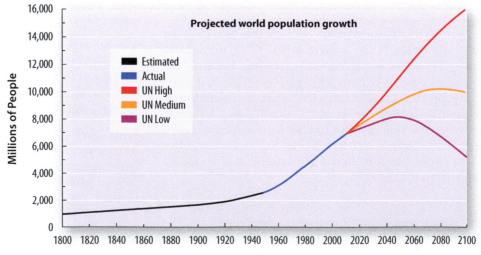

Finding an Economic Solution in the Developing World

Climate change is an international problem that requires an international solution that addresses the legitimate concerns of many nations. It will take strong political will and firm leadership to shape these divergent and competing interests into an effective international policy to address the problem of climate change. Knowledge can turn into action only if there is the political will to take action, but political will is notoriously ephemeral.

Political leaders around the world are often criticized for shortsightedness, but consider the enormous challenges they face. They must manage budgets, combat recession, create jobs, advance health care, tackle poverty, foster education, defend the nation, and (in democratic nations) face reelection every few years. It is easy to see how the cost of preventing climate change can seem too high in the developing world if it means taking funding away from more immediate priorities—such as women's and children's health, education, and economic growth—that are clearly defined international priorities under the International Covenant on Economic, Social and Cultural Rights and as stated in the United Nations Millennium Goals.

CHECKPOINT 8.7 ▶ Is near-term economic growth in developing nations a greater priority than the long-term threat of climate change? Explain.

In developing nations, tens of millions of people are living in poverty, with limited access to **potable** water, food, electricity, health care, and education. Politicians in those nations are continually forced to balance their interest in growth today against legitimate fears for the future, and they often resent pressure from developed nations to take action against a problem that they did not create. Their most obvious and immediate solution is to pursue strong economic growth, based on the use of cheap fuel resources. **Energy poverty** (when people lack access to modern energy services) remains one of the greatest barriers to social and economic growth in the developing world, with levels of energy consumption for millions below the base level necessary for survival. For many developing countries, the easiest and cheapest way to achieve these goals is to invest further in the extraction of fossil fuels. They lose concern about the impact on global emissions when they see developed countries using more energy than required to maintain a luxuriant standard of living. They (correctly) see the excessive level of emissions as a primary cause of global warming and look to the developed world to reduce these "luxurious" emissions first.

To many, it seems that capitalism, globalization, and the free movement of people, goods, money, and services

are the best routes for economic development in developing countries. But as the global demand for energy increases and as energy prices soar, there is a growing awareness that these nations need to invest in alternative energy out of economic self-interest. Many governments face growing pressure from indigenous environmental social movements that promote sustainable practices. China and India, the two biggest economies in the developing world, have already acknowledged the need for responsible and sustainable growth, and they have committed to enhancing energy conservation and efficiency and reducing the intensity of greenhouse gas emissions.

CHECKPOINT 8.8 ▶ Both climate and economic models are very complex, and their output can be only as good as the data we feed them. Why, then, is it reasonable to believe that their conclusions are reliable enough to act as a foundation for government policy?

Is There a Capitalist Solution to Climate Change?

Many environmentalists oppose a capitalist solution to climate change on principle. However, international competition for markets, resources, and influence is stronger than ever, and any action to address the problem of global climate change that is not based on the free market will have difficulty gaining global consensus. Those who imagine that a solution can still be found through the egalitarian pursuit of common international goals should consider the political fate and discouraging impact of the Kyoto Protocol discussed later in this chapter. Developing nations are simply not prepared to sacrifice economic growth when they have not yet reached the base level of per capita emissions necessary for survival; meanwhile, the developed world is not going to "go it alone" when they are faced with rising economic competition from countries such as Brazil, China, and India.

There are, however, plausible capitalist solutions that try to balance strong government intervention and regulation with entrepreneurship and the creative strength of the free market. We need to find a balance between an "enabling state" that restricts its role to stimulating free enterprise and an "ensuring state" that imposes strict regulations to ensure public health and safety. How much can industry be trusted to self regulate? How can governments avoid excess regulation that becomes punitive and hinders economic development? The nature of this balance is the subject of intense political debate.

Fiscal policies should encourage the use of alternative and cleaner sources of energy through tax incentives for innovation and technological development, tax breaks for alternative energy, carbon taxes, emissions caps, and emissions standards. Subsidies that support the fossil fuel industries should be phased out and replaced by new subsidies to support the development of a new green economy that will create employment, enhance competitiveness, generate tax revenues, and deliver energy security. In this way, the free market and government can work to mutual benefit.

In addition to promoting climate action at home, many economists also propose investing substantially in human capital and technological development overseas to help reduce global emissions. Overseas aid can help offset domestic emissions during the difficult transition from a high- to low-carbon economy. Properly invested, this capital will also create future markets for domestic goods and services.

Some of these proposals go too far for corporate, industrial, and energy stakeholders who want to limit the role of government and enable free market development. But many companies have already taken independent action to increase profitability by enhancing energy conservation and efficiency with no intervention from government. Others have diversified and restructured rapidly to meet the growing market for low-carbon goods and services.

While it is true that government intervention may not always be necessary, it is also true that the free market has not always responded to economic and social change in a timely manner. Powerful industrial and union stakeholders in energy-producing and energy-dependent states in the United States have, in the past, acted to stifle both innovation and the adoption of new technologies in key sectors such as the auto and steel industries. At different times, and under different political administrations, both failing industries were successfully restructured when government, industry, and unions worked together to meet the challenge. Now both the auto and steel industries are world leaders in technology and innovation.

Thus, government legislation can be used to create a political framework that enables economic change. Legislation to reduce overall greenhouse gas emissions has already been successful in many European countries where companies have invested in clean energy to reduce overall costs and enhance profits. Taxes (applied as a scalpel and not a hammer) have been used successfully to create the steady financial pressure necessary to catalyze social change. In the United States, efforts to increase energy conservation and efficiency still lag behind those of other developed nations. It will take time and patience to transform the high-throughput and high-waste economy of the United States into a more energy-efficient and environmentally aware economy, but the revenue generated though taxation and the money saved by reducing the cost of importing energy can be used to build or upgrade the energy infrastructure necessary to support sustainable economic growth and job creation.

We need a capitalist solution that creates new jobs and maintains the international competitiveness of U.S. industry. Corporate stakeholders will inevitably see any government regulation as a barrier to growth, but unregulated industry has repeatedly failed to offer the level of environmental protection that society needs and requires. Some government oversight is clearly necessary to maintain a balance between the public good and the demands of the free market. Past experience suggests that overregulation and punitive taxation will slow the development of a globally competitive low-carbon economy, but recent history also shows us that too much deregulation and the removal of critical government oversight can end in economic disaster. Government agencies, such as the U.S. EPA, must continue to play a vital role in maintaining effective oversight of regulations that are intended to avoid further damage to the global environment by reducing the level of heat-trapping greenhouse gas emissions.

CHECKPOINT 8.9 ▶ Are market forces alone capable of limiting the impact of greenhouse gas emissions? Explain.

The most viable solution lies in the application of free market solutions that are balanced by an appropriate regulatory framework and guiding fiscal control. This is an achievable solution, but the legislation required is not likely to succeed in the United States as long as the political debate remains polarized. The prospects of success are greater in Europe, where environmental social movements hold more political power. But at a time of economic crisis, it is not at all clear how this will work in practice (**Figure 8.7**).

CHECKPOINT 8.10 ▶ A capitalist solution will inevitably create both winners and losers, as the focus of the economy shifts to renewable energy, energy conservation, and energy efficiency. What, if any, social responsibility do we have for the losers?

8.2 **PAUSE FOR... THOUGHT** | *Who cares about science?*

Of the 535 members of the U.S. Congress, only a few are ever scientists and engineers. The 112th Congress, for example, has just one physicist, one chemist, and a single microbiologist. This is not uncommon for Western democracies. In China, the reverse is true: The majority of leaders have a science or engineering background. Could this be why the Chinese are forging ahead in the development of alternative energy and low-emissions technologies? It is not surprising that scientists have so much trouble communicating with policymakers in the United States. Even when politicians are interested, there is a chasm of communication to surmount before any meaningful dialogue can occur. When mostly male lawyers and businesspeople run a country, any scientist who wants to promote climate action must learn to speak the languages of both law and business.

▲ **Figure 8.7:** We can expect major advances in solar, wind, and other clean energy technologies over the next 50 years. If advances in energy technology over the past 50 years tell us anything, it is that future technology will likely have an even greater impact on society. But can we wait in the hope that one of these unforeseen advances will solve the energy problem?

The International Response to Climate Change

Despite growing scientific and economic consensus about the threat of climate change, there is little international political agreement about what needs to be done. We will start by considering the negotiations that led to ratification of the Kyoto Protocol, the first major international treaty to address the issue of climate change. This case study illuminates some of the advantages and many of the frustrations of international negotiation. Pay particular attention to the way that the United Nations uses scientific evidence to guide policy and compare this to the actions of your own government.

The Kyoto Protocol

The **Kyoto Protocol** was the first major international treaty to address the issue of global warming and climate change. This section explores how political negotiations led to its eventual ratification, despite protracted and often heated negotiations where many stakeholders vied for influence. Ratification of the Kyoto Protocol was a major achievement, but it was only the first faltering step toward a more complete political solution, and its final outcome still hangs in the balance.

This process unfolded very slowly and illustrates how it is difficult to arrive at an international consensus when the outcome of negotiations may result in significant global social and economic change. When the stakes are this high, the most powerful nations tend to dominate negotiations and overlook the interests of the smaller nations. This is a major problem for global governance and one reason the United Nations was the chosen venue for climate negotiations.

The Kyoto Protocol illustrates how agreement was achieved despite these competing interests, but the resulting treaty is less demanding than first envisioned. Try to imagine what the chief climate negotiators for the United States, European Union (EU), and China must have been thinking. What domestic pressures drove their decisions? They brought very different perspectives to the negotiating table, but remember that each had to sell any agreed solution to the people back home. When I teach a course on climate change, students often come up with idealistic solutions that are critical of the U.S. negotiating position. However, even if the U.S. representatives at Kyoto wanted Kyoto to succeed in every aspect, they were still confined by the political reality of what was acceptable in Congress. As you read the next section, consider what was politically achievable, given growing divisions in U.S. politics over climate change and the changing balance of international relations. What would you have done under the same circumstances?

The UN and Climate Change

When global warming became a subject of international concern in the late 1980s, the UN provided a neutral venue where competing governments could seek consensus. The impact of global climate change is not distributed equally around the world, and this

▲ **Figure 8.8:** The United Nations General Assembly hall in New York. There are now 193 member states in the UN (192 when the Kyoto Protocol was negotiated), which was formed after World War II in 1945. The UN strives to enable dialogue between nations that are often deeply divided politically, socially, and economically. Nations cooperate in the fields of international law, international security, economic development, social progress, human rights, and achieving world peace.

inequality is reflected in the imbalance of representation at regional and international levels. Many of the nations most affected are small, poor, and politically weak. The statutes of the UN are intended to prevent the larger nations from completely dominating the debate by allowing the smaller nations at least some influence on negotiations (Figure 8.8).

CHECKPOINT 8.11▶ Why is the United Nations a good venue for climate change negotiations?

A first step toward international agreement was taken when the World Meteorological Organization (WMO) and the United Nations Environment Program (UNEP) established the IPCC in 1988. The Intergovernmental Panel on Climate Change (IPCC) was to provide governments with the scientific, technical, and socioeconomic information needed to guide climate policy. The IPCC reports are *policy neutral* because they do not recommend any specific course of action, and they are *politically neutral* because they are approved by government representatives from all participating nations. The IPCC provides the information, and governments then decide how to use it.

CHECKPOINT 8.12▶ Why is it important that the IPCC provides information but does not determine policy?

First Steps: The IPCC's *First Assessment Report* and the UNFCCC

The IPCC focused worldwide attention on climate change and catalyzed UN efforts to organize an international response. The IPCC published the *First*

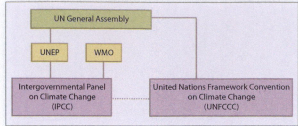

▲ **Figure 8.9:** The IPCC and UNFCCC institutional framework. In 1988, the WMO and the UNEP jointly established the IPCC, which led to the establishment of the United Nations Framework Convention on Climate Change (UNFCCC). The IPCC is organized into three working groups plus a task force on national greenhouse gas inventories.

Assessment Report on the status of climate change in 1990. This groundbreaking report presented the results of scientific research from around the world and concluded that global warming was taking place, with anthropogenic greenhouse gas emissions partly to blame. These shocking conclusions prompted the UN to propose a new international legal agreement to combat global warming. The resulting **United Nations Framework Convention on Climate Change (UNFCCC)** was opened for signature at the Rio de Janeiro **Earth Summit** in 1992 (**Figure 8.9**).

The initial goal of the UNFCCC was to stabilize greenhouse gas emissions in the atmosphere at 1990 levels by the year 2000. This was considered a "safe level" that could avoid significant damage to the environment and would be sufficient to ensure that changes already locked into the Earth's climate system would occur so slowly that threatened ecosystems would have time to adapt.

CHECKPOINT 8.13▶ How successful has the UNFCCC been at reaching its initial goal? Explain.

There was widespread international concern about the economic implications of the UNFCCC's goals. New "battle lines" were quickly drawn between the EU, which wanted strong, binding commitments to reduce emissions, and the United States, which argued that action was premature and lacked sufficient scientific basis. At the same time, there was also growing concern in the developing world about the impact that universal mandatory cuts in greenhouse gas emissions would have on economic growth and prosperity (**Figure 8.10**).

Despite these concerns, 192 governments, including the EU and United States, agreed to share information about national greenhouse gas emissions, publish national action plans, and take voluntary action to reduce greenhouse gas emissions to 1990 levels by the year 2000.

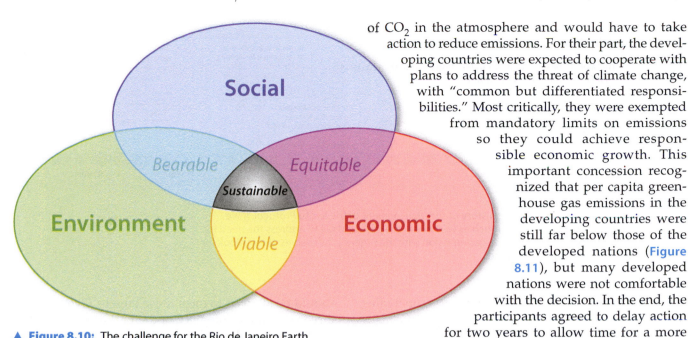

▲ Figure 8.10: The challenge for the Rio de Janeiro Earth Summit was to find a political framework that could balance the competing social, environmental, and economic needs of developed and developing countries while reducing greenhouse gas emissions to a sustainable level. This diagram illustrates how a sustainable solution that takes account of socioeconomic and environmental issues must be bearable, equitable, and viable.

CHECKPOINT 8.14 ▶ In your own words, compare the negotiating positions of the United States and the EU.

Defining the Role of Developing Countries: The Berlin Mandate The lead body of the UNFCCC is called the Conference of the Parties. By the time of its first meeting in Berlin in the spring of 1995, it was already clear that the developed countries would not reach their initial voluntary emissions targets. There was also growing disagreement over how much the developing countries should also reduce greenhouse gas emissions.

In response to this disappointing news, the UN issued a "ministerial declaration" now called the **Berlin Mandate.** The Berlin Mandate reluctantly accepted the failure of the developed countries to achieve voluntary targets and determined that mandatory limits on greenhouse gas emissions were going to be necessary to drive the process forward.

As an important part of the Berlin Mandate, the developed countries also accepted for the first time that they were responsible for historically high levels

of CO_2 in the atmosphere and would have to take action to reduce emissions. For their part, the developing countries were expected to cooperate with plans to address the threat of climate change, with "common but differentiated responsibilities." Most critically, they were exempted from mandatory limits on emissions so they could achieve responsible economic growth. This important concession recognized that per capita greenhouse gas emissions in the developing countries were still far below those of the developed nations (**Figure 8.11**), but many developed nations were not comfortable with the decision. In the end, the participants agreed to delay action for two years to allow time for a more detailed analysis and assessment.

CHECKPOINT 8.15 ▶ Why was the Berlin Mandate a groundbreaking agreement?

An Even Greater Risk of Warming: The IPCC's Second Assessment Report As the Conference of the Parties deliberated over the problem of emissions reduction, further evidence accumulated about the risks of unchecked global warming. The IPCC's *Second Assessment Report* was published in 1995. This report

▼ Figure 8.11: Per capita responsibility for current anthropogenic CO_2 in the atmosphere. Although China now emits more greenhouse gases in total than the United States, per capita emissions in China and the rest of the developing world are still far below those in the United States and most of the developed world.

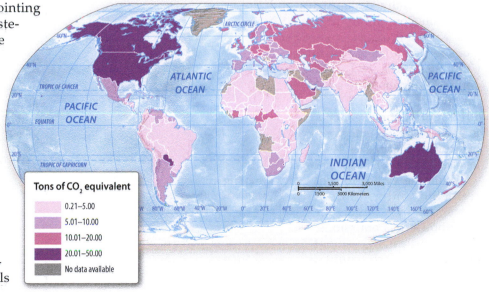

Tons of CO_2 equivalent

0.21–5.00
5.01–10.00
10.01–20.00
20.01–50.00
No data available

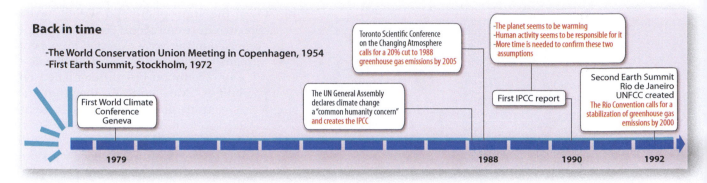

Back in time

-The World Conservation Union Meeting in Copenhagen, 1954
-First Earth Summit, Stockholm, 1972

First World Climate Conference Geneva

Toronto Scientific Conference on the Changing Atmosphere calls for a 20% cut to 1988 greenhouse gas emissions by 2005

The UN General Assembly declares climate change a "common humanity concern" and creates the IPCC

-The planet seems to be warming
-Human activity seems to be responsible for it
-More time is needed to confirm these two assumptions

First IPCC report

Second Earth Summit Rio de Janeiro UNFCC created
The Rio Convention calls for a stabilization of greenhouse gas emissions by 2000

1979   1988   1990   1992

▲ **Figure 8.12:** A schematic history of the development of the Kyoto Protocol. If we procrastinate in the future as much as we have in the past, and fail to address the need for global reduction of greenhouse gas emission with the urgency it deserves, the outlook for global sustainability is not good.

went much further than the first report, concluding that a discernible human influence on global climate was likely and would soon have a negative impact on society and the environment. The scientific confidence of this report increased international pressure for strong and immediate action against climate change, culminating in a new proposal under the UNFCCC for a treaty that would impose legally binding cuts to reduce greenhouse gas emissions to "safe" levels. This envisioned treaty emerged as the new Kyoto Protocol (Figure 8.12).

CHECKPOINT 8.16 ▶ In what ways have the IPCC scientific reports been successful at guiding policy?

Final Steps: Shaping the Kyoto Protocol Several dozen nations finally agreed to the future goals of the Kyoto Protocol in 1997 in Kyoto, Japan. This represented nearly a decade of work mobilizing the global community to combat climate change.

The Kyoto Protocol set specific and binding targets on greenhouse gas emissions for 37 participating industrialized countries and the European Community (Figure 8.13). The Kyoto Protocol covers six gases related to global warming: CO_2, SF_6, HCF, NO_x, PFCs, and CH_4. Targets were set for the five-year period 2008 to 2012, representing real cuts in greenhouse gas emissions of around 5% below 1990 levels (Figure 8.14).

While the Kyoto Protocol reflected international agreement over targets for greenhouse gas emissions, there was no agreement about how each country

should be allowed to meet those targets. Over the next three years the Conference of the Parties met to define all the rules and regulations that were needed before full implementation of the new protocol could take place.

The Challenge of Carbon Sinks One of the biggest obstacles to progress was the issue of **carbon offsets**. Each country had agreed to cut greenhouse gas emissions by a specific amount, but the Kyoto Protocol allowed them to offset part of this obligation by removing carbon from the atmosphere by alternative means.

Of particular interest was a proposal by the United States to offset some of its emissions by trapping greenhouse gases in so-called **carbon sinks**, such as reforestation and agriculture practices that

▼ **Figure 8.13:** Signatory countries to the Kyoto Protocol. This map shows very clearly that the United States stood alone in the world by refusing to ratify the Kyoto Protocol.

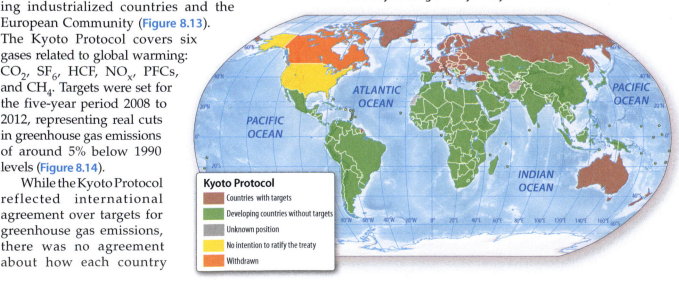

Kyoto Protocol
- Countries with targets
- Developing countries without targets
- Unknown position
- No intention to ratify the treaty
- Withdrawn

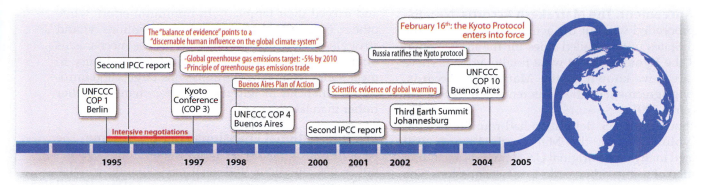

could sequester (lock away) large amounts of atmospheric carbon in the form of vegetative growth and biomass within the soil. This was a significant and controversial proposal, as it would allow the United States and some other industrialized nations to satisfy much of their emissions requirements without actually making any cuts in existing greenhouse gas emissions. Many in the developing world and in the EU saw this as "cheating" and resisted the proposal very strongly.

Climate negotiations faced a serious setback at this time, when the United States formally withdrew from the process following the election of President George W. Bush. This was a great disappointment for those who looked to the United States for political leadership. It was not entirely unexpected, as the Bush administration's policy challenged the science of climate change from the start. Some, however, felt that the United States had been holding back progress toward a final agreement and were glad the United States had stepped away from the negotiations.

Despite its formal disengagement, the United States sent observers to attend negotiations, as it wanted to maintain some influence on negotiations. No longer held back by the demands of the U.S. delegation, negotiations proceeded rapidly, and the parties soon reached agreement on the key issues. Some saw this as progress, but the absence of the United States, at that time the world's largest producer of greenhouse gases, placed the entire future of the Kyoto Protocol in doubt.

CHECKPOINT 8.17 ▶ What impact did the withdrawal of the U.S. delegation from the Kyoto Protocol have on climate negotiations?

Hot and Hotter Still: The *Third Assessment Report* The IPCC published its *Third Assessment Report* in July 2001, while the parties were still deep in discussion over the future of the Kyoto Protocol. This report concluded that there was "clear evidence" for a "discernible human influence" on global climate and warned of significant dangers in the near future.

This strong reaffirmation of the scientific evidence should have catalyzed international action, but it actually deepened division. The geopolitical implications and economic cost of the Kyoto Protocol concerned many industrial and political stakeholders, and they chose instead to weaken the agreement and cast doubt on the science rather than deal with the reality of climate change.

▼ **Figure 8.14:** Emissions targets for selected countries under the Kyoto Protocol. The results from the Kyoto Protocol are mixed. While the European Union and Australia were largely successful at meeting emissions targets, Canada with its burgeoning hydrocarbon industry fell far behind and eventually left the treaty.

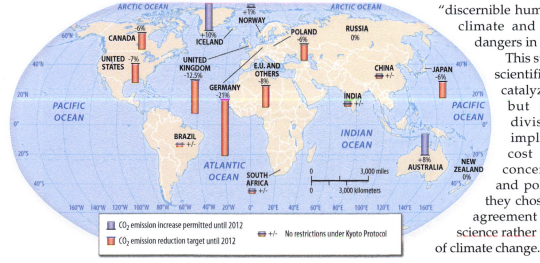

Final Agreement: The Marrakesh Accords Negotiations proceeded without the active engagement of the United States until detailed rules for the implementation of the Kyoto Protocol were finally agreed. These rules are now known as the **Marrakesh Accords,** following agreement by the Conference of the Parties in Morocco in 2001 (**Figure 8.15**).

Although the U.S. delegation had roundly rejected the Kyoto Protocol, the Marrakesh Accords still contained many of the original U.S. demands, including the option of using carbon sinks to offset emissions. The influence of the United States was also reflected in the inclusion of "flexible mechanisms"—free market–based solutions that would add financial incentives to the developed nations to meet their emissions targets and promote the transfer of technology and human capital to developing nations. This was done to leave open a door for the United States to ratify the Kyoto Protocol at a later stage.

The Policy Mechanisms of the Kyoto Protocol

Three major programs—the Emissions Trading Scheme (ETS), the Joint Implementation Mechanism (JI), and the Clean Development Mechanism (CDM)—were established to offer some flexibility in the way that nations could meet their Kyoto Protocol targets. It was clear from the start that some countries could meet and exceed their targets with minimum effort, while others, especially the more industrial nations, would struggle to meet their obligations. These mechanisms offered alternative pathways countries could use to meet their greenhouse gas emissions targets and introduced financial incentives to make changes earlier rather than later.

The International Emissions Trading Scheme The Kyoto Protocol placed a cap on the amount of greenhouse gas emissions that each participating country could produce. According to the **International Emissions Trading Scheme (IET),** countries that bettered their targets could profit by selling (trading) this excess to countries that failed to meet their obligation and needed to buy extra credits to make up their shortfall.

This **cap-and-trade** mechanism turned carbon into a global commodity that is now bought and sold on the international market. Carbon is tracked and traded like any other commodity, with traders buying and selling carbon credits like stocks in the financial markets.

This new carbon market expanded rapidly, as individual governments imposed their own limits on emissions in order to meet their Kyoto Protocol targets. The first round of cuts was targeted at the largest emitters, such as energy utilities, oil refineries, cement manufacturers, and the iron, steel, and paper industries. **National allowances** (caps) were simply

▼ **Figure 8.15:** A concept map that outlines how the different elements of the Kyoto Protocol work together to help reduce overall greenhouse gas emissions. At first glance, this diagram seems very complex, but it is worth taking time to follow the different links that connect the components of the Kyoto Protocol.

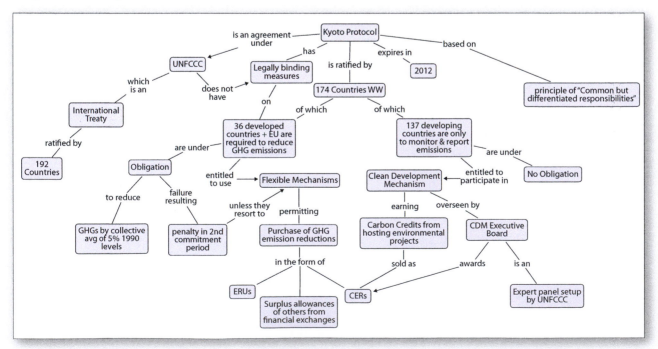

divided up among companies based on their emissions history. A large power plant, for example, would have been given an allowance based on its emissions over previous years.

This first round of national emissions reduction planning soon ran into problems when some countries overestimated demand and allocated too many credits for use. As a result, everyone met their targets, and the carbon market was flooded with excess credits, causing the price of carbon to fall dramatically due to lack of demand.

In the second round, national caps were set at a much more demanding level that required companies to make real reductions in greenhouse gas emissions. Allowances were no longer allocated on past performance; instead, they were sold to the highest bidder. This time companies were forced to either cut emissions or pay a lot of extra money to buy the rights to produce them. The outcome of this round was much more successful, and by 2007, carbon traders had exchanged over $21.9 billion. The prospect for a global market in carbon looked good.

This promising future came to a sudden and dramatic end with the economic crash in 2008, when there was a significant downturn in global energy consumption. As the level of CO_2 emissions fell, there was less demand for carbon credits, and the price of carbon collapsed. The carbon market will recover along with the global economy, but as of 2013, the price of carbon is still less than half what it was in 2007 (**Figure 8.16**).

The Joint Implementation Mechanism The **Joint Implementation Mechanism (JI)** allows developed countries to meet some of their legally binding emissions targets by investing in emission-reduction or emission-removal projects in another developed country. For example, Spain offsets some of its excess emissions by investing in a wind energy project in Poland. Each metric tonne (1 metric tonne = 1,000 kilograms [2,204 pounds]) of carbon dioxide offset by this investment earned Spain an emission reduction unit (ERU) that the country could count toward its overall greenhouse gas emissions target or sell to another country.

Some countries use the JI mechanism because it allows them to plan a more ordered transition to a lower-carbon economy. Buying the ERU credits they need to meet their emissions targets gives them more time to restructure and upgrade old and inefficient industry. ERU credits purchased this way may be cheaper than carbon credits bought directly on the open market and also may have a more immediate environmental impact. The overall advantage of the JI is that it allows countries like Spain to manage emission reductions more strategically, and others, like Poland, to benefit from the transfer of investment and technology (**Figure 8.17**).

The Clean Development Mechanism The **Clean Development Mechanism (CDM)** offers an alternative way for developed countries to offset emissions. By

▼ **Figure 8.16:** Diagram illustrating how emissions trading can work to establish a global market in carbon emissions. Every industrialized company that is required to reduce emissions by its national government receives a certain number of emissions certificates, in accordance with national emissions targets. Each certificate covers 1 metric tonne of CO_2. If one company does not use all of its allocated capacity, it can sell that capacity to another company that has failed to meet its targets and produced more CO_2 than it was allocated. The transfer of emissions certificates from one company to another takes place in a carbon market, where carbon is traded like other commodities and the price of the certificates is determined by supply and demand.

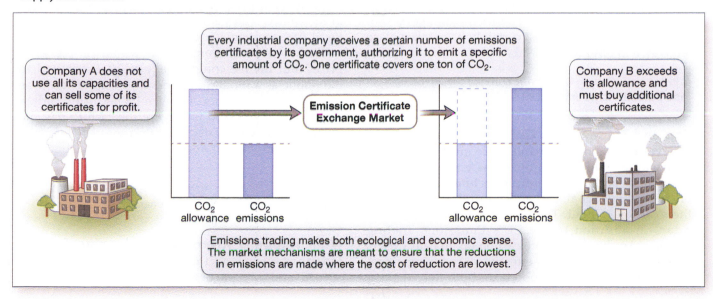

Company A does not use all its capacities and can sell some of its certificates for profit.

Every industrial company receives a certain number of emissions certificates by its government, authorizing it to emit a specific amount of CO_2. One certificate covers one ton of CO_2.

Emission Certificate Exchange Market

Company B exceeds its allowance and must buy additional certificates.

CO_2 allowance CO_2 emissions

CO_2 allowance CO_2 emissions

Emissions trading makes both ecological and economic sense. The market mechanisms are meant to ensure that the reductions in emissions are made where the cost of reduction are lowest.

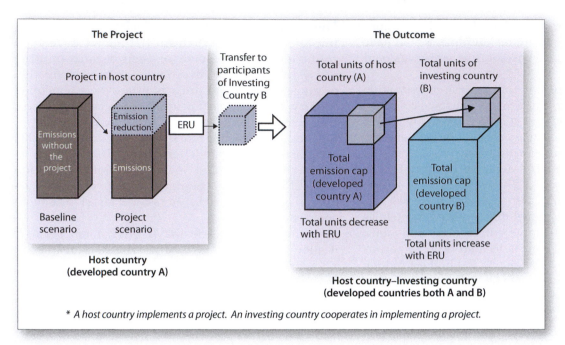

The Project

The Outcome

Transfer to
participants
of Investing
Country B

Project in host country

Emission
reduction

ERU

Emissions
without
the
project

Emissions

Baseline
scenario

Project
scenario

Host country
(developed country A)

Total units of host
country (A)

Total units of
investing country
(B)

Total
emission cap
(developed
country A)

Total
emission cap
(developed
country B)

Total units decrease
with ERU

Total units increase
with ERU

Host country–Investing country
(developed countries both A and B)

*A host country implements a project. An investing country cooperates in implementing a project.

▲ **Figure 8.17:** Diagram illustrating how JI works under the Kyoto Protocol. If country B invests in an emissions reduction project in country A, it will leave that country with excess emissions reduction credits that it can then use to add to its own capacity. Country B does end up producing more emissions than originally allocated, but the overall limit on carbon emissions is preserved.

financing projects that reduce global emissions in the developing world, countries earn Certified Emission Reduction (CER) credits that they can then use to meet Kyoto Protocol emissions targets or sell to countries that need them.

An example of a CDM activity might be a rural electrification project using solar panels or wind turbines. Since 2006, countries have registered more than 1,000 CDM projects that have reduced carbon dioxide emissions by more than 2.7 billion tonnes (2.98 billion tons). The advantage of CDM projects is that they are independently certified and can promote sustainable development by reducing emissions targets in the developing world. The disadvantage is that some projects would almost certainly have taken place anyway without funding (**Figure 8.18**).

CHECKPOINT 8.18 ▶ Describe some of the advantages and potential flaws in the CDM and JI.

Cap-and-Trade Versus Carbon Tax Not everyone believes that using a cap-and-trade mechanism is the best way to limit greenhouse gas emissions. Many economists still favor a simple **carbon tax**—a charge on fossil fuels (coal, oil, and natural gas) based on their carbon content. The United Kingdom uses such a tax, in addition to the EU's Emissions Trading Scheme. Applying a simple carbon tax to heavy users of fossil fuels can be very effective at reducing greenhouse gas emissions. Carbon taxes both raise the cost of energy from fossil fuels and make energy from carbon-free alternative sources more competitive. A carbon tax can also generate revenue that can be reinvested to support a growing green economy. Taxes, however, need to be set at a high level to ensure the required reduction in emissions, and this translates to a major hike in the price of energy for consumers. Politicians seeking reelection do not favor this strategy, especially during a global recession. But in a growing world economy, enacting carbon taxes is an effective and proven means to further reduce emissions.

A novel approach championed by James Hansen in the United States is to collect a **carbon fee** from the fossil fuel companies upon the first sale at the mine, wellhead, or port of entry. This would raise energy prices, but the fee could be equitably distributed to the public as a monthly dividend to help offset those costs and create support for even higher rising carbon fees. The public could either use the cash to pay the higher price of energy or pocket the money by reducing their consumption of energy through efficiency and conservation.

CHECKPOINT 8.19 ▶ How does the carbon market encourage the use of alternative energy?

What Happens After Kyoto: Planning for the Future As soon as the Marrakesh Accords were agreed and the participating countries started to

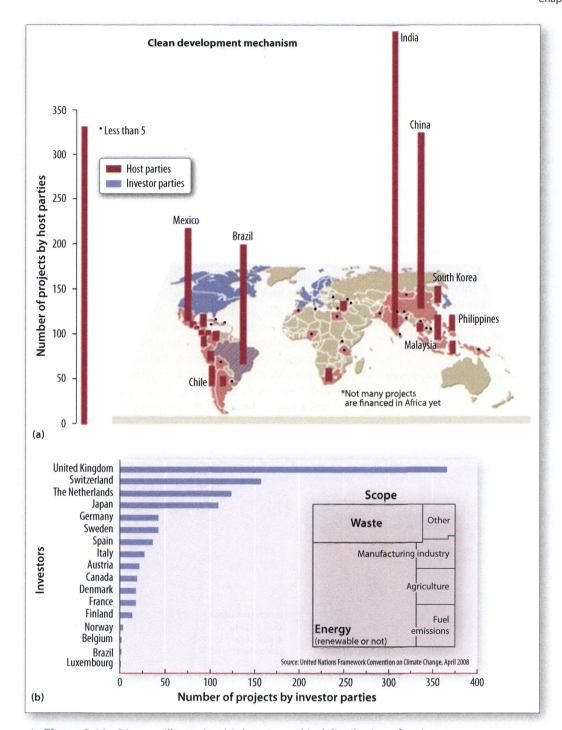

▲ Figure 8.18: Diagram illustrating (a) the geographical distribution of projects implemented in 2008 under the Kyoto Protocol's CDM, and (b) a graph to indicate the sectors affected. Notice that the majority of projects are related to energy in the more advanced developing nations, with the European Union and Japan leading as investor parties.

implement the Kyoto Protocol, negotiations started to focus on the future. In 2007, the parties met in Canada and published the **Montreal Action Plan,** a proposal to extend the Kyoto Protocol beyond 2012.

This was the first time that many people in the United States fully understood that the Kyoto Protocol was just a first step toward a much larger goal of reversing the trend of global emissions and that they would soon be required to accept even deeper cuts. As the debate intensified, the countermovement that opposed climate change legislation became very active in the United States. Lobbyists became more vocal, the press more critical, and politicians more conflicted between the demand for action based on scientific evidence and the economic consequences of the Kyoto Protocol on their constituents.

How Hot Does It Have to Get? By 2007, the scientific and physical evidence for climate change was stronger than ever before, but the economic and social costs of mitigation were so high and the consequences of implementing the Kyoto Protocol so evident that many stakeholders in the United States were not willing to pay the price. Driven by concern that the environmental movement might yet succeed in its efforts to have the United States ratify the Kyoto Protocol, corporate, industrial, and energy stakeholders engaged climate skeptics to embark on a preemptive and coordinated social, political, and media campaign to challenge the UN, the IPCC, and climate science. When the IPCC published the *Fourth Assessment Report* in February 2007, it concluded, as widely anticipated, that anthropogenic global warming was real and dangerous (**Figure 8.19**). However, in the United States, it met only a mixed response because climate skeptics had undermined the authority of the IPCC and climate science among conservative voters and their representatives.

CHECKPOINT 8.20 ▶ In your opinion, is there any future for the Kyoto Protocol? Explain.

The Bali Road Map, 2007: First Steps for the Future The long-anticipated *Fourth Assessment Report* and growing opposition to the Kyoto Protocol set a serious tone for negotiations later in 2007 in Indonesia. There was broad agreement that a plan to renew the Kyoto Protocol must include deeper cuts in global greenhouse gas emissions. The **Bali Action Plan** that emerged had two main tracks to keep negotiations moving—and still left an option open for the United States to participate in the future:

- The first track included all parties signed up to the UNFCCC, including the United States, and focused on climate action though mitigation, adaptation, financing, and development of new technologies. It called on developed countries to take immediate action to mitigate the impact of climate change and on developing countries to commit to future mitigation and also take immediate action to reduce emissions due to deforestation and forest degradation. In a major step forward, developing countries, including China, India, and Brazil, agreed to consider "measureable, reportable and verifiable" action at a national level to ensure emission reductions. However, they agreed to do so only with

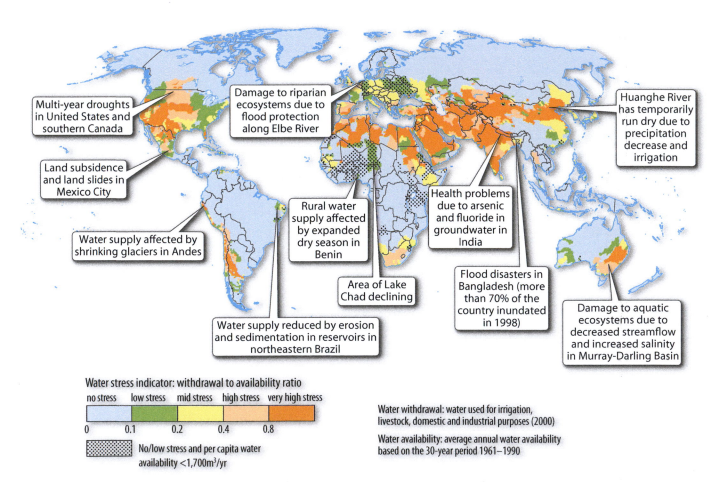

Multi-year droughts in United States and southern Canada

Damage to riparian ecosystems due to flood protection along Elbe River

Huanghe River has temporarily run dry due to precipitation decrease and irrigation

Land subsidence and land slides in Mexico City

Water supply affected by shrinking glaciers in Andes

Rural water supply affected by expanded dry season in Benin

Health problems due to arsenic and fluoride in India

Water supply reduced by erosion and sedimentation in reservoirs in northeastern Brazil

Area of Lake Chad declining

Flood disasters in Bangladesh (more than 70% of the country inundated in 1998)

Damage to aquatic ecosystems due to decreased streamflow and increased salinity in Murray-Darling Basin

Water stress indicator: withdrawal to availability ratio

no stress	low stress	mid stress	high stress	very high stress

0   0.1   0.2   0.4   0.8

No/low stress and per capita water availability <1,700m³/yr

Water withdrawal: water used for irrigation, livestock, domestic and industrial purposes (2000)

Water availability: average annual water availability based on the 30-year period 1961–1990

▲ **Figure 8.19:** The impact of climate change. A global increase in water stress is just one of the threats of climate change highlighted by the IPCC. The withdrawal to availability ratio is a measure of how much water is used each year with respect to total annual water availability.

assistance from developed countries in the form of technology, financing, and capacity building.

- The second track applied only to countries that had signed the Kyoto Protocol. It set targets for the reduction of greenhouse gas emissions of between 25% and 40% below 1990 levels for the period beyond 2012—a significant further reduction.

The Rise of the Skeptics: Climate Negotiations Stall in Copenhagen The Conference of the Parties that met in Denmark in 2009 was supposed to build on the Bali Action Plan and establish both post–Kyoto Protocol emissions reduction targets and additional mechanisms to provide financial support for the least developed countries under the greatest threat from climate change.

The meeting was a disaster. The economic crisis that started in 2008 became a major issue, and many stakeholders were concerned that any commitment to enforce emissions standards at this stage could delay economic recovery. A well-orchestrated, anti-science firestorm erupted shortly before the conference, based on the willful misinterpretation of emails stolen from highly reputable climate scientists in the United Kingdom. The "Climategate" headlines that accompanied these false allegations helped to derail the process. By the time the scientists were vindicated and cleared of all accusation of improper scientific practice, the world had moved on. The damage was done.

During the Denmark negotiations, Brazil, South Africa, India, and China (the **BASIC countries**) refused to accept mandatory cuts in emissions unless the United States took the lead. The United States insisted it would not act alone until the agreement was "symmetric," with measurable, reportable, and verifiable commitments from all the developing countries. Stalemate.

At the conclusion of these very difficult negotiations, the participating countries could only agree to share how they intended to voluntarily reduce greenhouse gas emissions at a national level. The United States proposed to cut emissions to 17% below 2005 levels by 2050, and China promised to reduce its energy intensity 40%, to 45% below a 2005 baseline.[1] No mandatory reductions were agreed, and no reporting measures were finalized. Even the level of voluntary cuts proposed was woefully inadequate. A UN report made it clear that if all the countries that promised to cut emissions actually fulfilled their goals (which was historically unlikely), the cuts would represent only 60% of the cuts necessary to prevent a dangerous level of global warming.

The position of the EU delegation was more conciliatory. They agreed to deeper mandatory cuts if the

developing world agreed to reduce emissions intensity. But with the United States and China at loggerheads, there was no chance of making any progress. The two countries that emit the largest amounts of greenhouse gases globally had to be involved for any substantive agreement, but they remained as far apart as ever. Even if the EU went ahead as proposed and voluntarily reduced its emissions between 20% and 30% by 2020, that amount would still only equal two weeks of greenhouse gas emissions from China.

The resulting nonbinding Copenhagen Accord made little headway along the Bali road map. There was some progress on the **reduction of emissions from deforestation and forest degradation (REDD)**, and developed countries made vague additional promises to provide "adequate, predictable and sustainable financial resources, technology and capacity-building to support the implementation of the adaptation action in the developing countries."

The Copenhagen Accord was to be achieved through a new technology mechanism with financial commitments to provide up to $30 billion for the period 2010–2012 and a further $100 billion per year by 2020 to address the needs of the most vulnerable developing countries. The new Copenhagen Green Climate Fund was established as a financial mechanism to deliver these funds, but there were no binding commitments regarding who was to pay into the fund. There was some hope that a new cap-and-trade program in the United States would raise some of the money required, but the rise of the **Tea Party** and other political factors in the United States dashed any hopes of cross-party support for that proposal.

CHECKPOINT 8.21 ▶ What is the economic basis of the current impasse in climate negotiations?

Is the Kyoto Protocol a Sinking Ship? At the following meeting of the Conference of the Parties in Cancun, Mexico, in 2010, and in Durban, South Africa, in 2011, expectations were low. In the depths of recession, Canada, Japan, and Russia declared that they would not sign any extension of the Kyoto Protocol. (Canada never really tried in the first place, and it subsequently decided to withdraw from the Kyoto Protocol because its position had become untenable due to plans for the exploitation of massive reserves of shale gas and oil sands.)

China and India made it clear that they would not sign up to reduce energy intensity unless the United States agreed to mandatory cuts in greenhouse gas emissions and made a firm commitment to a second Kyoto Protocol period. This was an additional demand beyond the scope of the Copenhagen Accord, and it was clear that the U.S. delegation could not accept it. The meeting in Durban in 2011 made some progress, and finally all countries agreed to take action to address climate change. But however much the Conference

---

[1]*Energy intensity* is a function of how much greenhouse gas emissions increase relative to the growth of the GDP. It is an important indication of how efficiently energy is used but does not mean that overall greenhouse gas emissions are reduced.

of the Parties tried to spin the positive side of the agreement, the results were still nonbinding, and no specific targets were identified. During the 18th annual summit in Doha Qatar in 2012, delegates agreed to work towards a new universal climate change agreement, the Doha Climate Gateway, covering all countries by 2020. If successful, this might keep the world on target to stay below a 2°C (3.6°F) rise in global temperature. They also agreed to allow all of the Kyoto special mechanisms (CDM, JI, IET) to continue, pending a new agreement and outlined how new market mechanisms outside the UNFCCC might be recognized as part of national or bilateral offset programs. There was some progress towards finalizing the operation of the Green Climate Fund by 2013, and the developed countries reiterated their earlier promise to deliver 100 billion dollars to help with climate adaptation and mitigation in the developing countries by 2020. In reality little was achieved beyond kicking the ball down the road in hope of better economic and political times to come.

At this writing in early 2013, it is hard to see what the future holds. The UN is the only venue where the voices of all countries can be heard and valued, but negotiations have reached such an impasse that there is little prospect of immediate progress. If, as seems likely, the impasse continues, the larger nations may choose to bypass negotiations at the UN and instead make direct multilateral agreements that aim to balance emissions, growth, and economic competition—but that end up delivering nothing. The marginalization of the UN from multinational climate negotiations would be a major loss for global governance and equity among nations. But in a world controlled by a handful of powerful economic giants—among them the United States, the EU, Brazil, China, India, Japan, Russia, and South Africa—this seems increasingly likely.

CHECKPOINT 8.22 ▶ Why is the United States not willing to take the lead in emissions reduction?

What Should the Developing Nations Be Required to Do? The Berlin Mandate assigned responsibility for the reduction of greenhouse gas emissions to developed industrialized nations. These countries built their economies by burning fossil fuels, and they have much higher per capita greenhouse gas emissions than do other nations. While this per capita difference will remain for some time to come, emissions from many of the developing countries, such as Brazil, China, India, South Africa, and Mexico, now exceed those of many developed countries. China, for example, already exceeds the United States in total greenhouse gas emissions, even though it remains far behind the United States and other developed countries in its per capita emissions (**Figures 8.11** and **8.20**).

China's growth is exceptional, but it is representative of the rapid economic growth in many economies across the developing world. With an average growth rate of 10% over the past 30 years, China has overtaken the United States as the world's largest manufacturer. China has a population of 1.3 billion and is becoming increasingly urbanized and energy dependent. With a labor force of over 820 million, consumerism is rapidly changing expectations in a society that understandably demands the same standard of living as the fully developed nations.

China is already a world leader in the development of renewable resources as it struggles to meet this demand for energy, but it will be decades before the supply of clean energy can possibly meet the accelerating demand. To sustain growth, China will have to increase its dependence on cheap coal as a major fuel source well beyond 2050.

Strong economic growth in China threatens many corporate interests in the United States and Europe that want greenhouse gas emissions restrictions applied more evenly and "fairly." Without such evenly applied restrictions, China would have an enormous industrial advantage in the world market.

It is now widely agreed among the signatory nations to the Kyoto Protocol that the economically stronger developing nations must make measurable, reportable, and verifiable reductions in greenhouse gas emissions intensity as part of any new international agreement to reduce greenhouse gas emissions

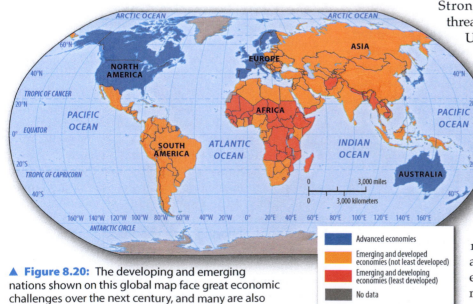

▲ **Figure 8.20:** The developing and emerging nations shown on this global map face great economic challenges over the next century, and many are also under great threat from climate change.

Advanced economies

Emerging and developed economies (not least developed)

Emerging and developing economies (least developed)

No data

and to limit the impact of climate change. It is equally accepted that emissions targets for economically weaker nations in the developing world must be set at responsible levels that will not inhibit economic growth and national prosperity.

The moral and ethical issues raised by the competing interests of the developed and developing worlds are considered in more detail in Chapter 9. Reflect for a moment on the political and economic pressures that the rapid economic growth in China and India now places on countries such as the United States, Canada, Japan, Australia, and the countries of the EU. How can developed countries compete with these rising economic giants if they carry the primary responsibility for limiting greenhouse gas emissions? If countries such as China, India, Brazil, Mexico, South Africa, and Indonesia are allowed to fuel their growth with the unrestricted use of fossil fuels, it is hard to see how the world can hope to limit the severity of climate change. Should citizens of developed nations accept responsibility for the unintentional actions of past generations because they continue to benefit and prosper as a consequence of those actions? If so, shouldn't the developing world share some of this burden as they now benefit from access to these technologies and do not face the enormous cost of replacing an aging energy infrastructure? Where does the "buck stop"? This is an issue that crosses the interface between climate policy and international development, and it is a long-term challenge for the UN.

Climate negotiations highlight the deep divisions between developed and developing nations and also between the strong emerging economies and small nations most at risk from climate change. Most developed nations are reluctant to take action to limit greenhouse gas emissions in the absence of firm reduction commitments from (at the very least) China, India, Brazil, Mexico, South Africa, and Indonesia. Most developing nations want the developed nations to cut emissions first, as a sign of good faith. The more vulnerable nations are left wondering what is going to happen to them when the climate "storm" strikes.

8.3 PAUSE FOR... THOUGHT | *Why was Vice President Al Gore not able to get the U.S. Senate to approve a proposal to ratify the Kyoto Protocol?*

Such a proposal must get a two-thirds majority in the Senate in order to be approved. When Vice President Gore considered his chance of success, he found little support from either side in the Senate, and he found strong resistance from both sides in the House of Representatives, where there was strong cross-party support for legislation to ensure that the treaty would fail. Why do you think Congress was so united against the Kyoto Protocol?

The Social Impact of Climate Change in the United States

Climate policy in the United States has reached an impasse, and there is little prospect of any progress. Public and political opinion is deeply divided, despite clear evidence of climate change from satellite data and ground-based observations. We must understand the social factors that lie behind these divisions if we want the cross-party support necessary to promote effective and timely climate change legislation.

The Six Americas Report: Weighing Public Opinion

A 2012 survey by Yale University identified "six Americas" that clearly define how national attitudes to climate change are deeply divided. Participants were asked about their attitude toward climate. The results: 13% were alarmed, 26% concerned, 29% cautious, 6% disengaged, 15% doubtful, and 10% dismissive. The majority of responses divide along traditional party lines. Only 9% of those who expressed alarm were Republicans, compared to 44% for Democrats, and only 6% of those who were dismissive were Democrats, compared to 43% for Republicans. Respondents associated with very conservative or somewhat conservative views comprised 76% of those dismissive of climate change. What social factors lie behind these divisions?

The Environmental Movement in the United States

The environmental movement in the United States is strong and relatively well funded, but it lacks the unified leadership necessary for successful political action. There is solid grassroots support for climate action, and activist stakeholders such as Greenpeace, Friends of the Earth, and the World Wildlife Fund do an excellent job educating the public and increasing general knowledge about climate change. Unfortunately, the political elite in Congress who can turn this knowledge into action are divided by the need to win, and retain control of, key swing states that either produce energy or need cheap energy for manufacturing. It might be possible for some of these politicians to pass climate change legislation, but only at the cost of losing political power in the next election. As a result, there is debate, heated argument, and political posturing, but in the end, the environmental movement has made little progress toward the reduction of national U.S. greenhouse gas emissions to date.

The Climate Change Countermovement in the United States

When climate change was declared a major global problem at the Rio Earth Summit in 1992, President George H. W. Bush responded by stating that the

"American way of life is not negotiable." Many in the United States still share this opinion, opposing all climate action and legislation. This opposition grew stronger as negotiations made progress toward ratification of the Kyoto Protocol. When it became clear that the United States would be required to make mandatory cuts in greenhouse gas emissions, a groundswell of opposition gave birth to a climate change countermovement that sought to delegitimize both the environmental movement and climate change science. Led by influential social, political, and cultural elites, the countermovement mounted a highly effective media campaign and has since grown to become one the strongest lobbying forces in the United States. Using contrarian scientists and their supporters, the countermovement successfully fostered doubt about the reliability of the science of climate change and changed the minds of many U.S. citizens.

CHECKPOINT 8.23 ▶ Should the American way of life be negotiable? Explain.

Twenty years after the Rio Summit, the United States is still deeply divided over climate change, as prospects fade for any effective renewal of the Kyoto Protocol. The climate countermovement has grown powerful, through the support of powerful stakeholders who feel threatened by prospective climate change legislation. Public and political debate have reached an impasse. Most Democrats and many independent voters support action to limit climate change on the grounds that the risk of inaction is too high. Republicans, and particularly social conservatives, oppose any action on the grounds that it will reduce economic competitiveness, increase government regulation, raise taxes, and undermine national sovereignty. With strong political leadership, passionate grassroots support, substantial financial backing, and some cross-party support, the climate change countermovement continues to block most climate change legislation that comes before Congress.

The Social Conservative Perspective

Founded on a rich tradition of individualism, self-reliance, self-determination, and deeply held Evangelical Christian values, social conservatism is strongest in the Republican Party and is an influential force in U.S. politics. Social conservatives hold a sincere conviction that climate change legislation will harm the United States, diminish personal freedom, and decrease personal liberty. Many are convinced that climate change is not happening at all and mistrust the scientific research community. Even those who accept that climate change is real believe it is natural or unavoidable. From this political perspective, the regulation of greenhouse gas emissions is a barrier to economic growth, and the Kyoto Protocol is an egregious example of big government interference in the economy. Social conservatives consider fossil fuels a vital natural resource that should be exploited in the general interests of the economy and energy security.

A More Liberal Perspective

In contrast to social conservatives, social (progressive) liberals are mostly Democrats, and they are much more likely to support the global environmental movement. Most trust the scientific research community and believe that climate change is a real and imminent global threat. They want immediate action to reduce greenhouse gas emissions and believe that the state has a valid and important role to play in securing sustainable economic growth and development. Most social liberals mistrust the free market to guide the economy in the absence of strong regulation and oversight. Social liberals have a long record of supporting global issues related to social justice and environmental protection, and they believe that the United States has a moral duty to take leadership in the global response to climate change.

CHECKPOINT 8.24 ▶ In what ways is climate change a threat to religious, political, and personal freedom?

▲ **Figure 8.21:** The countermovement against political action to address the risk of climate change uses the media to cast doubt on the science behind climate change. However, there are good examples of investigative journalism that highlight the weakness of climate change denial machine.

Climate Skeptics

There have been climate change skeptics as long as there has been a debate about climate change. Skeptics do not believe in the accumulated evidence for anthropogenic climate change, and they challenge the scientific consensus. But skepticism is an important and integral part of the scientific method. Healthy skepticism challenges the prevailing scientific paradigm and continually tests the quality and consistency of the evidence. Skeptics do not deny the evidence; they test it, move with the debate, and adjust their criticism in the light of new evidence. Many scholarly and respected skeptics make valuable contributions to the climate debate that are both important and constructive. For this reason, it is important to draw a distinction between the constructive climate skepticism of many mainstream conservatives and an emerging, and very vocal, community that deny the very foundations of climate science. Many of these voices identify closely with the Tea Party movement on the right wing of the Republican party.

In a 2011 survey, the majority of self-identified Tea Party members denied that climate change is happening and expressed no concern about climate change. They revealed a strong mistrust of academic and social elites who espouse the dangers of climate change and were critical of both scientists and the scientific method. Paradoxically, most believed that they were well informed about climate science and needed no further information. This is in marked contrast to the mainstream Republican Party, where a small majority still believes that climate change is happening, that humans are at least partly responsible, and that most scientists are in agreement.

Denialism at Work

Opposition to the mandatory reduction of greenhouse gas emissions in the United States is hardening around a rather extreme conservative countermovement that questions the authenticity of science and the motivation of leading climate scientists. Conservative blogs, cable television, and published media now describe climate science as a hoax, or "con job," and they have leveled lawsuits and ad hominem attacks against major climate scientists such as James Hansen and Michael Mann. Both Hansen and Mann have been vindicated of all accusations, but their critics continue to attack them. Extreme rhetoric like this is certainly not a new phenomenon. Since the early days of the Founding Fathers, politics in the United States has involved obfuscation, exaggeration, and duplicity. Anyone who doubts this should read accounts of the first elections that followed independence. At its best, this is just the normal "cut and thrust" of politics in a modern democracy; it is a hard game. But at its worst, it has the potential to motivate dangerous elements in society to take violent action. The most aggressive anti–climate change blogs have gone as far as publishing the names and addresses of climate researchers who are then subjected to threats and verbal abuse. Donors to major universities that support climate research are contacted and asked to stop their donations. University presidents are pressed to investigate the conduct of climate scientists and to dismiss faculty engaged in climate research. These coordinated email, blog, and ad hominem (personal) attacks seem determined to frighten scientists and drive them from the public debate.

The extremist rhetoric has created a growing chasm of mistrust between politicians and the scientific community, and it has succeeded in blocking the prospect of climate change legislation in the immediate future. Critics may delight in their successful efforts to undermine the integrity of science, but this may become a Pyrrhic victory if climate change becomes irreversible. It is hard not to agree with Pulitzer Prize–winning columnist Leonard Pitts, Jr., who wrote, "We are a people estranged from critical thinking, divorced from logic, alienated from even objective truth." But we can all hope he is mistaken.

CHECKPOINT 8.25 ▶ In your own words, clarify the distinction between skepticism and denialism.

Turning Knowledge into Action

To turn knowledge about climate change into action, we must consider how to break down the divisions that exist in U.S. society over the issue of climate change. The majority of the population is concerned about the risk, but why does a large minority still dismiss the evidence as a hoax or irrelevant? What social forces determine that intelligent and highly responsible citizens, who share many common cultural values, end up with diametrically opposite interpretations of the risks involved with climate change? This deepening division is a barrier that prevents our knowledge about changing climate from turning into effective political action, and the solution to this problem must be a priority for anyone concerned about the impact of global warming. The obvious answer is to find people of good will on both sides of the political divide who are prepared to work together on cross-party initiatives. So far, this tactic has been ineffective, and there is little sign of improvement. If a solution exists, it must lie in furthering our understanding of how people perceive risk and how to better communicate the risk of climate change to a skeptical audience.

CHECKPOINT 8.26 ▶ Before reading this text, where did you get most of your information about climate change?

The Role of the Traditional Media

The press and other media play a pivotal role in the process of turning knowledge into action by presenting complex scientific, economic, social, and political problems in a way that the general public can understand. The majority of environmental journalists base their work on sound research and have great depth of knowledge, but many are not sufficiently well informed and, despite overwhelming scientific evidence to the contrary, they try to offer "balanced" coverage to both sides of the climate debate. But where is there "balance" between a large majority of climate scientists on one hand and a tiny minority of climate skeptics on the other? Both views deserve respect, thought, and attention, but do they deserve equal weighting?

It is easy to blame journalists for giving too much credibility to climate skeptics. But scientists generally lack the rhetorical skills of climate skeptics and avoid their crisp well-targeted sound bites. Scientists feel at home with Socratic debate, not the adversarial antics of the courtroom, and they are simply out of their depth when faced by professional and erudite skeptics. Over the past few years, more scientists have learned to come out fighting like lawyers rather than professors, and while this may be anathema to many in the scientific community, it may be necessary to achieve political progress in a world dominated by sound bites and new social media.

CHECKPOINT 8.27 ▶ Why is it difficult for scientists to present their case through the traditional media?

The Impact of Online and Social Media

Websites, Facebook, Twitter, phone apps, blogs, and YouTube now offer skeptic and scientist alike opportunities to bypass the traditional media and speak directly to the public.

Social media in general have expanded to become the primary source of information and political opinion for the younger generation. Anyone wanting to address the issue of climate and policy in the future must learn to engage with these new media and surpass their rivals at using them. This is not an easy task; a simple Google search for "climate change" uncovers scores of attractive blogs and wikis written by climate contrarians who confuse the issues, cherry-pick the data, and boldly misinterpret the science. In contrast, sites written by climate scientists are often boring and unappealing. A celebrated success of the climate skeptics, however, is in forcing mainstream climate scientists to emerge from their Socratic cocoons and join with a growing and increasingly professional community of bloggers and videographers who make their arguments directly to the electorate. It is a new world.

CHECKPOINT 8.28 ▶ How much of the news and information you listen to is modulated through cultural filters? Explain.

▲ **Figure 8.22:** The rapid expansion of online social networking using tools such as Facebook, Twitter, and blogs has had a major impact on the way people access, process, and communicate information about a broad spectrum of political issues, including climate change.

Talking About Risk

Sociologists and political scientists study how risk is perceived to help minimize the impact of future disasters. People's perception of risk determines what actions they are prepared to take to protect themselves. Scientists and government agencies take an empirical approach to risk, providing facts, data, and safety advice. But it turns out that the perception of risk is much more of a social construction than a scientific formula, and many people ignore a clear warning of imminent danger.

The empirical definition of **risk** can be expressed as: Risk = Hazard × Exposure × Longevity × Probability (HELP). An example is a potential hurricane on the Florida Atlantic Coast (the **hazard**) that could have a damaging impact on millions living close to the coast (**exposure**); this impact could last for weeks (**longevity**) and can be expected every few years (**probability**). In a case like this, where the risk of a disaster is large, the appropriate response is to take immediate action to prepare and develop a disaster readiness plan to protect the population affected. If the hazard has a low probability, such as the chance of a Superstorm striking New York City, the appropriate response would be very different. In the New York case, even though the product of the hazard, exposure, and longevity is enormous, the very low probability made the overall risk very small and not worth the cost of mitigation (until the day after Hurricane Sandy actually occurred).

CHECKPOINT 8.29 ▶ How can we quantify the level of risk related to climate change?

When using the HELP formula, the risk from climate change appears to be very high. Millions are exposed globally to multiple hazards, the effect will last for millennia, and the probability of a rise in global temperature of at least 3°C (5.4°F) is very high. The need for action appears obvious to the scientific community, but the message is somehow lost by the time it reaches the different interpretive communities that comprise the nation. The two major interpretive communities that dominate U.S. politics, Democrats and Republicans, have used these same data to reach diametrically opposite conclusions based on their social construction of risk; this psychological phenomenon is a major barrier to turning our knowledge about climate change into effective political action. To solve this problem, it is important to understand how perceptions of risk are acquired and what strategies might be effective in promoting better communication about climate-related hazards.

Cultural Risk Theory

Each person faces decisions that involve risk throughout their lives. This could be as simple as determining whether to move from New York to Florida or whether it is worth the thrill to jump off a high bridge with a rubber rope strapped around your ankles. (My son tells me it is worth it, but I remain to be convinced.) People are known to assess personal risk more accurately if the impact of the hazard lies within their personal experience or has affected someone in their family or close circle of friends. The statistical data presented by most climate scientists are much less persuasive, and fewer people pay attention. The death of a loved one due to lung cancer is much more likely to end a smoking habit than is the statistic that 33% of smokers at age 35 die before they reach 80 years of age.

Social scientists who have considered this problem suggest that, over time, we all construct small-scale mental models of the world around us, based on our life experience, cultural factors, and social group norms. We use these models as a framework against which we test and validate new ideas, observations, and perceived risks. If information about a risk is consistent with our world model, we are likely to accept the threat seriously and take action. If the information we receive in inconsistent with our world model, we tend to ignore the risk, discredit the source, and take no action. In general, we respond quickly to risks that create a sense of fear or dread and disregard even greater risks that may seem more normal or less threatening. Many Americans accept the very high risk of an accident on the highway almost daily but live in dread of an improbable terrorist attack. As a result, the government spends billions of dollars on homeland security but allows many roads and bridges across the nation to fall into disrepair.

This **cultural risk theory** suggests that we must understand the cultural context of a target audience if we want to successfully communicate information

about the risks of climate change. Analytical reasoning plays a part in the process of personal risk assessment, but it does so alongside emotional responses, the impact of social networks, the influence of political elites, the news media, cultural groups, social aspirations, peer norms, and religious beliefs. Together these have created behavioral inertia that has delayed action to mitigate climate change. In this context, it is not surprising that many people in the United States conclude that the risk from climate change is low. The data are largely statistical and impersonal, few disasters generate fear or dread, and climate is often conflated with weather, which seems familiar, controllable, and much less threatening (this may change in the aftermath of Hurricane Sandy).

> **CHECKPOINT 8.30** ▶ In your own words, explain the concept of *cultural risk theory*.

The perception of risk can be amplified or attenuated through the influence of psychological, institutional, cultural, and political processes that influence our worldview—a process known to social scientists as the *social amplification of risk*. The 2011 Fukushima nuclear disaster in Japan is an example of how risk can be amplified. The death toll of this nuclear disaster was very small and had minimal impact compared to the devastation of the tsunami that triggered it, yet the ripples of nuclear dread spread around the world, and countries such as Germany have now completely reversed their prior energy policy and banned nuclear energy.

> **CHECKPOINT 8.31** ▶ Why do so many of us discount the level of risk from immediate hazards and overestimate the risk from less probable events?

In the United States, climate change provides an example of how risk perception can be attenuated. For a majority of the population, the scientific evidence is heavily filtered. At each stage of processing, the message is received, modulated, and re-emitted to the next stage with added interpretation and symbolic meaning. Both conservatives and liberals receive their information about climate change though social filters that are well established, widely recognized, and often highly respected. For conservatives, the risks associated with climate change are attenuated by a deep mistrust of science and big government and a strong social value system that focuses on national sovereignty, independence, self-reliance, faith, ethics, personal freedom, the economy, national security, and financial independence. The risks associated with climate change are tested against the consequences of climate action on each part of this value system, and the community is faced with the easy choice of rejecting the science of climate change or turning their complete mental model of the world inside out.

The First Steps Towards More Effective Communication

When every paper and report that highlights a new aspect of climate change is filtered through the same communications pathways, the 24% who are doubtful or dismissive of climate change are unlikely to change their minds. This kind of **belief persistence** is well known to politicians, who focus election campaigns on small groups of swing voters who are capable of changing their minds.

The **Six Americas Survey** offers some hope for the future and draws attention to key target groups that may be receptive to more information about climate change. These include the 35% who are currently cautious or disengaged about the risk of climate change, and the 65% who already believe that global warming is affecting weather in the United States. This is a large majority of the population, and it is powerful enough to demand political action if convinced by the evidence. It is encouraging that climate scientists are still the most trusted source of information about climate change for 74% of the population, despite the concerted campaign to undermine their credibility by extreme skeptical elements of the climate change countermovement. There is a sufficiently large subset of the population to make sure climate legislation is enacted if we can only find the right message to convince them it is necessary.

CHECKPOINT 8.32 ▶ What is *belief persistence*, and how does it impact the climate change debate?

8.4 **PAUSE FOR... THOUGHT** | *"Seeing is believing"?*

Most scientists think that we should see before we believe. But many of us are taught to believe first, and this has a profound effect on how we perceive the world. Is climate change the only scientific issue that is currently affected by this doubtful epistemology?

The Six Americas Survey also highlights topics that still need to be addressed. These include the 47% of respondents who say they would be more concerned if climate scientists stated publically that global warming is happening, the 42% who think that individual action can make no difference, and the 47% who still believe a technological solution in the near future will save them from making changes in lifestyle. By addressing these audiences first, it should be possible to convince many more people about the real dangers of climate change and maybe start a new social movement that has enough political influence to actually reverse the stagnation of climate policy. If

more people from all sides of the political spectrum can emerge from the entrenched positions they have assumed over the past decade, the future may yet look promising.

Developing an Effective Communication Strategy

Audiences need scientific information about the nature of risks from climate change, but they need it in ways they can understand and from sources they can trust. Social scientists suggest that a community-based approach is more likely to be successful than national campaigns that are often rejected and even resented by target audiences. The message should come from a trusted source and should be clear, concise, and not cluttered with too much information. The effectiveness of a message is enhanced when it is personalized, takes account of the local context, and focuses on the values, needs, and social norms of the target audience. Uncertainty should be acknowledged and addressed early in the presentation to build trust for the most important parts of the message.

Scientists need to understand the circumstances and acknowledge the current beliefs of the audience. They should appeal to emotions when describing the actual hazards, as these are closely tied to our understanding of risk, and introduce an appropriate level of fear and dread, but always remember that worry is a finite resource, and people will eventually discount false or exaggerated claims.

CHECKPOINT 8.33 ▶ Should scientists engage with new communications media to engage a skeptical public? Or should they focus solely on publication through peer-reviewed journals?

A presentation to major stakeholders must first be framed by a strategic analysis that recognizes their concerns and acknowledges any uncertainty in the science or data. Major stakeholders almost certainly understand these problems, so it is no use fudging the issue when honesty can build trust and dialogue. The aim is to reframe the question into a balance of pros and cons that clarify how climate action is actually in the interests of the free market. For more conservative communities, this will mean less emphasis on the environmental and social issues that appeal to liberals and more emphasis on clean energy, energy security, energy independence, national security, green jobs, and international competitiveness—all issues that attract strong cross-party political support.

The final message must be positive. It is not ethical to introduce fear into people's lives without offering them a constructive way to change the situation. We should focus on permanent behaviors that promote slow "buy in" to create gradual change rather than the kind of emotional enthusiasm that too often dies out after a few weeks. This is a game of chess, not checkers, and the objective is to introduce viable options that promote lasting change in the lifestyle of communities that can be replicated on a larger scale. We should offer ideas for personal and group action that can lead to tangible, observable, and rewarding changes in lifestyle, such as energy conservation and efficiency. We should stress the benefits of change that best match the concerns of the audience and recognize barriers to progress by suggesting ways to surmount them. We should use a variety of media to present the message, especially the **social networks** that people look to for information and guidance and **community-based social marketing** to access the communications spines that inform local communities. We should offer hope, not doom, and stress the long-term economic advantages of change.

The slogan used in Virginia to promote blue crab conservation was funny and pointed: "Save the crabs—and eat them!" What slogans would you recommend to promote climate action in your community?

8.5 PAUSE FOR... THOUGHT | *Political Minefields.*

In the 2012 selection of the Republican presidential candidate, only one candidate was in favor of climate change legislation. All the other candidates were diametrically opposed to any regulation, even though some had expressed serious concern about climate change in the past. The scientific evidence for climate change has only improved over this time. What political and social forces do you think caused these politicians' rhetoric to change?

▼ **Figure 8.23:** According to the latest statistics from NOAA's National Climatic Data Center, the average temperature for the contiguous United States for 2012 was 12.9°C (55.3°F), which was 1.8°C (3.2°F) above the twentieth-century average and 0.6°C (1.0°F) above the previous record from 1998. This map shows where the 2012 temperatures were different from the 1981–2010 average. Shades of red indicate temperatures up to 4.5 °C (8°F) warmer than average, and shades of blue indicate temperatures up to 4.5°C (8°F) cooler than average. The darker the color, the larger the difference from average temperature. As the direct impacts of climate change within the United States become increasingly apparent, they may help break down some of the political barriers to progress towards climate mitigation.

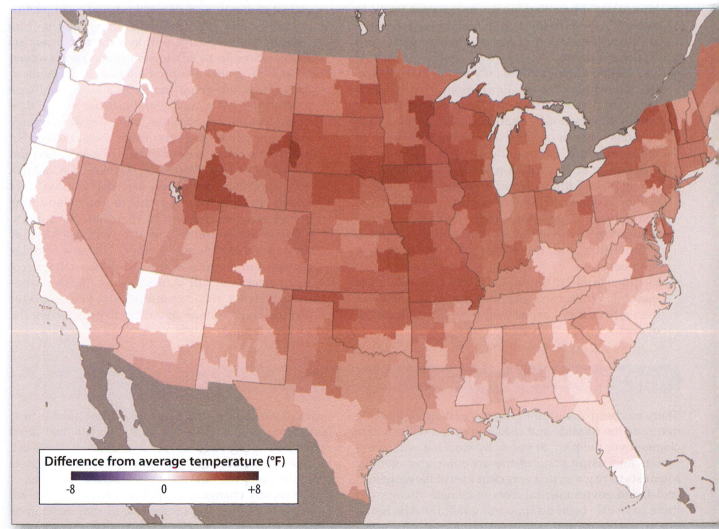

Difference from average temperature (°F)

-8   0   +8

Summary

- There are many different stakeholders in the climate debate who continue to influence policy. Some accept the science of climate change and are strong advocates for action to reduce greenhouse gas emissions. Others block progress and remain skeptical of climate science because they are concerned that the social and economic costs of action are too high.
- Scientists and economists have developed robust and accurate models that are able to reproduce past climate change and project scenarios of future change that depend on the actions we take today.
- Turning scientific knowledge of climate change into political action has not been straightforward. The fact that most countries in the world eventually ratified the Kyoto Protocol shows that progress is possible.
- The Kyoto Protocol was the first step towards limiting carbon emissions and the damaging impact of climate change. The capping of emissions, carbon trading, and other free market mechanisms encourage companies to plan ahead towards a time when energy consumption will generate few carbon emissions.
- International negotiations towards a new treaty to extend the aims of the Kyoto Protocol have reached an impasse due to competing economic and strategic interests.
- The United Nations is the only venue that protects the interests of the smaller nations, but it seems increasingly likely that the large developed and advanced developing nations will set their own agendas without considering the broader needs of the international community.
- One reason progress is slow is that negotiations are taking place within a larger geopolitical framework where major stakeholders such as multinational companies, corporations, and conglomerates are increasingly worried about the impacts of global competition and the threat of deep economic recession.
- Skepticism is an important part of the scientific method. While some skeptics are more tenacious than others, they change their perspective when the evidence becomes unambiguous.
- When scientific data challenges the intrinsic beliefs of a large section of the population it is pointless to keep appealing to reason and provide yet more data. New strategies are required to reach out and convince these stakeholders that change is important and consistent with their worldview and self-interest.
- The news media will continue to play a critical role in this process but must do a better job of holding all stakeholders accountable. Climate change is no longer a debate, and there is no longer any balance to maintain. The facts and data speak for themselves, but journalists still let too many unsubstantiated claims slip past without redress. They need to check each crispy sound bite for substance before broadcasting it, and they need to be prepared to name and shame deception wherever they find it. Unsubstantiated claims that linger unchallenged in public memory do immeasurable harm.
- Scientists must learn to wield their influence more effectively and need to speak more directly to the public. Taking the message to children and young adults in schools, community colleges, and universities should be a priority. The youth of today will almost certainly experience the negative impact of climate change within their lifetimes and will be a strong political force within a decade. Scientists should also start to talk to energy companies—even oil and gas giants—to promote new ideas on alternative fuels and green energy as a way to enhance profits, grow the economy, and increase competitiveness. Most importantly, they should start to engage with reasonable blogs and other social media that social conservatives use as sources of information and advice. Pick a blog, work on it, stay with it, establish a good reputation, and people may start to listen.
- Finally, it is vitally important to understand how people evaluate risk and to appreciate the cultural values that determine how they will react to new messages about climate change. This is a complex process, but analysis of the social forces that drive risk perception will lead to more effective ways of identifying and communicating key messages about climate change to ensure that they have the greatest impact.

Why Should We Care?

There has been significant progress in defining the true nature of the hazards and risks associated with climate change over the past 20 years. By working with economists and using state-of-the-art computer models, scientists have given us a very clear idea of the economic, social, and environmental costs of climate change. But there have also been disappointing failures. The Kyoto Protocol has been a qualified success, but there is little current prospect of effective future action. In the United States, cross-party support for climate action does exist, but progress will be slow as long as the denial of climate change remains a litmus test for social conservatives.

The rapidly changing geopolitical environment will continue to play an important role in shaping climate negotiations in the near future, and the key industrial stakeholders will retain their strong influence on climate policy. All nations, and especially those under threat from the first impact of climate change, should have a say in the search for a solution to climate change, and the United Nations remains the only viable forum that can give them both voice and vote. Negotiations will not be easy as a reluctant United States starts to navigate the recalcitrant suspicions of China and India. At this writing in 2012, the outcome is still very uncertain.

The media and public climate education sit at the core of the debate. Political solutions require political will, and lobbyists and grassroots constituents generate that will. As long as the public remains poorly informed, lobbyists who oppose climate action can dominate negotiations, and there can be no political progress. Quite simply, the public has to care. The scientific evidence for climate change is not balanced between pro- and anti-climate change factions. The previous seven chapters make it very clear that the evidence is overwhelming and underlines an urgent need to reduce heat-trapping greenhouse gas emissions. To convince the majority of voting Americans that the hazards and risks from climate change are real, scientists need to emerge from their academic cocoons and add their collective and varied voices to the public debate. Those who raise their heads above the academic parapet can expect some ad hominem abuse and even political censure. Such is the state of play in a game where the stakes are very high.

Looking Ahead . . .

In Chapter 9 you will consider fundamental social problems that create the problem of climate change and drive the global demand for cheap and abundant energy. You will investigate why we are likely to remain dependent on fossil fuels well into this century and investigate whether new sources of energy offer some hope for the future.

Critical Thinking Questions

The social and political issues that are introduced in this chapter raise many questions that are suitable for class discussions. The following scenarios are meant to stimulate discussion and the exchange of views between students and faculty.

1. As a U.S. Senate committee chairman, you understand that climate change should be avoided and that your government will have to make many important economic and social decisions over the next few years. The cost of action will be measured in billions, or even trillions, of dollars. The cost of inaction may be the loss of thousands of lives and the destruction of the environment.

   Scientific and economic models are good and getting better all the time, but they are still imperfect. As a U.S. politician, how much uncertainty are you willing to accept before you are prepared to support a new climate bill that will almost certainly cost your own state thousands of jobs? Explain.

2. As an economic consultant who advises the White House, you know that economic forecast models become less reliable as time passes. The vice president has taken charge of energy policy and demands to know why we should worry about climate change 100 to 200 years in the future, when new technologies are very likely to be available then that can reverse or mitigate the impact. He wants to spend more U.S. dollars on projects that support the United Nations Millennium Goals and win international recognition and support for the United States. What do you tell the vice president?

3. You are a senior advisor on the White House Energy Task Force. Every day you receive phone calls and invitations to lunch with executives and lobbyists from the major power utilities and fossil fuel industries who do not want you to change greenhouse gas emissions standards. These are the same people who helped finance a recent hard-fought election campaign that won your party a slim majority in Congress. You know your administration is much more environmentally friendly than the opposition, and if you lose your slim majority in the midterm elections, a number of vitally important conservation, health, and welfare bills are likely to fail in the House. How do you weigh the contrasting opinions of these stakeholder groups, in light of these external pressures?

4. You are a responsible media consultant for an international NGO that campaigns for legislation to combat climate change. You like to listen to all sides of an argument, but the opposition is now filling the airwaves with half-truths and all sorts of bogus claims, and these communications appear to be affecting public opinion. You are losing the argument. You are about to launch a new campaign to support new carbon tax legislation that could have lasting impact on greenhouse gas emissions. How much are you prepared to play the same game and lead with half-truths and bogus claims to support an argument that you firmly believe to be for the greater good? Explain.

5. During a major economic recession, you are the Secretary of State for Work and Pensions in the UK

Cabinet, and you are responsible for employment, social security, and education. You are in conference with the Secretary of State for Energy and Climate Change, who wants to reallocate £600 million from your budget to promote a new wind energy project in northern England. Unemployment is rising around the nation, tax revenues are falling, the demand for social security is skyrocketing, and members of your party are demanding new education programs for retraining redundant workers. Do you agree to the request of the Secretary of State for Energy and Climate Change? Explain.

6. You are once more the Secretary of State for Work and Pensions in the UK Cabinet, and you are responsible for employment, social security, and education. You are once more in conference with the Secretary of State for Energy and Climate Change. Nothing much has changed since your last conference, except that the national economy has shrunk by 3.2% over the past eight months, and the national debt is rising rapidly. This time, the Secretary of State for Energy and Climate Change wants you to support new legislation that would impose new greenhouse gas emissions standards on small to medium-sized manufacturing enterprises across the nation. You fear that at a time of recession, this legislation will lead to more unemployment, loss of manufacturing capacity, and even less tax revenue. Do you offer your support? Explain.

7. You are (against the odds) still Secretary of State for Work and Pensions in the UK Cabinet, and you are responsible for employment, social security, and education. Under increasing pressure from the prime minister, and thinking you might soon lose your position on the Cabinet, you are confronted by the Secretary of State for International Development, who wants the prime minister to allocate £800 million of development aid from his new budget to support clean energy projects in rural India. You like this idea, but you suspect that this money will also be reallocated from your budgets. Should you let the needs of those you represent politically outweigh the greater long-term need of people who are not UK citizens and who will have no say in your reelection? Explain.

8. You are Secretary for Energy in the U.S. Cabinet. You are under increasing public pressure from extreme environmentalists who have a record of cherry-picking and misrepresenting the published results of NASA and NOAA research. When the issues are critical and public support is imperative, how much latitude should you have to restrict and control the flow of information from government-sponsored research to the general public?

9. You are a scientist working for NOAA, and you discover some disturbing evidence that the West Antarctic Ice Sheet may collapse much sooner than previously expected. This is a sensitive issue, as the government is in the middle of important negotiations with small island nations under threat from rising sea level. Your supervisor seems to be holding back from publishing this data, and you think he is under political pressure to hide the information. You and the rest of NOAA have been told not to communicate directly with the press about your research. Your job is on the line, you are up for promotion, and you are hoping for a new $2 million grant for new monitoring equipment for your Antarctic research. What do you do? When is it ethical for a government-funded scientist to break the rules of confidentiality for what he or she believes to be the public interest? Given that the truth about of your concern may not become apparent for 50 to 100 years, should you be punished for divulging information or protected as a whistleblower?

10. You are senior environmental reporter for CNN, and you have a keen interest in climate change but no special expertise in climate research. As a good journalist, you want to be balanced in your representation of the issues, and almost daily you get calls from the Cato Institute and the American Enterprise Institute, demanding that the views of climate contrarians continue to be well represented on your program. Some of their arguments sound plausible, but your contacts in NOAA are scathing and critical of everything they say. You decide to invite representatives from both organizations to debate the issues. When does the need to achieve journalistic balance end up actually distorting the true balance of a scientific argument?

11. Was Al Gore's timely and well-staged arrival at the Rio World Summit pure political posturing or a determined effort to gain public support back in the United States? Did he have the desired effect? Explain.

12. The United States was determined to include forestry and agriculture in the balance of carbon offsets as part of the Kyoto Protocol. Is this position justified? Explain.

13. Compared to the media in the United States, the media in Western Europe is more biased in support of the Kyoto Protocol in its coverage of the climate change debate. Why is this so? At what point should the established U.S. media shift from a balanced approach to actively promoting change?

14. Was the United States government justified in delaying and prevaricating during the Kyoto negotiations, given widely perceived uncertainties in the validity of the scientific and economic models? Explain your position.

15. Do you think that a voluntary carbon emissions reduction scheme in the United States, as initially preferred by the George W. Bush administration, stood any chance of success? Explain your position.

16. To what extent was the position of the EU on the Kyoto Protocol driven by political ideology rather

than a considered argument driven by the scientific evidence? Explain.

17. As an EU representative at the Kyoto negotiations, you want to see mandatory targets for reducing carbon emissions. What arguments would you lead with to persuade your colleagues from the United States?

18. You are an investment banker in Germany and proudly European. You want to see the EU seize a competitive advantage from the initial reluctance of the United States to engage with the Kyoto Protocol. Where should your bank encourage investment to maximize profits while promoting the overall competitiveness of the EU?

19. Skeptics claim that EU environmental policy has been marked by a significant contrast between policy and practice at national and regional levels. As the EU commissioner with responsibility for the environment, how would you ensure that the member states comply with legislation to limit greenhouse gas emissions?

20. You sit on the executive board that oversees the administration of CDM programs under the Kyoto Protocol. What international guidelines would you put in place to prevent widespread fraud during the implementation of CDM programs?

21. To what extent could a carbon tax applied to all goods and services complement a carbon cap on emissions from major producers of greenhouse gas emissions?

22. You lead the U.S. delegation to implement the recommendations of the Copenhagen Agreement for the next stage of the Kyoto Protocol. The senior delegate from China meets with you in private and asks how you will react to a new proposal that countries that outsource the manufacturing of domestic goods overseas should pay for most of the greenhouse gas emissions released. How do you answer?

23. You work for an international development agency, and you are meeting with the president of a small nation that has a large population living on a coastal delta. He is concerned with economic and social welfare and is a strong advocate of the United Nations Millennium Goals. How do you persuade the president that his country must also address the urgent problem of climate change by investing in clean energy and changing agricultural practices?

24. You have just been appointed as the minister for social justice in a developing country in Africa. The ambassador from the United Kingdom, a key friend and ally, is asking you to support legislation that will encourage investment in clean energy. You know that your country has access to cheap coal, and you are concerned that this investment will detract from the urgent need to tackle energy poverty and will also limit funds that are available for education and health reform. How do you respond to the ambassador?

Key Terms

Please make sure you are familiar with the key terms introduced and highlighted in this chapter. A full glossary is available at the end of the book.

Bali Action Plan, p. 272
BASIC countries, p. 273
Belief persistence, p. 280
Berlin Mandate, p. 265
Cap-and-trade, p. 268
Carbon fee, p. 270
Carbon offsets, p. 266
Carbon sinks, p. 266
Carbon tax, p. 270
Clean Development Mechanism (CDM), p. 269
Community-based social marketing, p. 281
Cultural risk theory, p. 279
Department of Energy, p. 258

Earth Summit, p. 264
Emissions Trading Scheme, p. 268
Energy poverty, p. 261
Environmental Protection Agency (EPA), p. 258
Exposure, p. 278
Hazard, p. 278
International Emissions Trading Scheme (IET), p. 268
Joint Implementation Mechanism (JI), p. 269
Kyoto Protocol, p. 263
Longevity, p. 278

Marrakesh Accords, p. 268
Montreal Action Plan, p. 271
National Aeronautics and Space Administration (NASA), p. 258
National allowances, p. 268
National Oceanic and Atmospheric Administration (NOAA), p. 258
Potable, p. 261
Probability, p. 278
Reduction of emissions from deforestation and forest degradation (REDD), p. 273

Risk, p. 278
Six Americas Survey, p. 280
Social network, p. 281
Stakeholders, 256
Tea Party, p. 273
United Nations Framework Convention on Climate Change (UNFCCC), p. 264
U.S. Department of Agriculture (USDA), p. 258
U.S. Geological Survey (USGS), p. 258

PART 5

GLOBAL SOLUTIONS: MANAGING THE CRISIS

CHAPTER 9: THE ENERGY CRISIS

CHAPTER 10: TURNING KNOWLEDGE INTO ACTION

Wind turbines like these have the potential to supply clean energy to many parts of the world and help to reduce the global emission of heat trapping greenhouse gases.

With global population now over 7 billion, there is an ever-increasing demand for the supply of cheap energy. The rapid economic growth needed to sustain this population relies on the supply of cheap and plentiful energy from fossil fuels, which increases the risk of damaging climate change. Alternative sources of energy offer a solution to this problem, especially if introduced with energy conservation and efficiency measures. The goal of reducing emissions to a level where climate change is no longer a threat to society is achievable, but it is important to consider alternative strategies such as geoengineering in case current political strategies to limit emissions are not successful. Part 5 reviews the growing global demand for energy, considers the impact of different sources of energy on greenhouse gas emissions, and proposes a range of actions that can reduce the risk of climate change. Chapter 9 considers problems of energy supply and demand, and Chapter 10 reviews possible actions that could significantly reduce the risk of climate change.

The application of solar technology is not always complex and expensive. Simple parabolic mirrors like this one in Tibet can focus enough free solar energy to cook food and boil water; simple necessities that can save millions of lives in the developing world.

CHAPTER 9

The Energy Crisis

Introduction

Rapid population growth and economic development are increasing the global demand for energy. Demand is greatest in the developing world, where many communities lack sufficient energy to meet the basic demands of modern life. In these countries, limited access to potable water, sanitation, refrigeration, sterilization, and smoke-free cooking create unacceptably high levels of mortality in both adults and children. Most of our energy is produced by burning cheap fossil fuels that pollute the atmosphere and damage our health. Clean energy alternatives exist, but they are still expensive and require new investment in energy infrastructure.

To meet the rising demand for energy at an affordable price, many developing countries are investing in coal to generate power, but this increases the rate of greenhouse gas emissions and heightens the risks of climate change. The Kyoto Protocol set specific targets to limit the emission of greenhouse gases, but the continuing use of coal makes it almost certain that the level of CO_2 in the atmosphere will still exceed 560 parts per million (ppm) by 2050. If fossil fuels are used to supply most of the energy we need to feed economic growth over the next decade, a serious gap will develop between what needs to be achieved and what is likely to be accomplished. Energy is not our enemy, it feeds economic growth and prosperity, but the climate crisis will deepen until the world discovers how to decouple the generation of energy from the creation of greenhouse gases.

Learning Outcomes | *When you finish this chapter you should be able to:*

- Understand the economic factors that drive the global demand for energy
- Discuss the main political and ethical issues involved in the use of different sources of energy
- Discuss the important role of natural gas in managing the transition between coal and alternative sources of energy for generating electrical power
- Suggest how innovation in the transport sector can reduce the global demand for oil
- Consider the role of nuclear energy as an emissions-free alternative to coal and gas for power generation
- Discuss how it is possible to meet the rising demand for energy with clean and renewable sources of power
- Debate the role of government in regulating and directing the energy market
- Explain why a new energy infrastructure is vital for the efficient use of alternative sources of energy such as wind and solar power

The Energy Problem

The United Nations Intergovernmental Panel on Climate Change calls on the developed countries to reduce GHG emissions by 25–40% below 1990 levels by 2020, and 80–95% below 1990 levels by 2050, in order to keep the level of CO_2 in the atmosphere below 450 ppm and limit global warming to less than 2°C (3.6°F) This reduction must take place at the same time the developing countries[1] continue to increase emissions as their economies grow. This will not be easy. Greenhouse gas emissions from all sources have already grown from 7.6 billion tonnes of carbon per year in 1990 to over 10 billion tonnes of carbon per year in 2012, and if unchecked, they may double the level of 1990 emissions by 2050. *Fundamental change must take place in energy production if we hope to reach these important emission targets.*

CHECKPOINT 9.1 ▶ What emissions target did the IPCC set for 2050?

The Challenge of Energy Poverty

The world needs more energy to bring heat, light, and electrical power to billions of people who are still living in **energy poverty** without access to electricity or other modern energy services. The expected birth of 2 *billion* more people by 2050 makes this a critical problem. The chronic injustice of energy poverty will continue to raise issues of equality and human rights as long as the per capita energy consumption in developed nations remains so far ahead of the rest of the world (**Figure 9.1**). Does a developing nation have the right to exploit local fossil energy sources and increase the prospect of global climate change if by doing so it delivers cheap and reliable power to the people who need it? Solving the problem of energy poverty is a global priority that runs in parallel with the need to reduce greenhouse gas emissions, and any energy policy that does not address this problem will lose the support of the developing nations.

CHECKPOINT 9.2 ▶ Explain the challenge of energy poverty.

World Energy Demand

World energy demand has grown significantly over the past 50 years, and it continues to increase at around 2% to 3% per year (**Figure 9.2** and **Figure 9.3**). Sustained economic growth in **Organisation for Economic Co-operation and Development (OECD)** countries and rapid economic growth in the developing world is placing enormous demands on conventional energy resources.

---

[1]*Developing country* is not an exact term but refers generally to a country with low per capita income. One metric is whether a country qualifies for World Bank assistance. Developing countries in this book are non-OECD (Organization for Economic Co-operation and Development) countries plus the OECD members Mexico and Turkey, but excluding Russia and other formerly planned economies in transition.

The global consumption of power (energy consumed per year) in 2008 was 473×10^{18} **joules** or 448×10^{15} British thermal units (Btu). This is equivalent to over 15 terawatts, 15×10^{12} TWs^{-1} (a Watt is equivalent to 1 joule per second). By 2013 consumption is expected to exceed 570×10^{18} joules per year (or 540×10^{15} Btu per year) at a rate of 18 terawatts. Consumption is greatest in the United States (3.13 terawatts), but demand is also high from China (2.47 terawatts) and the EU (2.32 terawatts). Over the next 20 years, the demand for energy is expected to increase by 19% in developed nations and by a staggering 84% in developing nations, with China and India alone responsible for over 56% of demand in 2050 (Figure 9.3 and **Figure 9.4**). Over the past 15 years, China's gross domestic product (GDP) has risen by over 300%, and India's has risen by 140%; most of this growth has been powered by fossil fuels. During this time, greenhouse gas emissions have increased by over 50% and placed a heavy demand on fuel, power, and mineral resources worldwide (**Figure 9.5**).

CHECKPOINT 9.3 ▶ How much more energy do China and India need by the end of the century in order to keep pace with the developed world?

The economic miracle in Asia has improved the standard of living of many millions of people—but it has come at a price. By 2030, the global demand for electrical power will be 50% higher than it is today, and it may possibly be twice as high by 2050. The demand for liquid petroleum is likely to increase by 35%, natural gas by 53%, and coal by as much as 65%. Growth based on energy from fossil fuels is good news for the millions relieved of poverty and disease, but it is certainly not good news for the future of global climate.

CHECKPOINT 9.4 ▶ How much of the growth in the demand for energy is likely to be met by fossil fuels?

In order to meet IPCC targets, the world will have to meet the demand for energy in a fundamentally different way than it does today, and the task is daunting. When 2050 arrives, the global demand for electrical power is expected to be more than 26 terawatts. Some of this demand will be met from renewable sources—maybe even as much as 30%—but that still leaves 18.2 terawatts that has to be generated from coal, natural gas, and oil. Renewable energy must be the answer to this problem in the long term, but to have any chance of success in the short term, the world needs to focus on changing the fuel mix to more natural gas and less coal; it also needs to focus on the generation of more carbon-neutral energy from conventional fuels (**Figure 9.6**).

CHECKPOINT 9.5 ▶ What is the current global demand for electrical power? What is the demand likely to be in 2050?

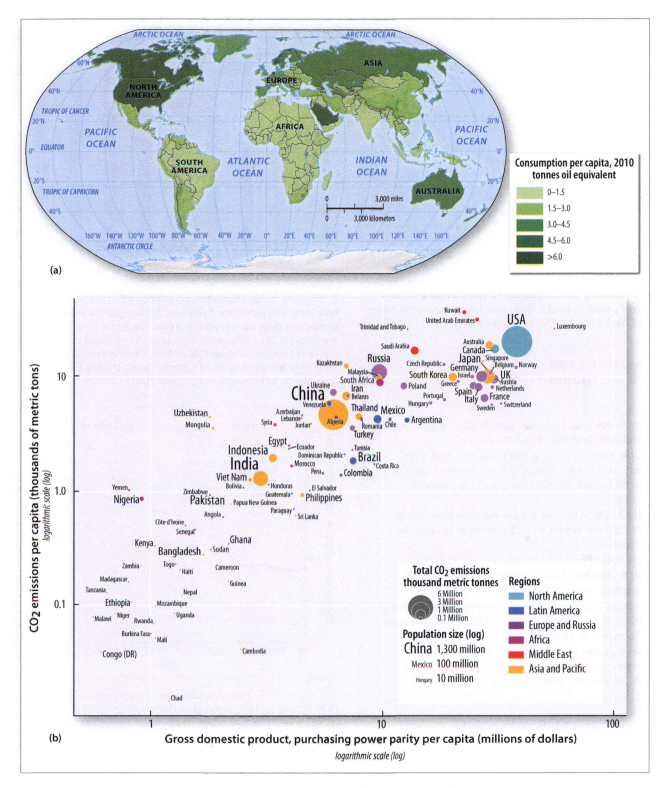

▲ **Figure 9.1:** (a) Map of world primary energy consumption per capita. Units are tonnes of oil equivalent (1 toe = 42×10^9 joules). Consider the impact on global emissions in 30 years, when all the countries with per capita emissions below 1.5 metric tonnes per year start to catch up with Australia, Canada, Europe, Russia, and the United States. (b) The relationship between carbon dioxide emissions and wealth, measured by comparing CO_2 emissions per capita (in thousands of metric tonnes) and gross domestic product (GDP in Purchasing Power Parity per Capita (basically GDP divided by population.)). Both scales are logarithmic, to allow many nations to be included on this chart. The separation between the United States and China would be much more apparent on a linear scale. The larger the text of the country, the larger the population, and the larger the bubble, the greater the level of carbon dioxide emissions. Logarithmic scales are used on both axes to show the large numerical range of values.

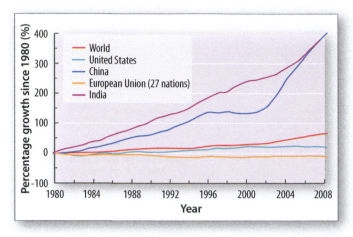

▲ **Figure 9.2:** National CO_2 emissions, 1998–2008. The rate of growth of CO_2 emissions from India accelerated slowly over this decade. China's emissions stalled in the 1990s, until economic reforms gave a major boost to industrial production and the generation of power to feed it. Both countries now produce over 400% more heat-trapping CO_2 emissions than they did in 1980.

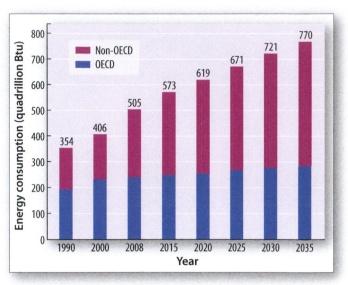

▲ **Figure 9.3:** World energy consumption (OECD and non-OECD) from 1990, with projection to the year 2035 in quadrillion Btu. (Each terawatt of power consumed over a year uses about 30 quadrillion Btu of energy [31.78 x 10^{18} joules].) If this growth in demand for energy is met by burning coal and other fossil fuels, the world will soon start to experience an increased risk of extreme climate change. OECD is the Organization for Economic Cooperation and Development.

9.1 PAUSE FOR... THOUGHT | *Some key facts about energy poverty from the World Health Organization.*

- Today as many as 1.4 billion people do not have access to electricity, and 85% live in rural areas.
- Around 3 billion people still cook and heat their homes using open fires and leaky stoves burning biomass (wood, animal dung, and crop waste) and coal.
- Nearly 2 million people die prematurely from illness attributable to indoor air pollution from household solid fuel use.
- Nearly 50% of pneumonia deaths among children under 5 years are due to particulate matter inhaled from indoor air pollution.
- More than 1 million people die each year from chronic obstructive pulmonary disease (COPD) that develops due to exposure to indoor air pollution.
- Adults exposed to heavy indoor smoke are two to three times more likely to develop COPD than are those who are not exposed to indoor smoke.

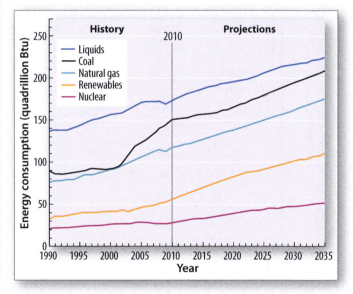

▲ **Figure 9.4:** Growth in world energy consumption from 1990 to 2010 and projected growth through 2035 for five primary energy sources: oil and other liquid petroleum products, coal, gas, nuclear, and renewable energy. The growth in renewable energy is remarkable but is not enough to offset the increase in the use of fossil fuels. Units are in quadrillion (10^{15}) British thermal units (Btu), with 1 quadrillion Btu equivalent to 1.05 x 10^{18} joules.

Conventional Sources of Power: Coal

Coal is used to create more electrical power in the world than any other fuel. It powered the Industrial Revolution and revolutionized transportation on land and sea. Coal remains the cheapest, most abundant, and most readily available fuel in the world. It is also toxic when burned, destructive to mine, and pollutes the atmosphere with acidic aerosols and carbon dioxide.

Origins of Coal

Coal forms deep underground, when plant material that once collected in surface swamps and bogs is buried, dehydrated, heated, and compressed. The quality of coal varies considerably from brown lignite, which is little more than hardened peat, to shiny **anthracite** with a very high proportion of carbon. Coal currently provides 27% of the world's demand for energy, and it will meet as much as 29% by 2030. It is hard to overstate the enduring importance of coal. Coal is dirty and inefficient, and it produces more CO_2 and toxic gases per tonne used and kWh produced than any other major fuel (**Figure 9.7**), but it is also cheap, abundant, locally

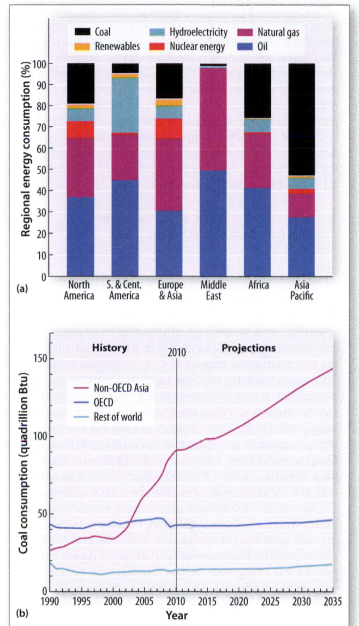

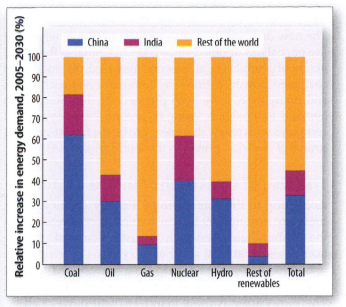

▲ Figure 9.6: Graph of projected increase in energy demand from different primary sources, 2005–2030. Note the very large increase in natural gas and renewables in the rest of the world and of the combined use of coal in China and India.

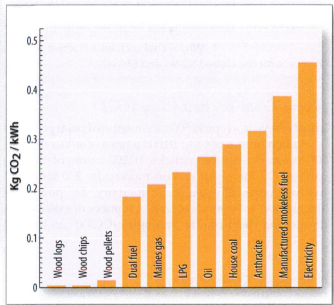

▲ Figure 9.7: Graph of CO_2 emissions per kilowatt-hour of power from different primary energy sources. A kilowatt-hour is the sustained production of power at the level of 1,000 Watts (joules per second) for an hour. It is the energy used to heat a typical single heating element bar of an electric fire.

▲ Figure 9.5: (a) Regional energy consumption pattern, 2010. The Asia-Pacific region continues to lead global energy consumption, accounting for 38.1% of the world total and for 67.1% of global coal consumption. Within the Asia-Pacific countries, coal is the dominant fuel, accounting for 52.1% of energy consumption. Oil is the dominant fuel for all other regions except Europe and Eurasia, where natural gas is the leading fuel. In addition to being the largest consumer of coal, the Asia-Pacific countries are also the leading users of oil and hydroelectric generation. Europe and Eurasia are leading consumers of natural gas, nuclear power, and renewables in power generation. (b) World coal consumption by region, 1990–2010 and projected to 2035, measured in quadrillion Btu. Compare these data with Figure 9.1 and notice how the rapid increase in coal consumption around 2001–2002 corresponds with the increase in CO_2 emissions from China. The emissions associated with this rapid increase in the use of coal are unsustainable because they will push the world toward more rapid and damaging climate change.

available, and supported by powerful stakeholders. Although there are many reasons to wish it were not so, coal is an essential fuel for the foreseeable future. The only way to deal with this unpalatable truth is to find ways to capture and sequester the CO_2 coal produces so that coal can become more efficient and as carbon neutral as possible.

CHECKPOINT 9.6 ▶ In your own words, describe how coal is formed.

The Use of Coal in China

The new demand for coal is greatest in China, where the use of coal has doubled since 2000. In China, at the recent height of production, at least one midsized, 500-megawatt coal-fired power station was commissioned each week, and many more have been planned for the future. With demand for coal now rising by around 2% per year, coal competes with renewable energy to be the fastest-growing source of electrical power in the world (Figure 9.8).

The Use of Coal in the United States

According to the **International Energy Agency (IEA),** there are around 909 billion tonnes of coal reserves in the United States—enough to sustain current production levels for at least another 155 years. (1 metric tonne = 1,000 kilograms = ~1 ton.) Coal is expected to meet as much as 46% of the demand for electricity in 2030. There are currently more than 1,440 coal-power stations in the United States, each with an average operating life of 47 years, and plans are to build at least 280 new 500-megawatt coal power stations before 2030. Carbon caps and carbon taxes would have a significant impact on the use of coal, but whatever happens, coal will remain a major source of energy well into the next century.

CHECKPOINT 9.7 ▶ Why is coal such an attractive resource for the United States and China?

Clean Coal Technology (CCT)

In just one year, a typical 500 megawatt coal plant produces 3.7 million tonnes of CO_2, 10,000 tonnes of sulfur dioxide, 500 tonnes of airborne particles, 10,200 tonnes of nitrogen oxides, 720 tonnes of carbon monoxide, 220 tonnes of hydrocarbons, 170 pounds of mercury, 225 pounds of arsenic, and 114 pounds of lead. A number of existing and developing **clean coal technologies (CCTs)** can be used

to capture, clean, and sequester greenhouse gases and toxic emissions from coal. One way is to capture these harmful gases and metals from the exhaust fumes before they reach the atmosphere. This can be done by using absorption towers where CO_2, toxic gases and **particulate matter** are captured (scrubbed) and stored. This process readily removes sulfur dioxide, nitrogen oxides, and particulate matter from combustion gases, but the high volume of CO_2 makes it difficult and expensive to remove completely. If coal must be a major source of energy into the next century, then the development of more efficient CTTs must be a global priority.

CHECKPOINT 9.8 ▶ Why is coal considered a "dirty" fuel, and how can it be "cleaned up"?

The Political Power of Big Coal

The power of clean coal technology is worthless without the political power to support it. A concerted legislative, scientific, and technological effort is now required to enforce the use of clean coal technology and minimize the environmental impact of CO_2 emissions from coal. This would not be the first time that government has demanded action from the energy utility companies to reduce emissions. When the Clean Air Act Amendments were passed in the United States in 1990, the law required utility companies to reduce toxic sulfur dioxide emissions. Despite vociferous lobbying and public protestations, they somehow found the $50 billion needed to comply with the regulations and continued to make large profits.

As carbon dioxide is not a toxic gas, it was not included in this initial clean air legislation, but in a groundbreaking and hard-fought decision in 2008, the U.S. Supreme Court gave the **Environmental Protection Agency (EPA)** permission to regulate CO_2 emissions *from vehicles* as if CO_2 were a pollutant. This is a first step—a thin wedge that may soon expand beyond the transport sector and

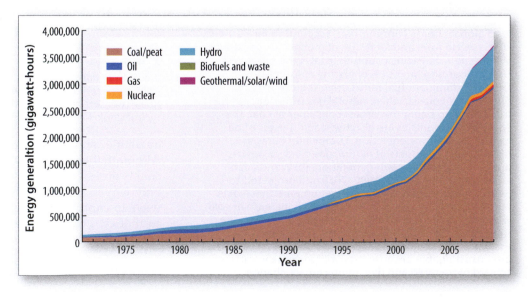

▶ **Figure 9.8:** China's electricity production in gigawatt-hours by source, 1971–2009. Gigawatt-hour is a unit of energy equal to $3.6 \times 1,012$ Joules.

have a significant impact on the ability of the energy utility companies to keep polluting "for free." Of course they know this and are lobbying hard to oppose the EPA and reverse or slow the progress of any such legislation.

In Europe, the impact of legislation to limit CO_2 emissions has already had a positive impact, following the introduction of mandatory emissions caps that forced the energy utilities to radically improve CO_2 emissions standards or face severe financial penalties.

CHECKPOINT 9.9 ▶ Why did the original Clean Air Act Amendments not cover carbon dioxide?

How Efficient Is Coal Power?

The generation of electricity is responsible for over 27% of global greenhouse gas emissions. Any increase in efficiency that can create more electrical power from every tonne of coal that is burned is a critical step toward reaching CO_2 emission targets. Power stations first pulverize and then burn coal in air, using the heat of combustion to drive steam turbines. This is a very inefficient process that converts as little as 30% to 40% of the available energy into electricity. However, it is possible to improve efficiency to as much as 50% by burning the coal at higher temperatures and pressures. The technologies already exist to do this (**Figure 9.9**) but require new and more expensive materials. In the absence of legislation to force compliance or of incentives to encourage it, there is little prospect that the utility companies will make much progress in this area.

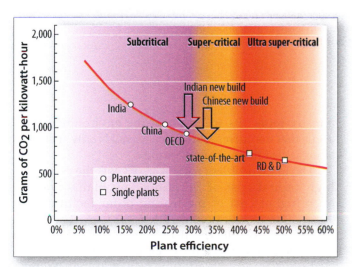

▲ **Figure 9.9:** Coal-fired plant efficiency and CO_2 emissions. Note that many of the new coal-fired power plants in China are to be constructed with higher-efficiency standards than the average of the OECD countries. However, old, very inefficient power plants are still a major problem in countries such as India and China. Only research and development programs have been able to push efficiency as far as 50%. The terms subcritical, supercritical, and ultra supercritical relate to the increasing temperature of combustion and refer to a physical property of water.

CHECKPOINT 9.10 ▶ What opportunities are there to increase the efficiency of coal combustion?

The Integrated Gasification Combined Cycle (IGCC) System

Coal has been used to produce hydrogen gas since the first public piped gas supply was introduced in London in 1807. The **Integrated Gasification Combined Cycle (IGCC) system** (**Figure 9.10**) works with up to 55% efficiency by burning pulverized coal in the controlled presence of oxygen and steam to produce a mixture of hydrogen and carbon monoxide known as syngas (synthetic gas). Fewer sulfur compounds, ammonia, mercury, and other metals are left in the syngas compared to the exhaust from a conventional coal-fired power plant, and these can be removed before the hydrogen is pumped to feed a gas turbine.

The hydrogen produced by this process is a clean, flammable gas that, when burned, produces water and heat. During the IGCC process, hydrogen is used to power a high-temperature gas turbine, and the residual heat is used to generate steam that drives a second conventional steam turbine. The pure carbon monoxide that was earlier removed, along with other impurities, is converted to CO_2 and removed for sequestration in a process known as carbon capture and storage (CCS), which is discussed next.

CHECKPOINT 9.11 ▶ Are there any advantages to producing hydrogen gas from coal? Explain.

Carbon Capture and Storage (CCS)

The theory behind **carbon capture and storage (CCS)** is simple, but the practice is proving difficult. It is possible to collect and compress the CO_2 produced during combustion and transport it by pipeline up to a few hundred kilometers away. The CO_2 can be pumped underground and stored in geological formations such as saline aquifers, gas fields, and coal beds, or it can even be taken out to sea and pumped into the deep ocean (**Figure 9.11**). Oil companies have successfully used the technology to pump some of the CO_2 collected during oil and gas production back down special wells, where it is injected into porous and permeable reservoirs to displace even more oil and gas into production wells.

The challenge for CCS technology is the quantity of gas involved. With so much gas needing to be stored, it is important to ensure that none of it can leak back to the surface, where it can cause suffocation and death. The search is now on for geological formations around the world that are suitable repositories for CO_2 sequestration (**Figure 9.12**).

FutureGen Alliance is an exploratory project that was launched by the George W. Bush administration to test new clean fuel technologies, including IGCC

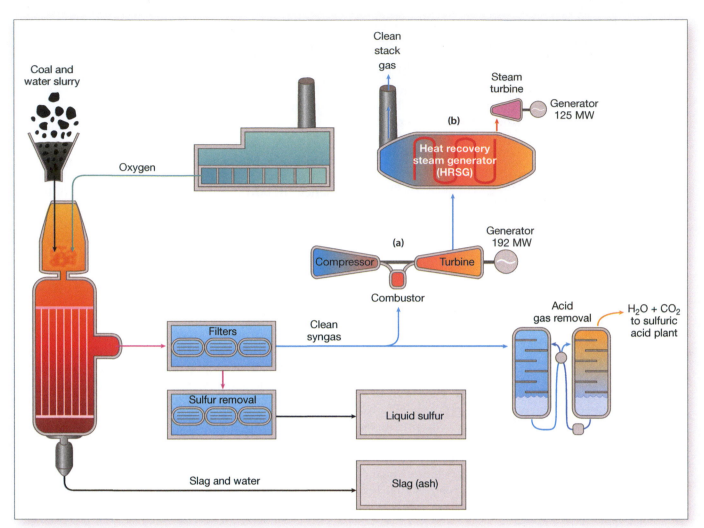

▲ **Figure 9.10:** Block diagram of an IGCC (Integrated Gasification Combined Cycle) coal-fired power station. The process is complex, as the diagram illustrates but at a basic. level, the hot syngas produced by burning coal is used to power a gas turbine (a) and generate electricity. Excess heat from the gas turbine exhaust is used to turn water into steam that drives a second (steam) turbine (b) to generate additional electrical power. Solid residue (slag) and gases are captured and removed for disposal.

and CCS. The storage capacity of CO_2 in saline rock formations in the United States is estimated to be around 2,970 gigatonnes. This is enough to meet the current demand from coal power of 2.48 gigatonnes of CO_2 per year for a long time. The success of this technology is a critical step toward achieving the goal of zero emissions of CO_2 from coal-fired power plants. At this writing in 2012, the project is planning to have the first power station completed in 2015. At that time, more than 1.3 million tonnes of CO_2 each year—more than 90% of the coal plant's carbon emissions—will be sequestered on site in underground reservoirs.

Carbon sequestration and storage will make coal power more expensive, but the higher price will encourage investment in alternative energy strategies. Many economists argue that the price of coal should already reflect its true cost to the environment, and argue that the current low price of coal is only sustained at great cost to future generations.

The main problem for IGCC/CCS is the extremely high capital cost of building new plants and the even higher cost of modifying older facilities that were seldom built on sites ideal for CCS. The world needs coal for electrical power, and it needs CCS technologies to meet IPCC emissions targets. If the cost of CCS implementation proves to be too high, the inefficient use of coal will continue and bring an end to any hope of managing greenhouse gas emissions. The prognosis using current technologies is not good.

CHECKPOINT 9.12 ▶ Describe the main barriers to the effective use of carbon capture and storage.

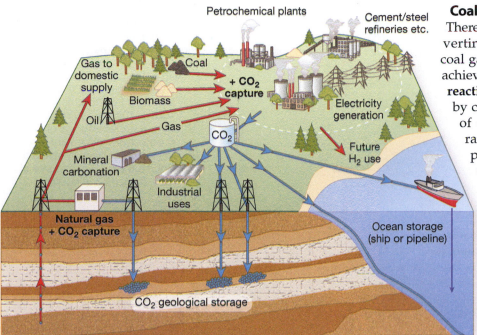

Petrochemical plants

Cement/steel refineries etc.

Gas to domestic supply

Coal

Biomass

Oil

Gas

+ CO$_2$ capture

Electricity generation

CO$_2$

Mineral carbonation

Future H$_2$ use

Industrial uses

Natural gas + CO$_2$ capture

Ocean storage (ship or pipeline)

CO$_2$ geological storage

▲ **Figure 9.11:** Schematic of possible Carbon Capture and Storage (CCS) systems. Proximity to the source of emissions is very important to keep the risks and costs of storage and transportation as low as possible.

Coal as a Source of Liquid Petroleum

There has been renewed interest in converting the syngas produced during coal gasification into petroleum. This is achieved by using the **Fischer–Tropsch reaction,** where the hydrogen produced by combustion is catalyzed in a series of chemical reactions to produce a range of petroleum products. This process was first used by coal-rich Germany in the 1920s, and it was used extensively during World War II. When hydrogen (a clean-burning gas) is sourced from coal, however, it does not reduce the overall level of CO$_2$ emissions to the atmosphere. A more carbon-neutral way of using the Fischer–Tropsch reaction to produce petroleum would be to generate hydrogen from water instead of coal, using nuclear or renewable energy sources.

CHECKPOINT 9.13 ▶ Why will the production of oil from coal not help to reduce emissions?

The "Coal Power in a Warming World" Report

A 2008 report titled "Coal Power in a Warming World," by the Union of Concerned Scientists, outlined eight reasonable ways to ensure a cleaner future for coal:

1. Invest more in research and development for clean coal technology.
2. Stop building new coal-fired power stations without CCS.
3. Stop investing in new coal-to-liquid petroleum plants.
4. Ensure that any coal-to-gas plants employ CCS and that the resulting fuel is used to offset coal use rather than natural gas use.
5. Significantly increase both deployment of and research and development for energy efficiency and renewable energy.
6. Adopt statutes and stronger regulations to reduce the environmental and societal costs of coal use.
7. Put a price on CO$_2$ emissions by adopting a strong economy-wide cap-and-trade program (as in the EU).
8. Ensure the transfer of low-carbon technologies to other countries (especially China and India).

How many of these methods do you think are viable?

CHECKPOINT 9.14 ▶ Is the price of coal too low, considering the environmental and health costs associated with its uses? Explain.

9.2 PAUSE FOR... THOUGHT | *How do we calculate how much carbon dioxide a typical 500 megawatt coal power station produces?*

From Figure 9.9 you can see that for every kilowatt hour of energy produced, 1kg of carbon dioxide is typically released. How can we work out how many kilowatt hours a 500 gigawatt power station produces in a year?

As with all energy calculations it helps to reduce the data to basic units of energy, in this case joules.

The power of 1 watt is defined as 1 joule of energy per second, so 1 kilowatt hour is the same as 1,000 joules per second over the period of an hour = 1,000 watts x 60 seconds x 60 minutes = 3.6×10^6 joules.

We now know that we emit 1 kg of carbon dioxide for every 3.6×10^6 joules of energy released for coal.

We need to find out how many joules a 500 gigawatt power station produces in a year. We can calculate this as: 500×10^6 joules/second x 3,600 seconds per hour x 24 hours per day x 365 days per year = 15.8×10^{15} joules per year. As each kilowatt hour is equivalent to 3.6×10^6 joules (above), we can calculate the number of kilowatt hours produced from a 500 gigawatt power station as 15.8×10^{15} joules divided by 3.6×10^6 joules = 4.4×10^9 kilowatt hours per year.

And finally, as each kilowatt hour produces 1kg of carbon dioxide, the total amount of carbon dioxide released each year from a 500 gigawatt power station will be: 4.4×10^9 kilowatt hours x 1kg = 4.4×10^9 kilograms or (as a tonne is 1,000 kg) 4.4×10^6 tonnes or carbon dioxide.

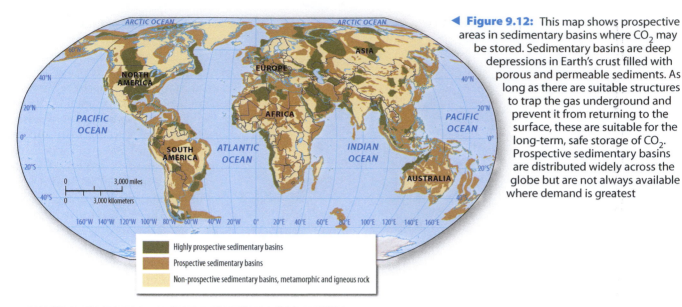

◀ **Figure 9.12:** This map shows prospective areas in sedimentary basins where CO_2 may be stored. Sedimentary basins are deep depressions in Earth's crust filled with porous and permeable sediments. As long as there are suitable structures to trap the gas underground and prevent it from returning to the surface, these are suitable for the long-term, safe storage of CO_2. Prospective sedimentary basins are distributed widely across the globe but are not always available where demand is greatest

Legend:
- Highly prospective sedimentary basins
- Prospective sedimentary basins
- Non-prospective sedimentary basins, metamorphic and igneous rock

CHECKPOINT 9.15 ▶ Why is the coal lobby so strong in the United States?

9.3 PAUSE FOR... THOUGHT | *Environmental impacts of mountaintop removal.*

It is possible that as much as 28.5 billion tonnes of high-quality coal remain beneath the mountains of the Appalachia coal-mining region. Before miners can gain access to these reserves, the mining companies may remove up to 122 meters (400 feet) of waste rock and earth from the top of a mountain, much of which they simply dump into the surrounding valleys. Despite some effort by the coal industry to restore the land after mining, the EPA noted the following environmental issues in a study of more than 1,200 streams that were impacted by mountaintop mining:

- An increase in the concentration of minerals such as zinc, sodium, selenium, and sulfate in the water can negatively impact wildlife.

- Mining activity can have a severe impact on aquatic communities.

- Streams in watersheds below valley fills tend to receive more input from the seepage of groundwater that can be contaminated.

- Streams are sometimes completely buried by the debris that is dumped into the valleys.

- The wetlands that often result from the dumping of waste are of low environmental quality.

- The remaining forests are often fragmented (broken into sections).

- The regrowth of trees and woody plants on restored land can be slowed due to soil compaction.

- The cumulative environmental costs have not been identified.

- The social and economic impact and loss of local heritage are significant.

None of this is accounted for in the price we pay for coal.

The Future of Coal

Clean coal technologies have the potential to meet the demand for energy from the developing world and still achieve global greenhouse gas emission targets. Finding these CCTs will require intensive research and development, massive investment in infrastructure, and strong legislative pressure. The long-term future for coal is not bright for the industry. From extraction to sequestration, coal is damaging to the environment and public health. We should replace it with new, cleaner ways of producing energy as quickly as possible, but we will be forced to use it for some time to come to meet the demands of energy poverty and economic growth.

Conventional Sources of Power: Natural Gas

Gas can be used as a replacement for coal to generate electricity (fuel switching). Compared to coal and oil, natural gas is cleaner, richer in energy, and more efficient to burn. It is not a renewable fuel, but it produces much less CO_2 during combustion than coal, and it has the potential to replace coal in existing power stations over the short term. This could significantly reduce the overall level of CO_2 emissions from electrical power generation over the next decade and help the world meet IPCC emissions targets.

Origins of Natural Gas

Natural gas (methane [CH_4]) is a product of the slow breakdown of organic material trapped in rocks, usually coals or organic-rich shales, deep under Earth's surface. The gas is slowly released over geological time and gradually rises up toward the surface, where most of it escapes naturally into the atmosphere. Some of the gas, however, is trapped in porous and permeable reservoirs beneath the surface. From there it can be extracted and sent by pipeline for uses in chemical processing and as fuel.

Recently, the commercial extraction of gas directly from its source rocks—referred to as hydraulic fracturing, or **fracking**—has also become possible and could significantly increase the potential of global reserves if it can be shown to be environmentally safe. A 2012 report from the Energy Information Administration (EIA) estimated that around 0.48 trillion cubic meters (17 trillion cubic feet) of methane could be extracted directly from coal beds and 13.65 trillion cubic meters (482 trillion cubic feet) from gas shales within the United States. British Columbia, Canada, is estimated to have approximately 2.8 trillion cubic meters (100 trillion cubic feet) of coal bed gas, and 2.2 trillion cubic meters (78 trillion cubic feet) of shale gas. Alberta is estimated to have at least 6.31 trillion cubic meters (223 trillion cubic feet) of recoverable shale gas reserves and up to 500 trillion cubic feet (14.16 trillion cubic meters) of coal bed gas. In addition to these major reserves, North America has much smaller reserves in the form of biogas (also methane) from the **anaerobic fermentation** of animal waste and methane from landfill sites. The sum of total natural gas available for use is considerable.

CHECKPOINT 9.16 ▶ What is the origin of natural gas?

Is Gas More Efficient than Coal?

Natural gas (methane) produces 43% fewer greenhouse gas emissions than coal and 30% fewer than oil. In addition, when burned, methane doesn't leave any of the toxic solid wastes associated with coal, and although it does produce CO_2 and nitrogen oxides, existing technologies can remove them relatively cheaply and effectively.

CHECKPOINT 9.17 ▶ Why is the use of natural gas considered preferable to the use of coal and oil?

World Demand for Natural Gas

World demand for natural gas is expected to increase by 52%, from 2.9 to 4.5 trillion cubic meters (104 to 158 trillion cubic feet) by 2030. New natural gas reserves have made gas much more price competitive, and if the price of coal rises further due to EPA regulations, natural gas may soon replace coal as the primary fuel for the generation of electrical power in the United States.

The greatest growth in demand for natural gas is expected in the developing world. In 2005, developed and developing countries consumed about the same amount of natural gas, but by 2030, developing countries are expected to have outpaced the rest of the world, accounting for 74% of the total increase in demand (**Figure 9.13**). China and India use relatively little natural gas at the moment, but their usage is expected to grow by around 4% to 5% per year until 2030. Neither country has large gas reserves, and

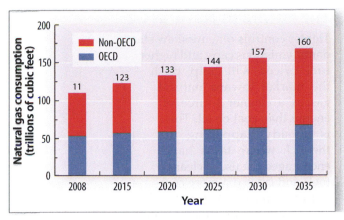

▲ **Figure 9.13:** Predicted changes in world natural gas consumption from 2008 to 2035, in trillions of cubic feet of gas (1 trillion cubic feet = 1.027 quadrillion Btu = 1.123×10^{15} joules).

both countries need to invest in basic infrastructure to get the gas they have from its sources to their growing cities.

CHECKPOINT 9.18 ▶ By how much is the demand for natural gas expected to increase by 2030?

Natural Gas Consumption

Natural gas has many uses apart from power generation. In 2007, the United States consumed 23 trillion cubic feet of natural gas, which provided 23% of total U.S. energy demand. Only 26% of this gas was used to generate electricity, and the rest was divided between the demands from residential (22%), commercial (14%), industrial (35%), and transportation (3%) sources. By 2030, the United States is likely to make much greater use of gas as a cleaner source of energy for electricity generation. Some estimates anticipate that natural gas production will grow by more than 35% by 2035 (**Figure 9.14**).

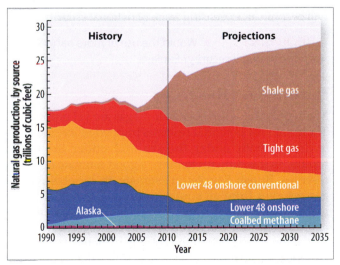

▲ **Figure 9.14:** Natural gas will meet a greater proportion of new generating capacity in the United States between 2009 and 2035 due to the recovery of both shale gas and **tight gas** (tight gas is gas contained within reservoirs with low permeability i.e. gas that cannot flow readily through the formation).

The Future of Natural Gas

Tighter controls on emissions standards, emissions caps, and higher fuel prices will help push the market toward natural gas as a cleaner source of energy than coal or oil. This is already happening in the EU, where taxes have encouraged the use of cleaner fuels and created a more competitive market for natural gas. With the continued strong growth of renewable and nuclear energy, natural gas may eventually become less competitive, but that is not likely to occur before the end of the 21st century.

Energy Security and Natural Gas in the European Union

The EU depends heavily on natural gas for the generation of electricity, and the demand for gas is expected to increase annually by as much as 3.9%. The Russian Federation currently supplies as much as 50% of this gas and has twice seriously disrupted gas supply to the EU following financial and political disputes with Ukraine. Energy security has thus become a key strategic issue, and Europe's response to this threat has been to explore alternative pipeline routes to other gas-rich states. Russia's use of its energy reserves to exert political pressure is not a new idea—this has been the practice of the **Organization of the Petroleum Exporting Countries (OPEC)** for some time—but it has forced the EU to seek more energy independence through the use of renewable energy and nuclear energy. As energy resources become more expensive and in greater demand, access to reliable and secure sources will become an increasing issue of national security.

CHECKPOINT 9.19 ▶ Would the use of more natural gas increase national security? Explain.

Natural Gas Reserves

Many potential sources of natural gas are yet untouched. There are vast potential reserves trapped in ice (methane hydrates) both under the sea and under the Arctic tundra, so it is difficult to make a good estimate of global reserves. British Petroleum estimate, that global proven gas reserves from all sources reached 208 trillion cubic meters (7,345 trillion cubic feet) in 2011, with 10.8 trillion cubic meters (381 trillion cubic feet) from North America, 7.6 trillion cubic (268 trillion cubic feet) meters from South America, 14.5 trillion cubic meters (512 trillion cubic feet) from Africa, 78.7 trillion cubic meters (2,779 trillion cubic feet) from Europe and Eurasia, 16.8 trillion cubic meters (593 trillion cubic feet) from the Asia Pacific Region, and 80 trillion cubic meters (2,825 trillion cubic feet) from

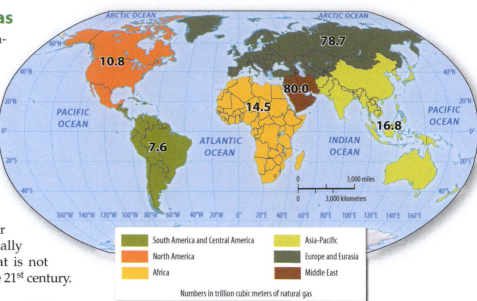

▲ **Figure 9.15:** Proven natural gas reserves as of 2011. Numbers are in trillion cubic meters of natural gas. Reserves in North America have risen recently due to the discovery of shale gas reserves.

the Middle-East. These are just the proven reserves. Some estimates place the total amount of gas available as high as 600 trillion cubic meters (21,188 trillion cubic feet). Based on the level of gas consumption in 2008 this is enough to last for over 130 years and equivalent to $23,000 \times 10^{18}$ joules of energy (**Figure 9.15**).

9.4 PAUSE FOR... THOUGHT | *Hydraulic fracturing.*

The Marcellus Gas Shale that lies beneath parts of New York and Pennsylvania is typical of an abundant type of fine-grained sedimentary rock that is rich in organic material. As the rock is buried and heated, the organic material it contains turns into natural gas that is trapped by the low-permeability shale. The total amount of gas available for extraction from such shale is immense. The Marcellus Shale alone may contain as much as 13.6 trillion cubic meters (482 trillion cubic feet) of gas, of which 2.38 trillion cubic meters (84 trillion cubic feet) can be recovered using current technologies. There are a number of such large shale formations in the United States.

The use of natural gas to replace coal and oil as a primary source of energy is an important part of any national strategy to reduce heat-trapping greenhouse gas emissions. However, the technology used to extract the gas is causing a lot of concern. Hydraulic fracturing (fracking) involves pumping large volumes of water into the ground under so much pressure that the shale fractures, creating pathways that allow the trapped gas to be extracted. Fracking is not a new process; it has been used to enhance oil and gas recovery for decades. But on the scale involved with shale gas, there is growing concern that wastewater and natural gas will escape, contaminating drinking water supplies, or that natural gas may leak to the surface, increasing the risk of explosions.

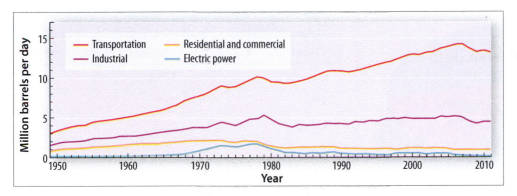

◀ **Figure 9.16:** U.S. petroleum consumption by sector, in millions of barrels per day (1 barrel = 43 U.S. gallons = 0.16 cubic meters). Note the steady increase of petroleum consumption in transport compared to other sectors.

in as little as 30 to 40 years. As the resource becomes scarce, prices will rise, and the cheap and abundant resource that fed the modern transportation revolution will come to a stuttering end.

CHECKPOINT 9.20 ▶ The extraction of natural gas by fracking may threaten the environment and human health. Explain why many people appear to believe that it is worth the risk.

Conventional Sources of Power: Oil

Oil forms deep within Earth, from source rocks that are rich in the remains of marine plankton. Because oil is less dense than the briny fluids that surround it, it migrates toward the surface, but some gets trapped as liquid in the pores of sedimentary rocks. Sometimes the oil loses so much of its volatile content that it becomes thick and difficult to extract by conventional means. Under these circumstances oil-rich deposits that lie close to the surface may be mined and processed as "tar sands." Oil can also be recovered from organic-rich shales deep under the surface in the same way that natural gas is extracted from gas shale deposits.

Oil is not a renewable resource, and it is running out fast. The observed depletion of liquid reserves has followed a theoretical model known as the Hubbert curve. This suggests we have already moved past the point of peak world production of oil and that new reserves will be harder to find and smaller in scale.

Oil is the dominant fuel of the transportation sector. Over 70% of the oil consumed in the United States is used to produce gasoline (**Figure 9.16**), which is responsible for as much as 25% of the nation's greenhouse gas emissions. Global demand for oil is expected to increase rapidly with the growth of car ownership in the developing world, especially in China and India.

CHECKPOINT 9.21 ▶ In your own words, describe how oil is formed.

Oil Reserves

Oil and liquid natural gas reserves total around 1,653 billion barrels, most of which is in Brazil, Canada, Europe, Kazakhstan, Nigeria, Mexico, the Middle East, Russia, the United States, and Venezuela. The rate at which new reserves are added may have already passed its peak (**Figure 9.17**), and global reserves will struggle to keep up with demand

Oil Sands, Heavy Oil, and Extra-Heavy Oil

The lifetime of oil may be extended using new technologies to extract it from unconventional sources, such as oil sands that contain heavy and extra-heavy thick viscous oil deposits (bitumen) (**Figure 9.18**). For example, Canada has conventional reserves of around 28 billion barrels of oil, and if the potential from oil sands is included, this could reach as high as 179 billion barrels—as large as the oil giants of Iran, Iraq, and Kuwait. The problem with oil sands and heavy oils is that extracting the oil is much more expensive than with traditional deposits. The process is also very harmful to the natural environment. Surface mining is destructive, releasing contaminants into the atmosphere and ground water, and large quantities of water are required for oil extraction and processing. These reserves, however, will be of immense future value as the price of oil rises towards $120 a barrel.. This is good news for the economy of United States because access to large oil reserves across a friendly border enhances energy security considerably.

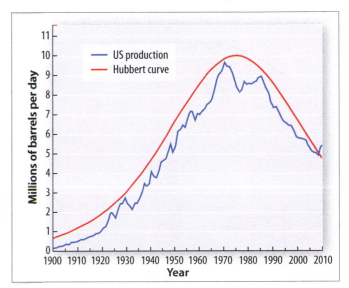

▲ **Figure 9.17:** Peak oil production and decline in the United States, compared to a theoretical model known as the Hubbert curve. One million barrels of oil is equivalent to 5.86×10^{15} joules, so 10 million barrels per day for a year is equivalent to 21×10^{18} joules per year.

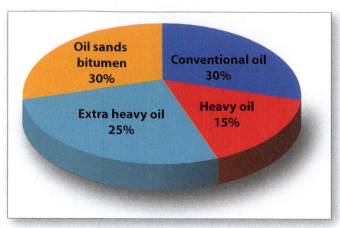

▲ **Figure 9.18:** Total world oil reserves were estimated to be around 1,653 billion barrels in 2012. This is equivalent to 9,689 x 10^{18} joules of energy (compare to 23,000 x 10^{18} joules of energy for gas reserves). Note that there is as much oil contained in tar sands, such as those discovered in Alberta, Canada, as contained within conventional reservoirs.

CHECKPOINT 9.22 ▶ Why is the extraction of oil from oil sands a greater threat to the environment than production from conventional sources?

Oil and Energy Security

Current world demand for oil is around 88 million barrels (1 barrel = 158.98 liters [42 gallons]) per day (bpd), and this is expected to grow at a rate of 2.5% per year to around 106 bpd by 2030. Today, the United States consumes 20 million bpd, Europe 16 million bpd, China 7.6 million bpd, and India 2.8 million bpd. Domestic oil resources in these same countries are very small compared to this demand and could meet 100% of current of demand for only as little as 1,000 to 2,000 days (**Figure 9.19** and **Table 9.1**). This is a significant shortfall for most countries and a major threat to national energy security.

Table 9.1 Select Oil Reserves as of 2011	
Country	**Thousand Million Barrels**
Canada	175.2
US	30.9
China	14.7
Europe	12.5
Mexico	11.4
India	5.7

The United States has to meet its 60% shortfall in domestic supply with imports from Canada (18.2%), Mexico (11.4%), Saudi Arabia (11.0%), Nigeria (8.4%), and Venezuela (10.1%). The strategic risk of this dependency on foreign oil was highlighted during the **oil crisis** of the 1970s when the global economy was severely affected by conflict in the Middle East. The risk is even greater today for China and India, where growth in demand for oil over the next 10 years will average around 3% to 4% per year, leaving these two countries almost totally reliant on imported oil. In 2006, China alone accounted for 38% of total world growth in demand for oil.

CHECKPOINT 9.23 ▶ Where does most of the oil used in the United States come from?

Oil and the Transportation Crisis

Transportation will account for 74% of the total projected increase in demand for liquid fuels up to 2030. There is only one way to break free from the power of oil, and that is to have a new transportation revolution, based on renewable energy and more efficient internal combustion engines.

In 1990, China had fewer than 250,000 passenger vehicles on the roads, but now there are more than 11 million, with nearly 1.8 million new cars sold each year (**Figure 9.20**). Naturally, the Chinese want to benefit from the freedom of individual transportation. But there are a lot of people in China, and even with so many cars already on the road, only 1% of the population owns a car. If everyone who wants a car in China and India buys a car as efficient as those sold in the United States, there is no way that the world will reach its target for greenhouse gas emission reduction. Developing nations will need to insist on higher **vehicle emission standards** and alternative fuels to realize the freedom of transportation-on-demand without critically altering the course of future climate change.

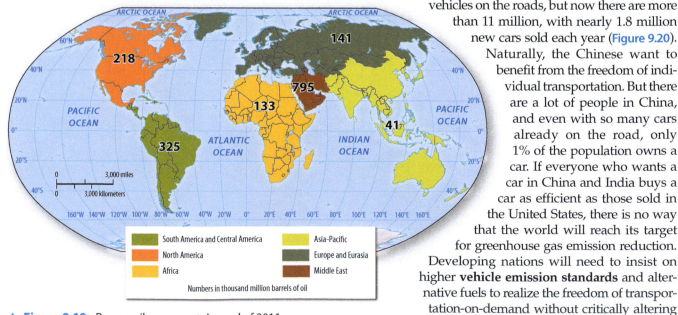

▲ **Figure 9.19:** Proven oil reserves at the end of 2011, in thousand million barrels.

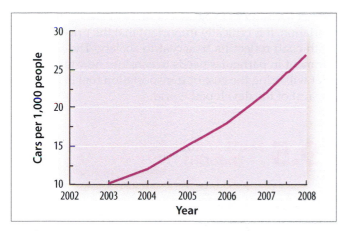

▲ **Figure 9.20:** Car ownership in China. Note the rapid rise in truck and car registration at the start of the transportation revolution in China.

CHECKPOINT 9.24 ▶ What impact will the per capita economic growth in China have on the global demand for and price of oil?

Fuel Efficiency and Mileage Standards

Both energy security and environmental security are best served if countries reduce their reliance on imported oil. One of the best ways to do this is to increase vehicle mileage standards. The EU and China have already introduced tough mileage standards and emission control standards that leave the United States far behind. Some cars now manufactured to permissible standards in the United States are no longer legal in many other countries of the world (**Figure 9.21**).

The United States has been reluctant to address the mileage question because of effective political lobbying on behalf of the oil and auto industries. Unlike many other countries around the world, and notably those in the EU, low taxation on gasoline in the United States continued to benefit the auto industry. The unusually low cost of fuel in the United States exerted little pressure on consumers to demand higher mileage standards. Buoyed by higher profit margins, the auto industry was happy to manufacture inefficient vehicles and missed an opportunity to take an early lead in the manufacture of energy-efficient vehicles. The auto industry in the United States now faces intense competition on its home turf from foreign companies that had early vision and better technology, but recent government bailouts and deep restructuring have made significant progress towards restoring international competitiveness. Standing still has never been an option in a truly global market.

The United States is not the only country to subsidize the cost of fuel. China and India also subsidize the price of fuel, but unlike the United States, the main reason is that the average per capita income is still too low for most people to pay the full price. In the future, as per capita income rises, all countries need to phase out fuel subsidies that only serve to encourage growth in the number of personal vehicles. The money could be better invested in an enhanced public transportation system, especially in large cities.

Future Fuel Efficieny and Mileage Standards

In 2007, the U.S. Congress somewhat reluctantly changed the nation's automotive mileage standard for the first time since 1985. It raised the **Corporate Average Fuel Economy (CAFE) standard** to 27.5 miles per gallon (mpg) by 2011 and to 35 mpg by 2020. In 2010, Congress amended this to 35.5 mpg by 2016. The current proposal is to increase the standard to 55 mpg as early as 2025.

Most other countries already meet or exceed existing U.S. standards and have plans for even higher standards in the near future. China has a new target of 43 mpg, while the EU recently set a new limit at 40 mpg. Many auto manufacturers

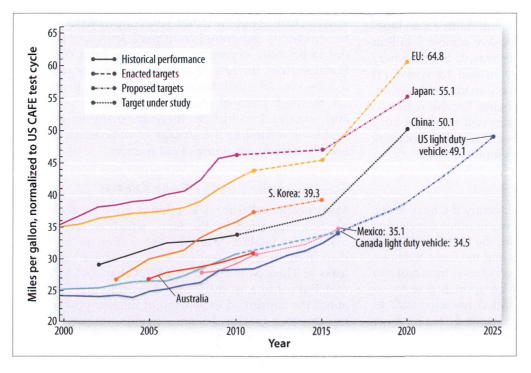

◀ **Figure 9.21:** Actual and predicted miles per gallon fuel economy for passenger vehicles, by country. The lines represent historical performance of existing targets, enacted targets, proposed targets, and potential targets currently under study.

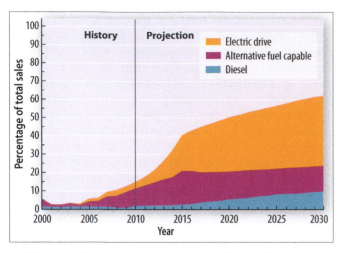

▲ **Figure 9.22:** Projected new alternative light vehicle sales in the United States, 2000 to 2030.

in Europe have met this challenge and prospered, with Ford planning a 65 mpg vehicle, Audi a 39-mpg vehicle, Seat a 62-mpg vehicle, and Volkswagen a 62-mpg vehicle. When gasoline costs more than $8 a gallon, as it does in the EU, such vehicles will not be a hard sell.

The standards auto manufacturers in Europe are meeting are well within the technical ability of automotive manufacturers in the United States. Switching to producing such efficient vehicles would significantly reduce overall CO_2 emissions and reduce the nation's dependency on imported oil. The EIA in the United States predicts that by 2030, more than 60% of vehicles sold in the United States will be fueled by alternative sources of energy (**Figure 9.22**).

CHECKPOINT 9.25 ▶ What impact could increased fuel efficiency standards have on the demand for oil?

The Cost of Fuel Inefficiency

If the United States set the same automotive standards as China and the EU, it would save almost 2 billion barrels of oil a year. Put another way, U.S. consumers would have as much as $200 billion to spend on something other than making the Canadians, Mexican, Saudis, and Venezuelans much richer. For this reason, it is very likely that the United States will soon introduce mileage standards and fuel taxes that will come more into line with those of other developed nations.

The Future of Oil

Global climate will change significantly if China, India, Mexico, and Brazil, along with many other developing nations, try to achieve parity with the developed world by using the same primary fuels and the same technologies. This is as true for electricity generation from coal and gas as it is for the use of oil in the transportation sector.

Increasing demand for a limited resource such as oil has caused prices to rise to well over $100 a barrel as the global economy recovers from a deep recession. To

attract investment to cleaner sources of energy for transportation, it is critically important that the price of oil (as with coal) reflect its true cost to society. The U.S. government in particular needs to consider new tax policies that will bring the cost of transportation fuel in line with the rest of the developed world.

9.5 PAUSE FOR... THOUGHT | *Is offshore gas worth the risks associated with exploration and recovery?*

More than 27 trillion cubic feet of natural gas may lie beneath the seabed off the East Coast of the United States, but a disastrous explosion aboard a BP oil rig in the Gulf of Mexico in April 2010 led to a ban on further offshore exploration. The production of gas is much less damaging than the use of coal, it is much less polluting when burned, and it produces less carbon dioxide. Given these factors, a decision to ban exploration for gas while continuing to subsidize the mountaintop removal of coal appears to serve neither the strategic energy interests of the nation nor the greater needs of the environment.

The continued burning of any fossil fuel presents a significant threat to the environment, but it appears very unlikely that alternative sources of energy can take their place by the middle of the 21st century. To keep emissions at a minimum, it is essential that, as much as possible, we use natural gas in place of coal in the production of electricity. If offshore reserves exist, it is hard to see how we can avoid exploiting them to the fullest.

Conventional Sources of Power: Nuclear Power

Nuclear power was initially seen as a clean and cheap solution to many of the world's energy problems. When the electricity generated from nuclear power turned out to be more expensive than power from conventional sources, the rate of construction of new reactors in most countries fell dramatically. The main problems are the high costs of planning, construction, safety, and security. In addition, there are many long-term problems related to the storage of radioactive waste and the decommissioning of old reactors.

The Nuclear Power Fear Factor

Many people still fear nuclear power, but these concerns are not supported by the facts, which clearly show many more fatalities from the use of coal, oil, and gas than from nuclear power (**Figure 9.23**). Major radiation leaks at Three Mile Island in 1979, Chernobyl in 1986, and Fukushima in 2011 raised worldwide concern about the continued expansion of nuclear power, but new modern reactor designs can significantly reduce the chance of a major radiation leak.

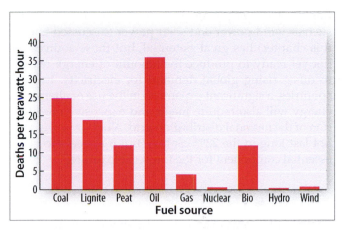

▲ **Figure 9.23:** Mean values of health effects, presented as deaths/TWh (the number of deaths recorded for each energy source relative to the amount of energy produced by that source), for the respective forms of electricity generation throughout the European Union

New Interest in Nuclear Power

International attention is once again focused on how nuclear power might help reduce global greenhouse gas emissions and increase energy security. Nuclear power offers many advantages. The raw fuel for nuclear power, **uranium oxide,** is widely available. Also, unlike using coal, gas, and oil, the generating power from nuclear energy releases no CO_2 into the atmosphere. Every 22 tonnes of uranium oxide used to generate electricity in place of coal prevents 1 million tonnes of CO_2 from entering the atmosphere. Existing reserves of uranium oxide could sustain demand for several hundred years. With the advent of a new generation of **fast breeder reactors** now operational in the United Kingdom, France, Russia, Germany, China, and India, 95% of spent fuel capacity can be recovered and recycled, and in this way the supply of nuclear energy could be sustained for thousands of years.

CHECKPOINT 9.26 ▶ Is our fear of nuclear power justified by history? Explain.

Did France Get Nuclear Power Right?

France produces power from 59 nuclear reactors that meet over 78% of the country's demand for electricity, with enough left over to earn €3 billion by exporting excess power to the rest of Europe. After the oil crisis of the 1970s, France realized that it was not rich in energy resources and decided to reach for energy independence through the development of alternative energy (**Figure 9.24**). At the moment, France meets 90% of its national demand for electricity from a combination of hydroelectric and nuclear power, and it stands out from the other large nations of Europe by producing only 6.5 tonnes of CO_2 per person compared to 20 tonnes per person in the United States and 10 tonnes in both Germany and the United Kingdom. France is a good example of the strategic independence and clean energy that nuclear power could offer the rest of the world.

CHECKPOINT 9.27 ▶ Why is France so committed to the use of nuclear energy? Has its strategy proved successful?

The United States and Nuclear Power

The United States has recently approved the construction of two new nuclear plants in the state of Georgia, the first approved since 1996, and is still the largest overall producer of nuclear power in the world. More than 20% of electricity in the United States is generated from over 100 nuclear reactors. This compares well with the average of 24% for the OECD countries but is still less than the EU, which can supply 34% of demand from nuclear sources. There have been 17 license applications to build 26 new nuclear reactors since mid-2007.

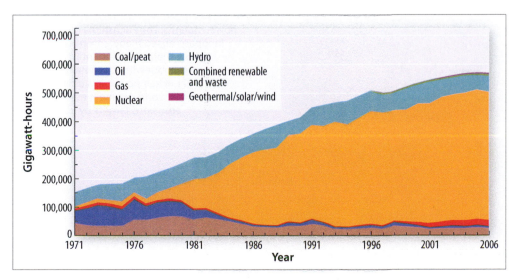

▲ **Figure 9.24:** Electricity production from various sources of power in France.

Both government and industry envision significant new nuclear capacity by 2020. These new plants will be economically competitive because they will be equipped with new technologies that extend the lifetime of a reactor from 40 to 60 years and get more return on investment. Tax credits for new nuclear plants were included in the Energy Policy Act of 2005 and, along with high fossil fuel prices, these should support the addition of 16.6 gigawatts of new capacity—equivalent to at least 16 large coal-powered plants and reducing carbon dioxide emissions by as much as 140 million tonnes per year.

CHECKPOINT 9.28 ▶ What are the main advantages of nuclear power?

The Rest of the World and Nuclear Power

The greatest expansion of nuclear power capacity is expected to take place in the developing world. China, India, and Russia will account for two-thirds of the projected increase in world nuclear power capacity before 2030. Other countries that once rejected nuclear power, such as Sweden, have also reconsidered their position, although the Fukushima disaster in 2011 has created renewed doubt. Many governments still believe that the advantages of nuclear power outweigh the disadvantages of fossil fuels and global warming. The world has changed, and any national threat assessment must put climate change high on the list of priorities.

Is Nuclear Power Really Necessary?

Nuclear energy can increase output quickly and help reduce greenhouse gas emissions within the critical time period between now and 2050. During this time, global demand for electricity will double; at the same time, greenhouse gas emissions must fall by 50%. The genera-tion of electrical power from clean, renewable sources such as wind power and solar power (discussed later in this chapter) has great potential, but those sources are not yet ready to produce the amount of energy needed to offset rising global demand for electricity. In most countries, a significant expansion in the use of alternative energy will also require major and expensive modification of the national distribution grid. Nuclear power may not last long in the 22nd century, but it appears to be an essential component for the foreseeable future.

The Future Is Glowing?

Despite a long history of concern about nuclear energy, there are now 439 nuclear reactors in 31 countries worldwide, generating 6% of world power and meeting 16% of the total demand for electricity (Figure 9.25). By 2050, global nuclear capacity is projected to increase by as much as 150% to 350%, but the proportion of electricity produced by nuclear energy may not actually increase due to the unprecedented demand for conventional power from the developing world. During the 1980s, one new nuclear reactor started up every 17 days on average; by 2015, this rate will have to increase to one reactor every 5 days to meet the rise in global demand for power. Today there are more than 33 nuclear power stations under construction around the world, with a further 94 planned and 222 proposed, ranging in size from 500 megawatts to 2 gigawatts. This is impressive, but not enough to create any significant reduction in the expected demand for fossil fuels (Figure 2.7).

CHECKPOINT 9.29 ▶ Discuss whether nuclear power can be anything more than a stopgap that sustains the demand for low-carbon energy, while the world develops alternative technologies.

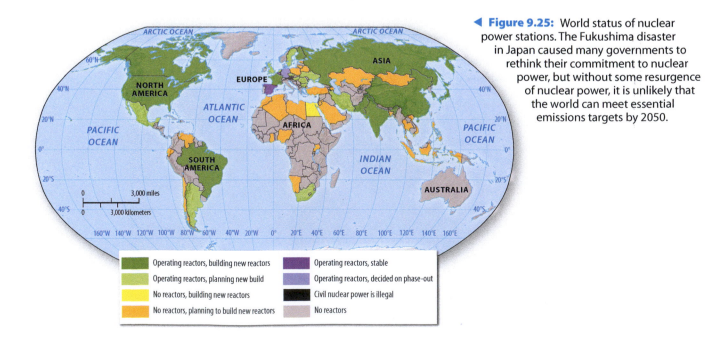

◀ **Figure 9.25:** World status of nuclear power stations. The Fukushima disaster in Japan caused many governments to rethink their commitment to nuclear power, but without some resurgence of nuclear power, it is unlikely that the world can meet essential emissions targets by 2050.

Legend:
- Operating reactors, building new reactors
- Operating reactors, planning new build
- No reactors, building new reactors
- No reactors, planning to build new reactors
- Operating reactors, stable
- Operating reactors, decided on phase-out
- Civil nuclear power is illegal
- No reactors

9.6 PAUSE FOR... THOUGHT

Was the Fukushima Daiichi nuclear disaster a fatal setback for the nuclear industry?

The terrible tsunami that struck Japan in March 2011 was so large that it overwhelmed the reactor's defenses and, with the loss of all coolant, the reactor underwent a partial meltdown that released radiation directly into the environment. Following the disaster, increased antinuclear sentiment was soon evident in many countries, including Germany, India, Italy, Spain, Switzerland, Taiwan, the United Kingdom, and the United States. Germany, which obtained one-quarter of its electricity from nuclear energy prior to 2011, decided to close all of its nuclear reactors by 2022. This rush to judgment in the immediate aftermath of the disaster may have made political sense in some instances, but it makes little long-term strategic sense. Despite the disaster at Fukushima, nuclear power will remain an important component in the production of energy, especially in the developing world.

Renewable Energy

The Renewable Energy Working Party of the IEA set down the following broad definition of **renewable energy**:

Renewable Energy is derived from natural processes that are replenished constantly. In its various forms, it derives directly or indirectly from the Sun, or from heat generated deep within the Earth. Included in the definition is energy generated from solar, wind, biomass, geothermal, hydropower and ocean resources, and biofuels and hydrogen derived from renewable resources.

CHECKPOINT 9.30 ▶ In your own words, describe the meaning of *renewable energy*.

Renewable energy attracted investments of more than $257 billion in 2011. It is the fastest-growing energy sector and of great importance to both sustainable development in the developing world and the long-term prospect of reversing climate change. It is carbon-neutral energy that makes little net contribution to the global production of greenhouse gases, and it can be used to displace coal and oil as primary fuels to help meet global greenhouse gas emissions targets (**Figure 9.26**).

CHECKPOINT 9.31 ▶ What is a "carbon neutral" source of energy?

Global Use of Renewable Energy

Renewable energy currently provides 12.7% of global **total primary energy supply (TPES),** but most of this (78%) is still from traditional solid biomass, such as wood and dung, used primarily by rural communities in the developing world. Other forms of renewable energy together generate only 3.2% of TPES, with hydroelectric energy providing 2.2% and geothermal energy 0.4%. Much to the surprise of many eager environmentalists, this leaves just 0.6% of TPES that is produced from all other kinds of renewable energy (**Figure 9.27**).

CHECKPOINT 9.32 ▶ What is the total contribution to the TPES of solar, wind, and tidal power?

Hydroelectric Power

Large hydroelectric power plants are the lowest-cost alternative energy solution. They meet 3% of the overall global demand for power and use this power to meet as much as 15% of electricity demand—almost as much as nuclear power. However, new large programs, such as the 22.5-gigawatt Three Gorges Project on the Yangtze River in China, are not appearing at the same rate as other alternative energy projects. This is partly due to their very high construction costs but is also due to increasing concern about the environmental damage and danger of catastrophic flooding. China, Canada, Brazil, the United States, and Russia together account for more than 54% of global hydroelectric production. China has more than tripled hydroelectric power production from 220 billion kilowatt-hours in the year 2000 to 720 billion kilowatt-hours in 2010. In 2011 hydroelectric power accounted for 15% of China's total electricity

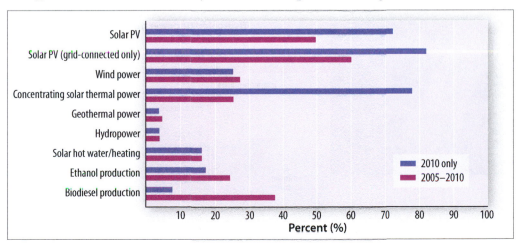

◀ **Figure 9.26:** Average global growth rates of renewable energy capacity, 2005–2010. The purple bars indicate the average growth over the period 2005–2010 and the blue bars the level of growth during 2010. Notice the remarkable growth of solar energy.

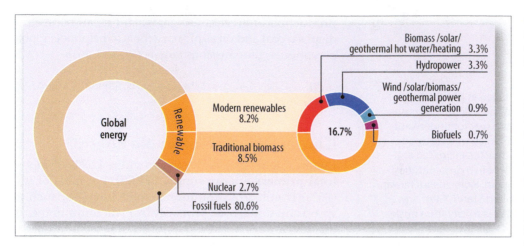

growth can also be expected from small projects that store energy from solar arrays and wind farms. During times of peak power generation, when supply exceeds demand, water can be pumped to higher elevations and stored until needed, when it is then allowed to flow back through a turbine to generate power. This is certainly not the most efficient way to store energy, but at the moment, it is cheap and uses tested and reliable technology. Conventional hydroelectric energy has been the low-hanging fruit of the renewable energy world, and its place in the future portfolio of clean energy is assured.

generation, and it is reported to have plans to increase capacity by 50% by 2015. Brazil is the second-largest producer of hydroelectric power worldwide. It produces 86% of its electricity from hydroelectric power stations including the Itaipu Dam, which generates over 92 billion kilowatt-hours per year (**Figure 9.28**).

Small Hydroelectric Schemes

The future of smaller hydroelectric programs, those less than 10 megawatts in size, is more promising than the future of larger projects. In 2006, China developed just as much power from its small projects as it did from its large projects, and much of the future growth globally is likely to be in small projects that meet local needs. In developing countries, small hydroelectric generators can replace reliance on old diesel generators. Some additional

CHECKPOINT 9.33 ▶ Why does the future of hydroelectric power lie with much smaller schemes than the infamous Three Gorges Project?

Geothermal Power

Geothermal power is derived from heat flowing naturally from Earth's interior. The rate at which temperature increases with depth is called the **geothermal gradient,**

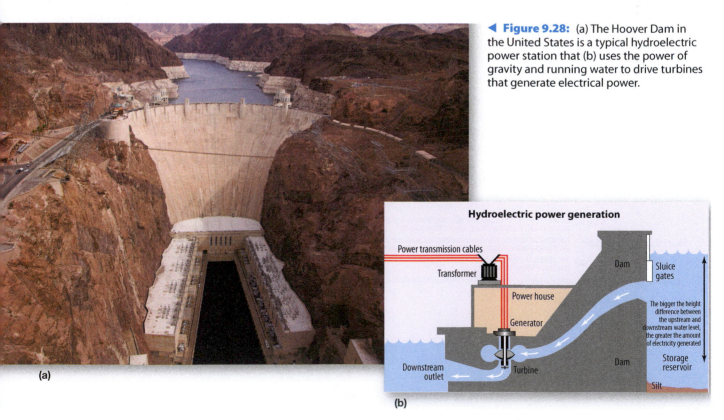

◄ **Figure 9.28:** (a) The Hoover Dam in the United States is a typical hydroelectric power station that (b) uses the power of gravity and running water to drive turbines that generate electrical power.

(a)

Hydroelectric power generation

Power transmission cables

Transformer

Power house

Generator

Downstream outlet

Turbine

Dam

Sluice gates

The bigger the height difference between the upstream and downstream water level, the greater the amount of electricity generated

Storage reservoir

Dam

Silt

(b)

is included, the total is equivalent to more than 100 gigawatts of power. Geothermal power is expensive—usually twice as expensive as coal—but it is CO_2 free and widely available around the world, and it deserves more support for future development.

> **CHECKPOINT 9.34** ▶ Is the potential of geothermal power confined to areas of the world where there is active volcanism? Explain.

"New" Renewable Energy

The world is focused on finding new clean, renewable energy as a solution to the climate crisis. The confluence of innovation, demand, funding, and political will has encouraged rapid development of many new technologies that will now compete for prime places in the growing renewable energy market. The potential for these forms of energy is enormous. Solar and wind alone have the potential to produce over $2,000 \times 10^{18}$ joules of energy per year, enough to meet 3.5 times the current demand for energy of around 504×10^{18} joules per year.

Wind Energy

Wind energy currently generates just 1% of global demand for electricity, but wind energy capacity is growing rapidly. In 2008, world capacity reached 94 gigawatts, after an astounding 10 years with growth at over 28% per year. The EU (especially Denmark, Germany, and Spain) dominates this expanding market, with 57 gigawatts, or 43%, of global energy capacity; the United States generates 16.9 gigawatts of wind energy, India 8 gigawatts, and China 6 gigawatts, all of these countries having invested heavily in the development of wind energy. In the United States, demand for wind power increased by 500% between 2000 and 2007, growing 28% in 2007 alone. Mexico and Brazil increased their demand more than 800% from 30 megawatts each to almost 300 megawatts each during 2006. Worldwide generating capacity is now growing faster than ever before, with installations in more than 70 countries.

A modern wind turbine can generate up to 5 gigawatts of power, but its capacity factor, a measure of the percentage of time it reaches optimal performance, can be as low as 25%. The wind does not always blow. However, by linking wind farms in different locations, often hundreds of miles apart, it is possible to offset less-than-optimal performance in one location with peak performance at another. But to do this we will need a new way to monitor and distribute energy that is beyond the capability of most existing electricity grids. Other possibilities are encouraging. Wind farms located offshore (as in the United Kingdom) have much higher capacity factors than land-based installations, and more constant higher-altitude winds await new airborne technologies that could revolutionize the delivery of wind power (Figure 9.30).

▲ **Figure 9.29:** A geothermal power plant in Nesjavellir, Iceland. The station produces approximately 120 megawatts of electrical power and delivers around 1,800 liters (476 gallons) of hot water per second.

and it varies from place to place depending on the nature of the rocks under the surface. A typical value of 20°C to 30°C per kilometer (58°F to 87°F per mile) can reach as high as 200°C per kilometer (580°F per mile) or more in places where molten rock (magma) lies close to the surface. In areas of known volcanic activity, such as in Iceland (Figure 9.29) and in Yellowstone National Park in Wyoming, the geothermal gradient is highest, and water is often boiling close to the surface.

Such spectacular geothermal gradients are not necessary for geothermal energy to be viable for the generation of power. Geothermal energy can be used directly to warm water that can be piped in to heat buildings, and it can also be used to convert water into steam to drive turbines. If the water is not hot enough for steam, it can be used to vaporize lower-boiling-point liquids that are then used to drive turbines. On a smaller scale, heat pumps use natural changes in near-surface temperature to warm buildings in the winter and cool them in the summer. Geothermal power is currently used in 70 countries and generates over 9.3 gigawatts of electricity and 28 gigawatts of energy for direct heating. If the energy saved by heat pumps

▲ **Figure 9.30:** Wind turbines, like these off the shores of the United Kingdom, can be installed offshore, where wind energy is often higher and the flow of wind is uninterrupted.

CHECKPOINT 9.35 ▶ Discuss the environmental advantages and disadvantages of wind power.

Solar Power

Heat from the Sun is certainly one of the most tangible and seemingly inexhaustible sources of natural energy, and it has already been used for thousands of years. Growing at rates as high as 35% per year, solar energy has become an essential part of the renewable energy strategy in the EU (especially in Germany and Spain), Japan, the United States, China, and India.

Two major and rapidly developing solar technologies that are available today could have a significant impact on global greenhouse gas emissions: concentrated solar power (CSP) and photovoltaic power plants. Other imaginative ways of capturing solar energy—such as using solar updraft towers—are also attracting growing interest.

Concentrated Solar Power With **concentrated solar power (CSP)**, arrays of movable mirrors focus the Sun's energy on receivers that store heat energy and drive a steam turbine that converts heat into electricity. Two examples are the Nevada Solar One generator in the United States, which will generate 64 megawatts of

power, and the PS10 Solar Power Plant in Spain, which will generate 11 gigawatts (**Figure 9.31**). Most hot, sunny places are not that close to large population centers, so, as with wind power, the future of this technology depends on the rapid growth of a new power distribution infrastructure that can give solar energy affordable access to national grids.

Photovoltaic Power Plants Some of the most rapid development in renewable energy technology is taking place in the exciting field of **photovoltaic (PV) solar cells**. PV solar cells convert radiation from the Sun directly into electricity. Solar cells have been around for a long time, but they have been bulky, inefficient, and expensive. New technologies are much cheaper, reliable, durable, and adaptable, and they are able to capture a broad spectrum of the Sun's energy. New thin-film solar cells can even be used to coat the roofing tiles and exterior walls of buildings, and they could herald a revolution in home energy supply. This is especially promising because of the advent of a new generation of electric meters that allow the average consumer to sell excess capacity back to the grid.

The Waldpolenz Solar Park in Germany is the world's largest thin-film PV power system, generating up to 40 megawatts of power. The Topaz Solar Farm in California, currently under construction, will generate 550 megawatts of power (**Figure 9.32**). Solar energy from PV cells has increased at a staggering 36.1% per year, and it has unprecedented potential application in the developing world. The growth of PV power in China is second only to the growth in Japan.

▼ **Figure 9.31:** The PS10 Solar Power Plant in Spain collects the Sun's energy from an array of surrounding reflectors.

▲ **Figure 9.32:** Solar power is a reliable source of energy and may soon be cheap enough to engage consumers eager to reduce domestic residential and commercial energy costs.

CHECKPOINT 9.36 ▶ What barriers might prevent the widespread consumer adoption of photovoltaic technologies in the United States?

Solar Updraft Towers A **solar updraft tower** is a tall tower surrounded by an apron of clear plastic that traps and heats air close to the ground. Heat can be stored in heat collectors filled with water that absorb heat during the day and then emit heat during the night to keep generating power (**Figure 9.33**). The apron acts as a funnel that directs warming air to the bottom of a hollow tower, where its natural buoyancy allows it to rise quickly with enough energy to drive a wind turbine that generates electricity.

A number of solar updraft tower prototypes are under construction. The proposed Ciudad Real Torre Solar in Spain would be the first of its kind in the EU. The plan calls for this tower to stand 750 meters (2,460 feet) tall and use air collected under a covered area of 350 hectares (1.35 square miles); it is expected to generate as much as 40 megawatts of electricity.

Wave Power

Several early attempts to generate power from wave and tidal energy following the oil crisis of the 1970s were disappointing, but modern technology is breathing new life into an old aspiration. There are now numerous competing designs that first turn the motion of waves into mechanical energy and then convert that energy

into electricity and transmit it to shore via submarine cables. Small facilities that will test the viability of commercial production are under construction in Oregon in the United States, in Scotland, and in Portugal. Output from these early commercial facilities is small, just 2 to 4 megawatts, but the future potential is enormous (**Figure 9.34**).

Tidal Power

Unlike wind, solar, or wave energy, tidal energy is always predictable and dependable. There are literally thousands of suitable locations around the world where the tidal range is large enough to drive turbines that generate electrical power from the twice-daily cycle of the tides (**Figure 9.35**). The technical challenges with tidal energy are considerable, and there is increasing concern about the potential environmental impact of large tidal energy programs. However, new systems are planned around the world, from the Severn Estuary in England to the Torres Strait in northern Australia to the East River in New York. These schemes can typically generate as much as 6 gigawatts of power, but the giant Penzhin Bay scheme in Russia is predicted to generate up to 87 gigawatts.

CHECKPOINT 9.37 ▶ In what ways is tidal power more attractive than solar power and wind power?

Solid Biomass

Solid biomass is composed of materials such as wood waste left over from forest management, paper mill residues, urban wood waste, wood chips,

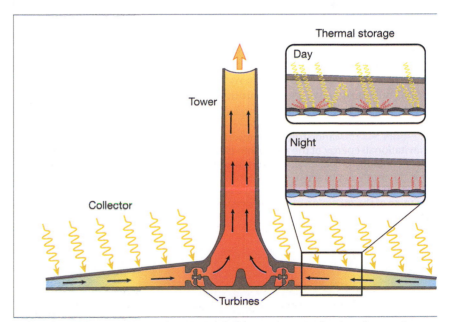

▲ **Figure 9.33:** The design of a solar updraft tower. During the day, direct solar heating drives convection. At night, heat collectors that absorbed energy from the sun during the day are still able to drive convection.

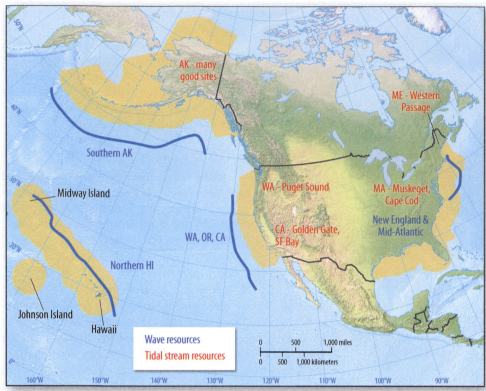

AK - many good sites

ME - Western Passage

Southern AK

Midway Island

WA - Puget Sound

MA - Muskeget, Cape Cod

New England & Mid-Atlantic

WA, OR, CA

CA - Golden Gate, SF Bay

Northern HI

Johnson Island

Hawaii

Wave resources
Tidal stream resources

0   500   1,000 miles
0   500   1,000 kilometers

◄ **Figure 9.34:** Locations of U.S. wave and tidal stream energy resources with the most electricity generation potential.

transportation. Biofuels may be that alternative. The conversion of **biomass** into liquid fuel is growing at 8% per year and could have a long-term impact on the global consumption of oil. The engines of most modern vehicles can be converted easily to run on bioethanol or run with a mix of gasoline and bioethanol, and diesel engines require almost no modification to run on biodiesel.

The United States expects to increase the use of biofuels from a baseline of 9 billion gallons in 2008 to 36 billion gallons by 2022, using new federal renewable fuel guidelines. In 2009, the **Energy Information Administration (EIA)** forecast that domestic oil demand might grow by as little as 0.2% over the ensuing two decades. This was the first time in more than 20 years that the EIA anticipated such limited growth, and it was largely due to both higher vehicle fuel standards and the use of biofuels.

Brazil produces as much as 6.5 billion gallons of bioethanol from sugarcane per year, and it is almost independent of foreign oil. The EU and China also produced as much as 1.2 billion gallons in 2008, and although production is still low in most other countries, especially in

and agricultural crop residues such as stalks. Solid biomass can be burned directly to produce heat or electrical power, used in **combined-cycle gasification systems** to generate electricity, or even burned in coal-fired power plants (in a process called *co-firing*) to reduce the overall demand for coal. Burning solid biomass produces large volumes of CO_2, but unlike coal, biomass is a renewable resource, and the CO_2 produced when solid biomass is burned is recaptured later, when the original forest or agricultural product is regrown.

Biofuels

The generation of greenhouse gas emissions–free electrical power is only half the answer to the problem of global warming. Any global solution must also offer a viable alternative to the use of gasoline and diesel for

▼ **Figure 9.35:** Marine turbines convert the tidal (gravitational) energy of the moon into electrical energy. (a) The technology is challenging as turbines like this one on its first journey out to sea need to be robust and reliable. (b) Electricity can also be generated by the action of waves. Here the force of the incoming wave compresses air that drives a generator.

(a)

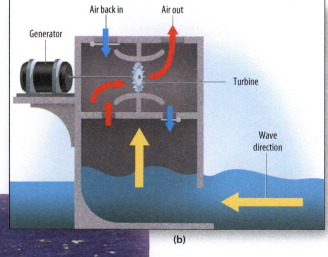

Air back in   Air out

Generator

Turbine

Wave direction

(b)

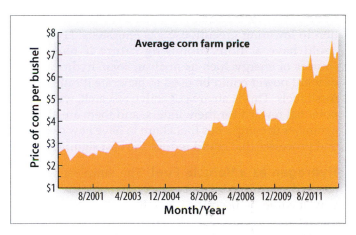

▲ **Figure 9.36:** The rapid rise in corn prices starting in 2008 was due to the competing demand from bioethanol manufacturers.

the developing world, the use of biofuels can be expected to increase significantly over the next 20 years.

Bioethanol In the United States and the EU, most gasoline sold today contains between 5% and 10% ethanol. Roughly 60% of world's **bioethanol** is produced from sugar crops, mainly sugarcane and sugar beet, with most of the remainder from corn.

Recently, the use of corn to produce bioethanol has become controversial. Rising demand for corn encouraged too many farmers to switch from food production to fuel production and led to much higher prices for corn on world food markets (**Figure 9.36**). The impact of higher corn prices in the developing world could be catastrophic. The "food for fuel" controversy deepened after new research cast doubts on claims that corn-based biofuels actually reduce greenhouse gas emissions. When the additional energy involved in the planting, harvesting, transportation, and processing of corn is considered, there may be little added value to using corn over conventional fuels at all, especially when the potential social impact is so negative.

CHECKPOINT 9.38 ▶ Why did using corn to generate biofuels become controversial?

Cellulosic Bioethanol There are many viable and more efficient alternatives to corn as a primary source of biomass for the production of ethanol. Inedible, perennial, fast-growing, cellulose-rich crops such as switchgrass and miscanthus, and even plant waste such as wood chips and stalks, can be processed to generate **cellulosic bioethanol.** This type of bioethanol can reduce greenhouse gas emissions by as much as 85% compared to gasoline—much more than corn-based bioethanol. Cellulosic bioethanol is produced using specially designed enzymes (biological catalysts) that break up the cellulose into sugars that are then fermented to produce ethanol. The world should probably preserve corn and sugar capacity to feed its growing population and focus on the use of cellulosic bioethanol as a primary source of biomass for fuel.

Biodiesel Many plants and animals produce oils and fats that can be used directly to produce **biodiesel**. In the United States, most biodiesel is made from soybeans, but biodiesel can also be made from canola oil, waste cooking oils, or animal fats. In the developing world, palm oil, Jatropha, and sunflowers can all be used to produce biodiesel. As with the cultivation of corn for bioethanol, there could be unexpected and unwanted consequences from the poorly planned expansion of this market. In Indonesia, there is growing concern that pressure on farmers to produce more corn oil for biodiesel could lead to an increase in the rate of deforestation, as well as a possible net increase in greenhouse gas emissions. In a world where people and markets are interconnected at multiple levels, there is a lot of room for unintended consequences from well-intended action.

CHECKPOINT 9.39 ▶ What are the primary alternatives to corn for the production of biofuels?

Algal Bioreactors Recent research has shown great potential for producing commercial volumes of biodiesel from natural oils extracted from algae. Algae have much higher oil content than crops such as corn, canola, and soy. Algal **bioreactors** are, at their most simple, just large aerated water tanks full of algae. They can be built anywhere there is enough light and water, even close to coal-fired power stations, where they could make productive use of the CO_2-rich exhaust stream to boost yields. Using algae as a source for biodiesel avoids the problem of using food crops for fuel, and unlike conventional crops, algae can be productive year-round if given enough sunlight and CO_2 (**Figure 9.37**).

Biogas Organic waste and wood can be converted into gas in other ways besides those already discussed. Waste that is properly disposed of in well-planned and well-constructed disposal sites generates **biogas** (methane) as the buried organic material rots under anaerobic conditions. The biogas can be collected, stored, and used to generate heat or electrical power.

Animal wastes on cattle and pig farms can also be composted to produce biogas that is used to generate power for the facility, or even feed excess energy back into the grid and make money for the farmers.

Hydrogen

Hydrogen is a very attractive source of energy. It is the ultimate clean fuel that burns in air to produce only heat and pure water. Unlike fossil fuels and solar and wind power, hydrogen is not a primary source of energy, as it must be manufactured using energy from other fuels. In this way, it is like electricity that can be used to both store energy (batteries) and transport energy (**Figure 9.38**).

The United States already uses over 10 million tonnes of hydrogen each year for industrial purposes, but 95% of this is produced from natural gas. Large volumes of

▲ **Figure 9.37:** A high-density vertical-growth algal bioreactor run by the Swedish energy group Vattenfall at an experimental greenhouse in Senftenberg, eastern Germany. Biofuels have great future potential, but there are still many technological barriers to overcome before they can make a major contribution towards meeting the demand for fuels.

CO_2 are emitted during this process, so although hydrogen has three times the energy content of methane, it does not help to reduce greenhouse gas emissions unless the CO_2 is captured and sequestered at or near the site of production. Hydrogen is also difficult to store, as it is a very light gas and difficult to contain. Distribution is also difficult because hydrogen is a reactive and corrosive gas.

Hydrogen can be produced by electrolysis, a process by which water is split into hydrogen and oxygen, using electricity. Because 70% of electricity in the United States is still produced by fossil fuels, using

hydrogen as a fuel source will not make a large contribution to lowering greenhouse gas emissions. Hydrogen could have a more promising future if renewable sources of energy such as nuclear, solar, hydroelectric, or wind power could be used to generate it rather than fossil fuels. This is an area of active research, with great potential reward for the winners, and there are already a number of companies that claim to have new catalysts and processes that could reduce the cost significantly.

Hydrogen as a Vehicle Fuel   The simplest way to use hydrogen is to burn it as a combustible gas. In this way, it could be used for home heating or to generate electrical power, but the further conversion of hydrogen to electricity is so inefficient that it is not yet economically viable. Hydrogen has more potential as a substitute for gasoline. Hydrogen-fueled engines are 25% more efficient than traditional gasoline-powered engines, but the large volume of gas required needs either very high levels of compression or expensive liquification.

Unless the local generation of hydrogen from clean sources of energy becomes commonplace or unless governments spend trillions of dollars on a national distribution and storage system, there is little prospect of the general population making the switch from gasoline to hydrogen. The future of hydrogen-powered vehicles is made less likely by the growing success of hybrid and electric cars that already have a national electric grid to connect to.

Hydrogen Fuel Cells   **Hydrogen fuel cells** have been used for years in specialized applications such as in space vehicles. Fuel cells combine hydrogen and

▼ **Figure 9.38:** An overview of the hydrogen economy indicating the different primary fuels used, and the means by which hydrogen is generated and distributed as a source of power for the generation of electricity and transport.

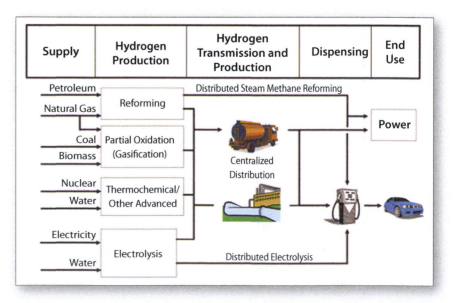

oxygen within a battery cell to produce water and electricity. These efficient and powerful batteries have great potential as a primary source of power that creates no greenhouse gas emissions. However, fuel cells still need fresh hydrogen when depleted, and they would require the same fueling infrastructure as gasoline to be competitive. It is highly unlikely that such an infrastructure will develop within the next 40 years.

> **CHECKPOINT 9.40 ▶** When will the use of hydrogen as a fuel really start to contribute to the reduction of greenhouse gas emissions?

The Hydrogen Fuel Initiative A national hydrogen energy roadmap was launched in the United States in 2002. The aim to replace fossil fuels used in passenger vehicles by 2040 was ambitious but deeply flawed. Just producing the fuel would require 150 million tonnes of hydrogen a year from clean energy sources. This amount of energy is equivalent to the output of 500 mid- to large-sized (500-megawatt) coal-fired power stations, or an area the size of Nevada covered in PV solar cells. And then there is the cost of the necessary distribution infrastructure. There is little doubt that hydrogen has a great future as a fuel, but new technologies are required to address the problems of greenhouse gas emissions, storage, and delivery. End-user technologies are also required that are accessible, affordable, and durable. Hydrogen is not likely to help reach the target of reducing greenhouse gas emissions by 2050.

> **CHECKPOINT 9.41 ▶** Examine some of the problems that may prevent the widespread adoption of hydrogen as a fuel.

9.7 PAUSE FOR... THOUGHT | *Is renewable energy really so green?*

Climate change will be devastating in many parts of the world, but the replacement of fossil fuels with renewable energy is not without some cost. Nuclear power requires the mining of uranium, carries the risk of contamination, and leaves behind the long-term problem of radioactive waste. Large dam projects uproot communities, flood valleys, and disrupt river ecology. PV cells have toxic components that require careful disposal, and solar arrays need vast areas of land. Wind farms can be dangerous to birds and bats, and they disrupt the landscape. Biofuels have resulted in the reallocation of agricultural land to the production of biomass, resulting in higher prices and food shortages. Even in combination, though, these factors do not rival the destructive capacity of fossil fuels. Still, we must recognize that no option is without some cost.

The Politics of Change

The technology of renewable energy is still in its infancy and needs a lot more investment from both governments and the private sector before it reaches maturity. The EU has shown how political and commercial interests can work together to accelerate this process. The EU agreed to meet 20% of energy demand from renewable sources by 2020 and to invest in collaborative research and development. In response, many EU countries started asking energy utilities to get a proportion of energy from renewable sources. This created a fiscal, legislative, and competitive environment that favors the development of new energy.

Renewable Portfolio Standards The United Kingdom and 17 other EU countries, along with Brazil, China, Indonesia, Israel, Nicaragua, Norway, South Korea, Sri Lanka, Switzerland, and Turkey, have all adopted mandatory renewable energy targets, called **Renewable Portfolio Standards (RPSs)**. An RPS requires a certain percentage of a utility's power to come from renewable sources by a given date. Failure to comply with an RPS normally leads to some significant sanction against the utility company. RPSs are powerful tools to encourage investment in renewable energy.

> **CHECKPOINT 9.42 ▶** In your own words, explain how RPSs aim to encourage the use of alternative technologies.

Renewable Energy Standards in the United States

In the United States, 27 states and the District of Columbia (DC) have adopted some kind of RPS. Texas, for example, is expected to avoid 3.3 million tonnes of CO_2 emissions annually with an RPS that required 2 gigawatts of new renewable generation by 2009. The fact that these standards apply only within state boundaries makes them much less effective than if a national standard were adopted. The United States will catch up with the EU over the next 5 to 10 years, but by that time, the Europeans will have extended their lead in new energy technology.

Renewable Energy Standards in China

The government in China passed a new renewable energy law in 2006 that aims to have at least 15% of energy generated from renewable sources by 2020. Using low-interest loans and tax incentives to encourage investment, the government also guarantees prices and requires the power grid utilities to purchase all the electricity generated by approved renewable energy facilities. The cost of purchasing this power is spread across all customers on the grid. In 2011, China announced plans to enhance wind capacity by 90 gigawatts to

100 gigawatts, hydroelectric power by 250 gigawatts to 260 gigawatts, and solar power by 10 gigawatts by 2015. These plans should challenge the United States to engage seriously with renewable energy.

A New Energy Infrastructure

One of the greatest assets of alternative energy is its flexibility and adaptability to simultaneously meet both large-scale and small-scale demand. When individual consumers—however big or small—have the ability to generate some of their own electrical power and even have spare capacity to give back to the grid, the world becomes one large distributed power network that is no longer focused on big energy utilities. Standing in the way of this vision is an outdated and almost obsolete system of power grids that emerged as poorly interconnected, unreliable networks founded on the assumption that power would continue to be in the hands of a small number of powerful utilities.

Existing power grids were simply not designed to respond to the changes that have taken place, and trillions of dollars of investment will be needed before these grids are able to make efficient use of the new distributed sources of alternative energy. The transition from old, poorly connected networks to modern, high-technology, intelligent networks will take time. The new networks must be able to communicate across national and international borders and must be able to balance the changing demand for power on a daily basis by using all available sources of energy. They will be more robust and less likely to suffer from prolonged power blackouts, and they will even be able to communicate directly with home appliances through intelligent networks that can manage the demand for power in real-time. Such a grid will eliminate the need for expensive backup (usually gas-powered) facilities that are always on standby and emit CO_2 even when they make no contribution to the grid.

Without this kind of modern network, renewable power will remain too expensive for too long, and it will not be able to compete on a level playing field with the large utilities. The wind does not always blow, and the Sun does not always shine, but at the moment, wind and solar facilities are still required to pay for access to the grid even when they are not using it. This adds to the price of wind and solar energy, making it less competitive. Countries in the developing world where national energy networks are still under development have an opportunity to take the lead in the use of alternative energy by investing directly in these new technologies. As governments look to extend national power grids over the next decade, they should invest in the development of new international networks that are ready for the future of energy and not just base them on EU and U.S. models that are as fossilized as the fuels they serve.

CHECKPOINT 9.43 ▶ What major changes need to be made to the U.S. electricity grid before alternative energy can reach its full potential?

Are We Subsidizing Greenhouse Gas Emissions?

A 2005 report by the United Nations Environment Program estimated that worldwide energy subsidies exceeded $300 billion, or as much as 0.7% of global GDP. Russia ($40 billion) and Iran ($37 billion) offered the largest subsidies for natural gas, and China, Saudi Arabia, India, Indonesia, Ukraine, and Egypt all gave fossil energy subsidies in excess of $10 billion. The Stern Report published in the United Kingdom in 2006 estimated that global subsidies for clean energy were just over $26 billion, but $16 billion was for nuclear energy, and only $10 billion was for renewable energy. In 2007, the EIA estimated that the U.S. government subsidized the energy industry by at least $16.6 billion. Some of the largest subsidies went to the coal industry for the development of clean coal technologies, but renewable energy was also well supported, especially considering the level of subsidy per kilowatt-hour of energy generated.

It is unclear how much the EU subsidizes energy, as there is no comprehensive official record. A 2004 report by the European Environmental Agency estimated that the 15 members of the EU at the time (Austria, Belgium, Denmark, Finland, France, Germany, Greece, Ireland, Italy, Luxembourg, the Netherlands, Portugal, Spain, Sweden, and the United Kingdom) spent over €29 billion, nearly 75% of it on fossil fuels (45% to support coal and 30% for oil and gas), 7% on nuclear power, and 18% on renewable energy. As in the United States, the amount of money spent per kilowatt-hour of energy generated was much higher for renewable energy than for fossil fuels, but the overall level of support for traditional fuels was still much higher.

These data make it abundantly clear that most governments still play a very active in role in subsidizing fossil fuel energy. Directly or indirectly, these subsidies decrease the price of fossil fuel energy in the local market. Not all subsidies are necessarily bad. Many developing countries have to subsidize the cost of fuel when the general population is not wealthy enough to pay the full market price. But if energy from fossil fuels is supported at the expense of renewable energy, the results will be higher greenhouse gas emissions, less competitive renewable energy, and failure to meet greenhouse gas emissions targets.

CHECKPOINT 9.44 ▶ What political justification is there for the large financial subsidies that governments give the fossil fuel industry?

The economic slump that began in 2008 highlights a significant weakness in the completely free market approach to energy. With oil prices close to $150 a barrel, renewable energy was economically viable and able to attract new investment. The collapse of the price of oil in 2008 wiped this slate clean, undercut the prospect of future investment, and placed the funds already invested at very high risk. Renewable energy innovation is essential if the world is to limit the long-term impact of climate change, but it cannot flourish in an environment of financial uncertainty. Long-term investment requires long-term planning and consistent policies if the private sector is to lead the world out of the climate crisis.

Governments need to establish investment-friendly tax policies that encourage new startup companies and give innovative and entrepreneurial companies a chance to establish a firm foothold in the energy market, without fear that some sudden change in the financial markets is going to wipe their financial slate clean every few years. Subsidies for fossil fuels need to be phased out as soon as possible. A guaranteed national price floor for oil, of say $100 a barrel, would encourage widespread energy conservation and efficiency in the transportation sector. This became evident in the United States when gas prices hit $4 per gallon and people rushed to trade their SUVs for more economical family sedans. In the EU, where government taxes on fuel are much higher than in the United States, the flight to vehicle efficiency has been in progress for some time.

Just a small rise in the price of electricity could make renewable energy much more competitive in the power sector, especially if the energy utilities are also required to buy a proportion of their energy from approved renewable sources. As long as traditional energy prices remain artificially low, whether due to short-term economic recession, conflict among OPEC producers, or government subsidy, the vision, innovation, and creativity of companies willing to risk capital on the growth of a green economy will remain stagnant or (worse) will be invested overseas.

The EU understood this dynamic from the earliest days of the climate change crisis and has acquired a commanding lead in the industry. Through a combination of tax incentives, price guarantees, and carbon taxes, combined with local innovation and the acquisition of abandoned know-how from the United States, Europe is well on its way toward the goal of meeting 20% of its energy needs from renewable sources by 2020.

Energy is not our enemy, but an essential part of the solution to global poverty and the challenges of population growth. The most recent NASA images of Earth at night highlight the stark difference between the amount of energy used and wasted in the developed and developing worlds (**Figure 9.39**).

CHECKPOINT 9.45 ▶ Is it government's responsibility to facilitate the growth of the alternative energy industry? Explain.

▼ Figure 9.39: Earth at night. This image was created from a series of NASA spacecraft images taken over a 22-day period. While some light is natural (from events such as wildfires, reflected moonlight, and aurorae), most of the visible light comes from cities- a dramatic reminder of our widespread and growing use of energy resources.

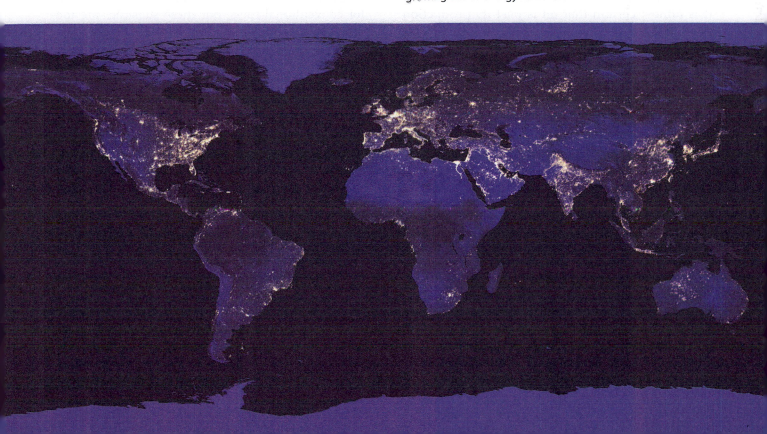

Summary

- The world is in the middle of an energy crisis, and there is no easy fix. The demand for energy is rising exponentially as population growth and economic development drive the demand for cheap energy to meet basic needs.

- The demand for energy is greatest in the developing world, where many communities lack sufficient energy to meet basic needs. In these countries, limited access to potable water, refrigeration, and smoke-free cooking leads to high levels of mortality in both adults and children. Many of these countries have turned to coal as a cheap and plentiful source of energy to feed economic growth, and the result is an increase in greenhouse gas emissions that heighten the risks of damaging climate change.

- Vast reserves of natural gas have the potential to replace some of this coal as a power source for the generation of electricity. Natural gas produces less CO_2 than coal when burned, but until recently was more expensive than coal to extract and distribute. Price barriers prevented its widespread use for the generation of electricity, as consumers were unwilling to pay for the full environmental cost of coal, and lobbyists convinced governments not to reduce energy subsidies. With the rapid exploitation of shale-gas reserves it is now possible that natural gas can replace coal as the major source of energy for the generation of electrical power and thus make a major contribution towards the reduction of greenhouse gas emissions.

- It is almost certain that oil is past its production peak, and yet the demand for fuel is rising faster than ever before. Unless production from oil sands and oil-rich shales increases rapidly, the price of oil will rise to levels of $140 a barrel and higher in the near future due to increasing demand and decreasing availability. This will make it difficult for countries such as India and China to keep subsidizing the cost of gasoline for consumers at a time when a transportation revolution is taking place and vehicle ownership is rising quickly. Alternative fuels, high-efficiency engines, and more efficient public transportation are all answers to this problem. There are signs that the necessary technologies can arrive in time, but high vehicle emission and mileage standards are essential to encourage manufacturers and consumers to adopt these technologies.

- Although many people mistrust nuclear power, it must be part of the solution. It is almost too late to look to nuclear power to help meet the 2050 greenhouse gas emissions targets, because the time required for planning and approval is so long. Waste disposal and security are critical problems, and acceptable solutions must be found, but energy from nuclear power has been proven safe, and it is ready to help fill the immediate gap in demand for electricity.

- There is great enthusiasm for alternative clean energy among environmentalists, and the future of energy is undoubtedly green, but despite the impressive growth in clean energy production, it is highly unlikely that new forms of alternative energy can make more than a modest contribution toward meeting global energy demand between 2050 and 2100.

- Alternative energy sources have great potential and have proven their ability to compete with fossil fuels on a level playing field. The problem is that the field is far from level. Government subsidies for renewable energy are more per kilowatt-hour than for conventional fuels, but these subsidies are too little and too late.

- The nature of the energy market will change fundamentally when consumers are able to contribute to the grid as well as buy from it. This is also a long way from happening, and a massive investment will be required, so there is little incentive for the utilities to meet this demand unless governments require them to produce a certain amount of energy from clean energy sources.

Why Should We Care?

Over the next 50 to 100 years, the world will see exceptional growth in the capacity of wind, solar, wave, and tidal energy and the ability of renewable energy sources to displace coal as the primary fuel for electrical power generation. Parallel growth in the biofuels industry can also meet the growing shortfall in oil. There is little doubt that clean energy will revolutionize the way that power is generated and consumed, but it is hard to avoid the disappointing conclusion that coal, gas, and oil will remain the dominant sources of primary energy well into the 21st century—and certainly well past the point where the growth of greenhouse gas emissions must be halted and then reversed. The continued emission of all greenhouse gases, and especially CO_2, is a continuing crisis that demands a political solution.

Chapter 10 will discuss this emissions crisis and consider how greenhouse gas emissions targets may still be met. No single technology has any prospect of success on its own, but success is attainable.

Looking Ahead . . .

The expansion of the global population is placing ever-increasing demands on energy to feed economic growth. The prospects of reducing greenhouse gas emissions to a level that will avoid the damaging impact of climate change seem increasingly remote unless we take rapid action very soon. Chapter 10 considers some of the options for such action and the political and social contexts that will determine their success or failure.

Critical Thinking Questions

1. Say that President Reagan had not rejected President Carter's energy policy that aimed to achieve energy independence. Say that President Reagan had instead continued to support investment in renewable energy after the return of cheap oil on the world market. Would the United States still have achieved the outstanding economic growth that occurred during the 1980s and helped bring a rapid end to the Cold War? Explain.

2. You are a representative for the nuclear power industry, sitting at a congressional hearing on clean energy. You believe that nuclear power is a "green glow" on the energy horizon. How do you persuade Congress that the advantages associated with nuclear power outweigh the problems?

3. You are sitting around the table with the UK Secretary of State for Energy and Climate Change. As a representative of the British coal industry, you have always believed it was a mistake for Margaret Thatcher to decimate the UK coal industry. You now believe that clean coal technologies offer a path to revitalize a failing industry and create thousands of new mining jobs in areas still struggling to recover from the recent economic downturn. What do you say to the Secretary of State to persuade him that clean coal is not an oxymoron but a concept to be embraced by necessity, at least until alternative energy sources become more viable alternatives?

4. At a UNFCCC hearing, you are a representative for a small biofuel energy company that is struggling to compete with the petroleum giants. You want to encourage investment in your company, but people are reluctant because you lost a lot of money the last time the price of oil fell to under $100 a barrel. What do you say to the UNFCCC officials to encourage them to adopt your proposal for Kyoto Protocol signatory countries to set a minimum price for oil at $100 a barrel?

5. As president of a power distribution company, you understand that a new modern distributed electricity grid is essential for the growth and efficiency of wind and solar energy. You also understand that it will cost your company hundreds of millions of dollars to make the necessary changes. Most of your senior board members are hostile to any such investment. They prefer to upgrade the existing grid in cooperation with the large energy utilities. How would you persuade the board that it is in their financial interests to invest in the development of a new energy infrastructure in your state?

6. You are the UK Secretary of State for Energy and Climate Change, and you are in a meeting with junior colleagues to formulate a new preelection public campaign to save energy. There is strong disagreement between two members of your team. One wants to lead with a campaign based on energy conservation, and the other wants to emphasize energy efficiency. You understand that both concepts are important, but you need to decide which idea is best to lead the campaign. What do you decide?

7. You are a celebrated environmental campaigner in Sweden and a strong advocate for wind energy. Your government has invited a senior representative of the French nuclear energy industry to testify at a hearing on the future of Swedish energy, and she argues convincingly that using nuclear power is the best way for nations with few natural resources to meet rising energy demand. Spend five minutes responding to this information.

8. As you sit over lunch with a delegate from China at an international symposium on climate change, the conversation turns to clean energy. Your new friend states categorically that there is no realistic prospect of China achieving economic growth without the use of coal as a major source of energy over the next 50 years. As he munches on his lunch, you get your chance to respond. You have three minutes before the bell rings to mark the start of the next conference session. What do you say?

9. You believe that the press puts too much emphasis on CO_2 emissions alone as the main source of global warming. You know that methane is also partially responsible, and you think that the public needs to be more aware of this. Devise a short media campaign to highlight the key points of your argument.

10. Some say that the willingness of one generation to sacrifice wealth, health, and education for a future generation—a generation that is already expected to be richer, healthier, and better educated—can be justified only if the continued action of the first generation will so fundamentally undermine the economic prospects of the next that the very future of humanity may be placed at risk. Explain whether you think this is true.

11. As a representative of The Pew Center on Global Climate Change, you are asked to attend a meeting organized by the Copenhagen Consensus. The keynote speaker argues that the immediate needs of the UN Millennium Goals must take priority over the more distant need to address global climate change. He argues that any reasonable discount rate shows that the cost of action today is much greater than the cost of action in the future. At a press conference afterward you are asked for your opinion. Explain your position.

12. You work for the governor of a large state and are in charge of energy policy. The governor wants you to prepare legislation that will give your state substantial control over the development, testing, and installation of a new modern power distribution grid. You consider this to be a vitally important strategic issue for your state, and there are no second chances: You must get it right first time. You meet with representatives of a large utility company, and they tell you that the future of clean energy must rest in the hands of private enterprise and not with government. They want legislative interference to be minimal. What will the bill you finally write contain?

13. As a very wealthy leader in the Middle East, you face the development of energy independence in the West with some trepidation. You know you will be able to produce oil at current levels for 20 to 30 more years. However, you are worried that the price of oil will start to fall just as you sign contracts for some major development projects that are designed to bring new life to the desert. You decide to summon your OPEC colleagues and present a plan that will maintain the flow of revenue for some time to come. What is your plan?

14. Mining tar sands and oil shales is highly destructive to the environment. As public relations officer for a large mining company, it is your job to argue that despite this, tar sands and oil shales are a cleaner alternative to coal. What do you say?

15. Recent disagreement over tariffs and fees between Russia and Ukraine led to significant short-term disruption of gas supply to much of central and Eastern Europe. You are minister for energy in one of those countries, which also happens to be sitting on large coal reserves. The minister for national security is demanding that the nation switch energy production from gas to coal power as a more reliable source of independent energy. How do you balance national security needs against global climate change?

Key Terms

Please make sure you are familiar with the key terms introduced and highlighted in this chapter. A full glossary is available at the end of the book.

Anaerobic fermentation, p. 299

Anthracite, p. 292

Biodiesel, p. 313

Bioethanol, p. 313

Biogas, p. 313

Biomass, p. 312

Bioreactor, p. 313

Carbon capture and storage (CCS), p. 295

Cellulosic bioethanol, p. 313

Clean coal technology (CCT), p. 294

Combined-cycle gasification system, p. 312

Concentrated solar power (CSP), p. 310

Corporate Average Fuel Economy (CAFE) standard, p. 303

Energy Information Administration (EIA), p. 312

Energy poverty, p. 290

Environmental Protection Agency (EPA), p. 294

Fast breeder reactor, p. 305

Fischer–Tropsch reaction, p. 297

Fracking, p. 299

FutureGen Alliance, p. 295

Geothermal power, p. 308

Geothermal gradient, p. 308

Hydrogen, p. 313

Hydrogen fuel cell, p. 314

Integrated Gasification Combined Cycle (IGCC) system, p. 295

International Energy Agency (IEA), p. 294

Joule, p. 290

Oil crisis, p. 302

Organisation for Economic Co-operation and Development (OECD), p. 290

Organization of the Petroleum Exporting Countries (OPEC), p. 300

Particulate matter, p. 294

Photovoltaic (PV) solar cells, p. 310

Renewable energy, p. 307

Renewable Portfolio Standard (RPS), p. 315

Solar updraft tower, p. 311

Solid biomass, p. 311

Total primary energy supply (TPES), p. 307

Uranium oxide, p. 305

Vehicle emission standards, p. 302

Workers install solar panels on a house. To mitigate the impact of climate change, we need immediate action at national and international levels to limit the emission of greenhouse gases, but we must also make personal choices about our own lifestyles. Simple changes that conserve and make better use of the energy we consume, such as better insulation, car sharing, the installation of solar panels, and the use of energy efficient appliances, may make a small individual contribution, but collectively they have a significant impact on the level of greenhouse gas emissions.

Turning Knowledge into Action

Introduction

This chapter investigates the socioeconomic factors that drive climate change and examines some potential solutions. The problem of climate change has its roots in rapid population growth and the rising demand for energy, space, and food. The combustion of fossil fuels, unsustainable deforestation, and inefficient agriculture have increased the concentration of greenhouse gases in the atmosphere to a level not experienced for millions of years. The risk of climate change is particularly high in the developing world, where many critical and endemic problems are related to energy poverty and hunger, but access to cheap energy, food, and economic development takes priority over climate change. What choice is there when millions of children are dying from preventable disease and hunger today? These countries look to the rest of the world for a solution that will provide them with the resources they need to grow today and preserve a sustainable future for their children. *If nothing is done to break the links between population growth, poverty, energy use, and greenhouse gas emissions, the world will soon experience changes in climate that present a major threat to society and the environment.*

Learning Outcomes | *When you finish this chapter you should be able to:*

- Understand why the demand for energy will continue to grow
- Consider who is responsible for the current energy crisis
- Discuss who should take responsibility for solving the energy crisis
- Identify nine target areas for international action
- Explain the concept of carbon wedges
- Discuss geoengineering as a potential solution to the climate crisis
- Demonstrate that it is still possible to meet greenhouse gas emissions targets

The Deep Roots of Climate Change

The population of the world reached 7 billion in 2011 and is growing at over 1% a year. By 2050, there may be 9 billion people on Earth, and they will all want cheap and plentiful energy to power their lives (**Figure 10.1**). The highest growth rates will be in countries such as India, China, Africa, Brazil, and Indonesia, but immigration will also maintain population growth in the United States, Canada, and Australia (**Figure 10.2**). Only the European Union, Russia, and Japan are likely to experience decreases in their birth rates with static or declining population. The world in 2050 will be hot and crowded, and the demand for energy will be more than double what it is today. This is clearly a problem.

Population Growth and Poverty as Drivers of Climate Change

The population of many developing countries is growing so rapidly that billions of people do not have enough energy to sustain even the most basic standard of modern living, and are considered to be living in energy poverty. Energy is needed for personal necessities such as potable water, light, and sanitation and to combat major social problems such as infant mortality, disease, unemployment, and education.

Only nations with strong economies can afford the infrastructure needed to bring energy to the people who need it most. Nations struggling to meet basic human needs are forced to pursue economic growth by all means available, and too often the energy used to drive this growth comes from biomass and cheap fossil fuels that accelerate the rate at which heat-trapping greenhouse gases are added to the atmosphere. Global climate change is likely to accelerate in the near future as this demand for cheap power increases throughout the century. Our challenge is to find mechanisms that can tackle all these issues at the same time.

The increased use of energy will eventually foster deep changes in social structure that can make society more sustainable and reduce the intensity of energy consumption—but not before millions of tonnes of carbon are added to the atmosphere. Social change is already taking place in China and India, where the rate of population growth is falling as the standard of living rises and large families become more of a liability than an asset. But it is happening too slowly to stabilize global emissions by 2050 at the level proposed by the IPCC. There is still time to meet the challenge of energy poverty before damaging climate change becomes inevitable, but as the search for an international solution grinds to a halt (Chapter 8), who will provide the leadership needed to reach a solution?

CHECKPOINT 10.1 ▶ Explain why population growth and poverty are root causes of the climate change problem.

Who Will Take the Lead?

Should the developed countries take the lead in cutting fossil fuel emissions? Under the Berlin Mandate, economically developed countries accepted responsibility for today's historically high levels of CO_2 in the atmosphere and acknowledged the urgent need to reduce global greenhouse gas emissions (see Chapter 8). Per capita greenhouse gas emissions in these countries are certainly much higher than the global average, and these countries have much more adaptive capacity to manage the impact of climate change.

Should developing nations share the responsibility for managing the impact of climate change? The Berlin Mandate made it clear that developing nations should accept common but differentiated responsibility for the

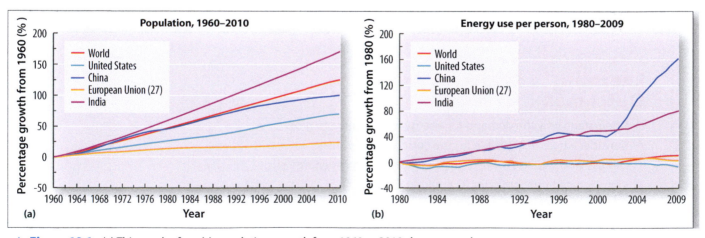

▲ **Figure 10.1:** (a) This graph of world population growth from 1960 to 2010 shows growth in the United States and Asia (mainly China and India); data are in millions of people. Note the decline in the rate of population increase in China that coincides with strong economic growth. (b) A graph of the growth in the energy use per person from 1980 to 2009, comparing different regions of the world. Can you explain the sudden rise in the rate of energy use per person in China after 2001?

▲ **Figure 10.2:** (a) On this cartogram map the size of each country is drawn in proportion to its population. (b) Population growth and urbanization in countries such as China place a greater demand on energy resources.

control of emissions. In other words, developed nations were never meant to "go it alone," and the reluctance of China and India to commit to a verifiable reduction in emissions intensity makes progress difficult. Their reluctance is understandable, as they did not create this problem, but if they fail to become part of the solution, they will not avoid the damaging impact of climate change (**Figure 10.3**).

So what is the way forward? Developed nations could act unilaterally and cut emissions without any commitment from countries like China, India, Brazil, Mexico, or Malaysia, but at a time of intense economic competition and rising concern over national security, this is a risk that few nations are likely to accept. It is increasingly hard to balance the rising economic and strategic concerns of developed nations with the legitimate aspirations of developing nations that demand the right to pursue economic growth (**Figure 10.4**).

CHECKPOINT 10.2 ▶ Do you think the developing countries should make any economic sacrifices to stop greenhouse gas emissions if the United States is unwilling to reciprocate? Explain.

Do Nations Have a Right to Pursue Economic Growth?

Developing nations claim that their right to pursue economic growth is an issue of social justice and equity. They are prepared to act sustainably but want the developed nations to accept responsibility and take the lead. There is substance to this argument, but any short-term economic advantage they gain by delaying action to reduce emissions today will be surpassed by the additional costs of mitigating climate change tomorrow. Action to reduce global greenhouse gas emissions has become a matter of national self-interest, and immediate action to reduce the greenhouse gas intensity of economic growth in the developing world should be a priority.

The developed nations have an obligation to assist and support the rest of the world. Most wealthy nations have outsourced at least some of their greenhouse gas–emitting industries to countries in the developing world. Who is accountable for all the CO_2 these factories emit? Is it the country that profits from the manufacturing or the country that uses the product? The nations of the world may be in competition for markets and resources, but their ultimate success is interwoven in the fabric of

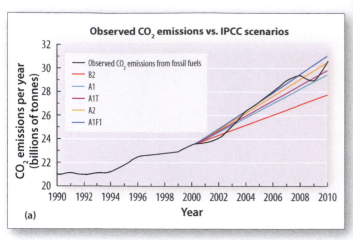

(a)

(b)

▲ **Figure 10.3:** (a) Carbon dioxide emissions from fossil fuels (excluding forestry and agriculture) from 1990 to 2010. Note that the dip in production during the economic recession that started in 2008 was followed by a rapid recovery. Compare the actual level of emissions to a number of the listed IPCC emissions scenarios. The B2 scenario is the best case, and the A1F1 scenario is the worst case that imagines the continued and widespread use of fossil fuels. Which scenario best matches the actual level of production? (b) Coal is cheap and can be found close to the surface, so that it can be extracted with a bucket excavator, but we can no longer sustain the level of CO_2 emissions this fuel produces without causing permanent changes to global climate.

international trade and commerce. In this context, helping your neighbor is just another way of helping yourself.

CHECKPOINT 10.3 ▶ Who should be responsible for carbon dioxide emissions produced by the outsourcing of manufacturing industry to the developing world? Is it the producer, the consumer of the good produced, or both? Explain.

The only viable solution is for developed nations to cut greenhouse gas emissions significantly and for developing nations to reduce **greenhouse gas intensity**—the rate at which greenhouse gases are emitted for each increment of economic growth. Models project that the level of greenhouse gases in the atmosphere will increase

by as much as 84% above preindustrial levels by 2050 if development continues at the current rate and nothing is done to limit emissions (**Figure 10.5**). This will almost certainly result in rapid and damaging climate change that is most severe in the least developed countries. It is clear that something has to be done.

The developing nations must weigh their fundamental right to economic growth against the consequence of economic failure and the damaging climate change that will follow.

CHECKPOINT 10.4 ▶ When considering actions to limit greenhouse gas emissions, explain why "helping your neighbor is just another way of helping yourself."

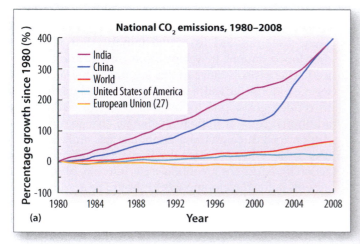

(a)

(b)

▲ **Figure 10.4:** (a) The growth of CO_2 emissions in the world, the United States, the EU, China, and India. The low growth rate of emissions for the United States looks good on this graph, but needs to be considered against a historic baseline of very high emissions. (b) The Beijing central business district skyline at night leaves a strong impression of just how much energy is used around the world as cities continue to grow in size.

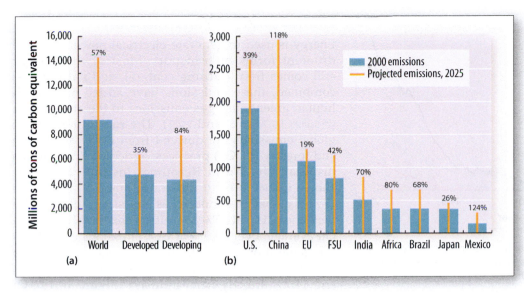

◀ **Figure 10.5:** Projected greenhouse gas emissions in 2025 and the percentage increase in emissions above levels in the year 2000 in US tons. Note that these figures are for total emissions, not just carbon dioxide, and that 1 tons (US) is equivalent to 1.023 tonnes (metric tons). (a) The developed and developing countries. Note the 84% increase in the level of emissions from the developing countries compared to 35% in the developed countries. (b) The growth in emissions of some key nations. Compare the projections for China, the EU, the United States, and India.

10.1 PAUSE FOR... THOUGHT | *Such a sensitive issue.*

The extent to which some countries will be affected by climate change over the next century depends on the sensitivity of global climate to greenhouse gas emissions. Low climate sensitivity may deliver neutral or even small net positive changes to many of the developed nations at high latitudes. Could this become a barrier to political action if the countries least affected by change come to believe it is not in their long-term strategic interest to help the countries most affected by change? Many of the developing countries will soon have enough economic and military influence to rival the traditional powers. History suggests that major powers do not yield their influence willingly.

Energy Demand and Greenhouse Gas Emissions

The steady emissions of anthropogenic greenhouse gases began when the first humans started to cut down trees and clear land for agriculture by burning the undergrowth. The rate of emissions accelerated rapidly in the 1950s, as cheap coal and oil were used to fuel postwar industrial development in the United States, the Soviet Union, Japan, and Europe. This trend intensified toward the end of the 20th century, following exceptional economic growth in China,

India, and other advanced developing countries (Figure 10.4). The growth in the level of emissions in developing countries is remarkable, with an expected 84% increase relative to 2000 levels by 2025 (Figure 10.5). Despite these advances in energy production and consumption, the per capita difference in energy consumption between developed and developing nations is very large, with the average person in the United States consuming as much as five times as much energy to sustain his or her standard of living as the average person in a developing nation (**Figure 10.6**).

▼ **Figure 10.6:** An illustration of per capita greenhouse gas emissions compared to a nation's overall percentage contribution to global emissions. Data are in metric tonnes (1,000 kg) of carbon dioxide. Emissions data may be presented as both tonnes of carbon dioxide (CO_2) or tonnes of the element carbon (C_E). To convert: Tonnes of CO_2 = Tonnes of C_E × 3.67. Compare the data from the United States with the UK data. The current world total of carbon dioxide emissions is ca. 33.5 gigatonnes (9.1 gigatonnes of carbon). Why is there such a large difference in the projected per capita emissions from these two developed nations?

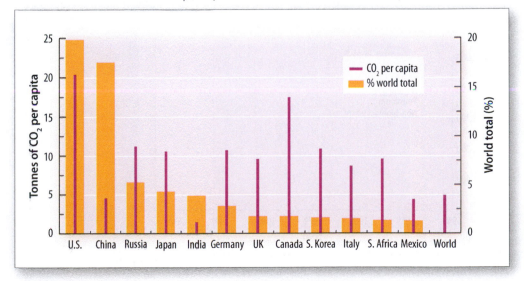

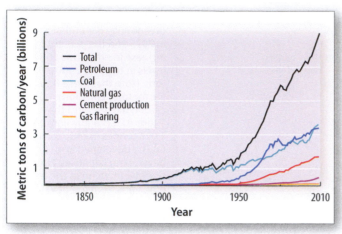

▲ **Figure 10.7:** Global carbon emissions from fossil fuels, broken down by sector. Why was there a sudden increase in the use of coal around the year 2000?

▲ **Figure 10.8:** A significant proportion of the demand for energy in developing countries is likely to be met using cheap and abundant local coal for much of the 21st century. This increases the risk of damaging climate change due to the continued emission of greenhouse gases.

Global Energy Consumption

Energy is used to generate electrical power and fuel different forms of transport. Most of this energy still comes from burning coal, oil, and natural gas; combined, these emissions have added 65% of all human greenhouse gas emissions to the atmosphere (**Figures 10.7, 10.8,** and **10.9**). The remaining 35% of global emissions are derived from land use change and forestry (LUCF) practices, agriculture, and waste disposal. The World Resources Institute flowchart shown in **Figure 10.10** illustrates that CO_2 makes up 77% of global greenhouse gas emissions, followed by methane (CH_4) at 15% and nitrous oxide (N_2O) at 7%. On average, over 24.9% of CO_2 emissions come from the generation of electricity and heat, 14.3% come from the use of fossil fuels used to power transport, and 14.7% directly from industry (Figure 10.10). Of the remainder, over 29% of CO_2 comes from a combination of waste, agriculture, and land use change.

CHECKPOINT 10.5 ▶ Discuss the advantages and disadvantages of the major sources of energy we use for the production of electrical power.

Emissions from the United States

The United States is a good example of a highly developed nation. With just 5% of the world's population, it supports 23% of the world's economy and produces 18% of global greenhouse gas emissions. Per capita emissions in the United States are the largest in the world, exceeding 20 tonnes of CO_2 per person. The United States is the largest global consumer of oil and the second-largest consumer of coal, after China. Over 27.2% of CO_2, twice the global

▼ **Figure 10.9:** (a) Trends in emissions from the electric power sector, 1949–2009. (b) The demand for power transmission is set to increase across the world, but unlike these high power electric transmission towers and lines from the Glen Canyon dam power plant, many other power grids are too old to make efficient use of alternative energy sources.

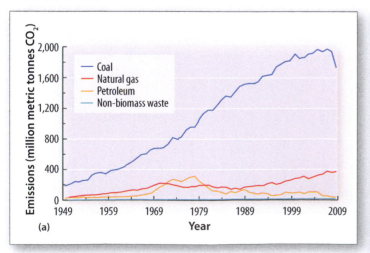

(a)

(b)

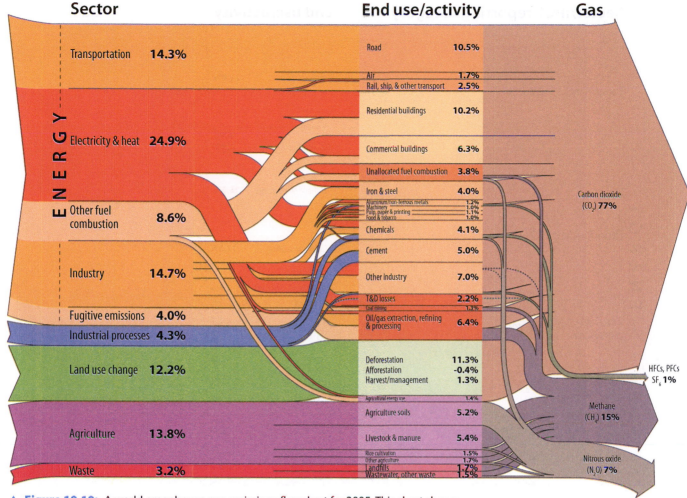

Sector | **End use/activity** | **Gas**

Sector		End use/activity		Gas	
ENERGY					
Transportation	14.3%	Road	10.5%		
		Air	1.7%		
		Rail, ship, & other transport	2.5%		
Electricity & heat	24.9%	Residential buildings	10.2%		
		Commercial buildings	6.3%		
		Unallocated fuel combustion	3.8%		
Other fuel combustion	8.6%	Iron & steel	4.0%	Carbon dioxide (CO_2) 77%	
		Aluminum/non-ferrous metals	1.2%		
		Machinery	1.0%		
		Pulp, paper & printing	1.1%		
		Food & tobacco	1.0%		
		Chemicals	4.1%		
Industry	14.7%	Cement	5.0%		
		Other industry	7.0%		
		T&D losses	2.2%		
Fugitive emissions	4.0%	Coal mining	1.3%		
Industrial processes	4.3%	Oil/gas extraction, refining & processing	6.4%		
Land use change	12.2%	Deforestation	11.3%	HFCs, PFCs SF$_6$ 1%	
		Afforestation	-0.4%		
		Harvest/management	1.3%		
		Agricultural energy use	1.4%		
		Agriculture soils	5.2%	Methane (CH_4) 15%	
Agriculture	13.8%	Livestock & manure	5.4%		
		Rice cultivation	1.5%		
		Other agriculture	1.7%	Nitrous oxide (N_2O) 7%	
Waste	3.2%	Landfills	1.7%		
		Wastewater, other waste	1.5%		

▲ **Figure 10.10:** A world greenhouse gas emissions flowchart for 2005. This chart shows the sources and activities across the global economy that produce greenhouse gas emissions. Energy use is responsible for the majority of greenhouse gases. Most activities produce greenhouse gases both directly, through on-site and transport use of fossil fuels, and indirectly, from heat and electricity that comes "from the grid."

average, comes from the use of fossil fuels to power transport, 32.4% from the generation of electricity and heat, and 12.4% from industry (**Figure 10.11**). The remaining 28% of emissions come from a combination of agriculture, land use, fugitive emissions (leaks), and waste.

Emissions from China

As an example of a rapidly developing nation, China has 20% of the world's population; it produces over 9% of the world's economy and 17% of global greenhouse gas emissions. Per capita emissions are equivalent to just 5.5 tonnes of CO_2. China is the world's largest producer and consumer of coal, accounting for almost half of the world's consumption. It is the second-largest importer of oil, after the United States.

Emissions from Bangladesh

As an example of a developing nation, Bangladesh has only 2% of the world's population but the highest population density on Earth (1,142.29 per square kilometer/441 square mile). Bangladesh's economy produces just 0.2% of global GDP, and the country is responsible for 0.15% of greenhouse gas emissions. The per capita emissions rate is just 0.36 tonnes, and yet Bangladesh may suffer first and greatest from the impact of climate change.

Facing an Uncertain Future

Population, poverty, energy and greenhouse gas emissions are closely linked. While the developed and strongly developing nations compete for strategic world dominance, the poorest nations strive for economic growth to overcome poverty and disease. Energy poverty is a major problem for many nations, and economic growth is a fully justified priority. In parallel with economic growth, however, the global level of greenhouse emissions will continue to rise. The use of fossil fuels to stimulate global economic development increases the probability of devastating climate change, and the poorest nations still lack the

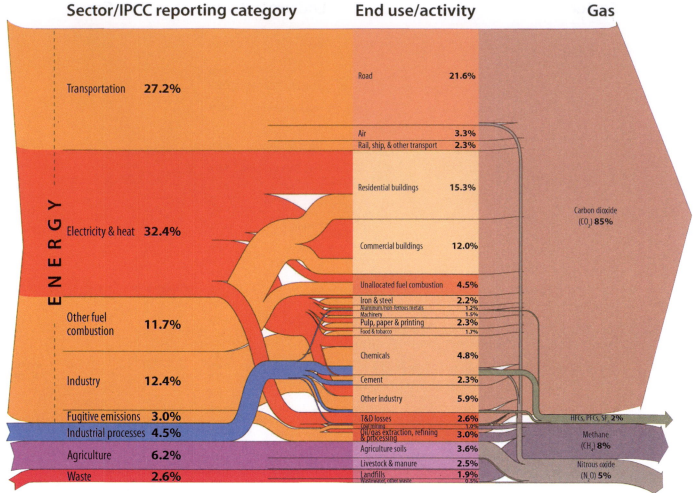

▲ **Figure 10.11:** A U.S. greenhouse gas emissions flowchart for 2003. This flow chart shows the sources and activities across the U.S. economy that produce greenhouse gas emissions. Energy use is by far responsible for the majority of greenhouse gases. Most activities produce greenhouse gases both directly, through on-site and transport use of fossil fuels, and indirectly, from heat and electricity that comes "from the grid."

resources to adapt to or mitigate its worst impacts. They have very good reasons to be worried about the future and resentful about the past. It is possible that the value of economic growth will offset the cost of climate change, but the prospects are not good.

Why Should We Worry About Emissions?

Human activity released around 10 gigatonnes of carbon (36.7 gigatonnes of CO_2) into the atmosphere in 2012, a figure expected to exceed 14 gigatonnes by 2050 (1 gigatonne = 1 billion metric tonnes = 1.1 tons). This is a major international and intergenerational problem. It is *international* because of the need to reduce emissions while maintaining social equity, and it is *intergenerational* because emissions added to the atmosphere today will affect global climate and ocean acidity for centuries to come.

The problem is compounded by the fact that the rate of change is not linear, and the threat of sudden and irreversible change in climate increases each year that greenhouse gas emissions continue unabated.

Can Emissions Decrease Without International Action?

There is some reason to believe that the world will eventually transition to a low-emissions economy without the need for an international agreement. By 2050, the tangible impacts of climate change will be more evident, and the price of oil and gas will be higher due to an increase in demand for diminishing resources. Renewable energy will become more competitive, and public opinion will demand change. By 2100, it is likely that the rate of population growth in China, India, and other developing economies will have started to level or even decrease, as their economies continue to grow. For the first time since the Industrial Revolution, the global level of emissions may start to fall.

But it will be too late. All climate models project that we have less than 50 years to achieve the goal of reducing greenhouse gas emissions in order to limit the damaging impact of climate change. Earth's climate is so sensitive to greenhouse gases that by 2050, irreversible changes will be locked into the climate system. Kyoto was a qualified success for the countries that participated, but it failed to meet its overall objective and emissions are still rising, and we are running out of time. We have the knowledge, but we are not turning it into action quickly enough to avoid disaster.

The IPCC set a safe target of limiting global warming to 2°C (3.6°F) by holding the level of CO_2 in the atmosphere below 570 ppm. To achieve this, the amount of carbon[1] added to the atmosphere each year would have to be limited to less than 7 gigatonnes per year, but it has already exceeded 10 gigatonnes in 2012 and is growing rapidly, with 52% of the growth coming from the use of coal in developing nations. The level of carbon dioxide in the atmosphere is now 39% above preindustrial levels and growing at 2.4 ppm per year. If this growth in carbon emissions is not limited, climate models project that global temperature will eventually rise between 5°C and 6°C (9°F and 11°F). A temperature change of this magnitude would likely exceed the ability of many natural and human-made systems to adapt.

If the developing world continues to expand its use of fossil fuels and does nothing to limit deforestation (there has been recent progress), there is little hope left. If China, India, and the United States refuse to accept that climate change is a danger to the environment, global economy, and national security, there is no hope left. But if the developing and developed worlds can cooperate to replace fossil fuels with clean and renewable sources of energy, if deforestation is reversed, and if agricultural practices are modernized, progress is still possible. Climate change will still occur, but it will be slower and less damaging, and most countries will have the time and capacity to adapt. Time is running out, and the world is looking to the United States, the EU, China, and India for strong and determined leadership.

The Way Forward: Finding a Solution

There is no single easy solution to the problem of greenhouse gas emissions, but the impact of hundreds of small changes can be dramatic. To borrow an idea from the world of strategic planning, we need to devise **SMART (specific, measureable, achievable, realistic, and deliverable on time) goals** to help reduce

green house gas emissions. Plans must be specific, stating exactly what needs to be done and how it will be done. They must be measureable, so that it is clear how to determine whether they are successful. They must be achievable using the resources available, and they must be financially and politically realistic. Most of all, the plans must be able to deliver the outcomes we need in the time we have left to achieve them. The data are clear: We must achieve significant greenhouse gas reduction by 2050 if we want to avoid the risk of damaging climate change by the end of the 21st century.

There are nine main target areas where SMART goals can reduce greenhouse gas emissions:

1. Enhance the sequestration of CO_2 from fossil fuels
2. Increase the generation of power from sources that produce fewer emissions
3. Enhance CO_2 sequestration during industrial production
4. Increase the efficiency of public and commercial transport by road
5. Decrease the use of cars and trucks for transport by road
6. Decrease the overall demand for power through conservation and efficiency
7. Reduce greenhouse gases from waste treatment
8. Stop unnecessary deforestation and loss of carbon from soils
9. Change agricultural practices to reduce greenhouse gas emissions

CHECKPOINT 10.6 ▶ In your own words, describe the meaning of a SMART goal.

Enhancing the Sequestration of CO_2 from Fossil Fuels

Burning fossil fuels to produce electricity produces the most anthropogenic CO_2, and once it escapes into the atmosphere, capturing it is difficult and expensive. We will have an almost unavoidable growth in the use of fossil fuels over the next 50 years. The capture and sequestration of CO_2 produced at its source is a crucial piece in managing emissions during the transition from a high-carbon to a low-carbon global economy, but there are significant barriers to progress. The technologies are expensive and still unproven on the scale required, and the operating costs of carbon capture are so high that sequestration may prove uneconomical using current technologies. This problem is compounded by major technical, safety, and legislative difficulties involved in transporting captured carbon dioxide to a disposal site for safe long-term storage. The capture and sequestration of CO_2 may be tomorrow's solution, but the technologies are needed today, and despite the enthusiastic claims of "clean coal" stakeholders, they are still far from proven (**Figure 10.12**).

---

[1] Other texts assess greenhouse gas emissions by using the mass of CO_2. Conversion is easy: $Mass_{(CO_2)} = (Mass_{(C)}/3) \times 11$.

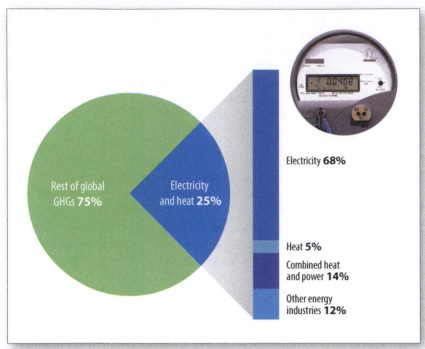

▲ **Figure 10.12:** Greenhouse gas emissions produced during the generation of electricity and heat as a proportion of total greenhouse gas emissions. Note: Data from different sources will show small differences that depend on the age of the data and the criteria used in their application.

Increasing the Generation of Power from Sources That Produce Fewer Emissions

The level of emissions can be reduced if governments limit the tax subsidies they give to the fossil fuel industry. This would account for the full social and environmental cost of greenhouse gas emissions, and help renewable energy to become much more competitive. Governments can further support the growth of alternative energy by investing in a renewable energy infrastructure that can support the growth of an alternative energy market. In order to reduce the level of greenhouse gas emissions as quickly as possible, we also need to take a fresh look at the potential of nuclear power. Despite reservations following the Fukushima disaster in 2011, nuclear power remains the safest and least damaging energy technology that is capable of producing power quickly and efficiently enough to meet emissions reduction targets.

Alternative energy is certainly a growing and vital part of the solution. By the end of the 21st century, these technologies will be in a strong position to replace fossil fuels as the primary source of electrical power and fuel, but they are not growing rapidly enough to reduce emissions to the desired level by 2050. These are tomorrow's technologies, they are the future, but they are not likely to reduce global emissions to a safe level on time. Nuclear power may be expensive, potentially hazardous, and unpopular, but it is a proven technology that can substitute for coal during the critical time period between now and 2050.

Enhancing CO_2 Sequestration During Industrial Production

Major industrial processes such as cement manufacturing and the petrochemical industry produce large volumes of CO_2 (**Figure 10.13**). Further regulations and emissions caps are required to impact these sectors and force them to reduce emissions through carbon

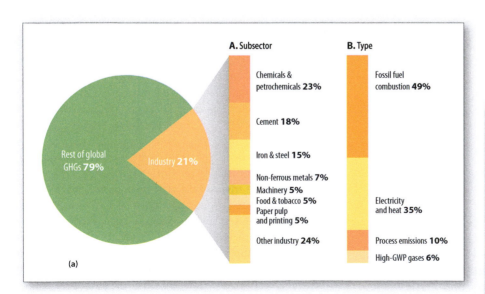

(a)

(b)

▲ **Figure 10.13:** (a) Greenhouse gas emissions produced by industry as a proportion of total greenhouse gas emissions. (b) The global demand for electricity for industry, like auto manufacturing, is increasing. Note: Energy data from different sources will show small differences that depend on the age of the data and the criteria used in their application.

capture and sequestration. These are major stakeholder industries with a lot of political power and legitimate concerns about strong competition from overseas. For this reason, any legislation to limit carbon emissions must have a global dimension to "level the playing field" and prevent companies from simply shifting production to a country with lower emissions standards. Such regulation will be contentious and unlikely to be agreed upon in time to meet the critical 2050 deadline.

Increasing the Efficiency of Public and Commercial Transport by Road

Mileage standards could be much higher than they are at the moment. There is no technological or price-based reason why modern engines cannot provide in excess of 55 miles per gallon (23.8 kilometers per litre or 66 miles per UK gallon). The success of hybrid vehicles and the future promise of efficient electrical and hydrogen-powered engines point toward a revolution in the future of transport. Higher gasoline prices will eventually push consumers toward this market, but the transition can be accelerated by a combination of stronger legislation, public education, and cheaper hybrid vehicles that allow the average consumer (and not just the environmentally conscious elite) to purchase a hybrid or electric vehicle.

Decreasing the Use of Public and Commercial Transport by Road

Over 72% of the emissions that come from the transport sector are from road transport (**Figure 10.14**). Any SMART action that can reduce the emissions intensity of vehicles on the road will have a big impact on emissions. This can be as simple as introducing higher tolls for traffic in the innermost zones around major cities and investing in clean-fuel public transport to manage the increase in demand. Mass transport systems have a proven track record of success in many cities, but they must be efficient, reliable, and affordable for this strategy to have an impact in the near future.

Decreasing the Overall Demand for Power Through Conservation and Efficiency

Residential and commercial buildings are responsible for 15% of greenhouse gas emissions (**Figure 10.15**). Most of this comes from the use of electricity and from heating. New and enhanced building standards that require proper insulation and make use of natural light and heat are a priority for the immediate future. The introduction of a whole new range of "smart" appliances that can be controlled automatically to reduce power consumption will also help to save energy. Conservation and efficiency result in increased comfort and decreased running costs, so consumers find these attributes appealing. But the financial threshold is often too high to promote widespread adoption, even for existing efficient technologies that are readily available, such as floor and roof insulation and double-glazing. One of the easiest and proven ways to get consumers to invest in energy conservation and efficiency is to offer tax breaks that soften the impact of the initial investment. These exist, but often for limited periods, and consumers tend to postpone action in the hope that tax cuts will appear sometime in the future. The United States needs a new tax policy that is consistent and applied nationally to ensure the rapid and widespread adoption of more efficient technologies.

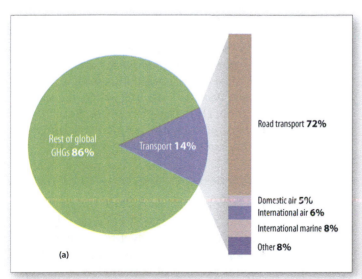

Rest of global GHGs **86%**

Transport **14%**

Road transport **72%**

Domestic air **5%**
International air **6%**
International marine **8%**
Other **8%**

(a)

(b)

▲ **Figure 10.14:** (a) Greenhouse gas emissions produced from the transport sector as a proportion of total greenhouse gas emissions. (b) The demand for personal transport, public transport, and air transport is growing rapidly in developing countries such as India and China.

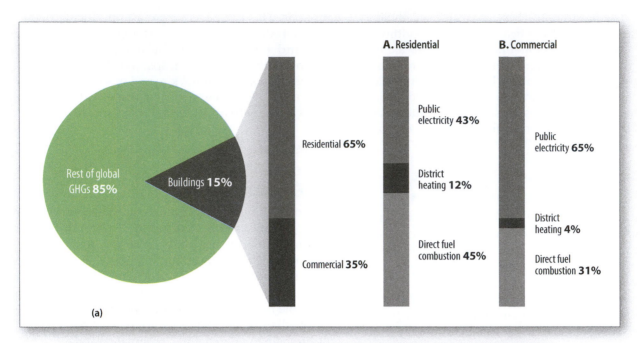

(a)

A. Residential

Residential **65%**

Public electricity **43%**

District heating **12%**

Direct fuel combustion **45%**

Commercial **35%**

B. Commercial

Public electricity **65%**

District heating **4%**

Direct fuel combustion **31%**

Rest of global GHGs **85%**

Buildings **15%**

(b)

◀ **Figure 10.15:**
(a) Carbon dioxide emissions from residential and commercial buildings as a proportion of total greenhouse gas emissions. (b) The Manhattan skyline in New York is a reminder of the high demand for electricity from commercial and residential premises.

Reducing Greenhouse Gases from Waste Treatment

Only 3.6% of greenhouse gases are generated by waste disposal in landfill sites, but 94% of this is in the form of methane, which is over 20 times as powerful a greenhouse gas as CO_2 (**Figure 10.16**). Gas from waste can be captured and sequestered or used onsite to generate power that can be fed into the grid and reduce the demand for coal. Tighter controls on the siting, design, and inspection of landfill sites are needed to ensure that these gases do not escape into the air or water supply.

A better strategy is to reduce the primary waste stream from commercial and domestic sources in the first place. Some locations charge consumers by the weight of domestic waste removed, but this almost always leads to an increase in the amount of illegal dumping. Public education, education in schools (children are a major source of influence in many households), and recycling can reduce the flow of waste to landfill sites. There is still a lot of public ignorance about what can and

can't be recycled. For example, most families discard biological waste that can be readily composted and used as a natural fertilizer. Most of this potentially useful waste is simply discarded and ends up as landfill, where it decays and releases CO_2 and CH_4 to the atmosphere. Instead of using a free and natural resource, people buy artificial fertilizers that further contaminate the environment.

Stopping Unnecessary Deforestation and Loss of Carbon from Soils

Land use change and forestry (LUCF) practices in the developing world account for nearly 18% of all global greenhouse gas emissions; this is more than the entire global transportation sector and is second only to the energy sector (**Figure 10.17**). Extensive deforestation, the conversion of grassland to pastureland, the use of fire to clear land, destructive logging, and the overall poor management of agricultural land all lead to the loss of carbon from the soil and its addition into the atmosphere. Deforestation can be balanced by sustainable forestry

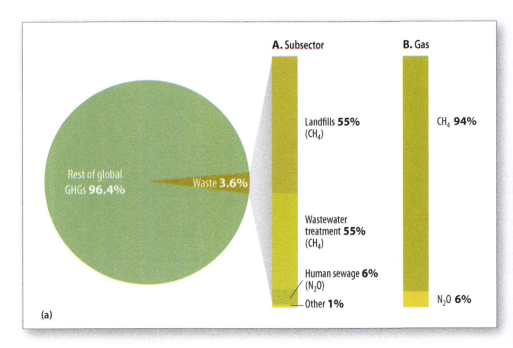

and new growth, and the transition to new farming techniques can cut down on the loss of carbon. However, each of these measures needs practical assistance and financial support. In the developing world, people need to see the immediate benefits of the changes they make. It is hard to see the advantage of change if that change means your family goes hungry.

The United Nations collaborative initiative **Reducing Emissions from Deforestation and Forest Degradation (REDD)** is designed to tackle this problem in developing countries. Launched in September 2008,

the program supports 42 partner countries spanning Africa, Asia-Pacific, and Latin America. According to the United Nations, REDD "focuses attention on the financial value of the carbon stored in forests and offers incentives for developing countries to reduce emissions from forested lands, and invest in low-carbon paths to sustainable development." A new REDD+ program aims beyond the problems of deforestation and forest degradation, promoting the sustainable management of forests and the enhancement of forest carbon stocks. Financial support from the international community for REDD+ could reach up to $30 billion a year if the participating parties fulfill their commitments. This should result in a significant reduction in carbon emissions and in other ecosystem-based benefits, such as conservation of forest biodiversity, water, soil, timber, forest

▼ **Figure 10.17:** (a) Carbon dioxide emissions from land use change and forestry as a proportion of total greenhouse gas emissions. (b) Smoldering forest following the application of slash and burn agriculture in the Yucatan Peninsula, Mexico. Note: Data from different sources will show small differences that depend on the age of the data and the criteria used in their application.

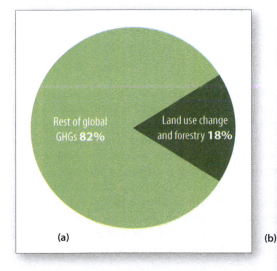

▶ **Figure 10.18:** (a) Greenhouse gases related to agriculture as a proportion of total greenhouse gas emissions. (b) Our taste for beef has a measurable impact on climate change through the emission of greenhouse gases from cattle and their organic waste.

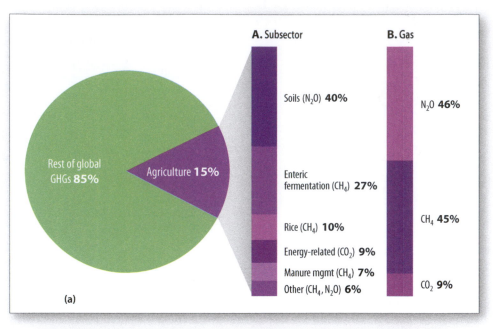

(a)

foods, and other non-timber forest products. For REDD to be really successful, however, all these activities must complement, and not substitute for, deep cuts in other sources of greenhouse gas emissions in developing nations.

Changing Agricultural Practices to Reduce Greenhouse Gas Emissions

Agriculture practices account for 14% of global greenhouse gas emissions and are the third-largest source of heat-trapping greenhouse gases, after fossil fuels. Both CO_2 and CH_4 are released directly into the atmosphere in approximately equal proportions (45% each). A further 9% of CO_2 is added from the use of fossil fuels for agriculture. This is a problem for both developed and

(b)

developing nations. China and India may now produce a staggering 29% of total agricultural emissions, but the United States, the EU, and Brazil together add a further 25%. Rapid population growth will increase the demand for higher crop yields, and this will exacerbate the problem of greenhouse gas emissions if nothing is done to break the link between more productive agriculture and increasing greenhouse gas emissions.

The largest sources of agricultural emissions are agricultural soil management, enteric fermentation (animal gas), the burning of biomass, rice cultivation, and the use of electricity and fuels (**Figure 10.18**). The expansion in crop area to meet the rising global demand for food and the increase in livestock to feed the growing demand for red meat are increasing pressure on global forests and grasslands. As more land is cleared for agriculture, the net effects are enhanced emissions, decreased global albedo, and accelerated global warming.

Agricultural Soil Management Soils artificially fertilized with nitrates emit 40% of agricultural emissions. Nitrogen is an important nutrient, but when soils become waterlogged due to heavy rain, flooding, or too much irrigation, anaerobic bacteria (those that need no oxygen for respiration) break down the nitrates into gaseous nitrogen oxides that are released from the soil. These powerful greenhouse gas emissions can be reduced significantly if the amount of fertilizer applied is matched to the specific needs of the crop grown and if crops are bred to make more efficient use of the nitrogen available. This is one area where action to prevent emissions can be successful without large financial investments.

Enteric Fermentation Almost 28% of agricultural emissions emanate from the assorted gaseous discharges of ruminants, and a further 7% come from manure management. There are 1.2 billion large ruminants (such as sheep, goats, and cattle) in the world, and each animal releases methane through a unique digestive process called *enteric fermentation*. Approximately four times as much comes from the front end than the back, but in combination, the total volume of gas produced is enormous. The solution to this problem is for the world to eat less red meat, but the demand is rising as the people in the more affluent developing nations change their diet.

Better grazing management and dietary supplementation can make a difference, but this is a problem without an immediate viable solution.

Burning of Biomass The burning of agricultural biomass generates up to 12% of total agricultural emissions. Burning releases CO_2, CO, NO_x, and even CH_4 into the atmosphere and reduces the albedo of agricultural land. Alternatives to burning are available. These include physically removing plant residues and mulching the residue before mixing it into the topsoil. Government regulations can be used to limit the amount of burning in the developed world but not in the developing world. The alternatives to burning are more expensive and time-consuming, and it is hard to convince farmers in developing nations to change practices they have used to maintain the land for generations. Progress can be made in this area, but it will not be easy.

CHECKPOINT 10.7 ▶ Some people believe that we should focus our efforts to lower greenhouse gas emissions on deforestation and agriculture rather than on industry, commerce, and transport. Explain why they think this is a good idea.

Rice Cultivation Around 10% of agricultural emissions come from methane produced in the water-saturated soils of paddy fields during rice production. The only solution to this problem is to drain these fields for part of the year, but this solution is economically and logistically prohibitive for many rice farmers. This will be a hard problem to solve, but it is a major issue, as rice will remain an important food crop for a long time to come.

Use of Electricity and Fuels The fuel and electricity that are used to support farming contribute 9% toward total agricultural emissions. As much as 34 liters (9 gallons) of diesel fuel can be used to manage each hectare of land. This amount of fuel can be reduced through simple changes to practices that require less time for machinery on task, such as shallow tilling, leaving more plant residue in the topsoil, and switching from annual to perennial crops when possible.

CHECKPOINT 10.8 ▶ If you had to prioritize the nine areas where SMART goals could reduce greenhouse gas emissions, which would you put at the top of the list and which at the bottom?

Driving in the Carbon Wedges: The Carbon Mitigation Initiative

By now you should understand why the IPCC goal of cutting 1990 emissions by 25–40% by 2020 and then by 80% by 2050 is a tremendous challenge. Physicist Robert Socolow and ecologist Stephen Pacala, co-directors of the Carbon Mitigation Initiative at Princeton University,

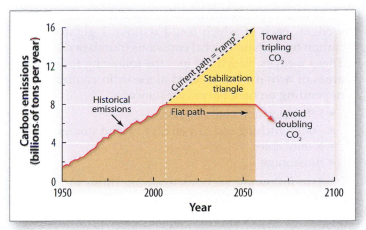

▲ **Figure 10.19:** How carbon wedges could be developed over time to either remove or prevent the additional 8 gigatonnes of CO_2 that will be added to the atmosphere by 2050 if no action is taken. The yellow area on this diagram is the stabilization triangle that is constructed by stacking carbon wedges (see Figure 10.20).

started a vigorous debate in 2004, following a publication in the journal *Science*. They suggested that the world can meet greenhouse gas emissions targets by creating between 7 and 18 "carbon wedges," depending on the rate of economic growth and global greenhouse gas emissions (Figure 10.19). A **carbon wedge** is a concrete action, a SMART goal that can reduce greenhouse gas emissions by 1 gigatonne of carbon a year by 2055. Starting as soon as possible, with small but increasing effect, carbon wedges are aimed at stabilizing greenhouse gas emissions at close to current levels and then actually decreasing the atmospheric concentration of CO_2 starting around 2055 (Figure 10.20). The gap between the predicted level of greenhouse gas emissions with no action and the flat path of emissions proposed by this plan opens up a so-called stabilization triangle over time that represents all the CO_2 that has been successfully removed and sequestered or never produced in the first place.

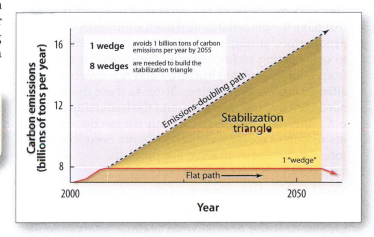

▲ **Figure 10.20:** The introduction of eight individual programs that reduce greenhouse gas emissions by at least one billion tons each (one gigatonne) by 2055 will create a stabilization triangle that can offset the expected increase in greenhouse gas emissions driven by economic growth, and maintain the level of greenhouse gas emissions at close to current levels.

In their paper, Socolow and Pacala identify 18 different wedges, each of which could remove 1 gigatonne of carbon from annual global emissions (numbered below). These wedges, which fall into the following four major areas of activity, can be used alone or in combination, depending on national circumstances:

- Efficiency and conservation
- Fuel switching and carbon capture and storage (CCS)
- Renewables
- Biostorage

The following sections examine these four areas and the carbon wedges outlined for each.

Efficiency and Conservation

Several carbon wedges are related to efficiency and conservation:

1. Double the fuel efficiency of 2 billion cars, from 30 to 60 mpg.
2. Decrease the number of car miles traveled by half.
3. Use best-efficiency practices in all residential and commercial buildings.
4. Produce current coal-based electricity with twice today's efficiency.

Conservation is learning how to use less. **Efficiency** is learning to do more with what we already use. Of these two measures, conservation is by far the most difficult. It is hard to get people to use less energy if it means changing their lifestyle. Remember how President George H. W. Bush was determined that the "American lifestyle was not negotiable"? Conservation and efficiency will become more attractive as the cost of energy rises and prompts people to make these changes.

The total amount of energy lost during the inefficient production, distribution, and consumption of power in the United States is more than 55%. Think about it. More than half the energy used and paid for in the United States goes to produce hot air that is vented into the atmosphere. We need to implement new technologies that have the potential to increase the efficiency of power production, storage, and distribution, and we need to use existing technologies to cut down on heat loss from poorly insulated buildings and manufacture more efficient combustion engines. To make these changes, we need both the will for change and the leadership to drive it. Change never comes easily, but we can do a lot with the resources we have today. We do not need to wait for future technologies to solve this problem.

Fuel Switching and CCS

If coal remains the primary fuel that powers economic growth over the next 50 years without intervention to capture and store the CO_2 produced, global temperatures will almost certainly increase by as much as 5°C to 6°C (9°F to 11°F). It does not have to happen like this. By changing the mix of fuels that are used to generate power and capturing the greenhouse gas emissions produced in the process, we could reduce emissions. Several carbon wedges are related to fuel switching and CCS:

5. Increase efficiency at 1,600 coal-fired plants between 40% and 60%.
6. Build 1,400 large (1-gigawatt) natural gas plants instead of coal plants. Natural gas produces only half the emissions of coal.
7. Apply CCS to 800 large (1-gigawatt) coal plants or 1,600 large gas plants.
8. Use CCS to generate hydrogen from coal as a fuel for 1.5 billion vehicles.
9. Use CCS to capture CO_2 from 180 large syngas (coal-to-gas)-generating facilities.
10. Triple the world's current nuclear capacity.

Renewables

Renewable energy currently provides just 7% of the demand for energy in the United States. Most of this is in the form of hydroelectricity and geothermal power, where there are limited options for significant growth. New clean commercial sources of energy such as wind power, solar power, hydrogen power, and the use of biofuels are on the threshold of exponential growth as new technologies increase the efficiency of energy conversion and decrease the cost. The wedges related to renewables must be part of any global plan to address the problem of greenhouse gas emissions:

11. Increase current wind capacity by a factor of 30.
12. Install photovoltaic cells over a total area of 20,000 square kilometers (7,700 square miles and just 1/5th the size of lower Michigan spread out all over the globe)—a 700-fold increase in capacity in just 50 years—to replace coal.
13. Increase wind power 80-fold to make hydrogen fuel for cars.
14. Increase the production of bioethanol by a factor of 30.
15. Fuel 2 billion cars with ethanol.

Biostorage

Our ability to change agricultural and forestry practices in order to keep more carbon locked away in the biosphere is known as **biostorage.** Over 18% of greenhouse gas emissions globally come from land use changes, mainly deforestation and agriculture. We need to increase biostorage while recognizing the rights of indigenous peoples to exploit local resources. The destruction of tropical forest by logging companies with no plans for management, conservation, sustainability, or restoration should be made illegal, and we need international support for enforcement and monitoring. The wedges here are quite simple:

16. Halt global deforestation in 50 years.
17. Plant new forest over an area the size of the United States.
18. Apply carbon management strategies—such as conservation tillage—to agricultural practices to keep more carbon trapped in the soil.

CHECKPOINT 10.9 ▶ In your own words, explain how carbon wedges can be used to plot a realistic path toward the reduction of greenhouse gas emissions.

10.2 PAUSE FOR... THOUGHT | *Can individuals really make a difference in the battle against climate change?*

Faced with the overwhelming threat of climate change, many people feel that individual action is pointless and give up without even trying. While it is true that an individual's action can have limited impact, the combined impact of millions of people, working towards a common goal, can have unimaginable impact. Perhaps the greatest impact occurs when individuals exercise the right to vote and elect politicians who enact legislative changes that can limit the overall production of greenhouse gases.

The United States: Regional and State Initiatives

Across the United States, a number of important regional initiatives and more than 30 individual states have proposed climate change action plans that hope to implement SMART changes and gradually reduce greenhouse gas emissions. The following are some examples:

- The *Northeast Regional Greenhouse Gas Initiative (RGGI)* is one of the most extensive and innovative independent programs. Ten northeastern and mid-Atlantic states (Connecticut, Delaware, Maine, Maryland, Massachusetts, New Hampshire, New Jersey, New York, Rhode Island, and Vermont) adopted a regional cap-and-trade program designed to curb CO_2 emissions from power plants by 10% by 2018. RGGI is an emissions trading scheme that requires organizations to purchase the right to emit CO_2. The first auction of allowances generated nearly $40 million for the six states that participated; this is money that can be invested in the future of renewable energy.
- The *New England Governors and Eastern Canadian Premiers (NEG-ECP) Climate Change Action Plan* is a parallel resolution adopted in 2001 by the New England governors and the eastern Canadian premiers. This resolution called for a reduction in greenhouse gas emissions to 10% below 1990 levels by 2020 and a 75% to 85% relative reduction below 1990 levels as a long-term goal.
- The *Midwestern Regional Greenhouse Gas Reduction Accord* set a long-term target to cut emissions by between 60% and 80% below 2007 levels, using a multisector cap-and-trade program. Illinois, Iowa, Kansas, Michigan, Minnesota, Wisconsin, and the Canadian province of Manitoba signed as full par-

ticipants; Indiana, Ohio, and South Dakota joined the agreement as observers to participate in the development of the cap-and-trade system.
- The *Energy Security and Climate Stewardship Platform for the Midwest* involves 12 states and 1 Canadian province: Wisconsin, Minnesota, South Dakota, Illinois, Indiana, Iowa, Kansas, Michigan, Missouri, Nebraska, North Dakota, Ohio, and Manitoba. Members have agreed to focus on renewable electricity, carbon capture and storage development, and biofuel development.
- The *Western Climate Initiative (WCI)* brings together the states of California, Montana, New Mexico, Oregon, Utah, and Washington, as well as the Canadian provinces of British Columbia, Manitoba, and Quebec. The WCI is an effort to reduce greenhouse gas emissions by setting joint emissions targets and establishing a cap-and-trade system. The WCI has set an emissions target of 15% below 2005 levels by 2020, or approximately 33% below business-as-usual levels. The WCI builds on work by two other regional agreements: the *Southwest Climate Change Initiative* and the *West Coast Governors' Global Warming Initiative*.
- California opened a new carbon trading market in 2013 as part of an innovative cap and trade policy that hopes to reduce emissions to 1990 levels by 2020 (a 15% reduction), and ultimately reduce emissions to 80% below 1990 levels by 2050. The scheme raised $233 million from the sale, by auction, of 23.1 million carbon permits for the year 2013. The program will cover 600 facilities starting with electric utilities and large industrial facilities in 2013 and then distributors of transportation, natural gas and other fuels in 2015. The program allows companies to meet up to 8% of their obligation to reduce emissions using carbon offsets in certain areas as long as the emissions reduction project takes place within the United States.

Geoengineering: A Last Resort?

According to the U.S. National Academy of Sciences, **geoengineering** is "the large-scale engineering of our environment in order to combat or counteract the effects of change in atmospheric chemistry." Geoengineering is not a solution to climate change; it is a pragmatic stopgap measure that we might have to use as a last resort to limit the extent of global warming while we transition to a carbon-free economy.

Geoengineering Technologies

The technologies involved in geoengineering are designed to enhance and extend natural processes that either remove carbon from the atmosphere or reflect more of the Sun's radiation. **Solar radiation management (SRM)** acts to reduce the amount of solar radiation absorbed by Earth's surface, and **carbon dioxide reduction (CDR)** actually captures CO_2 from

▶ **Figure 10.21:** The geoengineering technologies illustrated here, both by solar radiation management and carbon dioxide reduction, aim to reduce the amount of solar radiation reaching Earth and increase the rate at which carbon dioxide is removed from the atmosphere.

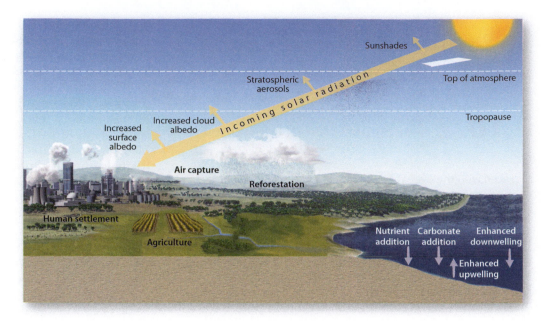

the atmosphere and stores it safely in deep natural reservoirs (Figure 10.21). There are many uncertainties about the long-term effects of these kinds of interventions in Earth's climate system, but a combination of technologies may be necessary if we fail to manage the growth of greenhouse gas emissions over the next 50 years.

CHECKPOINT 10.10 ▶ Explain why geoengineering is considered a stopgap measure rather than a solution to the problem of climate change.

Solar Radiation Management (SRM)

Solar radiation management aims to reduce the intensity of incoming solar radiation through a combination of technologies in space, in the atmosphere, and on land. A reduction in solar radiation of just 1.8%, or a change in global albedo from 0.31 to 0.32, would reduce the amount of energy entering Earth's climate system by 4 Wm^2—enough to offset the effect of doubling the level of CO_2 in the atmosphere.

Space-Based SRM Mechanisms An object in orbit around the sun at the **L1 (Lagrange) point** is balanced between the gravitational pull of Earth and the Sun. To achieve a reduction in solar intensity of 4 Wm^{-2}, an array of up to 16 trillion tiny solar wide reflectors covering 3 to 5 million square kilometers (1.2 to 1.9 million square miles) would need to be placed in an L1 orbit. This array would maintain a constant position and reduce the total amount of solar radiation reaching the top of the atmosphere. Just to keep pace with future emissions, the array would have to grow by as much as 30,000 to 40,000 square kilometers (11,500 to 15,400 square miles) each year.

Large solar reflectors could also be positioned in orbit around Earth to reduce the intensity of solar radiation over a large part of the atmosphere. One

estimate suggests that 55,000 reflectors, each 100 square meters (120 square yards) in size, would be required to reduce the global flow of radiation by 4 Wm^{-2}. Another suggestion is to pulverize and eject 2 billion tonnes of dust particles into orbit around Earth. The impact of either technology would be most effective if it were done over the poles, where warming is most intense and the albedo feedback from melting ice is greatest.

CHECKPOINT 10.11 ▶ Describe the main advantages of SRM.

These space-based geoengineering technologies do nothing to alleviate the problem of accumulating greenhouse gases in the atmosphere or to reverse the serious impact of ocean acidification. They simply offset the warming caused by greenhouse gas emissions, and their sudden failure would lead to a very rapid rise in global temperature. The environmental impact of these technologies is also uncertain. They would have no impact at high latitudes during the winter, but over the summer, they would reduce the contrast between seasons and the amount of light available for photosynthesis. At low latitudes, they could have a major impact on the Asian and African monsoons. Their deployment would be expensive and risky, and because our knowledge of their impact on the global ecosystem is limited, they are unlikely to be our first choice, using current technology.

CHECKPOINT 10.12 ▶ Describe the main disadvantages of SRM.

Atmosphere-Based SRM Mechanisms One atmosphere-based SRM mechanism would be to use aerosols a few tenths of a micron in size to increase the albedo of the stratosphere and reduce the amount of solar radiation reaching the ground. This would be an extension of the natural climate forcing that results from large volcanic eruptions. Aerosols are quickly and evenly distributed by stratospheric winds and are much cheaper to deploy and maintain than space-based reflectors. They are also removed naturally after just four or five years, so their effect can be easily reversed.

Many kinds of aerosols could be effectively used for SRM, but sulfate aerosols are the first choice as they are produced naturally by volcanic eruptions, and we have been able to observe their impact. More than 1 to 5 Mt sulfur y^{-1} would be required to achieve the cooling necessary, but generating this amount would not be impossible. Sulfate aerosols remain one of the most popular options for geoengineering. A disadvantage of using aerosols in this way is that we are uncertain of their long-term global impact. Much depends on their size and composition. If used improperly, they could accelerate the destruction of ozone at high latitudes by increasing the number of polar clouds that form in the winter.

Another atmosphere-based SRM mechanism is **marine cloud reflectivity.** Scientists have known for some time that ship exhaust ejects millions of small particles into the lower troposphere. These particles act as **cloud condensation nuclei (CCN)** that cause marine stratocumulus clouds to form in their wake and increase global **cloud albedo.** If this process could be replicated on a much larger scale in parts of the Southern Ocean where marine stratocumulus clouds are less common, their increased reflectivity would have a large impact on global albedo. One way to do this would be to use large unmanned wind-powered vessels that would spray a fine mist of seawater up into the air. This water would evaporate quickly and leave small salt particles that are very effective CCN. We would need between 500 and 1,500 vessels (presumably solar and wind-powered) pumping continually to double the CCN concentration in the Southern Ocean and offset the effect of doubling carbon dioxide in the atmosphere.

Surface-Based SRM Mechanisms It is arguably safer and cheaper to do as much as possible to reflect more of the Sun's energy on the ground than to attempt atmosphere- and space-based SRM mechanisms. Various suggestions have been made, including increasing the albedo of deserts by covering large areas with reflective materials and painting the roofs of buildings with white paint. It is even possible to biologically engineer changes in vegetation and crops to make them more reflective. Increasing the albedo of 50 million square kilometers (19.3 million square miles) (10%) of Earth's land surface from 0.15 to 0.8 would increase reflectivity by $4\,Wm^{-2}$. However, engineering an area this large in this way would be politically and economically impractical, so the use of large-scale surface reflectors is unlikely to happen. Still, a significant contribution from a large number of smaller projects is possible.

Carbon Dioxide Reduction (CDR)

CDR mechanisms aim to solve the greenhouse gas emissions crisis by "pumping" CO_2 directly from the atmosphere. This can be achieved directly from the air or through the enhancement of natural processes on land and in the oceans.

The Carbon Cycle Carbon is constantly exchanged between large reservoirs in the lithosphere, atmosphere, hydrosphere, biosphere, pedosphere (soils), and cryosphere as part of a complex cycle. It can take millions of years to move a single atom of carbon from the atmosphere into the terrestrial or marine biosphere, then through decay, excrement, water, soils, and sediment, before it is finally locked away in the lithosphere, where (in the absence of human intervention) it will stay for millions of years before volcanism or metamorphism finally releases it back into the atmosphere.

The amount of carbon involved in this cycle is very large, but the system is in a state of dynamic equilibrium, with as much as 400 gigatonnes of carbon exchanged each year between the atmosphere, land, and oceans. This equilibrium has been upset by the addition of anthropogenic greenhouse gases, mostly in the form of carbon dioxide. In 2012 over 10 gigatonnes of carbon were released into the atmosphere in this way, a figure that will grow with the global economy. The shift from equilibrium has been partially balanced by an increase in the rate of absorption of carbon in the oceans and on land, with an extra ~2.7 gigatonnes per year entering the ocean reservoir and ~3.3 gigatonnes per year entering the terrestrial biosphere. The remaining 4.0 gigatonnes per year is accumulating rapidly in the atmosphere, and this is the source of global warming (**Figure 10.22**).

The carbon cycle has been disrupted in the past through the sudden release of CO_2 by volcanism and changing climate. The large carbon reservoirs in the oceans, atmosphere, and biosphere slowly restore equilibrium over thousands of years. If we burned all of the 10,000 gigatonnes of carbon that remain as fossil fuel tomorrow it would be a sizable fraction of the ca. 40,000 gigatonnes of carbon already stored within the oceans, but Earth would slowly restore the planet to carbon equilibrium by moving the excess carbon into deep isolated reservoirs. It would take many thousands of years for this to happen, and by that time, surface temperatures would have risen by as much as 6°C (11°F), with devastating impact on humanity and global ecosystems.

CHECKPOINT 10.13 ▶ Describe the main advantages of CDR.

▶ **Figure 10.22:** The movement of carbon between land, atmosphere, and oceans based on global human emissions of 10 gigatonnes a year. Yellow numbers are natural fluxes, red numbers are human contributions, and white numbers indicate stored carbon. All numbers are in gigatonnes of carbon per year. Of the 10 gigatonnes of human emissions, ca. 3 gigatonnes are absorbed into the oceans, 3 gigatonnes are used in photosynthesis, and 4 gigatonnes accumulate in the atmosphere.

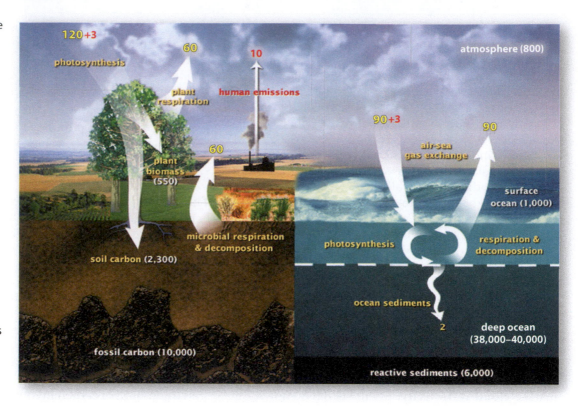

Air Capture Various engineering schemes have been proposed to capture CO_2 directly from the atmosphere by pumping air over relatively inexpensive chemicals. Tall air extractors could remove CO_2 for later extraction and sequestration. Often called "artificial trees," these towers could be distributed across the landscape in large arrays and could use solar, wind, or nuclear power to maintain a constant flow of air. Although it would take "forests" of these artificial trees to have any impact on global emissions, this practical technology has a promising future. One of the main problems with air capture lies in knowing what to do with the captured CO_2. Suggestions range from pumping it into deep porous rock formations to dumping it directly into the deep oceans.

Ocean Circulation: Enhancing the Solubility Pump
Ocean circulation plays a major role in removing carbon from the atmosphere. Close to the poles, the water is cold and can hold much more carbon dioxide in solution than at the equator. This carbon-enriched surface water sinks from the surface to form deep ocean conveyor currents that transport water around the globe. This powerful **solubility pump** has been removing CO_2 from the atmosphere for a long time and has built up an immense reservoir of 38,000 to 40,000 gigatonnes in the deep oceans. Some studies suggest that it might be possible to increase the formation and flow of deep water in the North Atlantic Ocean by cooling the surface water and enhancing the production of ice, but the price of doing this using current

technology is prohibitive. Increasing the flow by 1 million cubic meters per second (264,000 gallons per second) would increase the flow of carbon by only 0.01 gigatonnes per year.

Another suggestion is to construct large pipes to draw nutrient-rich water up from deep in the oceans toward the surface to enhance biological productivity. However, the cost of this solution would be high, the technology is unproven, and the environmental consequences are uncertain.

CHECKPOINT 10.14 ▶ What are the main disadvantages of CDR?

Oceans: The Soft Tissue and Carbonate Pumps
Large areas of the ocean are not very biologically productive due to a deficiency in common nutrients such as iron, phosphorus, and nitrogen. Field experiments have shown that **ocean fertilization** by iron, phosphorus, and/or nitrogen can lead to intense bursts of biological activity that can remove CO_2 from the atmosphere in living tissues and the carbonate shells of marine plankton. Most of this carbon is recycled in the upper layers of the ocean and then released back into the atmosphere. However, some of it reaches sea floor sediments, where it is locked in the lithosphere for millions of years. There is a lot of interest in using this technology as a way to generate carbon offsets that could be sold on the global carbon market, but there are many unsolved

problems. The 1 gigatonne per year of carbon that can be removed by using this technology is significant—about the same as can be removed by afforestation and reforestation—so it is worth experimenting further to monitor any side effects it might have on the marine ecosystem.

Enhancing Ocean Alkalinity The amount of CO_2 that can be removed from the ocean depends on the pH of the water. It is theoretically possible to increase the alkalinity of the ocean by adding lime to enhance the solution of CO_2. However, the amount of mining, processing, and transportation required to achieve this makes it an unattractive first alternative.

Enhanced Weathering The natural weathering of rock removes CO_2 from the atmosphere at the slow rate of 0.1 gigatonnes per year. It is theoretically possible to increase this rate through artificial weathering, but it takes at least 2 tonnes of rock to remove one tonne of CO_2, so the environmental cost of mining would be considerable. It is also possible to remove CO_2 by adding rocks rich in the common mineral olivine to soil, where it reacts to form carbonate and silica. However, this solution would require as much as 7 cubic kilometers (1.7 cubic miles) of rock per year, and that is twice the volume of current coal mining.

Changes in Land Use and Agriculture Deforestation is the primary target for land use change because it produces as much as 20% of all anthropogenic greenhouse gas emissions. However, other land use changes, including the following, could help mitigate the impact of climate change:

- The cultivation of natural and bioengineered crops that are selected to help increase global albedo.
- The combination of bioenergy production with **carbon capture and sequestration (BECS)** by growing biofuels and using them to generate power while collecting waste CO_2 for long-term storage in underground reservoirs.
- Burying biomass from waste leaves, wood, crops, straw, and even food in landfills can lock away carbon from the atmosphere for a long time. However, bacterial decomposition would still occur and release CO_2 and CH_4 that must be captured and stored.
- Burning organic waste in a low-oxygen environment to create **biochar**, a form of charcoal. Biochar is very stable and can retain carbon for thousands of years. It can be buried in landfill or added to soil to improve productivity.

The application of these additional changes in land use could reduce emissions by as much as 0.4 to 0.8 gigatonnes per year by 2030—an amount equaling

as much as 2% to 4% of projected emissions increases over that period. These are promising technologies and worth further investigation. The costs are low, and the benefits are high. However, to offset all greenhouse gas emissions by 2100 in this manner would require 10% of all global primary production. Clearly a combination of all technologies is needed to make the geoengineering dream come true.

CHECKPOINT 10.15 ▶ Why would an effective combination of SRM and CDR not solve many of the major environmental problems associated with the emission of greenhouse gases into the atmosphere?

Geopolitical Implications of Geoengineering

Some of the geoengineering technologies have potential, but they are not ready for deployment (**Figure 10.23**). There is still too much uncertainty about costs and benefits, as well as their state of technological readiness. There is no international agreement to monitor implementation, and there are no climate models with the regional resolution required to predict their impact. We need to do much more research and development before we can trust geoengineering, and we need to carefully assess each new technology in terms of its effectiveness, timeliness, safety, and cost. Geoengineering may not be our first choice, but in the absence of strong political leadership, it may become a pragmatic alternative.

Perhaps the greatest problem with geoengineering is not the uncertain science or unproven technology but the absence of a legal framework for governance. We need a regulatory framework that can determine who is liable if something goes wrong. Most of the proposed geoengineering solutions have an international dimension that raises questions about responsibility and accountability. At the moment, multinational alliances, individual states, corporations, and even rich individuals could act independently without consulting anyone about the impact. The United Nations is an obvious venue for negotiations, as it gives smaller nations a chance to be involved. But if the world cannot agree to reduce greenhouse gas emissions in the first place, it is unlikely to agree to a new framework for the implementation of unproven technology. In the end, social, political, religious, and ethical issues may delay action to deploy geoengineering as a "solution" until global temperatures have reached an unprecedented level and there is no longer an alternative.

CHECKPOINT 10.16 ▶ Discuss why an effective legal framework is necessary prior to the large-scale implementation of a major geoengineering project.

▶ **Figure 10.23:** All geoengineering technologies need to be evaluated in terms of their safety, timeliness, effectiveness, and affordability. This diagram illustrates how these factors vary on a scale of 1 (low) to 6 (high) among the different technologies discussed in the text. Although all the different methods are worth considering, the cluster of technologies in the middle of the diagram, including CCS, BECS, biochar, cloud albedo, and ocean fertilization, have the best chance of success, as they offer the best combination of all factors. The colors of the symbols relate to level of safety (red = low; yellow = medium; green = high) and timeliness (large = quick; small = slow).

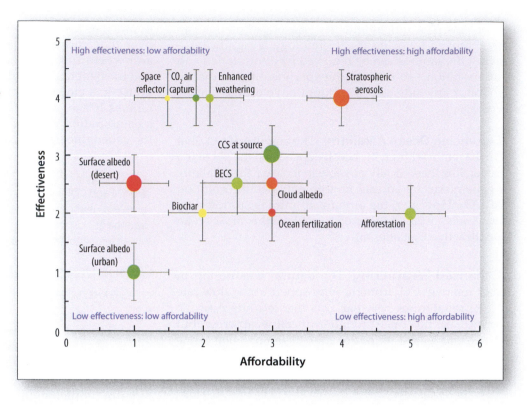

Epilogue

I hope you use the evidence presented in this book to reach your own conclusions about climate change and global warming. As new climate records are broken each year, we must recognize that any single event could be natural, but when the entire climate record is considered, the statistical probability that all recent changes in global climate are "natural" is vanishingly small.

Accepting Responsibility

The people of New Jersey and New York suffered terribly from the impact of Hurricane Sandy in 2012, reeling from the irreparable loss of life and stunned at the billions of dollars it will cost to repair businesses and property. This was a rare weather event, but the impact was typical of the widespread future devastation we can expect if Earth continues to warm. Climate change will continue to cost lives and jobs across the world if we use cheap carbon to drive economic growth and do nothing to reduce the level of greenhouse gas emissions. If we fail to take political action, we are simply passing on the cost of climate mitigation to our children and grandchildren. Do nothing today, and the world will warm by 3°C to 4°C (5°F to 7°F) by the end of the century. Do nothing by 2050, and parts of the world could warm by a devastating 6°C (11°F) by the end of the century.

▶ **Figure 10.24:** Recent elections in the United States have reminded us how important each vote can be in deciding the future of national policy. The future of climate change legislation will be decided in the ballot box. Get out and vote for what you believe in!

The Power of One

The political path towards success may seem impenetrable, but the solution lies within the ballot box and the integrity of the leaders we choose to elect (Figure 10.24). Every vote counts and the younger generation that will carry the burden of climate change still has a chance to use its influence as a catalyst for political change. It is time to find common ground among fellow citizens of all political persuasions who care deeply about the future of this country and our leadership in the world around us.

Time for Action

The time for action is now. It is time to apply our knowledge, determination and entrepreneurial spirit to find solutions that will moderate climate change while meeting the needs of a growing world population. Alternative energy is abundant and available, and the production of clean electrical power and fuels is growing. But we must invest in yet further growth, upgrade our infrastructure, and focus on the expansion of a new "green" economy that rewards energy conservation and efficiency. The time for talking is past. It is time to turn our knowledge into action.

10.3 PAUSE FOR... THOUGHT | *Should we wait for technological solutions to climate change?*

At the beginning of the 20th century, no one could have guessed at the technological advances that would occur over the next 100 years. The very first cars were just appearing, and regular transatlantic flights, moon landings, washing machines, personal computers, the Internet, and nuclear power were all unknown—in most cases not even imagined.

It is almost certain that our grandchildren will look back from the end of this century with great curiosity, wondering how we managed to live our lives without the advantages of the latest technologies. It is equally certain that some of those technological innovations will have potential to mitigate or reverse climate change. Is this an excuse to wait and do nothing today? Such a strategy would be dangerous. If, as seems possible, Earth's climate system reaches a critical tipping point before that time, we might experience climate change that human technologies are unable to reverse.

▶ **Figure 10.25:** Earth is a beautiful and resilient planet that we share with more than 7 billion people. Surely it is our responsibility to address global climate change and protect our only home for future generations?

Summary

- As the population of the world approaches 9 billion, the demand for energy to feed economic growth will escalate. The nations of the developed world have some choices to make. Energy is not the problem; it is the path to a basic standard of living for billions of people around the world. The problems are how we produce energy and the wasteful ways we consume it. We must change how we produce, consume, and conserve energy and transition from an old economy built on the use of cheap and abundant fossil fuels to a new and green economy based on energy generated from renewable resources.

- The current energy crisis has its roots in the industrial revolution when coal was first used to power machinery, but the impact was greatly accelerated by industrial expansion in the United States, Europe, the Soviet Union, and Japan during and following World War II.

- Countries that are large consumers of energy must assume leadership and take initiative to confront the growing social, economic, and environmental threat from climate change. The responsibility can no longer rest on the developed nations alone, but also rests with the richer developing nations such as China, India, Brazil, and South Africa. The smaller developing nations bear no responsibility for the problem of climate change and consume much less energy than the largest nations, but they should promote the use of clean energy globally, as they have the least capacity to adapt to climate change.

- Proposals to reduce greenhouse gas emissions must pass the SMART test. They should be specific, measureable, and achievable realistically by 2050. No single mechanism holds the answer, but small changes in conservation and efficiency alone can make a major difference.

- If we can identify a cluster of carbon wedges, each with the potential to reduce greenhouse gas emissions by as much as 1 gigatonne, the world still has a realistic chance of meeting greenhouse gas emissions targets by 2050. Identifying and implementing these wedges will require immediate and concerted action by all parties. These changes will take place within the lifetime of most students reading this text.

- Geoengineering may offer a partial solution if political action is delayed until rising damage and the high price of energy force governments into action. This would leave the future dependent on experimental, untested, and potentially dangerous solutions that lack an international legal framework. It would also fail to address the critical issue of ocean acidification.

- The global climate at the end of the 21st century will be very different from the climate today if no action is taken to limit emissions. There is increasing confidence in the depth of knowledge and projections of climate scientists, but that knowledge needs to be turned into action before it is too late. In the end, the world will transition to a low-carbon economy, but we need to take urgent action or it may only happen after climate change exacts such irreparable damage that economic growth in much of the developing world may be overwhelmed by the cost of climate mitigation.

Why Should We Care?

The international community must find a way to supply the developing world with the sustainable energy resources it needs for population and economic growth over the rest of the century. Specific and effective action appears to be realistic and achievable before 2050 if we take effective action immediately.

All nations that have the capacity to take action to reduce emissions must choose. Will they show leadership and engage with the process? Or will they simply retrench and protect their own domestic social and economic interests? There will be a tangible price to pay for progress. The price of failure remains uncertain, but we can be sure that future generations will hold us accountable for the decisions we make today. These challenges raise enormous ethical issues for political leaders, who must balance the welfare of their nations against the wider interests of humanity and the environment. In the transitory world of politics, it may be easier to procrastinate until there are no outstanding questions. By that time, it will be too late.

Looking Ahead . . .

It is hard to be optimistic about the immediate future when so many powerful stakeholders are determined to block action against climate change. The social, political, and economic barriers may seem insurmountable at times, but one lesson that democracy has taught us is that the impact of just one committed individual can be enormous. It is time for us all to take action, to make our voices heard both at home and in the ballot box. There is a legitimate debate about climate change that we need to engage with, there are real uncertainties that we need to address, and there are moral and ethical dilemmas that we cannot avoid . . . . and yet progress is achievable. It is not too late, and we still have options, but complacency is not one of them. We need informed, engaged, and active citizens who are willing to create change in society.

Critical Thinking Questions

1. Sitting in a bar one night, someone overhears your conversation and asks, "Why don't we just promote better birth control overseas rather than cut back on my standard of living? Why should I pay for other people's decision to have too many children?" How would you answer?

2. Do the rich nations have a moral right to tell developing nations to cut back on emissions if doing so impairs their economic growth and social welfare? Explain.

3. Do nations act ethically only when moral questions coincide with self-interest? Explain.

4. Does the United States have a right to consume so much power and generate so many greenhouse gas emissions per capita as long as it continues to produce 22% of global GDP? Explain.

5. The emissions wedges might be achievable, but are they realistic? Explain.

6. Suggest ways, besides those listed in this chapter, to reduce the demand for energy and cut down on greenhouse gas emissions.

7. As mayor of your city you are faced with the challenges of traffic congestion and air pollution and you understand the increasing threat from climate change. Suggest how you would try to persuade your electorate to make more use of public transport and pay higher tolls to take their cars into the city, and still hope to get re-elected?

8. You represent small island nations on a new committee of the United Nations that was established to approve new geoengineering projects. You have the power to veto projects. You are concerned that a new program to seed marine clouds is going to adversely affect the level of precipitation on many small Pacific islands, but the fact that this will also slow the rate at which global temperature is rising is proving a decisive factor for many representatives. How would you use the power of your veto?

9. Frustrated by slow progress to limit the rate of global warming, and after suffering from a series of disastrous grain harvests, the Russian government decides to act unilaterally and invests billions of dollars to place solar reflectors into space. The results are impressive at first, but problems soon arise as lower solar radiation over the boreal winter affects the position of the jet stream over the USA and Europe; Arctic winters start to take hold as early as November. The European Union, already incensed by a series of sharp increases in the price of natural gas from Russia, is demanding immediate action. As the US Ambassador to Russia you are asked to mediate. What would you do?

10. A new hard-line government in China refuses to cut emissions any further until India and the African nations agree to impose a one-child-only policy. They argue that only this can lower and then decrease population growth in time to limit the damage from climate change. As health minister in South Africa you are deeply opposed to the injustice of this proposal, but you desperately need Chinese investment in clean energy development. How would you respond with an alternative proposal?

Key Terms

Please make sure you are familiar with the key terms introduced and highlighted in this chapter. A full glossary is available at the end of the book.

Biochar, p. 343
Biostorage, p. 338
Carbon capture and sequestration (BECS), p. 343
Carbon dioxide reduction (CDR), p. 339
Carbon wedge, p. 337
Cloud albedo, p. 341

Cloud condensation nuclei (CCN), p. 341
Conservation, p. 338
Efficiency, p. 338
Geoengineering, p. 339
Greenhouse gas intensity, p. 326

L1 (Lagrange) point, p. 340
Marine cloud reflectivity, p. 341
Ocean fertilization, p. 342
Reducing Emissions from Deforestation and Forest Degradation (REDD), p. 335

SMART (specific, *m*easure-able, *a*chievable, *r*ealistic, and deliverable on *t*ime) goals, p. 331
Solar radiation management (SRM), p. 339
Solubility pump, p. 342

Climate Change and the Scientific Method

All scientific research is guided by a series of conventions known collectively as the *scientific method*. The scientific method describes how to plan and execute a research project; how to analyze, synthesize, and evaluate the data; and how to form a hypothesis that can explain the data and predict future results with some certainty (**Figure A.1**). The scientific method is easy to learn and applicable to most topics. It guides and directs research so that its conclusions truly extend our knowledge and understanding. The following sections review the scientific method and show how you can use it to guide your study of this text.

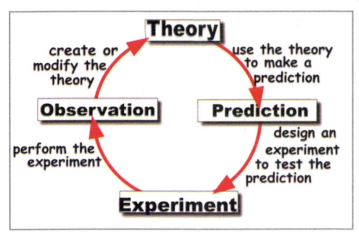

▲ **Figure A.1:** The steps and cycle of the scientific method.

The Question

All research starts with a simple question. Most scientific questions are routine and add to our knowledge and understanding of some existing theory, but the best questions are profound and can shake the foundations of prevailing science.

Paradigms in science are tenacious, but they can be wrong and history is littered with the shattered remains of scientific "certainties." We used to believe in a flat world, a biblical flood, and a **geocentric universe** and more recently we were quite sure that continents do not move and that much of the DNA in our genome is just "junk." The hardest part of any research is to accept that what you know for sure to be true may not, in fact, be true. This is exactly why a disproportionate amount of original research is produced by young minds that haven't yet learned that it "just can't be so."

The Role of Skepticism

All climate scientists are climate skeptics. Scientists must be prepared to consider an unpalatable truth, even if it upsets a favorite theory. Skepticism is the essence of the scientific method. Science focuses on the question, not the anticipated answer, and the scientific method guides scientists along a path with the best chance of advancing real knowledge and understanding. Climate science is a young science with few unshakable paradigms. There is considerable outstanding uncertainty, and there are many questions that still need to be asked.

The Scientific Method

As an example to illustrate the scientific method, let's explore the question "Why are the glaciers in Glacier National Park retreating?"

Four main factors could explain this:

1. There has been a decrease in winter precipitation.
2. More winter precipitation falls as rain than in the past.
3. More ice melts in the spring and summer than in the past.
4. More ice is lost by direct evaporation than in the past.

To find out what is actually happening, researchers must first decide what factors they want to investigate, what questions they need to ask, and what observations they need to make.

What observations and measurements would you make in order to answer the question "Why are the glaciers in Glacier National Park retreating?"

Literature Survey

Most research begins by checking to make sure the original premise is correct. To determine why the glaciers in Glacier National Park are retreating, you should start with a literature survey to become familiar with existing research. This will help you to understand the basic facts and direct you to sources where you can collect all the available data on regional precipitation, air temperature, ice thickness, evaporation, snowfall, rates of glacier retreat, and **meltwater** discharge. Following your literature survey and some follow-up communication with key researchers, you should be able to spot what data are missing and devise a plan to complete the record.

Planning

When planning research, you must think very carefully about the data required. Where do you need to go, and for how long? What instruments do you need, how much do they cost, and how much extra help is required? Climate research is seldom conducted under ideal conditions. Most research is conducted on a limited budget that forces you to distinguish between what is essential and what is desirable. A seasonal window often limits the time you have on-site for field observations and the installation of equipment. You have one chance to get it right because afterthoughts are prohibitively expensive.

Observing and Recording

Most climate research involves fieldwork to collect data. Data collection lies at the core of a research project and is often the most enjoyable and most challenging aspect of the research experience. In order to determine why the glaciers in Glacier National Park are retreating, you may need to extend the data collected from photographs, satellites, and meteorological records with direct ground-based readings of local temperature, the amount and form of precipitation, the flow rate of meltwater, and the variable thickness of ice. This is the point where careers are made and lost and each day scientists wonder what important field evidence might be staring them in the face if they could only just see it. Scientific observation requires us to see what is really there, as opposed to what we expect to be there, but the cards of perception are stacked against us. It is not easy to observe objectively. Most often we perceive what we expect to see and fit our observations into prevailing mental models of how the world works. Too often these models are incomplete.

Analysis

The word "analysis" comes from the Greek to "loosen up" and refers to the process where scientists start to look for statistically meaningful trends and relationships that exist within each data set. But how accurate are these data? Were the instruments calibrated properly? What were the physical conditions like during installation and on-site observation? (Warm, dry, well-fed researchers are much more reliable than cold, wet, hungry ones.) Are you sure the recorded data actually measured what you wanted to measure? What are the statistical limits of certainty? If the primary observations are not yours, how much are you willing to base your reputation on another researcher's data? In

this example your analysis should provide you with a reliable record of trends in temperature, precipitation, seasonal melting, surface evaporation, and meltwater flow that covers multiple seasons.

Data Synthesis

The word "synthesis" comes from the Greek for "pulling together" and is the act of combining all the different analytical data together from different sources to look for meaningful relationships. Scientists often wish they had more data at this point in their research, and this is exactly why the first steps in designing a research program are so important.

Synthesis requires careful attention to detail. It is all too easy to overlook a small subset of your data that is inconsistent with your expectations and just dismiss them as due to instrument or human error. U.S. scientists had data about the developing ozone hole over Antarctica before their British colleagues finally understood their implication. The U.S. team had the data, but misinterpreted readings as instrument error. In doing so, they missed one of the most interesting scientific observations of the century. Making significant scientific progress is not easy, not obvious, and most often not achieved by running with the crowd.

When your synthesis is complete, you should have some ideas of how fast the glaciers are retreating, how temperature, precipitation, evaporation, and melting have changed over time, and how these factors are interrelated.

Forming a Hypothesis

A hypothesis is a plausible explanation for all the observable data that allows scientists to make accurate future projections. Most often a good hypothesis starts with an educated guess. In the case of the receding glaciers, you will have observed that:

1. More winter precipitation does fall as rain than in the past.
2. More ice does melt in the spring and summer than in the past.
3. More ice is lost by direct evaporation than in the past.

You can therefore hypothesize that the reason the glaciers are retreating is that rising global temperature has resulted in more ice melting over the spring and summer months and less ice accumulating over the winter months. If just one of these observations turns out to be false, and, for example, glaciers continue to retreat when there is an increase in frozen winter precipitation, then it is back to the drawing board for a new hypothesis that will explain all observations.

Testing the Hypothesis

Testing a hypothesis is at the core of the scientific method. As scientists collect more data each year, they must check whether the data are consistent with previous observations. In the case of the receding glaciers, a further increase in global temperature should be followed by more winter rainfall, the earlier onset of spring, higher maximum summer temperatures, an increase in surface evaporation and melting, and an increase in meltwater runoff. If future observations match these projections, the hypothesis is supported. If even one observation fails to match the projected outcome, the hypothesis must go back to the drawing board for revision.

Forming a Theory

When all the strands of research are drawn together and the different hypotheses are reconciled, tested in the field, and proven to have predictive capability, a new theory is born. A unified theory for global climate change is within reach, but many competing hypotheses must still be tested. Sometimes it feels like we are getting close, and at other times it seems more distant, but we have a problem: If we wait to prove beyond doubt that all our hypotheses about climate change are correct, it may be too late to prevent some of the worst impacts of climate change on society. Science can take us only so far.

A Quick Science Primer:
KEY SCIENTIFIC CONCEPTS

Temperature

In the world of science, temperature is measured on the Kelvin (K) scale, starting from the point of **absolute zero temperature** (−273.15°C). The more commonly used centigrade scale is measured from 0°C, the freezing point of water, which is at 273.15 on the Kelvin scale. In most cases in this book, temperature is given in degrees centigrade, as °C. We use degrees as a defining unit when we discuss temperature in terms of Centigrade (e.g. 100°C) but we never use degrees when referring to the temperature in Kelvin (e.g. 100 K).

Pressure

Pressure is a measure of force acting on a unit area. Under the système international d'unités (the SI system), pressure is measured in newtons (the SI unit of force) per square meter and given the unit pascals (Pa). Many older books and web sources still express pressure in terms of bars and millibars, where standard atmospheric pressure is measured as 1,013.25 bars. Conversion is as easy as moving the decimal point, so that 1 standard atmosphere = 101,325 Pa = 1013.25 hPa (hectopascals).

The pressure of air falls off in the troposphere with increasing height, so that at 1,000 meters it falls to around 900 hPa, at 2,000 meters it falls to 800 hPa, at 6,000 meters it falls to 450 hPa, and at 10,000 meters it is just 250 hPa.

Energy, Force, Work, and Power

Energy is a very familiar concept to us all. We understand that it can be neither created nor destroyed, we can feel and measure its effects, and we observe how it can be converted from one form to another and back again, as long as the total amount of energy in a system is always conserved (that is, stays the same).

In our study of climate, we must consider many different forms of energy, such as electromagnetic (from the Sun), kinetic (from moving water, ice, and wind), potential (from the pull of gravity), electrical (lightning), and, of course, thermal (heat) energy.

Energy is used when we apply a *force* to do *work*. Think of the energy used when we do the work of moving a large apple across the top of a table. To start the apple moving, we use energy to apply a force (we push it) so that it starts to move faster and faster. As

long as this same force is applied, the apple will keep moving faster and faster (*accelerating*) as time passes. We measure the total force applied by taking the mass (m) of the apple and multiplying that by the rate of acceleration (a), as F = ma. This makes sense, as we need more force to move a big apple (mass) and still more if we want to move it faster (acceleration). Force is measured in units called *newtons (N)* ($1\ N = 1\ kg \cdot m/s^2$)

The total amount of energy used to move the apple depends on how far it is moved under the influence of this force. The total amount of energy, the work done, is calculated by multiplying the magnitude of the force applied by the distance over which it is applied (W = F × S). *Work* is measured in *joules (J)* ($1\ J = 1\ kg \cdot m^2/s^2$), the units of energy, and this is why you often see energy defined as the ability to do work.

So we can describe how large a push was applied to the apple and how far it was moved, but how long did it take for the apple to move across the table? Did it take just a second, or did it take hours? The time taken is a measure of the "power" applied. *Power* is the *work* done divided by the time (t) over which the force was applied (P = W/t) and is measured in joules per second, or watts ($1\ W = 1\ kg \cdot m^2/s^3$). To move the same apple the same distance more quickly requires more energy.

Solar Radiation

Energy from the Sun reaches Earth as electromagnetic radiation, and most of this is at the wavelength of visible light. If you think about it, this makes sense, because our eyes have evolved to make best use of as much of the available light as possible. Electromagnetic energy can be thought of in terms of traveling waves, with alternating peaks and troughs. The distance between successive peaks or troughs is known as the *wavelength*, the size of each peak as the *amplitude*, and the speed at which they travel as the *frequency*.

The electromagnetic spectrum is very broad, ranging from high-energy, very short-wavelength gamma rays at one end of the spectrum; through ultraviolet, visible, and infrared light in the middle; to very long-wavelength microwaves and radio waves at the other end of the spectrum. The stratosphere acts as a barrier against much of the short-wavelength, high-energy ultraviolet radiation from the Sun that never makes it to Earth's surface. Ultraviolet radiation is harmful and can cause irreparable damage to living organisms.

Blackbody Radiation

All objects above absolute zero temperature emit blackbody radiation. The amount of energy a body emits is in direct proportion to the fourth power of temperature (T^4). Cool objects emit radiation at infrared wavelengths, which are invisible to the naked eye but easily visible with infrared cameras. Objects hotter than 480°C emit visible light starting at the red end of the spectrum with dark red at 580°C, and becoming white hot at greater than 1,400°C.

Heat Capacity

The heat capacity of a substance is a measure of how much energy it takes (such as radiation from the Sun) to raise its temperature by 1°C. This is not the same for all materials, and it varies with changing temperature, but at around 20°C, it takes as much as 4,186 J kg^{-1} (joules per kilogram) to raise the temperature of pure water by 1°C, and it takes as little as 790 J kg^{-1} to raise the temperature of granite by 1°C. The following is a list of the heat capacities of some materials of interest:

Heat Capacity (J kg^{-1})	Substance
790	Granite
830	Quartz sand
1,005	Dry air
1,700	Wood
2,093	Ice
2,512	Mud
4,186	Water

The same amount of energy will raise the temperature of granite five times higher than it will raise the temperature of water. A high heat capacity means that water stores a lot of heat energy but warms slowly and cools slowly. In contrast, sand, with a low heat capacity, warms quickly and cools quickly. Think of a day at the beach: The Sun heats the surface of both sand and ocean with the same amount of energy, but you burn your feet on the sand and not the water. At night, the sand loses energy quickly and cools rapidly, but the sea stays warm. This very important physical property has a profound impact on weather and climate around the world.

Latent Heat

If heat capacity is a measure of the energy it takes to raise a mass of 1 kg by 1°C, then latent heat is a measure of the "extra" energy it takes to turn that substance from solid to liquid or liquid to gas. It takes around 4,186 Joules (4.186 kJ) of energy to raise water by 1°C but a further 2,400,000 Joules (2,400 kJ)—more than 500 times as much energy—to convert 1 kilogram of water into water vapor. When water heats up and evaporates, the water vapor stores all this extra energy internally as it convects in the atmosphere and moves with the wind. When that water vapor finally cools and condenses back into liquid water, to form clouds, all that energy is released back into the atmosphere, pumping a massive amount of heat into the troposphere. Next time you watch a thundercloud overhead, watch how fast the storm can build and consider how the thermodynamic engine that drives the storm is fueled by the release of latent heat. You can experience this effect for yourself. Blow on the palm of your hand and feel how cool the air feels. Now wet your palm and blow again. It feels cooler this time because the water is evaporating and taking the latent heat of vaporization from its surroundings, including your hand!

Heat Transfer

In the oceans and atmosphere, heat is mainly transferred between objects by conduction, convection, and radiation.

Conduction

Conduction occurs as heat is transferred between objects through direct contact. This is why a metal spoon soon feels too hot to the touch if left lying in hot coffee. Some materials are good thermal conductors, and some are poor conductors (and therefore great insulators). Water and air are generally poor thermal conductors.

Convection

Convection occurs due to the thermal expansion of materials when they are heated. As air and water expand, they can become less dense than their surrounding medium and gain buoyancy. In the atmosphere, convection sets in as the surface air warms and rises into the atmosphere, drawing cooler air in at the surface to be heated. Heat is also exchanged with the surrounding air during convection through turbulent mixing and conduction. This is exactly how a convective room heater or radiator works in your home.

Radiation

Heat can be transferred by electromagnetic radiation. Nonreflective materials such as the ocean surface, forests, fields, and mountains absorb incoming radiation from the Sun. As they warm, they start to emit more radiation at different (infrared) wavelengths to the surrounding environment. This is how you "feel" heat from the Sun or your fire or stove. Some may be due to convection heat transfer or conduction, but most is due the direct transfer of energy to your skin and clothes by infrared radiation.

Sensible Heat

Sensible heat, as the name suggests, is the heat you can feel in the air around you. When air near Earth's surface is heated by the sun it expands and becomes lighter and more buoyant than the surrounding air. This warm buoyant air rises into the troposphere transferring sensible heat into the troposphere and moves with the prevailing wind. This is an important factor in the distribution of the Sun's energy in the atmosphere, but it is not as important as the impact of latent heat.

Volumetric Thermal Expansion

Most materials expand and increase their volume as they warm and contract and decrease their volume as they cool. This is true of both air and water, although fresh water is unusual because it reverses this trend and starts to expand again when it gets close to its freezing point. This is good news for aquatic life because it means that the coldest water rises to the surface before it starts to freeze instead of freezing from the bottom up. This density reversal does not happen in the oceans, and seawater becomes progressively denser as the temperature falls toward the freezing point.

The amount of expansion/contraction varies with temperature. For water, the proportional increase in volume is on the order of 88×10^{-6} per degree centigrade, but for air it is much larger, around 3.43×10^{-3} per degree centigrade. These figures seem small, but they have a big impact on the ways that water and air behave. To illustrate this point, simply calculate what the effect would be of increasing the temperature of 1 cubic kilometer of water by 5°C. The water level would rise by as much as 30 to 40 centimeters. This is why much of the rise in sea level over the past century has had more to do with thermal expansion than with melting ice.

Convection

The lowest part of the atmosphere is called the *troposphere*, a term that comes from the Greek for "turning" or "mixing." Air in the troposphere is in a constant state of motion. As the Sun heats the surface of Earth, the air above soon warms, expands (see the section on thermal expansion above), and rises into the troposphere. As it ascends, it slowly cools, until it reaches the same density as the surrounding air. At this point, it comes to rest until is loses heat, becomes denser than the surrounding air, and starts to sink. Convection stops at the top of the troposphere (called the *tropopause*) because the stratosphere above is warmer due to the presence of ozone, a greenhouse gas.

Adiabatic Processes

Adiabatic processes are very important because they determine how the temperature of air changes as it rises and falls through the troposphere. The assumption is that a parcel of convecting dry air will not exchange heat directly with its surroundings (think of a well-insulated balloon). This is not exactly true, but it is a close approximation to real-life observations if the air moves rapidly.

A parcel of dry air that ascends through the troposphere expands because the confining pressure exerted by the surrounding atmosphere falls with altitude. To expand, it must push the surrounding atmosphere out of the way. This is a form of work that uses energy. Because this is an adiabatic process, the energy required must come from within the parcel of air, and as it uses energy to expand, it cools.

When a parcel of dry air descends through the troposphere, it is compressed by the growing weight of the overlying atmosphere. Such compression is a form of work done on the parcel of air by the atmosphere, and this causes it to warm and (ideally) return to the ground at its original temperature.

You can demonstrate the expansion process by breathing onto your hand with your mouth open, as if you were yawning. The air feels warm, as it is at body temperature. Next blow hard onto the palm of your hand. The air feels much colder, yet it is the same air. The reason for this is that when you blow hard, the air coming out of your mouth is expanding rapidly, pushing against the surrounding air and cooling as it uses thermal energy to expand. Anyone who has ever used a bicycle pump knows that the compression theory is also true: As you pump (compress) the air, it is heated, and you can feel the heat where the pump connects to the wheel. This is not just of academic interest. The next section considers how the atmosphere convects, with air rising and falling in the troposphere. From this point, you should consider that anywhere air is ascending, it will be cooling, and anywhere air is descending, it will be warming.

Lapse Rate

The lapse rate is a measure of how air temperature changes with altitude. The environmental lapse rate (ELR) is the temperature you could measure each day if you sailed up through the atmosphere with a thermometer; it is roughly 6.48°C per kilometer. To understand the stability of the atmosphere, it is important to consider what happens to an individual parcel of air as it rises or falls through the atmosphere by convection. If the parcel stays warmer than the surrounding air, it will continue to ascend; if it reaches the same temperature as the surrounding air, it will have neutral buoyancy and stay where it is, but if it becomes cooler than the surrounding air, it will start to descend toward the surface.

We have already seen how adiabatic processes cause air to cool as it rises and warm as it descends. The rate at which the temperature of a rising parcel of air changes due to expansion and compression is known as the dry adiabatic lapse rate (DALR). If we calculate what the temperature of this parcel of air will be at any one time and then compare that to the temperature of the surrounding air (remember that this depends on the environmental lapse rate), we can tell if the parcel is going to stay where it is, keep ascending, or descend toward the surface. The DALR is normally around 9.8°C per kilometer.

This calculation becomes more complex when a rising parcel of air cools to the point where condensation takes place and releases the latent heat of vaporization into the rising air. This helps the air to stay warmer for longer and allows it to rise higher into the troposphere before it cools to the point where it can rise no further. This slower rate of cooling is known as the moist (or saturated) adiabatic lapse rate (MALR) and is variable, at around 5°C per kilometer. If ascending air loses most of its moisture by precipitation, it can return to the surface warmer than it started if that moisture is not replaced due to the difference between moist and dry adiabatic lapse rates.

Condensation

Clouds play a very important role in regulating the temperature of Earth by reflecting up to 70% to 95% of the Sun's energy. Anyone who has been in an airplane and looked out the window at the clouds below will know just how bright this reflection can be. Anything that either promotes or inhibits the formation of clouds is a critical factor in determining global temperature. Before clouds can form, the air needs to be cooled to the point where it becomes saturated with water. Even then, water does not easily condense without some solid surface to act as a substrate for condensation.

The air is full of small particles that can act as nuclei for condensation, such as dust, clay, soot, salt, and sulfate aerosols. The average diameter of cloud condensation nuclei (CCN) is only 0.0001 millimeters (1 micron), compared to 0.02 millimeters for a cloud drop and 2 millimeters for a raindrop. Many common processes can change the number of CCN in the atmosphere, including volcanic eruptions, dust storms, forest fires, and industrial pollution.

Some research suggests that marine plankton and even cosmic rays can increase the number of CCN. This is critical research because clouds cover around 65% of the globe at any one time, and a decrease in cloud cover of just 1% could almost double the impact of anthropogenic warming. Of course, the opposite is also true: If cloud cover were to increase by 1% to 2% globally, this could offset any of the effects of warming due to greenhouse gases. This is so critical that there are now two NASA satellites, *TERRA* and *AQUA*, working to gather critical information on cloud cover.

Dew Points and Water Saturation

It is true that it can be too cold to snow! The amount of water vapor in the air around us depends on the temperature of that air and is known as the degree of water saturation. Water evaporating into the air is (in some ways) like sugar dissolving in coffee. You can add only so much before you reach a point where no matter how much you try to add, an equal amount will precipitate out at the bottom of the cup. The colder the air (coffee), the less water vapor (sugar) it can hold. Conversely, the warmer it is, the more it can hold.

Another way to think of air is as if it were a sponge. Imagine a dry sponge to which you slowly start to add water. With each drop added, more is absorbed until the point where the sponge becomes "full." From that point on, for every new drop added, a drop of equal amount simply leaks out from the bottom. If you then squeeze the water-saturated sponge, water will run out freely until you stop. In this way, a dry sponge is like warm, dry air that becomes progressively saturated as it moves over a lake or an ocean. As this air, now rich in water vapor (but not usually saturated), warms at the surface, it expands and rises into the troposphere, where it soon cools. Cooling this moisture-laden air is like wringing water out of the sponge. The air can no longer hold as much water vapor and the water condenses ("is squeezed out") to form clouds or precipitation. The temperature at which condensation first starts to take place is known as the *dew point*, a name taken from the temperature at which night air slowly cools, becomes saturated, and forms dew on the ground.

There are limitations to the analogy of "sponge air." One major breakdown is that a water-saturated sponge is heavier than a dry one. Water vapor is actually lighter than the molecules it displaces, so the buoyancy of moist air actually increases with water saturation, helping it to rise faster into the troposphere.

The link between water saturation and temperature is one of the main reasons that the Antarctic is a cold desert and one of the driest places on Earth, while the tropics are so wet. Tropical air can be warmer than 30°C, and at that temperature, air can hold as much as 40 to 50 grams of water per kilogram. When air with this much water vapor is cooled to the point of saturation, all the latent heat that was stored during evaporation is suddenly released, pumping vast amounts of energy into the atmosphere. This gives the air enough buoyancy to rise rapidly to the highest parts of the troposphere and sometimes enough momentum to rise past the troposphere and penetrate into the lower stratosphere for a while.

Albedo

Albedo is a measure of how much of the Sun's incident energy is reflected from any surface on Earth. This is an important calculation, as a decrease in albedo will lead to global warming. The following table shows the albedos of some common surfaces:

Natural Material	Albedo
Fresh snow	0.8–0.9
Ocean ice	0.5–0.7
Clouds	0.1–0.9
Desert sand	0.4
Bare sand	0.17
Trees	0.15–0.18
Conifer forest	0.08

Positive and Negative Feedback

One of the reasons that Earth's climate is difficult to model is that there are many feedbacks locked into the system. Many of these are well known, and some are less well known, and it is still quite possible that there are some critical feedback mechanisms that remain to be discovered. Feedback occurs when some change in the climate system is either accentuated or diminished because of other changes that take place as a consequence.

An example of a positive reinforcement would be when rising global temperatures melt polar ice and expose either dark earth or ocean underneath. These new surfaces are much less reflective than ice and absorb more of the Sun's energy, helping to melt even more ice. An example of a negative reinforcement would be when higher global temperatures lead to the evaporation of more water from the oceans, and this in turn leads to the formation of more clouds. As clouds reflect more of the Sun's heat, this would cool Earth and offset the effect of the original heating. There are many more examples of both positive and negative reinforcement that are covered in Chapters 3 and 4.

Online Resources

Government Agencies and NGOs

United States Government

www.whitehouse.gov/energy
The White House

www.weather.gov/organization
National Weather Service

www.climate.gov/#climateWatch
National Oceanic and Atmospheric Administration (NOAA)

www.cpc.ncep.noaa.gov
NOAA Climate Prediction Center

www.ncdc.noaa.gov/paleo/perspectives.html
NOAA National Climate Data Center, Paleo Perspectives

www.ncdc.noaa.gov/ghcnm/
NOAA National Center for Atmospheric Research (NCAR)

http://lwf.ncdc.noaa.gov/sotc/
NOAA State of the Climate

http://climate.nasa.gov
National Aeronautics and Space Administration (NASA) main climate change portal

www.giss.nasa.gov
NASA Goddard Institute for Space Studies (GISS)

http://science.nasa.gov/earth-science/missions/
NASA satellite missions related to climate change

www.epa.gov/climatechange/
Environmental Protection Agency (EPA) Climate Change

www.epa.gov/climatechange/glossary.html
Glossary of Climate Change Terms

www.nap.edu/catalog.php?record_id=12914
National Security Implications of Climate Change for U.S. Naval Forces

http://nsidc.org
National Snow & Ice Data Center (NSIDC)

www.usgs.gov/climate_landuse/
U.S. Geological Survey (USGS) climate portal

www.globalchange.gov
U.S. Global Change Research Program

www.cdc.gov/climateandhealth
U.S. Centers for Disease Control and Prevention (CDC)

www.hud.gov/offices/cpd/about/conplan/greenhomes.cfm
U.S. Department of Housing and Urban Development

www.state.gov/e/oes/climate/
U.S. Department of State: Global Climate Change

http://climate.dot.gov/about/index.html
U.S. Department of Transportation.

www.cna.org/reports/climate
Climate Change Clearinghouse

www.cna.org/reports/climate/
U.S. Center for Naval Analysis

https://www.llnl.gov
Lawrence Livermore National Laboratory

University and Private Research Units

www.cesm.ucar.edu
University Cooperation for Atmospheric Research (UCAR)

https://www.meted.ucar.edu
UCAR MetEd site (requires free registration)

http://ncar.ucar.edu
National Center for Atmospheric Research (NCAR)

www.comet.ucar.edu/index.php
UCAR/NOAA COMET Program

www.yaleclimatemediaforum.org
Yale University Forum on Climate Change & the Media

http://berkeleyearth.org
Berkeley Earth Surface Temperature

www.whoi.edu
The Woods Hole Oceanographic Institution

National Council for Science and the Environment

http://www.camelclimatechange.org/resources/
Climate Change Exercises

European Union

www.esa.int/esaEO/SEMTLZ2VQUD_planet_0.html
European Space Agency (ESA) climate portal

https://earth.esa.int/web/guest/missions
ESA satellite missions related to climate

http://ec.europa.eu/dgs/clima/mission/
index_en.htm
European Commission Climate Action

www.cru.uea.ac.uk/en
Climate Research Unit, University of East Anglia, UK

www.metoffice.gov.uk/climate-change
United Kingdom Meteorological Office (Met Office) Climate

United Nations and Other International Organizations

www.un.org/wcm/content/site/climatechange/
gateway/
United Nations climate change gateway

http://unfccc.int/2860.php
United Nations Framework Convention on Climate Change (UNFCCC)

http://unfccc.int/kyoto_protocol/items/2830.php/
UNFCCC: Kyoto Protocol

www.ipcc.ch
Intergovernmental Panel on Climate Change (IPCC)

www.ipcc.ch/publications_and_data/ar4/wg1/en/
contents.html
IPCC Working Group I (WGI) report

www.ipcc.ch/publications_and_data/ar4/wg2/en/
contents.html
IPCC Working Group II (WGII) report

www.ipcc.ch/publications_and_data/ar4/wg3/en/
contents.html
IPCC Working Group III (WGIII) report

www.ipcc.ch/ipccreports/sres/emission/index
.php?idp=0
IPCC Special Report Emissions Scenarios (SRES) report

www.un.org/en/globalissues/index.shtml
United Nations Global Issues

www.wmo.int/pages/themes/climate/index_en.php
World Meteorological Association (WMO) Climate

http://climatechange.worldbank.org
The World Bank Climate Change

Nongovernmental Organizations

www.ucsusa.org/global_warming/
Union of Concerned Scientists Global Warming

www.climatehotmap.org/index.html
Union of Concerned Scientists Climate Hot Map

www.worldwildlife.org/climate/index.html
World Wildlife Fund (WWF)

www.rtcc.org
Responding to Climate Change (RTCC)

www.nrdc.org/globalWarming/default.asp
National Resources Defense Council Global Warming

www.greenpeace.org/usa/en/
Greenpeace

www.worldwatch.org
Worldwatch Institute

www.c2es.org
Center for Climate and Energy Solutions (C2Es)

http://climateandsecurity.org
Center for Climate & Security

www.cfr.org/climate-change/climate-change-
national-security/p14862
U.S. Council on Foreign Relations

Energy Agencies/Organizations

www.eia.gov/environment/
U.S. Energy Information Administration (EIA)

www.wri.org
World Resources Institute

www.wri.org/project/cait/
World Resources Institute (WRI) Climate Analysis Indicators Tool (CAIT)

www.iaea.org
International Atomic Energy Agency

www.ren21.net
Renewable Energy Policy Network

www.bp.com/sectionbodycopy.do?categoryId=
7500&contentId=7068481
BP Statistical Review of world energy

www.worldenergyoutlook.org
International Energy Agency

http://energy.gov
U.S. Department of Energy (DOE)

www.ne.doe.gov/np2010/neScorecard/
neScorecard.html
U.S. DOE Nuclear Energy

www.altenergy.org
Alternative Energy

www.oecd-nea.org
Organisation for Economic Co-operation and Development (OECD) Nuclear Energy Agency

http://europa.eu/legislation_summaries/energy/european_energy_policy/index_en.htm
European Energy Policy

http://ec.europa.eu/energy/index_en.htm
Energy Strategy for Europe

http://europa.eu/legislation_summaries/energy/nuclear_energy/index_en.htm
Nuclear Energy

http://iet.jrc.ec.europa.eu
Joint Research Centre (JRC)

www.managenergy.net
European Commission, Energy

www.wcre.de/en/index.php
World Council for Renewable Energy

www.futuregenalliance.org
FutureGen Alliance

www.world-nuclear.org/info/inf16.html
World Nuclear Association

www.wavec.org/en
WaveEnergy Centre

www.worldenergy.org
World Energy Council

www.gwec.net
Global Wind Energy Council

http://cta.ornl.gov/cta/
Oak Ridge National Laboratory (ORNL) Center for Transportation Analysis

http://webarchive.nationalarchives.gov.uk/+/http://www.hm-treasury.gov.uk/sternreview_index.htm
Stern Review on the Economics of Climate Change

http://royalsociety.org/policy/publications/2009/geoengineering-climate/
The Royal Society report on geoengineering

www.copenhagenconsensus.com
Copenhagen Consensus Center

Selected Media Sources

http://news.bbc.co.uk/2/hi/in_depth/sci_tech/2004/climate_change/
British Broadcasting Cooperation Climate Change

http://www.pbs.org/search/?q=climate%20change
Public Broadcasting System Climate Change

http://dotearth.blogs.nytimes.com
New York Times Dot Earth

http://dsc.discovery.com/convergence/globalwarming/globalwarming.html
Discovery Channel Global Warming

http://environment.nationalgeographic.com/environment/global-warming/
National Geographic Global Warming

www.exploratorium.edu/climate/index.html
Global Climate Change Research Explorer

Examples of Climate Blogs

http://agwobserver.wordpress.com
An up-to-date list of recent publications related to climate change.

www.carbonbrief.org
Brief reports on the latest developments in climate science and fact-checking of stories about climate and energy online and in the press.

www.climatecentral.org/news/
Communicates the science and effects of climate change to the public and decision makers and inspires Americans to support action to stabilize the climate, prepare for impacts of climate change, or some combination of the two.

http://climateinc.org
Climate Inc. is a blog devoted to discussion of business and climate change. It examines how business is (or is not) responding, the economics of business action, and how public policy can support action on climate change.

http://thinkprogress.org/climate/issue/
ThinkProgress was voted "Best Liberal Blog" in the 2006 Weblog Awards and chosen as an Official Honoree in the 2009 and 2012 Webby awards.

www.desmogblog.com
The DeSmogBlog Project began in January 2006 and quickly became the world's number-one source for accurate, fact-based information regarding global warming misinformation campaigns.

www.drroyspencer.com
Dr. Spencer is the U.S. Science Team leader for the Advanced Microwave Scanning Radiometer flying on NASA's *Aqua* satellite and a noted climate skeptic.

www.skepticalscience.com
The goal of Skeptical Science is to explain what peer-reviewed science has to say about global warming.

www.wwfblogs.org/climate/
For 50 years, WWF has been protecting the future of nature. The world's leading conservation organization,

WWF works in 100 countries and is supported by 1.2 million members in the United States and close to 5 million globally.

http://dels.nas.edu/basc
Climate change at the National Academies: The National Academy of Sciences, National Academy of Engineering, Institute of Medicine, and National Research Council are the nation's preeminent source of high-quality, objective advice on science, engineering, and health matters.

http://zvon.org/eco/ipcc/ar4/#
Zvon.org provides a guide to the Fourth Assessment Report of the Intergovernmental Panel on Climate Change.

www.350.org
350.org is building a global grassroots movement to solve the climate crisis that is led from the bottom up by thousands of volunteer organizers in more than 188 countries.

www.energynow.com
energyNOW! is a half-hour weekly TV news-magazine and opinion program designed to inform and engage Americans on the most pressing energy issues of the day. It was created in response to two overlapping crises.

http://judithcurry.com
The Climate Etc. blog provides a forum for climate researchers, academics and technical experts from other fields, citizen scientists, and the interested public to engage in discussion on topics related to climate science and the science–policy interface.

www.worldclimatereport.com
This popular blog by noted climate skeptic Patrick J. Michaels points out what he considers to be weaknesses and outright fallacies in the science that is being touted as "proof" of disastrous warming.

http://grist.org
Grist has been dishing out environmental news and commentary with a "wry twist" since 1999.

www.realclimate.org
RealClimate is a commentary site on climate science by working climate scientists for the interested public and journalists.

Sources of Climate Data

www.ncdc.noaa.gov/ghcnm/
NOAA Global Historical Climatology Network (GHCN)

www.ncdc.noaa.gov/oa/ncdc.html
NOAA National Climatic Data Center

http://lwf.ncdc.noaa.gov/oa/climate/climatedata.html

NOAA Online Climate Data Directory

www.ngdc.noaa.gov/ngdc.html
NOAA National Geophysical Data Center

www.ncdc.noaa.gov/paleo/datalist.html#historical
NOAA World Data Center for Paleoclimatology—Data Sets Listing

http://data.giss.nasa.gov
Goddard Institute for Space Studies

http://gcmd.gsfc.nasa.gov
NASA Global Change Master Directory

http://aom.giss.nasa.gov/srorbpar.html
NASA Determination of Earth's Orbital Parameters

www.imcce.fr/Equipes/ASD/insola/earth/earth.html
www.imcce.fr/Equipes/ASD/insola/earth/online/index.php
Astronomical Solutions for Earth Paleoclimates

www.nationmaster.com/index.php
NationMaster

www.yaleclimatemediaforum.org
Yale University Forum on Climate Change & the Media

http://edgcm.columbia.edu
Educational Global Climate Modeling

http://berkeleyearth.org
Berkeley Earth Surface Temperature

www.cgd.ucar.edu/cas/guide/
National Center for Atmospheric Research An Informed Guide to Climate Data Sets

www.climatedata.info/index.html
Climate Data Information

www.cru.uea.ac.uk/cru/data/
Data Available from Climatic Research Unit

www.ipcc-data.org
IPPC Data Distribution Centre

www.remss.com
Remote Sensing Systems

http://discover.itsc.uah.edu/amsutemps/
Daily Satellite Temperatures from the University of Alabama Huntsville

www.iceandclimate.nbi.ku.dk/data/
Niels Bohr Institute Center for Ice and Climate

http://cdiac.ornl.gov/epubs/ndp/ushcn/ushcn.html
U.S. Historical Climatology Network

Collections of Images and Animations

http://svs.gsfc.nasa.gov
NASA Scientific Visualization Studio

www.youtube.com/user/NASAexplorer
NASA GISS: A selection of educational videos

www.gfdl.noaa.gov/visualizations-oceans
NOAA Geophysical Fluid Dynamics Laboratory: Many videos at this site

www.youtube.com/user/NOAAVisualizations/videos
NOAA's Visualization Lab

www.youtube.com/user/noaa
NOAA: Videos

www.youtube.com/user/jbnsidc/videos
National Snow and Ice Data Center (NSIDC): Videos

www.pbs.org/wgbh/nova/search/results/page/1?q=what%27s+up+with+the+weather&x=0&y=0&facet[]=dc.subject%3A%22Planet+Earth%22
PBS-NOVA: Videos on climate change

Chapter by Chapter Resources

Chapters 1 and 2

http://geography.uoregon.edu/envchange/clim_animations/index.html
Global climate animations, University of Oregon: The animations at this site show the climatology of the seasonal cycle for selected climate variables for the time period 1959–1997.

http://svs.gsfc.nasa.gov/vis/a010000/a011000/a011003/
Scientific Visualization Studio, Excerpt from "Dynamic Earth": Much of the radiation energy that makes it through is reflected back into space by clouds, ice, and snow, and the energy that remains helps to drive the Earth system, powering a remarkable planetary engine—climate. It becomes the energy that feeds swirling wind and ocean currents as cold air and surface waters move toward the equator and warm air and water moves toward the poles, all in an attempt to equalize temperatures around the world.

www.teachersdomain.org/asset/ess05_int_seawifs/
Global View of the Seasons

www.teachersdomain.org/ext/ess05_int_climatezones/index.html
Understanding different climate zones

http://youtu.be/jj0WsQYtT7M
NASA Center for Climate Simulation (NCCS): Goddard Space Flight Center is the home of a state-of-the-art supercomputing facility called the NCCS that is capable of running highly complex models to help scientists better understand Earth's climate. This short video introduces the NCCS and takes you behind the scenes into the fascinating field of climate modeling. Using supercomputers to process data from satellite observations, these models are used to predict weather

and give a picture of how the Earth's systems and climate are changing.

http://youtu.be/lrPS2HiYVp8
How Does the Climate System Work? UK Meteorological Office video.

http://youtu.be/363HhzYzJlA
Climate Feedbacks: Dr. Ben Booth, climate scientist at the Met Office, explains what climate feedbacks are and how they play a role in global warming.

http://youtu.be/bjwmrg__ZVw
What Is Climate? Met Office climate change guide.

www.youtube.com/user/greenman3610/videos
Climate Denial Crock of the Week

www.youtube.com/user/scienceonasphere
Science on a Sphere: Researchers at NOAA developed Science on a Sphere as an educational tool to help illustrate Earth system science to people of all ages.

Chapters 3 and 4

Satellites

http://youtu.be/ARc9D9Ofr1o
NASA's *Aqua* Satellite: One of the primary instruments on NASA's Aqua spacecraft is the Atmospheric Infrared Sounder (AIRS), which is providing a detailed three-dimensional view of the atmosphere. This new view is helping scientists to better understand the climate system and is proving of great value also in several practical applications, including weather forecasting.

www.teachersdomain.org/resource/ess05.sci.ess.eiu.earthorbit/
Satellites Orbiting Earth: In recent years, there has been a push to better understand how Earth works as a system—how land, oceans, air, and life all interact. Satellites in orbit around Earth provide a fast and efficient means of gathering remotely sensed data about the planet as a whole. This animation adapted from NASA shows the orbital paths of the satellites in the Earth observing system.

www.teachersdomain.org/resource/ess05.sci.ess.eiu.essatellites/
Earth System: Satellites: While the Moon is Earth's only natural satellite, there are thousands of artificial satellites circling our planet for navigation, communications, entertainment, and science. These satellites are an integral part of our everyday life, and they provide a source for scientific data unavailable from Earth's surface. This video segment adapted from NASA's Goddard Space Flight Center describes some of the different kinds of satellites that orbit Earth.

http://youtu.be/unlfchZaRo0
An Introduction to *Aqua*: The first in a series of episodes looking at the instruments and applications of the *Aqua* satellite.

The Energy System

www.teachersdomain.org/resource/ess05.sci.ess
.earthsys.infrared/

Infrared: More Than Your Eyes Can See: Infrared radiation, located in the range just beyond visible light, is emitted by all matter and is very useful in gathering data about Earth and space. In this video segment adapted from NASA, learn about infrared radiation, its properties, and some of its practical applications.

http://svs.gsfc.nasa.gov/goto?20142

Electromagnetic Spectrum: This animation shows a graphical representation of the electromagnetic spectrum and includes radio waves, infrared, visible, ultraviolet, X-rays, and gamma rays.

http://youtu.be/vIUqyYEuOz4

The Electromagnetic Spectrum: NASA Science—Best of Science about the electromagnetic spectrum (EMS), gamma rays, X-rays, ultraviolet rays, visible light waves, infrared waves, microwaves, and radio waves.

www.teachersdomain.org/resource/npe11.sci.phys
.energy.infrareden/

Infrared Energy: How does NASA "see" thermal radiation? This video, adapted from NASA, explores what infrared energy is and how NASA detects it to study Earth's systems more completely. The video also includes a look at the experiment Sir William Herschel conducted that led to the discovery of infrared.

www.teachersdomain.org/resource/phy03.sci.phys
.energy.emspectrum/

The Electromagnetic Spectrum: Physicists can classify different types of electromagnetic waves on the basis of their wavelengths and frequencies. What determines the wavelength and frequency of a given wave and distinguishes one wave on the spectrum from another is the amount of energy each wave carries—specifically the level of energy in the photons of each. This video segment adapted from FRONTLINE explores the different types of radiation represented by electromagnetic waves and how their properties vary.

www.teachersdomain.org/resource/ess05.sci.ess
.eiu.sunbasics/

Characteristics of the Sun: Our Sun is an ordinary star—just one of billions of stars in our galaxy alone. However, as our own star, the Sun holds special status for us and is essential to our existence. The Sun's gravity holds the solar system together, and nuclear fusion within the Sun supplies the energy for life on Earth. Without the Sun, Earth would be a drastically different place. In this video segment adapted from NASA, learn some basic facts about the Sun.

www.teachersdomain.org/resource/ess05.sci.ess
.eiu.solarwind/

Solar Wind's Effect on Earth: The Sun produces a solar wind—a continuous flow of charged particles—that can affect us on Earth. It can, for example, disrupt communications, navigation systems, and satellites. Solar activity can also cause power outages, such as the extensive Canadian blackout in 1989. In this video segment adapted from NASA, learn about solar storms and their effects on Earth.

www.teachersdomain.org/resource/ttv10.sci.ess
.land/

The Effect of Land Masses on Climate: This video produced by ThinkTV describes the complex connections between land (the lithosphere) and other parts of Earth's climate system. Animations from NOAA show how distance from the equator affects average temperature in the Northern and Southern Hemispheres. The video emphasizes five land factors that influence climate: latitude, elevation, topography, surface reflectivity, and land use. Landmasses are one part of the Earth system that affects climate.

http://youtu.be/uVkfh89iyeU

Tracking Earth's Heat Balance: Is the heat budget of the planet changing? Thermometers on the ground can give us a snapshot of a summer heat wave or winter cold spell, but it takes something like NASA's CERES instruments to give a long-term picture of whether the planet is keeping more of its heat than it loses back into space.

http://youtu.be/w9SGw75pVas

What We Know About Climate Change: The evidence for climate change.

www.teachersdomain.org/resource/phy03.sci.ess
.watcyc.co2/

Global Warming: Carbon Dioxide and the Greenhouse Effect: Human activities are causing increasing amounts of carbon dioxide to be pumped into the atmosphere. Is this increase resulting in global warming? Most climate experts see a distinct correlation between increased atmospheric CO_2 concentrations and rising global temperatures. This video segment adapted from NOVA/FRONTLINE provides a generally accepted explanation by illustrating the heat absorbing role carbon dioxide plays in our atmosphere.

www.teachersdomain.org/resource/ess05.sci.ess
.watcyc.maunaloadata/

CO_2 Concentrations at Mauna Loa Observatory, Hawai'i: The current practice of burning fossil fuels to provide energy for transportation, heating, cooking, electricity, and manufacturing adds carbon dioxide to the atmosphere more rapidly than it is removed. The result, most climate experts agree, is an intensification of the greenhouse effect and an increase in global temperatures. This document, adapted from the Carbon Dioxide Information Analysis Center, plots an increase in atmospheric CO_2 concentrations since 1958.

www.teachersdomain.org/resource/phy03.sci.phys.matter.greenhouse2/
Global Warming: The Physics of the Greenhouse Effect: Earth's relatively stable and hospitable average temperature is the result of a phenomenon called the greenhouse effect. The presence in the atmosphere of naturally occurring compounds, known as greenhouse gases, maintains Earth's temperature. This video segment adapted from NOVA/FRONTLINE describes how human activities are increasing greenhouse gas concentrations and explains what effect this might have on global temperatures.

www.teachersdomain.org/resource/kqedcl11.sci.ess.climatemodels/
Climate Models: In this lesson from Clue into Climate, produced by KQED, learn about climate models, experiment with your own climate model, and investigate how climate models are used to predict how species distributions may change as the planet warms.

www.teachersdomain.org/resource/ttv10.sci.ess.climatemodels/
Climate Models: In this video segment produced by ThinkTV, adapted from a video by University Corporation for Atmospheric Research, see how scientists construct computer-generated climate models to forecast weather, understand climate, and project climate change. Animations depict the way scientists divide the globe into a three-dimensional grid, with many layers reaching up into the atmosphere to capture many types of information. The computers running these models create a virtual Earth, helping scientists to discern patterns that may serve as a guide to climate in the real world.

http://youtu.be/DjILZWW6Ko0
Piecing Together the Temperature Puzzle: The decade from 2000 to 2009 was the warmest in the modern record. This video illustrates how NASA satellites enable us to study possible causes of climate change. The video explains what role fluctuations in the solar cycle, changes in snow and cloud cover, and rising levels of heat-trapping gases may play in contributing to climate change.

http://youtu.be/jj0WsQYtT7M
Supercomputing the Climate: Goddard Space Flight Center is the home of a state-of-the-art supercomputing facility called the NASA Center for Climate Simulation (NCCS) that is capable of running highly complex models to help scientists better understand Earth's climate.

http://youtu.be/LjFz1FCKfT8
A Warming World: This short video announces the launch of the A Warming World Web page on NASA's Global Climate Change website.

http://youtu.be/h88WF4wOqwl
Projection of Temperature Rise to 2100: Provides the latest results from the UK Met Office's climate change research. The data are based on a midrange IPCC emissions scenario A1B.

http://youtu.be/7KQ-cAqwtXs
Global Temperature Projections with Increasing and Decreasing Greenhouse Gas Emissions: A video from the UK Met Office.

http://youtu.be/EoOrtvYTKeE
Temperature Data: 1880–2011: From NASA.

http://climate.nasa.gov/ClimateReel/TakingEarthTemp640480/
Taking Earth's Temperature: Earth's climate is changing at an unprecedented rate. This video explores climate modeling and other tools that NASA scientists use to take the Earth's temperature.

http://youtu.be/sCxIqgZA7ag
This World Is Black and White: A look at how the historic Daisy World model illustrates Earth science concepts, such as albedo and feedback loops.

www.teachersdomain.org/resource/tdc02.sci.life.eco.earthstemp/
Taking the Earth's Temperature: This video segment from FRONTLINE/NOVA follows groups of climate researchers collecting temperature data from a wide range of locations in an effort to determine the current rate of global climate change relative to climate shifts of the distant past.

http://youtu.be/JAjjjtDY8UU
Over 15,000 Records Broken as March 2012 Becomes Warmest on Record: From NOAA

http://youtu.be/-Rr2Tflw9BE
The HadCRUT Global Temperature Dataset: Compiled by the UK Met Office and the University of East Anglia's Climatic Research Unit.

http://youtu.be/KK45yZqcZdg
Met Office Projections of Temperature Changes Due to Climate Change

www.teachersdomain.org/assets/wgbh/ipy07/ipy07_int_albedo/ipy07_int_albedo.html
Earth's Albedo and Global Warming: Which would keep you cooler in summer: a white shirt or a dark shirt? Consider this: Light-colored objects reflect more solar energy (i.e., heat), and they stay cooler as a result. In contrast, dark-colored objects absorb more solar energy, becoming warmer.

www.ucar.edu/webcasts/voices/3-uncertainty.html
What Is Meant by "Climate Uncertainty?" The more we understand the sources of uncertainty, the better the science. Sources in climate science include differences in the behavior of computer models and the lack of a

crystal ball to tell us how human beings will behave in the future (in terms of population growth, industrial activities, transportation, and so on). Even so, the scientists note, it's possible to act based on what we already know with considerable confidence about global warming and climate change.

The Oceans

http://climate.nasa.gov/ClimateReel/
OceansClimateChange640360/
Oceans of Climate Change: Oceanographer Josh Willis discusses the heat capacity of water, performs an experiment to demonstrate heat capacity using a water balloon, and describes how water's ability to store heat affects Earth's climate.

www.teachersdomain.org/resource/ttv10.sci.ess
.watcyc.currents/
The Role of Ocean Currents in Climate: Ocean surface currents have a major impact on regional climate around the world, bringing coastal fog to San Francisco and comfortable temperatures to the British Isles. This video segment produced by ThinkTV uses data-based visual representations adapted from NOAA to trace the path of surface ocean currents around the globe and explore their role in creating climate zones.

www.teachersdomain.org/ext/ess05_int_globalsurf/
index.html
Ocean currents: An animation examining surface ocean currents.

www.teachersdomain.org/resource/ess05.sci.ess
.watcyc.gulfstream/
What Causes the Gulf Stream? Even with the waves lapping at their feet, few people consider ocean currents and their importance to global climate. Although the Gulf Stream cannot be seen flowing by off North America's East Coast, in Western Europe, the current's warming effect is undeniable. This video segment adapted from NOVA uses satellite imagery to illustrate the Gulf Stream's path and animations to explain how atmospheric phenomena cause it to move.

http://youtu.be/6vgvTeuoDWY?hd=1
The Ocean: A Driving Force for Weather and Climate: This animation uses Earth science data from a variety of sensors on NASA Earth observing satellites to measure physical oceanography parameters such as ocean currents, ocean winds, sea surface height, and sea surface temperature.

http://youtu.be/lazg1F9hE_c
Thermohaline Circulation: The word *thermohaline* is a combination of terms referring to "heat" and "salinity." The movement of most mid and deep water in the ocean is driven by changes in heat and salinity, both of which affect density. This animation shows the general circulation pattern of the ocean, where water sinks in the North Pole, becomes cold, and travels around the globe, in some places rising to the surface as it warms.

http://youtu.be/FuOX23yXhZ8
Density Current: The salinity and temperature of water increases its density. This video describes how the different densities of water in the ocean create currents.

http://svs.gsfc.nasa.gov/vis/a000000/a003600/
a003658/
The Thermohaline Circulation—The Great Ocean Conveyor Belt

http://youtu.be/CCmTY0PKGDs
Perpetual Ocean: This visualization shows ocean surface currents around the world during the period from June 2005 through December 2007. The visualization does not include a narration or annotations; the goal was to use ocean flow data to create a simple, visceral experience.

http://youtu.be/ZVssbK0K4wc
Sea Surface Temperature Simulation (Indian Ocean): Sea surface temperature (SST) simulation from Geophysical Fluid Dynamics Laboratory's high-resolution coupled atmosphere–ocean model. As the animation focuses on various locations of the world ocean, we see the major current systems, such as the Agulhas Current, Brazil Current, Gulf Stream, Pacific Equatorial Current, and Kuroshio Current. The small-scale eddy structure is resolved and evident.

http://youtu.be/JniS83wKmNk
El Niño: A video about the 1997–1998 El Niño.

http://youtu.be/I6-l0flmHY4
Satellites Monitor La Niña in the Pacific: NOAA's POES satellites measure the temperature of the ocean surface—one of the primary indicators of the El Niño–Southern Oscillation.

http://climate.nasa.gov/ClimateReel/
KeepingCarbon640360/
Keeping Up with Carbon: Carbon forms living organisms, dissolves in the ocean, mixes in the atmosphere, and is stored in the crust of the planet. The ocean plays a critical role in the carbon cycle and is key to understanding Earth's changing climate.

http://youtu.be/pAIb8vhUxlY
Water Vapor Transport, June through November 2005 (NASA): The Atmospheric Infrared Sounder (AIRS) in conjunction with the Advanced Microwave Sounding Unit (AMSU) sense emitted infrared and microwave radiation from the Earth to provide a three-dimensional look at Earth's weather and climate.

The Atmosphere

http://youtu.be/QWeorlkJ39M
UK Met Office Cloud Spotting Guide: How do clouds form? How can you tell one cloud from another? In this video, James Chubb explains how clouds form and how to tell one cloud from another.

http://youtu.be/G7Ewqm0YHUI
What Are Weather Fronts? What kind of weather do weather fronts bring? In this UK Met Office video, James Chubb explains what weather fronts are and the kinds of weather they bring.

http://youtu.be/kvk-hBFnBTI
What Are Air Masses? How do air masses affect our weather? In this UK Met Office video, James Chubb explains what air masses are and how they bring all sorts of different weather to the United Kingdom.

http://youtu.be/Vsri2sOAjWo
Ship Tracks Reveal Pollution's Effects on Clouds: NASA's MODIS satellite instrument reveals how air pollution may alter clouds, affecting global temperatures.

http://svs.gsfc.nasa.gov/vis/a000000/a003600/a003658/
The Thermohaline Circulation: The Great Ocean Conveyor Belt: From NASA.

www.teachersdomain.org/resource/mck05.sci.ess.watcyc.fronts/
Compare and Contrast Warm and Cold Fronts: *Weather fronts* are defined as the boundaries between different air masses. Depending on the direction of movement and the characteristics of the air involved, different types of fronts form. In this visualization from McDougal Littell/TERC, see movement of warm and cold fronts as well as the characteristic clouds that are generated by each.

www.teachersdomain.org/resource/ess05.sci.ess.watcyc.convective/
Convective Cloud Systems: The transfer of energy between Earth's surface and the atmosphere causes all weather. Energy can be transferred through three main processes: convection, conduction, and radiation. This video adapted from the Atmospheric Radiation Program explains the differences between tropical convective cloud systems formed over land and those formed over oceans.

www.teachersdomain.org/resource/ess05.sci.ess.watcyc.cloudtype/
Cloud Types: The clouds that you see are more than just determinants of weather conditions. They also help to maintain the energy budget and climate on Earth and are an essential part of the water cycle. In this interactive resource, adapted from NASA's S'COOL Project Tutorial, learn about how clouds are named and identified.

www.teachersdomain.org/resource/ess05.sci.ess.watcyc.viewjet/
A Five-Day View of the Jet Stream: Although few weather phenomena have the ability to affect weather on a global scale, the jet streams are one such phenomenon. These bands of high-speed wind roar through the atmosphere 10 to 15 kilometers (6 to 10 miles) above Earth's surface. This animated image from NOVA Online illustrates a jet stream's wild undulations over North America during a five-day period in January 2001.

www.teachersdomain.org/resource/ess05.sci.ess.watcyc.risejet/
Giving Rise to the Jet Stream: The high-speed winds of jet streams, found near the top of the troposphere, play a major role in guiding weather systems. Many factors, including Earth's rotation and the Sun's uneven heating of Earth's surface, contribute to the formation of the powerful eastward flows of the jet streams. In this interactive resource from NOVA Online, learn about the factors that create these powerful meteorological forces.

www.teachersdomain.org/resource/ttv10.sci.ess.jet/
The Effect of Jet Streams on Climate: High above Earth, fast-moving rivers of air called jet streams play a major role in our weather systems and regional climates. In this video produced by ThinkTV, examine the jet streams in the Northern and Southern Hemispheres, and discover the ways in which a change in the established spatial patterns of the jet streams could lead to more severe weather events. Scientists are carefully tracking the jet streams because climate change may permanently change their spatial patterns.

http://climate.nasa.gov/ClimateReel/27Storms640360Atm/
27 Storms: Arlene to Zeta: During the 2005 Atlantic hurricane season, 27 named storms formed, breaking many records, including the most hurricanes, the most category 5 hurricanes, and the most intense hurricane ever recorded in the Atlantic.

http://youtu.be/lO848AiDszA
2008 Hurricane Season with Sea Surface Temperature: This animation depicts the 2008 hurricane season and the corresponding water temperature for the dates June 1, 2008, through November 30, 2008. The colors on the ocean represent the sea surface temperatures, and satellite images of the storm clouds are laid over the temperatures to clearly show the positions of the storms.

http://youtu.be/ugObc9NW1Lc
The Earth at Work Bringing Hot Air North: Notice how the circulation strongly brings hot air from near the equator, where the Sun shines bright and strong, to the frigid north, and at the same time cold air from the

North Pole moves south. A ridge of high pressure over the Four Corners area gives rise to south, southwesterly flow in Reno, bringing in warm air. A trough of low pressure sits off the coast, and brings in cooler air. Air is lifted along the boundary of the cold and warm air.

www.teachersdomain.org/resource/ess05.sci.ess.watcyc.eselnino/

Earth System: El Niño: The climatic phenomenon known as El Niño is a disruption of the ocean-atmosphere system in the tropical Pacific that impacts weather and climate around the globe. An El Niño occurs every 4 to 12 years, causes die-offs of plankton and fish, and affects Pacific jet-stream winds, altering storm tracks and creating unusual weather patterns around the world. This video segment adapted from NASA's Goddard Space Flight Center details some of El Niño's far-reaching effects on both marine life and humans.

www.teachersdomain.org/resource/ess05.sci.ess.watcyc.eshurricane/

Earth System: El Niño's Influence on Hurricane Formation: Warm water fuels the tropical storms that ultimately form hurricanes. In this video segment adapted from NASA's Goddard Space Flight Center, learn how El Niño events—climatic anomalies that occur periodically in the Pacific Ocean—alter the course of atmospheric circulation and lessen hurricane formation in the Atlantic Ocean.

www.teachersdomain.org/resource/ess05.sci.ess.watcyc.rainfall/

20-Year Map of Global Rainfall: The distribution of rainfall on Earth follows clear patterns that can be traced to factors that influence cloud formation, such as the amount of solar heating, surface temperatures, topography, and proximity to moisture. In this visualization from NASA, observe the monthly distribution of global rainfall from January 1979 to January 2001, as illustrated by data gathered with a combination of remote-sensing and ground-based methods.

www.teachersdomain.org/resource/ess05.sci.ess.watcyc.cloudprecip/

Water Vapor Circulation on Earth: Water vapor plays an important role in the water cycle and in the distribution of heat around the planet. By observing the movement of water vapor, scientists can study global wind patterns and the development of cyclonic storms. In this simulation from the National Center for Atmospheric Research, water vapor circulates around Earth over the course of a year.

www.teachersdomain.org/resource/ess05.sci.ess.watcyc.oceancur/

Ocean Temperatures and Climate Patterns: Interactions between Earth's atmosphere and oceans drive weather and climate patterns. Although these interactions and patterns are complex, they are also predictable. This animation from The New Media Studio explains precipitation patterns by illustrating how differences in ocean surface temperatures create wind, and how wind patterns can in turn affect ocean surface temperatures.

www.teachersdomain.org/resource/ttv10.sci.ess.watcyc.aerosols/

The Effects of Atmospheric Particles on Climate: In this video produced by ThinkTV, see how atmospheric particles, or aerosols, can affect regional climate change. It is commonly assumed that climate changes involve atmospheric warming, but some particles, such as sulfates, can actually cool the atmosphere. Dramatic images adapted from NOAA reveal the regions where atmospheric particles are concentrated and their patterns of movement around the globe. Researchers continue to investigate the role of aerosols in regional and, ultimately, global climate change.

www.teachersdomain.org/resource/ess05.sci.ess.watcyc.hurricane/

Hurricanes: New Tools for Predicting: Hurricane Katrina, which struck New Orleans and other Gulf Coast communities on August 29, 2005, provided the worst kind of reminder of the importance of accurate hurricane prediction—and of heeding those predictions. This video segment adapted from NOVA scienceNOW a year before Katrina struck describes the current state of research into what causes hurricanes and how scientists are now able to "see" inside the storms in their ongoing efforts to more accurately predict both the path and intensity of these powerful storms.

www.teachersdomain.org/resource/ess05.sci.ess.watcyc.hurrlife/

How Hurricanes Form: Meteorologists have gone to great lengths to identify the atmospheric conditions that trigger hurricanes. The list of ingredients is fairly short: warm ocean temperatures, lots of moisture in the middle and upper atmosphere, and light winds in the upper atmosphere. In this animation from NASA, learn about the conditions necessary for the creation of a hurricane.

www.teachersdomain.org/resource/ess05.sci.ess.earthsys.deserts/

Deserts: This still collage produced for Teachers' Domain shows different features of deserts across the globe. Some are hot, and others are cold. Some contain rippled sand dunes, while others have rocky landscapes. However, deserts share one trait: all are extremely dry. Their remarkable and unique natural features—including rolling dunes, wind-sculpted rock, and flat-topped mesas rising from barren plains—are the result of their dryness.

The Cryosphere

www.teachersdomain.org/resource/ipy07.sci.life
.eco.comparepoles/

Compare the Poles: In this interactive activity adapted from the Woods Hole Oceanographic Institution, compare characteristics of the Arctic and the Antarctic. Explore differences in physical features, weather, plants, wildlife, and human impact.

http://youtu.be/PfhuvkEUD64

Larsen B Ice Shelf: This movie shows the events of January, February, and March 2002, as recorded by NASA's Moderate Resolution Imaging Spectrometer (MODIS) satellite sensor. The images show the Larsen B Ice Shelf and parts of the Antarctic Peninsula (on left).

http://youtu.be/txpHVm-9rTg

Satellite Observation of Antarctic Sea Ice Concentration: Video of sea ice concentration December 12, 1991–January 24, 2012 (from NSIDC).

www.teachersdomain.org/asset/ipy09_int_lima/
Observing Antarctica: Interactive presentation.

http://youtu.be/qwvWH5tn1Tk

Glacial Ice Loss: Greenland and Antarctica: A University of Colorado, Boulder–led team used NASA data to calculate how much Earth's melting land ice is adding to global sea level rise.

www.teachersdomain.org/resource/ipy09.sci.ess
.watcyc.lima/

Observing Antarctica: In this interactive activity adapted from NASA, students are invited to view Antarctica in unsurpassed detail using a compilation of high-resolution images called the Landsat Image Mosaic of Antarctica (LIMA). Students can choose to investigate Antarctica's moving ice; get close-up views of icebergs, snow dunes, and other land and seascape features; learn how cracks in a glacier can be used to gauge its speed; see how ridges and troughs in the ice help scientists determine glacier flow patterns; and take a flying tour of the area surrounding McMurdo Station, a scientific research center operated by the United States.

www.teachersdomain.org/resource/ess05.sci.ess
.earthsys.icesheets/

Antarctic Ice Movement: Part I: With more than 29 million cubic kilometers (7 million cubic miles) of ice and snow, the Antarctic Ice Sheet is so massive that its weight depresses the underlying crust by 900 meters (nearly 3,000 feet). New snow that collects on the ice sheet's surface causes the ice beneath it to spread out and move along the slope of the land. In this video segment adapted from NOVA, a team of glaciologists carves into one glacier on the East Sheet to monitor the nature and speed of its movement.

www.teachersdomain.org/resource/ess05.sci.ess
.watcyc.seaice/

Antarctica: Sea Ice: Each winter, the ice apron that surrounds the continent of Antarctica expands from its summertime area of about 4 million square kilometers (1.5 million square miles) to 20 million square kilometers (7.5 million square miles). Although its presence has proven treacherous for would-be explorers and commercial shippers, sea ice provides essential hunting, feeding, and breeding habitats to polar bears, seals, and penguins. It also helps regulate temperature, moisture, and ocean salinity worldwide. In this video segment adapted from NOVA, learn how sea ice forms and how its seasonal fluctuation dramatically changes the continent of Antarctica.

www.teachersdomain.org/resource/ess05.sci.ess
.earthsys.icestreams/

Antarctic Ice Movement: Part II: For the most part, an ice sheet moves downslope slowly because the ice is in direct contact with underlying bedrock. In some places, however, ice races along much faster than the rest of the sheet. These areas of fast-moving flow, called ice streams, are believed to be caused by a thin, lubricating layer of water and mud between the ice and the land. In this video segment adapted from NOVA, a team of scientists seeks evidence to support their hypothesis that atmospheric warming—either now or in the past—may explain why water has formed beneath the ice sheet.

www.teachersdomain.org/resource/ipy07.sci.ess
.watcyc.cryoarctic/

Earth's Cryosphere: The Arctic: This video segment adapted from NASA uses satellite imagery to provide an overview of the cryosphere in the northern hemisphere, including the Arctic. Investigate the cryosphere's role in North America, the extent of permafrost in the Northern Hemisphere, and how the polar regions play a role in regulating global climate. Finally, observe changes in Arctic ice cover, such as the decrease in summer sea ice over the past few decades and changes in the flow rate of Greenland's Jakobshavn Glacier.

http://youtu.be/HBlLP6KTrsE

Climate Change: Arctic Warming Pushes Winter Weather Further South: The video series is part of a NASA-supported global climate change education project at the College of Environmental Science and Forestry in Syracuse, New York, to increase climate science literacy.

http://youtu.be/6j8SGs_gnFk

Arctic Sea Ice Timelapse from 1978 to 2009: This movie shows the complete 30-year history of the NSIDC satellite–derived Arctic sea ice extent in a single video. Brown is land, black is shoreline, and blue is water (except for the large blue dot in the center of the plot).

www.teachersdomain.org/resource/lsps07.sci.life
.eco.polarbear/
Polar Bears and Climate Change: This video from the World Wildlife Fund addresses the primary threat to polar bears in the Arctic today: global warming. Scientists monitor the effects of climate change on the large predator's activities and range, study the bears' physical condition, and explore why the melting of glaciers and reduction of sea ice in the Arctic region may ultimately have dire consequences for the polar bears.

http://youtu.be/uyXQWimVrfw
NASA Video Shows Greenland Melting: NASA has released a video showing time-lapse animated images of Greenland's ice sheet from 1979 to 2009.

http://youtu.be/8SoBF4vFArg
NASA: Increasing Rate of Ice Melt in Greenland: Report on the increasing rate of ice melt in Greenland (from the BBC in 2005).

www.teachersdomain.org/ext/ess05_int_glaciers/
index.htm
Introduction to Glaciers: An interactive presentation on glaciers.

http://youtu.be/_RVAilkTF3E
Greenland Ice Lakes Drain at Speed of Niagara Falls: Researchers have been studying the melting of glaciers and ice sheets in Greenland. Although global warming is affecting the region, the runaway effect isn't as extreme as was feared.

www.teachersdomain.org/resource/ess05.sci.ess
.earthsys.esglaciers/
Earth System: Ice and Global Warming: Scientific evidence strongly suggests that different regions on Earth do not respond equally to increased temperatures. Ice-covered regions appear to be particularly sensitive to even small changes in global temperature. This video segment adapted from NASA's Goddard Space Flight Center details how global warming may already be responsible for a significant reduction in glacial ice, which may in turn have significant consequences for the planet.

www.teachersdomain.org/resource/ipy07.sci.ess
.watcyc.fastglacier/
Fastest Glacier: This video segment adapted from NOVA scienceNOW features western Greenland's Jakobshavn Glacier, dubbed the world's fastest-flowing glacier. Scientists attempt to explain why this glacier is moving at a rate that far exceeds the average speed of glaciers and is contributing to a rise in global sea level. Their explanation revolves around the recent discovery that increased meltwater caused by global warming flows down holes in the glacier. Once it reaches the underlying rock, the theory goes, this water actually lubricates and lifts the glacier, allowing it to glide downhill with decreased resistance.

www.teachersdomain.org/resource/ttv10.sci.ess
.earthsys.fieldresearch/
Field Research on Glacial Change: Glacial meltwater is a natural resource critical to the survival of people in many parts of the world. Understanding the complex relationships between changing climate, populations, and water requires years of field research. In this video produced by ThinkTV, see how researchers solve the problems of making water measurements in rugged mountain locations—one small part of the efforts of an international climate research team.

http://youtu.be/n36EMqQ6k80
Global Warming Effects in the Himalayas: Experts fear that the effects of global warming may have disastrous results in Nepal in the near future. Taken from the documentary "Meltdown: In the Shadow of Nepal's Lost Glaciers," available at FirstScience.tv.

http://youtu.be/ZN8a-pP60wk
The Most Dangerous Lake in Nepal: Lake 464—so newly formed by glacial melt that it doesn't even have a name—may be a glacial lake outburst flood waiting to happen. This clip from the film *Outburst*, still in production, shows why.

http://youtu.be/UrRQBPyBaJA
Outburst **Trailer:** In the remote Hongu Valley, melting glaciers are forming new lakes due to rapid climate change—and they are starting to burst their terminal moraines, destroying lives and livelihood downstream. A featured National Geographic/Waitt Grant project.

http://youtu.be/6FTPPqe13b0
Glacial Ice Loss Around the World: A University of Colorado, Boulder–led team used NASA data to calculate how much Earth's melting land ice is adding to global sea level rise.

http://youtu.be/YYymZgqZLb4
Glacial Ice Loss: Himalayas: This animation shows ice loss over the Indian subcontinent, as measured by the GRACE mission.

www.teachersdomain.org/resource/ipy07.sci.ess
.watcyc.icesimulate/
Ice Shelf and Ice Sheet Simulation: This interactive activity adapted from Texas A&M University simulates how the melting of ice shelves and ice sheets affects sea level. In one video, observe a block of ice melt while it floats in the tank, simulating an ice shelf; the water level does not change. In a second video, see a block of ice melt as it rests on a plank of wood above the water, simulating an ice sheet; the water level rises.

www.teachersdomain.org/asset/ess05_int_waterdist/
Global Water Distribution: This is an interactive presentation of global water distribution.

www.sms.cam.ac.uk/media/1100246
Cambridge Ideas—This Icy World: Cambridge University glaciologist Professor Julian Dowdeswell has spent three years of his life in the polar regions.

http://youtu.be/oa3M4ou3kvw
Burning Methane from Frozen Lake: This video, recorded in November 2007, on a lake near Fairbanks, shows a research team drilling a hole through lake ice and then lighting the escaping methane.

Chapters 5 and 6:
Climate History

www.pbs.org/wgbh/evolution/change/deeptime/index.html
Deep Time: An explanation of the term deep time.

www.ucar.edu/webcasts/voices/2-tools.html
What Tools Are Used to Study Climate? Climate modelers and analysts explain the complexity and reliability of the observations, prehistoric records, and climate models used to study past, present, and future climate.

http://youtu.be/ZGFAWzjO378
The 8 Minute Epoch: 65 Million Years with James Hansen: In a speech at the University of Oregon, James Hansen, NASA's chief atmospheric scientist, walks listeners through the most recent geologic periods and speaks to the commonly heard climate canard, "It's been hotter in Earth's history before—so what's the big deal?"

http://youtu.be/mX3pHD7NH58
Snowball Earth: Scientists think there may have been a time when Earth was completely covered by glaciers. From National Geographic.

http://youtu.be/G80mlbF5yEg
What the Ice Cores Tell Us: This video explores ice cores and the Medieval Warm Period.

www.andrill.org/media
Andrill Videos: A selection of videos from Andrill.

www.teachersdomain.org/resource/ess05.sci.ess.watcyc.greenland/
Greenland Ice Sheet Project 2: A Record of Climate Change: An increase of 1°F in average global temperature sounds innocuous enough. Daily high and low temperatures can fluctuate far more than that on an average day. However, changes to global averages can alter conditions that most of us take for granted. This interactive activity adapted from materials from the Wright Center for Science Education at Tufts University describes how climatologists obtain and interpret evidence from the Greenland Ice Sheet in an effort to piece together a picture of Earth's distant climate history.

www.teachersdomain.org/resource/ess05.sci.ess.watcyc.climatechange/
Climate Change: Weather is notoriously unpredictable. From one moment to the next, any of dozens of atmospheric variables can change to create a new weather event. In contrast, climate descriptions, which identify average and normal temperatures and precipitation levels, tend to be perceived as stable, at least over time scales that humans can easily relate to. However, that hasn't always been the case. This video segment adapted from NOVA describes climate data that suggest the Earth has undergone dramatic climate shifts in relatively short spans of time.

http://youtu.be/H2mZyCblxS4
Time History of Atmospheric CO_2: This video shows the history of atmospheric carbon dioxide from 800,000 years before present until January 2009. Recommend full screen/HD to read titles.

www.teachersdomain.org/resource/tdc02.sci.ess.earthsys.dinokill/
What Killed the Dinosaurs? Definitive answers to some of life's most enduring questions are often difficult or impossible to come by. However, scientific processes piece together the information that is available as a way of providing alternative explanations for a wide variety of phenomena. This interactive activity from the Evolution website outlines the evidence gathered to explain what caused the extinction of the dinosaurs 65 million years ago. It proposes four possible hypotheses and invites you to consider the evidence and come to your own conclusion.

www.teachersdomain.org/resource/tdc02.sci.life.evo.permtriassext/
Permian-Triassic Extinction: In this video segment from Evolution's "Extinction!" geologist Peter Ward shows rock layers laid down during the Permian and Triassic periods. The Permian layers contain abundant animal fossils and fossilized traces of animals, while the Triassic layers are almost devoid of fossils, suggesting that a mass extinction event occurred 250 million years ago, at the end of the Permian.

www.teachersdomain.org/resource/oer09.sci.life.evo.massextinct/
Mass Extinction: This video segment adapted from NOVA scienceNOW examines a developing theory that might explain a 250-million-year-old "murder mystery." While several possible causes have been considered for the mass extinction that occurred at the end of the Permian period, the scientists featured in the video think a chain of events, beginning with massive volcanic eruptions, might have made Earth uninhabitable for a significant number of terrestrial and marine organisms. According to their theory, these eruptions released gases that warmed both the atmosphere and the oceans. This warming in turn created ideal conditions for killer bacteria to poison the environment.

www.teachersdomain.org/resource/clim10.sci.ess
.earthsys.diatom/
Diatoms Measure Climate Change: This video segment adapted from NOVA's "Becoming Human" examines one method scientists use to understand ancient climate conditions. To test the idea that eastern Africa had undergone rapid swings in climate—from wet to dry to wet again—over a period that began 10 million years ago, German scientists studied the fossils of tiny one-celled aquatic organisms called diatoms. Knowing that white layers of a rock formation consist of deep-water diatoms, and darker layers consist of shallow-water diatoms, the scientists have interpreted the alternating layers in the formation to mean that a massive lake appeared and disappeared many times in their study area. If this part of Africa indeed experienced wet and dry periods over time, this supports a new idea that suggests climatic variability may have shaped human evolution.

www.teachersdomain.org/resource/clim10.sci.ess
.watcyc.evoclimate/
Climate and Human Evolution: This video segment adapted from NOVA's "Becoming Human" examines a compelling theory that links climate change and human evolution.

www.teachersdomain.org/resource/ttv10.sci.ess
.clues/
Clues from Past Climates: How do scientists make judgments about past climates, when reliable instrumental records have only been available for the past few hundred years? What data do they rely on? In this interactive activity produced by ThinkTV, explore climate data sources from the natural world as well as human-made documentary evidence. Learn how scientists use data from tree rings, historical documents, coral records, pollen records, cave formations, and lake and seashores to better understand past climates.

Chapter 7
Impacts of Climate Change

http://youtu.be/UgNnuLpYywU
Changing Face of Planet Earth: From NASA.

www.ucar.edu/webcasts/voices/4-impacts.html
What Are the Likely Impacts of Climate Change? Scientists describe some of the major environmental impacts of global warming already being seen and expected in the future.

http://youtu.be/dS4ft5QTyxA
Projecting Future Climate: This video explores three development paths or scenarios that climate scientists commonly use to enhance understanding of potential future climate changes. Future precipitation and temperature projections vary in response to the emission pathway that humanity follows.

http://youtu.be/wF0KkTrwUu0
Climate Change—The Evidence and the Impacts: UK Met Office scientist Dr. Debbie Hemming gives an overview of climate change, the evidence that it is happening, and the impacts on our world.

http://youtu.be/O8qmaAMK4cM
6 Degrees Warmer: Mass Extinction? If the world warms by 6°F, oceans will turn into marine wastelands, and natural disasters will become common events. From National Geographic.

http://youtu.be/7nRf2RTqANg
5 Degrees Warmer: Civilization Collapse: If the world warms by 5°F, the planet will reach a nightmare vision of life on Earth, as traditional social systems break down. From National Geographic.

http://youtu.be/skFrR3g4BRQ
Rising Seas: If the world warms by four degrees oceans will rise and glaciers will disappear, cutting off fresh water to billions.

http://youtu.be/6rdLu7wiZOE
3 Degrees Warmer: Heat Wave Fatalities: If the world warms by 3°F, the Mediterranean and parts of Europe will wither in the summer's heat. From National Geographic.

http://youtu.be/P-0_gDXqYeQ
2 Degrees Warmer: Ocean Life in Danger: If the world warms by 2°F, some of the changes to the biosphere will no longer be gradual. From National Geographic.

www.teachersdomain.org/resource/ess05.sci.ess
.watcyc.biomemap/
Biomes: The distribution of plants and animals around the world is anything but random. Instead, it is a result of the interplay of individual environmental tolerances of species and the environmental conditions, especially variations in temperature and precipitation. These interactions result in biomes, the categories into which ecologists organize similar communities of plants, animals, and the environmental conditions in which they live. This interactive resource adapted from NASA features some of the physical and biological characteristics of seven of the world's biomes.

www.teachersdomain.org/resource/ess05.sci.ess
.watcyc.sealevel/
Antarctic Ice: Sea Level Change: In 1995 and again in 2002, large fragments of the Larsen Ice Shelf sheared away from the West Antarctic Ice Sheet. In the second event, an area the size of Rhode Island collapsed from the sheet. Although these dramatic events did not add to the modest 8-centimeter rise in global sea level experienced over the past 50 years, much of the predicted 25-centimeter or greater rise in the next century may

result from the incremental melting and growing instability of the world's glaciers. In this video segment adapted from NOVA, learn what might happen to the global sea level if atmospheric warming precipitated the collapse of West Antarctic Ice Sheet.

www.teachersdomain.org/resource/ipy07.sci.ess
.earthsys.glacierphoto/

Documenting Glacial Change: In this collection of images adapted from the National Snow and Ice Data Center, photo pairings demonstrate that glacier positions can change dramatically, even over a relatively short period in geological time of 60 to 100 years. The effect on the immediate environment once glaciers recede is readily apparent. As several of the images reveal, receding glaciers leave telltale signs of their presence in the landscape. With the ice gone, glaciated terrain is characterized by an eroded valley bed and debris piles. It is ripe for colonization by pioneer plant species.

www.teachersdomain.org/resource/ttv10.sci.ess
.watcyc.glacialwater/

Climate Change Affects Glacial Water Source: Water is a precious commodity in Peru. The populous areas west of the Andes are largely desert and rely on glacial meltwater as an important source of fresh water. But the Peruvian glaciers, high in the Andes, are in rapid retreat. In this video produced by ThinkTV, join scientists from the Byrd Polar Research Center who specialize in glaciers and water quality as they monitor the steadily shrinking glaciers and the impact on local populations.

www.teachersdomain.org/resource/ayv09.sci.life
.eco.riskwater/

Global Warming Threatens World Water Supply: This video segment adapted from FRONTLINE: "Heat" explains the important role glaciers play in storing and distributing fresh water in many parts of the world. Because of global warming, glaciers are melting more rapidly than normal. This is especially significant for Himalayan and Tibetan glaciers, which feed several major rivers that flow through China and the Indian subcontinent and sustain agriculture, livestock, and the human population. As glaciers disappear, consequences of water scarcity will likely include drought and political instability between nations.

www.teachersdomain.org/resource/ean08.sci.ess
.watcyc.shishmaref/

Global Warming Threatens Shishmaref: In this video segment adapted from Spanner Films, learn about how global warming and changing sea ice conditions affect the Alaska Native village of Shishmaref. Hear firsthand accounts about how climate change has altered the condition, extent, and freeze-up of sea ice. Understand how the local subsistence way of life relies on the presence of sea ice. Learn about how houses were relocated after a strong storm in 1997 and how erosion continues to threaten the village.

www.teachersdomain.org/resource/ean08.sci.ess
.watcyc.bakedalaska/

Losing Permafrost in Alaska: This video, adapted from Spanner Films, describes the effects of changes to permafrost, the frozen layer of soil and ice that underlies much of Alaska, caused by a warming climate. Permafrost, some of which has persisted since the last ice age, more than 10,000 years ago, is melting rapidly. In affected areas, this has led to building and road damage, shrinking lakes, rapidly eroding river banks, and the disappearance of some wildlife, including fish. Scientists such as Gunter Weller, who is featured in the video, suggest that these changes will continue to worsen unless society reduces its use of fossil fuels.

www.teachersdomain.org/resource/ean08.sci.ess
.earthsys.permafrost/

Melting Permafrost: This video, adapted from the International Institute for Sustainable Development, discusses the changes happening to permafrost in the Arctic landscape. Alaska Native peoples and Western scientists discuss both the causes of melting and its impact on the ecosystem. The video shows the consequences of erosion, including mudslides and inland lakes being drained of water. An Inuit expresses his uncertainty about the ultimate effect this will have on his community and culture.

www.teachersdomain.org/resource/echo07.sci.life
.coast.climate/

Arctic Climate Perspectives: This video, adapted from material provided by the ECHO partners, shows the changes now happening in Barrow, Alaska, due to global warming. The Iñupiaq people who live in Barrow present their observations of these changes based on their centuries-old knowledge of their environment, and describe how these changes are already affecting their lives. Scientists who have come to Barrow to study climate change also offer their perspective.

www.teachersdomain.org/resource/ean08.sci.life
.eco.arctichange/

How the Arctic Ecosystem Might Change: This video segment adapted from the National Film Board of Canada examines the unique, fragile Arctic marine ecosystem. A scientist discusses the connections that exist in the Arctic food web, whose overall lack of diversity makes many of its members prone to extinction as change occurs in their sea ice habitat. The video explains that as global warming continues and sea ice disappears from the Arctic Ocean, new species from warmer, more southern oceans will eventually replace those species unable to adapt to the changing conditions.

www.teachersdomain.org/resource/ean08.sci.ess
.earthsys.inpassage/

Climate Change Impacts Alaska glaciers: In this video adapted from KTOO, explore the historic cycle of advance and retreat of tidewater glaciers in southeast

Alaska. Over the past 250 years, most glaciers in this region have lost more of their mass through melting and calving than they have gained through the new accumulation of snow and ice. This signals the end of a cooling trend called the Little Ice Age, which lasted nearly 3,000 years. If this period of retreat is extended by global warming, scientists warn that it may result in outright loss of the glaciers.

www.teachersdomain.org/resource/ess05.sci.ess
.earthsys.esglaciers/

Earth System: Ice and Global Warming: Scientific evidence strongly suggests that different regions on Earth do not respond equally to increased temperatures. Ice-covered regions appear to be particularly sensitive to even small changes in global temperature. This video segment adapted from NASA's Goddard Space Flight Center details how global warming may already be responsible for a significant reduction in glacial ice, which may in turn have significant consequences for the planet.

www.teachersdomain.org/resource/ttv10.sci.ess
.earthsys.tropical/

High Altitude Glaciers in the Tropics: This video by ThinkTV takes us to the high-altitude glaciers in the Andes of Peru and explains that glaciers can exist even at tropical latitudes. Learn how the glaciers are formed and how glaciers in the Southern Hemisphere differ from glaciers in the Northern Hemisphere. See the evidence of shrinking glaciers and learn how scientists are trying to relate this evidence to the patterns of change in other glaciers around the world.

www.teachersdomain.org/resource/tdc02.sci.life
.eco.arctic/

Arctic Tundra: This video segment from Wild Europe: "Wild Arctic" describes some of the plants and animals that make up the tundra biome and captures the harshness of the treeless arctic environment and the adaptations organisms use to survive a year's worth of seasons there.

http://youtu.be/783uOqArcWw

Polar Bears in Decline: From the Discovery Channel, Frozen Planet.

www.teachersdomain.org/resource/ipy07.sci.ess
.watcyc.icesimulate/

Ice Shelf and Ice Sheet Simulation: This interactive activity adapted from Texas A&M University simulates how the melting of ice shelves and ice sheets affects sea level. In one video, observe a block of ice melt while it floats in the tank, simulating an ice shelf; the water level does not change. In a second video, see a block of ice melt as it rests on a plank of wood above the water, simulating an ice sheet; the water level rises.

www.teachersdomain.org/asset/ess05_int_icemelt/

Mountain of Ice: If the Ice Melts: An interactive presentation from NOVA.

www.teachersdomain.org/resource/ean08.sci.ess
.watcyc.samoa/

Samoa Under Threat: This video segment adapted from Bullfrog Films examines the possible effects of global warming on the Pacific island of Samoa. For many Samoans who grew up with a subsistence way of life, learning to cope with natural disasters is nothing new. But, as Penehuru Lefale—the climatologist interviewed in the video—asserts, extreme weather events appear to be on the rise, threatening the survival of a Polynesian culture that is thousands of years old.

www.teachersdomain.org/resource/tdc02.sci.life
.oate.rainforest/

Amazon Rainforest: The tropical rain forests of the world are home to thousands of species of plants and animals—including many that remain undiscovered by scientists. This video segment from the Race to Save the Planet teaching module "Saving the Diversity of Life" explores both the ecological importance and fragility of the rainforest biome and its organisms.

http://youtu.be/cBHsdjPYLsk

New Model Predicts Fire Season Severity in the Amazon: From NASA. By analyzing nearly a decade of satellite data, a team of scientists led by researchers from the University of California, Irvine and funded by NASA has created a model that can successfully predict the severity and geographic distribution of fires in the Amazon rain forest and the rest of South America months in advance.

www.teachersdomain.org/resource/kqedcl11.sci
.ess.regulatinggreenhousegases/

Regulating Greenhouse Gases: In this lesson from Clue into Climate, produced by KQED, students learn about how destruction of the world's largest rain forest affects global climate change. Prior to engaging in this lesson, students need to have a general knowledge of greenhouse gases (particularly carbon dioxide), the greenhouse effect, and global warming.

www.teachersdomain.org/resource/tdc02.sci.life
.eco.coralreefconnections/

Coral Reef Connections: In order to survive and reproduce, each organism must adapt to its physical environment, as well as to the other organisms in that environment it shares relationships with. As you dive through Australia's Great Barrier Reef in this Evolution Web feature, you'll discover relationships that have evolved between the resident organisms. Some are predators and prey; others compete for space, food, or mates; and still others are dependent or codependent on each other.

http://youtu.be/98ynSn_t3lM

Warming Oceans, Bleaching Corals: Recently, the frequency and severity of coral bleaching events have worsened, especially in the Caribbean. This animation shows data related to the increasing sea surface temperature of the ocean and its effect on coral reefs.

http://youtu.be/60jof35WuAo

Coral Bleaching: Corals get up to 90% of their energy supply from the zooxanthellae that live within them. Stressful conditions (including high water temperature) cause the corals to expel their zooxanthellae. Bleached corals begin to starve once they bleach.

www.teachersdomain.org/resource/ecb10.sci.ess .watcyc.phenology/

Climate Wisconsin: Phenology: In this multimedia video produced by the Wisconsin Educational Communications Board, Nina Leopold Bradley recounts how she learned as a child to record first flower blooms and the arrival of birds on her family's land along the Wisconsin River. She introduces the practice of phenology, or the study of the timing of life cycles of plants and animals, and explores evidence of climate change from records her family has kept since 1935.

www.teachersdomain.org/resource/ecb10.sci.ess .watcyc.uncertainty/

Uncertainty in Climate Change Modeling: Meet John Lyons, a fisheries research biologist with the Wisconsin Department of Natural Resources. He describes why trout are important indicators of stream health and climate change impacts in Wisconsin. This video, produced by the Wisconsin Educational Communications Board, also explores uncertainty in climate change modeling and the range of possibilities for temperature and precipitation that may result for different regions of Wisconsin.

www.teachersdomain.org/resource/ean08.sci.ess .watcyc.caribou/

Global Warming Threatens Caribou: In this video segment, adapted from the series "Arctic Mission," learn about climate change and its impact on the migration of Canada's barren-ground caribou. Caribou time their migration to benefit from advantageous feeding conditions, safety from predators, and relief from insects. Their routes are typically determined by weather factors. However, as global warming continues, traditional migration timings and pathways are leading more and more caribou to their death.

www.teachersdomain.org/resource/wds10.sci.life .eco.climig/

Climate Migrants in Bangladesh: In the South Asian nation of Bangladesh, climate changes are forcing people to leave their homes and farms; experts predict numbers of the displaced could rise to 13 million by midcentury, creating untold destabilization in the country. In this audio segment from PRI's The World Science podcast, hear first hand accounts from farmers on the leading edge of this huge migration. Learn how the Bangladeshi government is trying to help and some of the measures that politicians and policy experts suggest for countering the problem.

http://youtu.be/818TrQmBJ2A

Acid Oceans: New research is showing that carbon dioxide is not only causing global warming, but it's also causing sea water to become more acidic. As this ScienCentral News video explains, the implications of this on underwater life are becoming a hot topic for everyone from scientists to IMAX filmmakers.

http://youtu.be/5cqCvcX7buo

Acid Test: The Global Challenge of Ocean Acidification: This groundbreaking NRDC documentary explores the startling phenomenon of ocean acidification, which may soon challenge marine life on a scale not seen for tens of millions of years. The film, featuring Sigourney Weaver, originally aired on Discovery Planet Green.

http://youtu.be/xuttOKcTPQs

NOAA Ocean Acidification Demonstration: Ocean acidification is a global-scale change in the basic chemistry of oceans that is under way now as a direct result of the increased carbon dioxide in the atmosphere.

Chapter 8
The Politics of Climate Change

http://youtu.be/krXF9Icfa6k

Climate Scientist **Gavin Schmidt** Talks About the Politics of Climate Change

http://youtu.be/j5cGqzxlo28

Climate Change Speech: Science over Politics: Senator Sanders speaks out against a resolution that would dismantle strong environmental rules issued by the Environmental Protection Agency to address global climate change.

http://youtu.be/fWlnyaMWBY8

James Hansen: Why I Must Speak Out About Climate Change: Top climate scientist James Hansen tells the story of his involvement in the science of and debate over global climate change. In doing so, he outlines the overwhelming evidence that change is happening and why that makes him deeply worried about the future.

http://youtu.be/2T4UF_Rmlio

The American Denial of Global Warming—Perspectives on Ocean Science: Polls show that between one-third and one-half of Americans still believe that there is "no solid evidence" of global warming, or that if warming is happening, it can be attributed to natural variability. Others believe that scientists are still debating the point. Join scientist and renowned historian Naomi Oreskes as she describes her investigation into the reasons for such widespread mistrust and misunderstanding of scientific consensus and probes the history of organized campaigns designed to create public doubt and confusion about science.

http://youtu.be/hPWff2L25E0
Climatologist Michael Mann Headlines 2012 U.Va. EnviroDay Event: Climatologist Michael Mann of Penn State University speaks about scientific findings relating to global warming in this University of Virginia EnviroDay appearance. His talk, titled "The Hockey Stick and the Climate Wars: Dispatches from the Front Lines," shows how his findings have brought political heat from some Virginia and national politicians. Mann continues his research and outlines its scientific soundness during this packed public presentation. Prior to his move to Penn State in 2005, Mann served as a professor of environmental sciences at the University of Virginia.

http://youtu.be/8e2GlooAPkM
Michael Mann: The "Hockey Stick" Under Oath

http://youtu.be/cqBURjOdOG8
Climate Denial Crock of the Week: Climate Change and National Security–Part 1

http://youtu.be/vmv3NAO9sRc
Climate Denial Crock of the Week: Climate Change and National Security–Part 2

http://youtu.be/uHhLcoPT9KM
Global Warming: It's Not About the Hockey Stick

www.ucar.edu/webcasts/voices/1-causing.html
What Is Causing Climate Change? Atmospheric, environmental, and social scientists explain the human role in global warming, which is changing the climate in disturbing ways.

http://youtu.be/OfWDAmT1bh0
Richard Lindzen Copenhagen Climate Conference: Prior to the Copenhagen Climate Conference in December 2009, several climate scientists tried to tell the delegates that global warming is a hoax. This was Richard Lindzen's appeal.

http://youtu.be/Qp-nJKBwQR4
Communications expert George Marshall offers six strategies for talking to people who don't accept that climate change is happening. Drawing on his workshops in climate communications and the latest social research he proposes a respectful approach that responds to their interests and values.

http://youtu.be/21qPJ1xG_Ro
Climate Pt 2—Ice Cores, CFC Gases, the Greenhouse Effect and the Ozone Layer: Excerpt from a BBC climate program on climate change evidence, which takes into consideration the arguments of NASA and environmental experts and skeptic opposition.

http://youtu.be/52KLGqDSAjo
Climate Change: The Scientific Debate: A basic look at how climate scientists infer that human-made

carbon gases are changing the climate and how this view is contradicted by other climate scientists who are skeptics.

www.youtube.com/user/thecdmvideos
Clean Development Mechanism: A collection of videos from the United Nations on the Clean Development Mechanism of the Kyoto Protocol.

http://youtu.be/qKrFURSR5u0
The Clean Development Mechanism: This mechanism, defined in Article 12 of the Kyoto Protocol, allows a country with an emission-reduction or emission-limitation commitment under the Protocol to implement an emission-reduction project in developing countries. Such projects can earn saleable certified emission reduction credits, each equivalent to 1 ton of CO_2, which can be counted toward meeting Kyoto targets.

http://youtu.be/pl5tmN-_U7o
Carbon Trade Exchange: Next Generation in Carbon Trading: A global electronic exchange platform serves both the voluntary and regulatory carbon market. With members in 27 countries, the exchange provides a trusted marketplace for buying and selling carbon credits, with confidence in the quality and origin of the products being sold.

http://youtu.be/P6jIeXjadnM
Emissions Trading and Climate Finance: Is 2012 the Dead End? With the apparent end of the Kyoto Protocol era at hand, the role of market forces to drive emission mitigation is entering an uncertain era. Marc Stuart, founding partner at Allotrope Partners, discusses what Kyoto accomplished and the lessons to be learned.

http://youtu.be/YfQyPl6BkP4
Carbon Trading Simplified: This short video clearly explains concepts such as carbon financing, carbon offsetting, and carbon credits trading.

http://youtu.be/_xG8Hpnagg0
Dirty Trading: The Clean Development Mechanism: Kevin Smith is a campaigner and activist with Carbon Trade Watch. In this video, he talks to us about the UN's Clean Development Mechanism and how it is being used to profit polluters.

http://youtu.be/pA6FSy6EKrM
The Story of Cap & Trade: This 2009 video is a fast-paced, fact-filled look at the leading climate solutions discussed in Copenhagen and on Capitol Hill. Host Annie Leonard introduces the energy traders and Wall Street financiers at the heart of this scheme and reveals the "devils in the details" in current cap and trade proposals: free permits to big polluters, fake offsets, and distraction from what's really required to tackle the climate crisis.

http://youtu.be/Dtbn9zBfJSs
Bjorn Lomborg: Global Priorities Bigger Than Climate Change: Given $50 billion to spend, which would you solve first: AIDS or global warming? Danish political scientist Bjorn Lomborg comes up with surprising answers.

www.teachersdomain.org/resource/phy03.sci.phys
.energy.developing/
Global Warming: The Developing World: Thanks to technology, the world is changing faster than ever before. Unfortunately, many of these changes are probably having negative impacts on the global climate. This video segment adapted from NOVA/FRONTLINE takes a look at what the future might hold for the environment as a result of the expanding use of technology.

www.teachersdomain.org/resource/frnht10.guide
.china/
Corporations or Consumers: Who's Responsible for Climate Change? FRONTLINE Heat is a global report on what is being done by the world's largest corporations and governments to respond to the challenges posed by climate change and to the call for reductions in carbon emissions. In this video case study from Heat, correspondent Martin Smith goes to Shenua Energy Corp. in China to examine a leading source of carbon emissions and ask who is really responsible for climate change: corporations or consumers?

www.teachersdomain.org/resource/frnht10.guide
.india/
India: Is India's Growth the Planet's Doom? In this video case study and supplemental extended interview video with Anand Mahindra, vice chairman, Mahindra and Mahindra, Ltd., correspondent Martin Smith travels to India to investigate India's massive economic growth and its alarming rise in greenhouse gas emissions.

www.teachersdomain.org/resource/frnht10.guide
.leaders/
Leadership on Climate Change: Can America Summon the Political Will? In this video case study from Heat, correspondent Martin Smith examines whether the United States can summon the political will to address climate change.

http://youtu.be/FJUA4cm0Rck
Noam Chomsky: How Climate Change Became a "Liberal Hoax": Linguist, philosopher, and political activist Noam Chomsky talks about the Chamber of Commerce, the American Petroleum Institute, and other business lobbies enthusiastically carrying out campaigns "to try and convince the population that global warming is a liberal hoax."

www.teachersdomain.org/resource/ecb10.sci.ess
.watcyc.adaptation/
Climate Change Adaptation and Mitigation: This animated video produced by the Wisconsin Educational Communications Board distinguishes the roles of mitigation and adaptation in responding to climate change. The video offers examples of actions that humans can take as individuals and a society to adapt to and mitigate the impacts of climate change on natural and built environments.

Chapter 9
Energy

http://youtu.be/3_LT9nL4gVQ
BP Energy Outlook 2030—Global Energy Trends: By 2030, the world will have experienced great changes in the way we use and produce energy, according to BP's *Energy Outlook 2030*, released in January 2012. Emerging markets will generate major growth in energy consumption, and demands of a growing population will result in a greater reliance on renewable sources of energy. The report provides deeper insight into how our relationship with energy will evolve over the next 20 years.

http://youtu.be/6j2yzL0PdOw
BP's U.S. Energy Outlook 2030—America's Energy Trends: The United States plays a big part in BP's *2030 Energy Outlook* report, released in January 2012, which forecasts great changes in how America will fuel itself over the next 20 years. BP predicts that renewable energy will increase from 3% to 10% in the United States by 2030, as oil use declines from 36% to 28%. *2030 Energy Outlook* predicts a future defined by advances in power generation and renewable energy in the United States.

www.teachersdomain.org/resource/phy03.sci.phys
.energy.energysource/
Energy Sources: Innovation has allowed societies to tap into a wide variety of energy sources. This video segment, produced for Teachers' Domain, identifies some of these sources, discusses their benefits and drawbacks, and explains how they are used to produce energy that people can use.

www.teachersdomain.org/resource/kqedcl11.sci
.ess.energysources/
Energy Sources: With this video, diagram, and accompanying lesson from Clue into Climate, produced by KQED, students learn how using renewable energy sources to create electricity helps reduce fossil fuel consumption. Students also investigate which form of renewable energy has the potential to have the largest impact on climate change.

www.teachersdomain.org/resource/watsol.sci.ess
.water.geocoal/
The Geology of Coal: In this video, a geologist describes how coal, a sedimentary rock, was formed when organic materials piled up in swamps millions of years ago. Over time, heat and pressure transformed the buried materials into peat and into various forms of

coal. The geologist shows samples of low- and high-sulfur coal. High-sulfur coal contains a lot of pyrite, which is a mixture of sulfur, iron, and traces of other minerals. As long as it is underground, this mixture causes no environmental problems. But when it is exposed to air through mining, the pyrite rusts and forms sulfate salts on the coal's surface that can contaminate water.

www.teachersdomain.org/resource/frnht10.guide
.coal/
Coal in America: What Is the Strategy? In this video case study from Heat, correspondent Martin Smith investigates the challenges facing America's coal industry and examines its ability to thrive in a new carbon-constrained world.

www.teachersdomain.org/resource/ayv09.sci.life
.eco.cleancoal/
Clean Coal? This video segment adapted from FRONT-LINE: "Heat" examines clean coal technology. It provides statistics for overall annual U.S. consumption as well as average household usage, and then explains the need for developing a cleaner way to convert coal into energy. Visit a Florida energy plant that turns solid coal into a clean-burning fuel gas (syngas) and learn about a new approach to capture and store carbon dioxide gas, a by-product of coal burning. Carbon dioxide is an important greenhouse gas, and each year U.S. coal plants emit 2 billion tons of it into the atmosphere. While promising, most applications of clean coal technology are still not ready for implementation.

www.teachersdomain.org/resource/ket09.sci.phys
.energy.coal/
Making Electricity at a Coal-Burning Plant: In this video from KET, visit a coal burning power plant in rural Kentucky to see how electricity is produced in much of the United States. Follow coal's path from bulldozer to furnace and then trace the energy transformations that make electricity available to homes, schools, and businesses.

http://youtu.be/UGx5_2IYZ4Y
Alberta Oil Sands: Find out the real story about the Alberta oil sands, including the challenges we are facing and how the Government of Alberta is addressing them as a responsible energy producer.

http://youtu.be/mBePqKw7tyk
The Story of the Alberta Oil Sands: This video describes the impact of oil sand extraction on the environment.

http://youtu.be/MeRTCepmkqQ
Richard Heinberg: Peak Oil and the Globe's Limitations: Richard Heinberg, senior fellow with the Post Carbon Institute and the author of *The Party's Over*, *Peak Everything* and, most recently, *Blackout*, discusses the phenomenon of peak oil and how it will affect life on this planet.

http://youtu.be/BzLZnidztpI
Natural Gas: The Energy to Move Forward: Natural gas is clean, safe, and affordable, with abundant supplies that can help meet America's energy needs and spur economic growth.

http://youtu.be/p92SrIr2V3o
Special Report: Natural Gas: As new technology drives a natural gas revolution in America, our correspondents discuss the wider economic impact.

http://youtu.be/L4wMduY1__I
Natural Gas and Alternative Energy in China: Leslie Hook of the *Financial Times* looks at the use of alternative energy—specifically natural gas, coal bed gas, and shale gas—and its effect in Shanxi and Beijing, China.

http://youtu.be/-0Xn21q-yQk
Shale Gas Drilling: Pros & Cons: While some complain that extracting natural gas from shale rock formations is tainting their water supply, others who have allowed drilling on their property are getting wealthy.

www.teachersdomain.org/resource/envh10.sci
.phys.energy.fracking/
Fracking: In this video segment adapted from Need to Know, learn about health concerns regarding exposure to chemicals used in natural gas drilling. An animated diagram shows how the process of hydraulic fracturing—fracking—is used to extract natural gas. A reporter explains how the process may be polluting water resources with hazardous chemicals. Hear about health problems among people whose only common connection is proximity to gas drilling sites. In addition, learn about the challenges of investigating the possible link between the contaminants and illness because the gas industry is not required to disclose which chemicals they use in fracking fluids.

www.teachersdomain.org/resource/tdc02.sci.life
.eco.energyuse/
Snapshot of U.S. Energy Use: There are times when our role as energy consumers is clear. For instance, when we fill our cars' fuel tanks, the amount of money we spend makes it obvious how much gas we are using. But what we seldom think about is the energy we consume by simply living our lives in a developed society. In this video segment adapted from NOVA/FRONTLINE, experts estimate the amount of energy that is burned during daily activities, and how much CO_2 those activities contribute to the atmosphere.

www.pbs.org/wgbh/nova/tech/algae-fuel.html
Algae Fuel: What if you could generate useable biofuels that didn't have the problems associated with corn ethanol or plant-based processes, didn't use arable land or fresh water, didn't rely on foodstuffs, and also cost very little and could be "tuned" to produce hydrogen? Well, there is such a thing. It's algae, which just may be our biofuel future.

www.teachersdomain.org/resource/kqedcl11.sci
.ess.solarpower/

Solar Power: In this lesson from Clue into Climate, produced by KQED, learn how energy from the Sun is captured and the benefits and challenges of harnessing and using solar energy. The video compares traditional solar cells with new nanotechnologies.

http://youtu.be/nr-grdspEWQ

Solar Power Revolution—Here Comes the Sun—Documentary: Countries all over the world are leading the way toward a green economy. Unfortunately, lobbying by the oil, gas, coal, and nuclear industries is hindering progress. Very soon, perhaps even now, depending on the cost of electricity in your area, solar technology will be more economically cost-effective than traditional forms of electrical production.

www.teachersdomain.org/resource/klvx09.vid
.klvxsolar/

Solar Power: In this video segment from Vegas PBS, visit the largest solar power system installed in the United States as of 2009, at Nellis Air Force Base in Nevada. This photovoltaic array can generate up to 30% of the base's annual electricity needs. While solar power is clean and renewable, it also makes economic sense. Because Nevada receives abundant sunshine year-round, the base expects to save $1 million annually thanks to the new system.

www.teachersdomain.org/resource/psu06-e21.sci
.paphotovoltaics/

Photovoltaics: This animation shows how photovoltaic panels and their solar cells capture sunlight's energy and create electricity. Sunlight is made up of tiny energy packets called photons. Every minute, enough of this energy reaches Earth to meet a year's worth of Earth's energy needs. Photovoltaic panels consist of many solar cells, made of materials such as silicon, a commonly available element. Solar cells are designed to free electrons from absorbed photons with a positive and a negative layer that create an electric field, just like in a battery. Photovoltaic panels can be designed to meet specific energy needs.

www.teachersdomain.org/resource/wnet08.sci
.phys.energy.wnetsolar/

Solar House: We need energy to power up our technological world, from laptops to lamps, cell phones to cars. In this video segment from What's Up in the Environment? learn about the ways in which energy is created and how these processes impact the environment. After exploring the current research into producing energy from "green" sources, meet an electrical engineer and his 13-year-old son, who produce electricity in their home in Virginia.

www.teachersdomain.org/resource/idptv11.sci
.phys.energy.d4kgen/

Green Energy: This video segment from IdahoPTV's D4K defines green energy and renewable and nonrenewable energy sources. The video discusses pros and cons of each type of energy and provides suggestions about how you can help conserve energy.

www.teachersdomain.org/resource/envh10.sci.life
.eco.biodiesel/

Converting to Biodiesel: In this video from Earth Island Institute's New Leaders Initiative, meet Andrew Azman, a college student at the University of Colorado. When Azman suggested running the university's buses on biodiesel, people were skeptical. But he persevered. He built a processor to convert used french fry grease into biodiesel and convinced the university to run a campus bus on the fuel. Eventually, this led to all 13 campus buses converting to biodiesel, a commitment from the city of Boulder to use biodiesel in its fleet, and the opening of the first public biodiesel pump in Colorado.

www.teachersdomain.org/resource/vtl07.la.ws
.process.diffuel/

A Different Kind of Fuel: The increase in the world's population combined with declining fossil fuel supplies has created the need to develop an alternative form of fuel. In this video segment, adapted from Curious, scientists are developing ways to create fuel using Earth's greatest energy supplier, the Sun. In order to harness this energy, scientists are attempting to re-create photosynthesis and store the hydrogen fuel released during the process.

www.teachersdomain.org/resource/biot09.biotech
.app.synthfuel/

Engineering Biofuels: This video segment, adapted from KQED's QUEST, profiles the work of Jay Keasling, a synthetic biologist experimenting with ways to produce a cleaner-burning fuel from biological matter, using genetically modified microorganisms. Keasling's approach involves engineering microbes to eat simple sugars found in plant matter and then excrete fuel. The video explains the two techniques that may enable Keasling to accomplish this: metabolic engineering and directed evolution. It also suggests that creating biofuels from plants may actually contribute to global warming rather than help combat it.

www.teachersdomain.org/resource/psu06-e21.sci
.pabiomass/

Pennsylvania Energy: Biomass: One definition of *biomass* is biological activity that we can convert into energy. Many potential, renewable sources of biomass energy exist in Pennsylvania, especially on farms. This video profiles various facilities that make biomass energy. One facility in Pennsylvania converts wood to steam to generate electricity. Another company harvests forests selectively so as not to deplete its resources. The turnpike commission is fueling trucks with biodiesel made from soybean oil. Another company is supplementing natural gas with landfill gas created through heat and compaction with some bacteria. Still another company offers a furnace that runs on renewable grains, including corn.

www.teachersdomain.org/resource/biot09.sci
.engin.systems.ethanol/
Ethanol Biofuel: This video segment, adapted from NOVA, examines ethanol, a cleaner-burning fuel alternative to gasoline, and the efforts to produce it more efficiently. Today, most ethanol in the United States is made from corn kernels. But converting corn into ethanol requires lots of energy, as well as corn, which might otherwise be used to feed people and livestock. The video features research efforts to use less valuable plant matter, called cellulosic biomass, and microorganisms that may be able to accomplish the conversion from plant matter to fuel in a single step.

www.teachersdomain.org/resource/kqedq11.sci
.fromwastetowatts/
From Waste to Watts: In this video segment from QUEST, produced by KQED, explore how methane gas is converted into electricity, what a methane digester is used for, and how the scraps from food you eat every day can get converted into green energy.

www.teachersdomain.org/resource/kqedcl11.sci
.ess.geothermalenergy/
Geothermal Energy: Harnessing the Power of the Earth: In this lesson from Clue into Climate, produced by KQED, learn about how geothermal energy is produced and its limitations.

www.teachersdomain.org/resource/klvx09.vid
.klvxgeoth/
Geothermal Power: In this video segment from Vegas PBS, visit a construction site where a geothermal heat pump system is being installed to provide a new building with an energy-efficient way to meet its heating and cooling needs. Geothermal energy systems like this one circulate water through two pipes—one a supply line to the building and the other a return line from the building. Depending on the season, the water delivers heat to the building to warm it or removes heat from the building to cool it. Geothermal systems replace traditional heating and air conditioning systems that require considerably more fossil fuel to run.

www.teachersdomain.org/resource/eng06.sci
.engin.systems.fuelcells/
Fuel Cells: Imagine if we could find a nonpolluting, highly efficient source of power for our society's energy needs. One technology—the fuel cell—offers a strong and proven model. However, in order for the fuel cell to replace gas-powered engines and batteries, the fuel used to power it must be readily available. This video segment adapted from NOVA scienceNOW highlights the electrochemical reaction that takes place within a hydrogen fuel cell and explores the challenges of producing the pure hydrogen that fuels it.

www.teachersdomain.org/resource/phy03.sci.phys
.matter.hydrogencar/
Global Warming: The Hydrogen Car: Engines have been used in transportation and industry since the 18th century. Those powered by fossil fuels such as coal and gasoline emit harmful pollution, deplete the limited supply of natural resources, and contribute to global warming. In this video segment adapted from NOVA/FRONTLINE, see why hydrogen fuel cells are being touted as a possible replacement for the automobile's internal combustion engine and the challenges that surround this technology.

www.teachersdomain.org/resource/phy03.sci.phys
.energy.hooverelec/
Hoover Dam and Hydroelectric Power: Since its completion, Hoover Dam has wowed visitors with its massiveness and its ability to create and hold back a reservoir that covers nearly 650 square kilometers (250 square miles). Yet perhaps even more impressive is the dam's ability to harness the potential energy stored in the reservoir and convert it to electricity. In this video segment, adapted from Building Big, series host David Macaulay explores Hoover Dam's hydroelectric capabilities.

http://youtu.be/wWo_KVfr8Pw
Energy for the Future: The Nuclear Renaissance: This video showcases new nuclear power plants in Georgia. It was a 2012 EEI Edison Award Winner.

http://youtu.be/znO2cAvR5SM
Inside Story: The Future of Nuclear Power: Germany's ruling coalition says it has agreed to a date of 2022 for the shutdown of all of its nuclear power plants.

http://youtu.be/CQ55vYgbB1I
The Basics of Nuclear Energy

http://youtu.be/VJfIbBDR3e8
How Nuclear Energy Works

http://youtu.be/cvF6qQasMfg
Benefits of Nuclear Energy

http://youtu.be/UK8ccWSZkic
Debate: Does the World Need Nuclear Energy? The energy crisis has even die-hard environmentalists reconsidering it. In this first-ever TED debate, Stewart Brand and Mark Z. Jacobson square off over the pros and cons. This discussion will make you think—and might even change your mind.

http://youtu.be/AAFWeIp8JT0
The Future of Nuclear Power: Getting Rid of the Nuclear Waste: For the first time in decades, there are an abundance of new designs for nuclear power reactors—ones that are safer, more powerful, and more portable, and even ones that produce hardly any nuclear

waste. From the Department of Nuclear Science and Engineering at Massachusetts Institute of Technology.

http://youtu.be/4lq-h4ShZ8s
Tidal Energy Pty Ltd

http://youtu.be/ZcA3e8_j8XA
Harnessing the Power of Waves & Tides: UK energy and power experts meet to discuss the latest developments and initiatives in wave and tidal power technology across Britain.

http://youtu.be/tsZITSeQFR0
Energy 101: Wind Turbines: See how wind turbines generate clean electricity from the power of the wind. Highlighted are the various parts and mechanisms of a modern wind turbine.

http://youtu.be/uHTz8qdDPlM
For Wind Energy's Future, Researchers Look High in the Sky: The next major innovation in wind power might not involve big, white turbines dotting the countryside. KQED QUEST reports on research being done on "tethered airfoils" that could capture wind energy more efficiently than Earth-bound turbines.

www.teachersdomain.org/resource/nsn11.sci
.engin.systems.smartgrid/
Smart Power Grid: This video segment adapted from NOVA scienceNOW examines how new technology can help monitor and modernize the infrastructure of the century-old U.S. power grid, which is ill-equipped to handle the increasing demand for electricity. New technology is creating a "smart" grid that allows utility companies to monitor power needs and respond quickly to distribute it more efficiently, prevent outages, and repair problems. Nearly 50% of our electricity comes from burning coal, which emits greenhouse gases that fuel climate change. A smart grid would conserve energy and help integrate modern power plants that use cleaner, renewable energy sources such as wind, solar, tidal, and geothermal.

http://youtu.be/Cm4dHLflK08
A discussion of **Carbon Capture & Sequestration**

http://youtu.be/RQLR_L9vmLY
What Is Carbon Capture and Storage? The science of carbon capture and storage (CCS) could play a crucial part in reducing climate change, as well as open up valuable commercial opportunities.

Chapter 10
Solutions

www.ucar.edu/webcasts/voices/6-respond.html
How Can We Best Respond to the Problem? Scientists outline what it will take to face the realities of global warming and climate change head on.

www.teachersdomain.org/resource/frnht10.guide
.wind/
Wind: Investing in Carbon-Free Power: In this video case study from Heat, correspondent Martin Smith examines the role of entrepreneurship in climate change mitigation and how to provide incentives that will spark business innovation.

www.teachersdomain.org/resource/ecb10.sci.ess
.watcyc.adaptation/
Climate Change Adaptation and Mitigation: This animated video produced by the Wisconsin Educational Communications Board distinguishes the roles of mitigation and adaptation in responding to climate change. The video offers examples of actions that humans can take as individuals and a society to adapt to and mitigate the impacts of climate change on natural and built environments.

www.teachersdomain.org/resource/nsn08.sci.ess
.watcyc.capcarbonint/
Capturing Carbon: Where Do We Put It? This interactive activity from NOVA scienceNOW reviews several potential means of storing carbon dioxide (CO_2) captured from industrial sources. Among the featured ideas are technologies that deliver compressed CO_2 to underground cavities, saline aquifers, and the deep seabed. The benefit of storing, or sequestering, captured CO_2 could be significant in the fight to slow or limit global warming. However, the list of drawbacks associated with carbon sequestration includes high cost, storage capacity limitations, a still-incomplete understanding of the relevant Earth systems, and uncertainty as to whether the CO_2 can be safely and permanently contained.

www.teachersdomain.org/resource/nsn08.sci.ess
.watcyc.capcarbon/
Capturing Carbon: This video segment adapted from NOVA scienceNOW explains how geophysicist Klaus Lackner and two engineers, Allen and Burton Wright, teamed up to develop a technology to capture an important greenhouse gas, carbon dioxide (CO_2), in the air. Modeling their design after a tree—and one of Lackner's daughter's science experiments—the team tested different materials, configurations, and coatings that together would act as leaves do to remove CO_2 from the air. However, instead of mimicking photosynthesis, their process uses a manufactured fabric that attracts CO_2 and pulls it out of the air for subsequent storage.

http://youtu.be/vzMVfJKJK_c
Geoengineering Earth's Climate: Emergency preparedness is generally considered to be a good thing, yet there is no plan regarding what we might do should we be faced with a climate emergency. Such an emergency could take the form of a rapid shift in precipitation patterns, a collapse of the great ice sheets, the

imminent triggering of strong climate system feedbacks, or perhaps the loss of valuable ecosystems.

The Scientific Method

www.teachersdomain.org/resource/drey07.sci
.phys.matter.scprocess/
Scientific Processes: This interactive activity adapted from NOVA gives students a way to gain a more accurate understanding of how science is done. Two videos, which feature the work of renowned scientists Percy Julian and Judah Folkman, demonstrate that while scientists may use an orderly approach to learn new information and solve problems, they proceed along different paths in their quests. Instead of referring to a delineated set of steps, the term *scientific process* encompasses the true nature of scientific inquiry. Procedures are often refined as data are collected, and initial hypotheses can be modified or replaced in light of new evidence.

The Köppen Climate Classification System

The Köppen climate classification system was designed by Wladimir Köppen (1846–1940), a German climatologist and botanist, and is widely used for its ease of comprehension. The basis of any empirical classification system is the choice of criteria used to draw lines on a map to designate different climates. Köppen–Geiger climate classification uses *average monthly temperatures, average monthly precipitation*, and *total annual precipitation* to devise its spatial categories and boundaries. But we must remember that boundaries really are transition zones of gradual change. The trends and overall patterns of boundary lines are more important than their precise placement, especially with the small scales generally used on world maps.

Take a few minutes and examine the Köppen system, the criteria, and the considerations for each of the principal climate categories. The modified Köppen–Geiger system has its drawbacks, however. It does not consider winds, temperature extremes, precipitation intensity, quantity of sunshine, cloud cover, or net radiation. Yet the system is important because its correlations with the actual world are reasonable and the input data are standardized and readily available.

Köppen's Climatic Designations

Figure B1 shows the distribution of each of Köppen's six climate classifications on the land. This generalized map shows the spatial pattern of climate. The Köppen system uses capital letters (A, B, C, D, E, H) to designate climatic categories from the equator to the poles. The guidelines for each of these categories are in the margin of Figure B.1.

Five of the climate classifications are based on thermal criteria:

A. Tropical (equatorial regions)

C. Mesothermal (Mediterranean, humid subtropical, marine west coast regions)

D. Microthermal (humid continental, subarctic regions)

E. Polar (polar regions)

H. Highland (compared to lowlands at the same latitude, highlands have lower temperatures—recall the normal lapse rate—and more efficient precipitation due to lower moisture demand)

Only one climate classification is based on moisture as well:

B. Dry (deserts and semiarid steppes)

Within each climate classification additional lowercase letters are used to signify temperature and moisture conditions. For example, in a tropical rain forest *Af* climate, the *A* tells us that the average coolest month is above 18°C (64.4°F, average for the month), and the *f* indicates that the weather is constantly wet, with the driest month receiving at least 6 cm (2.4 in.) of precipitation. (The designation *f* is from the German *feucht*, for "moist.") As you can see on the climate map, the tropical rain forest *Af* climate dominates along the equator and equatorial rain forest.

In a *Dfa* climate, the *D* means that the average warmest month is above 10°C (50°F), with at least one month falling below 0°C (32°F); the *f* says that at least 3 cm (1.2 in.) of precipitation falls during every month; and the *a* indicates a warmest summer month averaging above 22°C (71.6°F). Thus, a *Dfa* climate is a humid-continental, hot-summer climate in the microthermal category.

Köppen Guidelines

The Köppen guidelines and map portrayal are in Figure B.1. First, check the guidelines for a climate type, then the color legend for the subdivisions of the type, then check out the distribution of that climate on the map.

▼ **Figure B.1:** World climates and their guidelines according to the Köppen classification system.

Köppen Guidelines
Tropical Climates — A

Consistently warm with all months averaging above 18°C (64.4°F); annual water supply exceeds water demand.

Af — Tropical rain forest:
 f = All months receive precipitation in excess of 6 cm (2.4 in.).

Am — Tropical monsoon:
 m = A marked short dry season with 1 or more months receiving less than 6 cm (2.4 in.) precipitation; an otherwise excessively wet rainy season. ITCZ 6–12 months dominant.

Aw — Tropical savanna:
 w = Summer wet season, winter dry season; ITCZ dominant 6 months or less, winter water-balance deficits.

Mesothermal Climates — C

Warmest month above 10°C (50°F); coldest month above 0°C (32°F) but below 18°C (64.4°F); seasonal climates.

Cfa, Cwa — Humid subtropical:
 a = Hot summer; warmest month above 22°C (71.6°F).
 f = Year-round precipitation.
 w = Winter drought, summer wettest month 10 times more precipitation than driest winter month.

Cfb, Cfc — Marine west coast, mild-to-cool summer:
 f = Receives year-round precipitation.
 b = Warmest month below 22°C (71.6°F) with 4 months above 10°C.
 c = 1–3 months above 10°C.

Csa, Csb — Mediterranean summer dry:
 s = Pronounced summer drought with 70% of precipitation in winter.
 a = Hot summer with warmest month above 22°C (71.6°F).
 b = Mild summer; warmest month below 22°C.

Microthermal Climates — D

Warmest month above 10°C (50°F); coldest month below 0°C (32°F); cool temperature-to-cold conditions; snow climates. In Southern Hemisphere, occurs only in highland climates.

Dfa, Dwa — Humid continental:
 a = Hot summer; warmest month above 22°C (71.6°F).
 f = Year-round precipitation.
 w = Winter drought.

Dfb, Dwb — Humid continental:
 b = Mild summer; warmest month below 22°C (71.6°F).
 f = Year-round precipitation.
 w = Winter drought.

Dfc, Dwc, Dwd — Subarctic:
Cool summers, cold winters.
 f = Year-round precipitation.
 w = Winter drought.
 c = 1–4 months above 10°C
 b = Coldest month below −38°C (−36.4°F), in Siberia only.

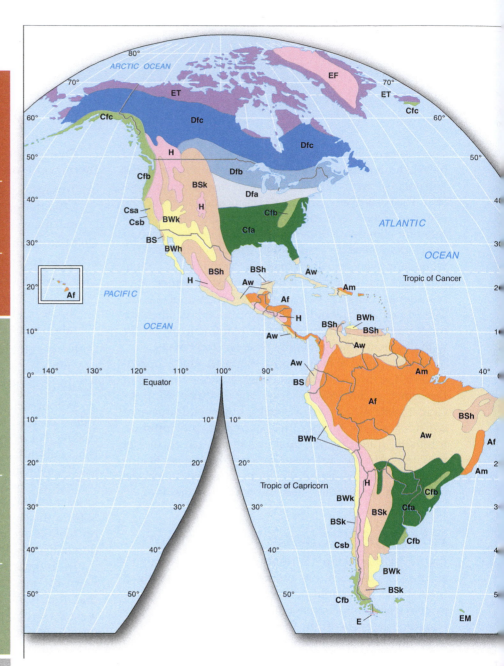

Dry Arid and Semiarid Climates — B

Potential evapotranspiration* (natural moisture demand) exceeds precipitation (natural moisture supply) in all B climates. Subdivisions based on precipitation timing and amount and mean annual temperature.

Earth's arid climates.

BWh — Hot low-latitude desert

BWk — Cold midlatitude desert
 BW = Precipitation less than 1/2 natural moisture demand.
 h = Mean annual temperature >18°C (64.4°F).
 k = Mean annual temperature <18°C.

Earth's semiarid climates.

BSh — Hot low-latitude steppe

BSk — Cold midlatitude steppe
 BS = Precipitation more than 1/2 natural moisture demand but not equal to it.
 h = Mean annual temperature >18°C.
 k = Mean annual temperature <18°C.

Polar Climates — E

Warmest month below 10°C (50°F); always cold; ice climates.

ET — Tundra:
 Warmest month 0–10°C (32–50°F); precipitation exceeds small potential evapotranspiration demand*; snow cover 8–10 months.

EF — Ice cap:
 Warmest month below 0°C (32°F); precipitation exceeds a very small potential evapotranspiration demand; the polar regions.

EM — Polar marine:
 All months above −7°C (20°F), warmest month above 0°C; annual temperature range <17 C° (30 F°).

*Potential evapotranspiration = the amount of water that would evaporate or transpire if it were available-the natural moisture demand in an environment.

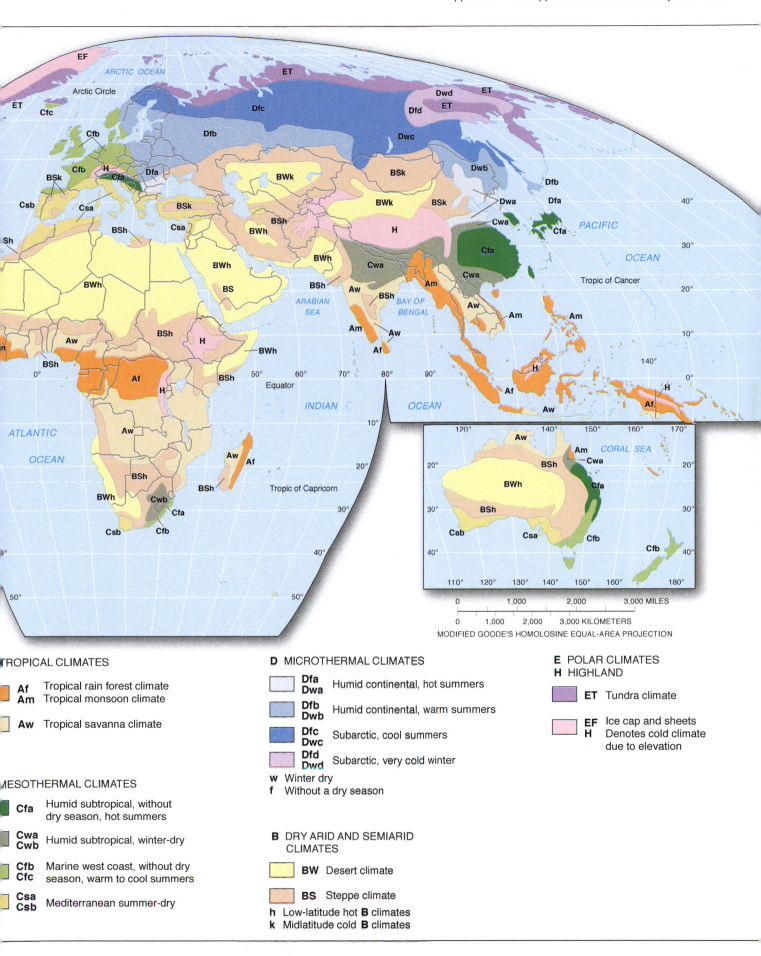

MODIFIED GOODE'S HOMOLOSINE EQUAL-AREA PROJECTION

TROPICAL CLIMATES

Af Tropical rain forest climate
Am Tropical monsoon climate

Aw Tropical savanna climate

MESOTHERMAL CLIMATES

Cfa Humid subtropical, without dry season, hot summers

Cwa
Cwb Humid subtropical, winter-dry

Cfb
Cfc Marine west coast, without dry season, warm to cool summers

Csa
Csb Mediterranean summer-dry

D MICROTHERMAL CLIMATES

Dfa
Dwa Humid continental, hot summers

Dfb
Dwb Humid continental, warm summers

Dfc
Dwc Subarctic, cool summers

Dfd
Dwd Subarctic, very cold winter

w Winter dry
f Without a dry season

B DRY ARID AND SEMIARID CLIMATES

BW Desert climate

BS Steppe climate

h Low-latitude hot **B** climates
k Midlatitude cold **B** climates

E POLAR CLIMATES
H HIGHLAND

ET Tundra climate

EF Ice cap and sheets
H Denotes cold climate due to elevation

Weather Extremes

This appendix presents various weather extremes and the geographical locales of the extremes on world maps. The weather extremes are presented on the following pages: Note that extreme weather is not a new phenomenon and that many records were established long before we were concerned about climate change and global warming. What is now different is the rate at which records are being broken, and the fact that these new records reflect climate change that is permanent and global and not just transient and regional.

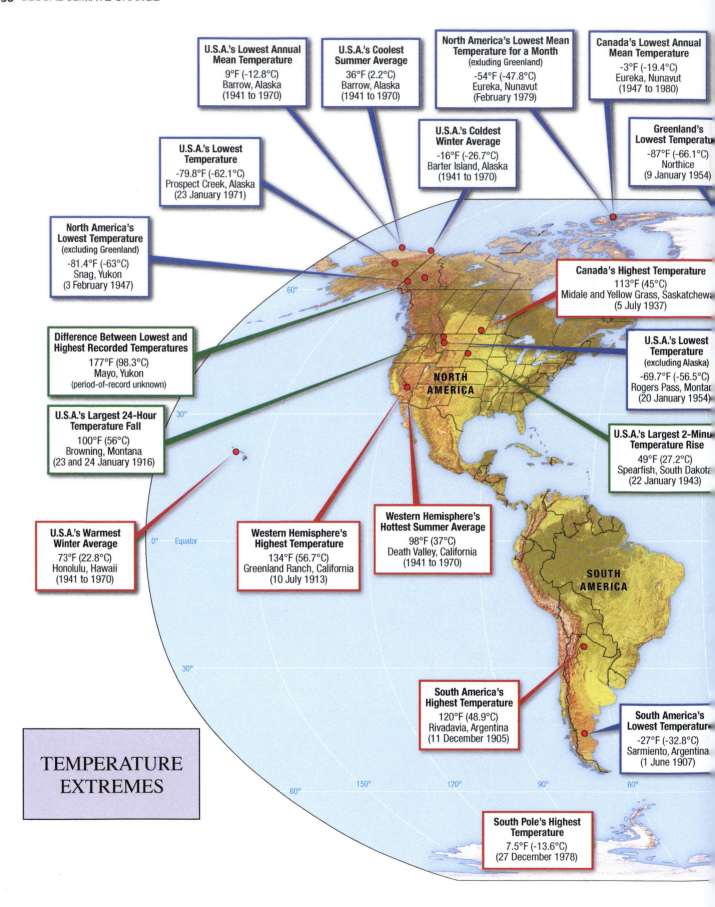

U.S.A.'s Lowest Annual Mean Temperature
9°F (-12.8°C)
Barrow, Alaska
(1941 to 1970)

U.S.A.'s Coolest Summer Average
36°F (2.2°C)
Barrow, Alaska
(1941 to 1970)

North America's Lowest Mean Temperature for a Month
(excluding Greenland)
-54°F (-47.8°C)
Eureka, Nunavut
(February 1979)

Canada's Lowest Annual Mean Temperature
-3°F (-19.4°C)
Eureka, Nunavut
(1947 to 1980)

U.S.A.'s Lowest Temperature
-79.8°F (-62.1°C)
Prospect Creek, Alaska
(23 January 1971)

U.S.A.'s Coldest Winter Average
-16°F (-26.7°C)
Barter Island, Alaska
(1941 to 1970)

Greenland's Lowest Temperature
-87°F (-66.1°C)
Northice
(9 January 1954)

North America's Lowest Temperature
(excluding Greenland)
-81.4°F (-63°C)
Snag, Yukon
(3 February 1947)

Canada's Highest Temperature
113°F (45°C)
Midale and Yellow Grass, Saskatchewan
(5 July 1937)

Difference Between Lowest and Highest Recorded Temperatures
177°F (98.3°C)
Mayo, Yukon
(period-of-record unknown)

U.S.A.'s Lowest Temperature
(excluding Alaska)
-69.7°F (-56.5°C)
Rogers Pass, Montana
(20 January 1954)

U.S.A.'s Largest 24-Hour Temperature Fall
100°F (56°C)
Browning, Montana
(23 and 24 January 1916)

U.S.A.'s Largest 2-Minute Temperature Rise
49°F (27.2°C)
Spearfish, South Dakota
(22 January 1943)

U.S.A.'s Warmest Winter Average
73°F (22.8°C)
Honolulu, Hawaii
(1941 to 1970)

Western Hemisphere's Highest Temperature
134°F (56.7°C)
Greenland Ranch, California
(10 July 1913)

Western Hemisphere's Hottest Summer Average
98°F (37°C)
Death Valley, California
(1941 to 1970)

NORTH AMERICA

SOUTH AMERICA

South America's Highest Temperature
120°F (48.9°C)
Rivadavia, Argentina
(11 December 1905)

South America's Lowest Temperature
-27°F (-32.8°C)
Sarmiento, Argentina
(1 June 1907)

South Pole's Highest Temperature
7.5°F (-13.6°C)
(27 December 1978)

TEMPERATURE EXTREMES

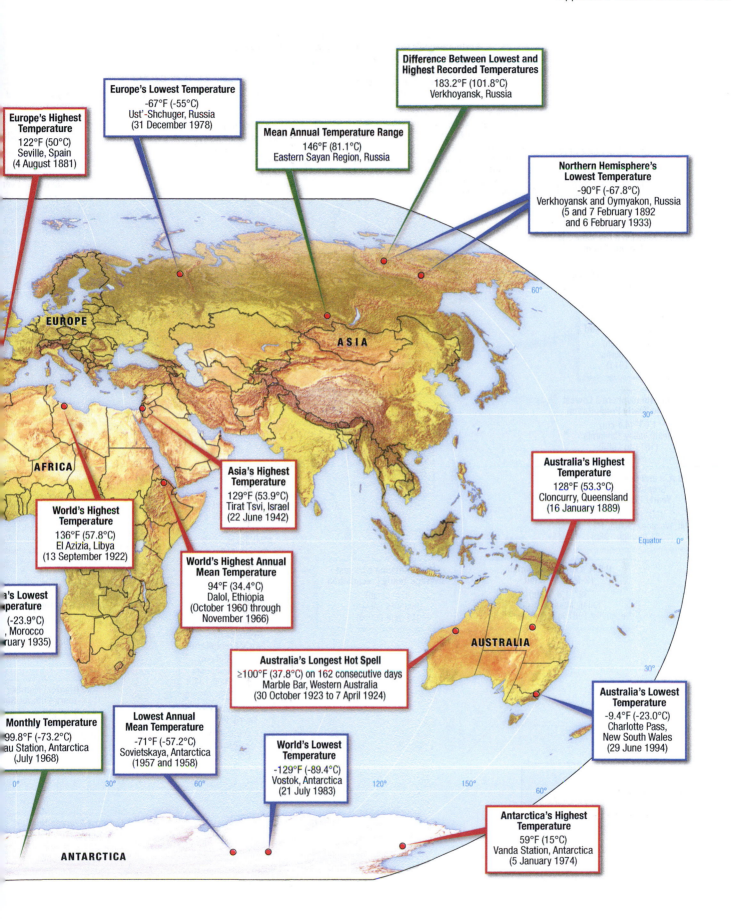

Difference Between Lowest and Highest Recorded Temperatures
183.2°F (101.8°C)
Verkhoyansk, Russia

Europe's Lowest Temperature
-67°F (-55°C)
Ust'-Shchuger, Russia
(31 December 1978)

Mean Annual Temperature Range
146°F (81.1°C)
Eastern Sayan Region, Russia

Northern Hemisphere's Lowest Temperature
-90°F (-67.8°C)
Verkhoyansk and Oymyakon, Russia
(5 and 7 February 1892
and 6 February 1933)

Europe's Highest Temperature
122°F (50°C)
Seville, Spain
(4 August 1881)

EUROPE

ASIA

AFRICA

Asia's Highest Temperature
129°F (53.9°C)
Tirat Tsvi, Israel
(22 June 1942)

Australia's Highest Temperature
128°F (53.3°C)
Cloncurry, Queensland
(16 January 1889)

World's Highest Temperature
136°F (57.8°C)
El Azizia, Libya
(13 September 1922)

World's Highest Annual Mean Temperature
94°F (34.4°C)
Dalol, Ethiopia
(October 1960 through November 1966)

's Lowest perature
(-23.9°C)
, Morocco
uary 1935)

Equator   0°

Australia's Longest Hot Spell
≥100°F (37.8°C) on 162 consecutive days
Marble Bar, Western Australia
(30 October 1923 to 7 April 1924)

AUSTRALIA

Australia's Lowest Temperature
-9.4°F (-23.0°C)
Charlotte Pass,
New South Wales
(29 June 1994)

Monthly Temperature
99.8°F (-73.2°C)
au Station, Antarctica
(July 1968)

Lowest Annual Mean Temperature
-71°F (-57.2°C)
Sovietskaya, Antarctica
(1957 and 1958)

World's Lowest Temperature
-129°F (-89.4°C)
Vostok, Antarctica
(21 July 1983)

Antarctica's Highest Temperature
59°F (15°C)
Vanda Station, Antarctica
(5 January 1974)

ANTARCTICA

North America's Greatest Average Yearly Precipitation

276" (701 cm)
Henderson Lake, British Columbia
(15-year period)

Canada's Greatest Snowfall in One Season

963" (2,446.5 cm)
Revelstoke, Mt. Copeland,
British Columbia
(1971 to 1972)

Canada's Heaviest Hailstone

10.23 oz (290 gm)
Cedoux, Saskatchewan
(27 August 1973)

U.S.A.'s Largest Hailstone Circumfer

18.75" (47.6 cm)
Aurora, Nebraska
(22 June 2003)

Canada's Greatest 24-Hour Rainfall

19.3" (49 cm)
Ucluelet Brynnor Mines,
British Columbia
(6 October 1967)

North America's Greatest 24-Hour Snowfall

76" (193 cm)
Silver Lake, Colorado
(14–15 April 1921)

Extreme 24-H Snowfall Eve

77" (195.6 cm
Montague, New
(11–12 January
(not an official rec

North America's Greatest Snowfall in One Season

1,122" (2,850 cm)
Rainier Paradise Ranger
Station, Washington
(1971 to 1972)

North America's Greatest Snowfall in One Storm

189" (480 cm)
Mt. Shasta Ski Bowl, California
(13–19 February 1959)

North America's Greatest Depth of Snow on Ground

451" (1,145.5 cm)
Tamarack, California
(11 March 1911)

World's Greatest 42-Minute Rainfa

12" (30.5 cm)
Holt, Missouri
(22 June 1947)

Western Hemisphere's Lowest Average Yearly Precipitation

1.63" (4.1 cm)
Death Valley, California
(1911 to 1953)

U.S.A.'s Longest Dry Period

767 days
Bagdad, California
(3 October 1912 to
8 November 1914)

Northern Hemisphere Greatest 24-Hour Rain

64.3" (163.4 cm)
Isla Mujeres, Mexico
(21–22 October 2005

U.S.A.'s and Possibly World's Greatest Average Yearly Precipitation

460" (1,168 cm)
Mt. Waialeale, Kauai, Hawaii
(1931 through 1960)

U.S.A.'s Greatest 24-Hour Rainfall

43" (109 cm)
Alvin, Texas
(25–26 July 1979)

Possibly World's Gre 1-Minute Rainfal

1.5" (3.12 cm)
Barot, Guadeloupe
(26 November 1970
(exact coordinates unkno

U.S.A.'s Greatest Rainfall 12-Month Period

739" (1,877 cm)
Kukui, Maui, Hawaii
(December 1981 to December 1982)

North America's Lowest Average Yearly Precipitation

1.2" (3 cm)
Batagues, Mexico
(14-year period)

Possibly World's Grea Average Yearly Precipit

523.6" (1,330 cm)
Lloro, Columbia
(1932 to 1960)

South America's Greatest Average Yearly Precipitation

354" (899 cm)
Quibdo, Columbia
(1931 through 1946)

Greatest Number Years Without Ra

>14 consecutive ye
Arica, Chile
(October 1903 to Januar

World's Lowest Average Yearly Precipitation

0.03" (0.08 cm)
Arica, Chile
(59-year period)

NORTH AMERICA

SOUTH AMERICA

60°

30°

0° Equator

30°

60°

150°

120°

90°

60°

PRECIPITATION EXTREMES

Greatest Average Number of Days per Year with Rain

325 days/year
Bahia Felix, Chile

rope's Greatest Average Yearly Precipitation

183" (465 cm)
vica, Bosnia-Herzegovina
(22-year period)

cord Snowfall

.7" (172 cm)
ssans, France
6 April 1969

World's Greatest 20-Minute Rainfall

8.10" (20.6 cm)
Curtea de Arges, Romania
(7 July 1889)

Europe's Lowest Average Yearly Precipitation

6.4" (16.3 cm)
Astrakhan, Russia
(25-year period)

Relative Variability of Annual Precipitation

94%
Themed, Israel
(1921 to 1947)

Relative Variability of Annual Precipitation

108%
Lhasa, Tibet, China
(1935 to 1939)

Asia's and Possibly World's Greatest Average Yearly Precipitation

467.4" (1,187.3 cm)
Mawsynram, India
(1941 to 1979)

World's Greatest 60-Minute Rainfall

15.78" (40.1 cm)
Shangdi, China
(3 July 1975)

World's Greatest 1-Month Rainfall

366" (930 cm)
Cherrapunji, India
(July 1861)

World's Greatest 12-Month Rainfall

1,042" (2,647 cm)
Cherrapunji, India
(August 1860 to July 1861)

World's Heaviest Hailstone

2.25 lbs (1.02 kg)
Gopalganj District, Bangladesh
(14 April 1986)

Asia's Lowest Average Yearly Precipitation

1.8" (4.6 cm)
Aden, Yemen
(50-year period)

Africa's Lowest Average Yearly Precipitation

<0.1" (<0.25 cm)
Wadi Halfa, Sudan
(39-year period)

ca's Greatest ge Variability f Annual ecipitation

" (191 cm)
bundscha,
ameroon

Africa's Greatest Average Yearly Precipitation

405" (1,029 cm)
Debundscha, Cameroon
(32-year period)

World's Greatest 12-Hour Rainfall

46" (117 cm)
Grand Ilet, Reunion
(26 January 1980)

World's Greatest 24-Hour Rainfall

72" (182.5 cm)
Foc-Foc, Reunion
(7–8 January 1966)

World's Greatest 5-Day Rainfall

196.06" (498 cm)
Cratere Commerson, Reunion
(24–28 February 2007)

Australia's Greatest Average Yearly Precipitation

340.16" (864 cm)
Bellenden Ker, Queensland
(9-year period)

Australia's Greatest 24-Hour Rainfall

44.9" (114 cm)
Bellenden Ker, Queensland
(4 January 1979)

Australia's Lowest Average Yearly Precipitation

4.05" (10 cm)
Troudaninna, South Australia
(42-year period)

ASIA

EUROPE

AFRICA

AUSTRALIA

ANTARCTICA

Equator 0°

60°

30°

30°

60°

0°   30°   60°   120°   150°   60°

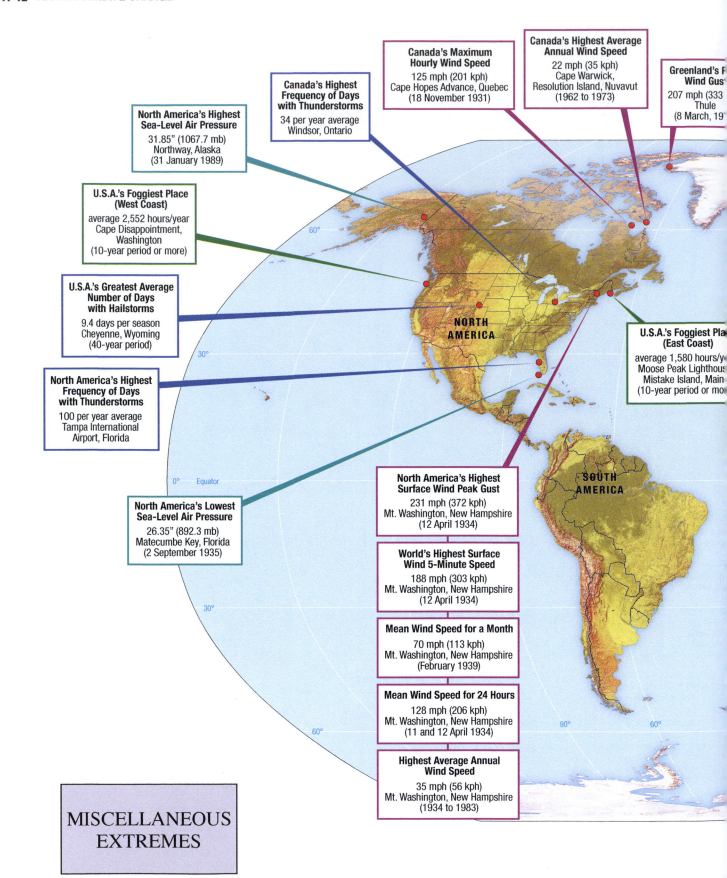

Canada's Highest Frequency of Days with Thunderstorms

34 per year average
Windsor, Ontario

Canada's Maximum Hourly Wind Speed

125 mph (201 kph)
Cape Hopes Advance, Quebec
(18 November 1931)

Canada's Highest Average Annual Wind Speed

22 mph (35 kph)
Cape Warwick,
Resolution Island, Nuvavut
(1962 to 1973)

Greenland's F Wind Gus

207 mph (333
Thule
(8 March, 19

North America's Highest Sea-Level Air Pressure

31.85" (1067.7 mb)
Northway, Alaska
(31 January 1989)

U.S.A.'s Foggiest Place (West Coast)

average 2,552 hours/year
Cape Disappointment,
Washington
(10-year period or more)

U.S.A.'s Greatest Average Number of Days with Hailstorms

9.4 days per season
Cheyenne, Wyoming
(40-year period)

North America's Highest Frequency of Days with Thunderstorms

100 per year average
Tampa International
Airport, Florida

North America's Lowest Sea-Level Air Pressure

26.35" (892.3 mb)
Matecumbe Key, Florida
(2 September 1935)

U.S.A.'s Foggiest Pla (East Coast)

average 1,580 hours/ye
Moose Peak Lighthous
Mistake Island, Main
(10-year period or mor

North America's Highest Surface Wind Peak Gust

231 mph (372 kph)
Mt. Washington, New Hampshire
(12 April 1934)

World's Highest Surface Wind 5-Minute Speed

188 mph (303 kph)
Mt. Washington, New Hampshire
(12 April 1934)

Mean Wind Speed for a Month

70 mph (113 kph)
Mt. Washington, New Hampshire
(February 1939)

Mean Wind Speed for 24 Hours

128 mph (206 kph)
Mt. Washington, New Hampshire
(11 and 12 April 1934)

Highest Average Annual Wind Speed

35 mph (56 kph)
Mt. Washington, New Hampshire
(1934 to 1983)

NORTH AMERICA

SOUTH AMERICA

60°

30°

0° Equator

30°

60°

90°

60°

MISCELLANEOUS EXTREMES

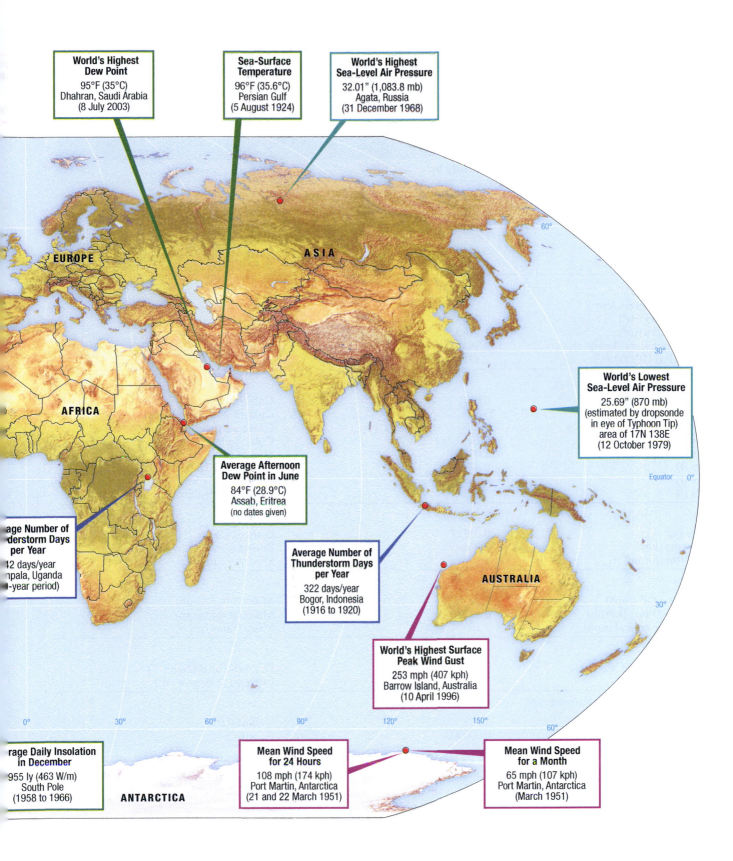

World's Highest Dew Point
95°F (35°C)
Dhahran, Saudi Arabia
(8 July 2003)

Sea-Surface Temperature
96°F (35.6°C)
Persian Gulf
(5 August 1924)

World's Highest Sea-Level Air Pressure
32.01" (1,083.8 mb)
Agata, Russia
(31 December 1968)

World's Lowest Sea-Level Air Pressure
25.69" (870 mb)
(estimated by dropsonde in eye of Typhoon Tip)
area of 17N 138E
(12 October 1979)

Average Afternoon Dew Point in June
84°F (28.9°C)
Assab, Eritrea
(no dates given)

Average Number of Thunderstorm Days per Year
322 days/year
Bogor, Indonesia
(1916 to 1920)

age Number of derstorm Days per Year
2 days/year
pala, Uganda
-year period)

World's Highest Surface Peak Wind Gust
253 mph (407 kph)
Barrow Island, Australia
(10 April 1996)

rage Daily Insolation in December
955 ly (463 W/m)
South Pole
(1958 to 1966)

Mean Wind Speed for 24 Hours
108 mph (174 kph)
Port Martin, Antarctica
(21 and 22 March 1951)

Mean Wind Speed for a Month
65 mph (107 kph)
Port Martin, Antarctica
(March 1951)

EUROPE
ASIA
AFRICA
AUSTRALIA
ANTARCTICA

Equator

60°
30°
0°
30°
60°

0° 30° 60° 90° 120° 150°

Common Conversions

Metric to English

Metric Measure	Multiply by	English Equivalent
Length		
Centimeters (cm)	0.3937	Inches (in.)
Meters (m)	3.2808	Feet (ft)
Meters (m)	1.0936	Yards (yd)
Kilometers (km)	0.6214	Miles (mi)
Nautical mile	1.15	Statute mile
Area		
Square centimeters (cm^2)	0.155	Square inches ($in.^2$)
Square meters (m^2)	10.7639	Square feet (ft^2)
Square meters (m^2)	1.1960	Square yards (yd^2)
Square kilometers (km^2)	0.3831	Square miles (mi^2)
Hectare (ha) (10,000 m^2)	2.4710	Acres (a)
Volume		
Cubic centimeters (cm^3)	0.06	Cubic inches ($in.^3$)
Cubic meters (m^3)	35.30	Cubic feet (ft^3)
Cubic meters (m^3)	1.3079	Cubic yards (yd^3)
Cubic kilometers (km^3)	0.24	Cubic miles (mi^3)
Liters (l)	1.0567	Quarts (qt), U.S.
Liters (l)	0.88	Quarts (qt), Imperial
Liters (l)	0.26	Gallons (gal), U.S.
Liters (l)	0.22	Gallons (gal), Imperial
Mass		
Grams (g)	0.03527	Ounces (oz)
Kilograms (kg)	2.2046	Pounds (lb)
Metric ton (tonne) (t)	1.10	Short ton (tn), U.S.
Velocity		
Meters/second (mps)	2.24	Miles/hour (mph)
Kilometers/hour (kmph)	0.62	Miles/hour (mph)
Knots (kn) (nautical mph)	1.15	Miles/hour (mph)
Temperature		
Degrees Celsius (°C)	1.80 (then add 32)	Degrees Fahrenheit (°F)
Celsius degree (C°)	1.80	Fahrenheit degree (F°)
Additional water measurements:		
Gallon (imperial)	1.201	Gallon (U.S.)
Gallons (gal)	0.000003	Acre-feet

1 cubic foot per second per day = 86,400 cubic feet = 1.98 acre-feet
Additional Energy and Power Measurements

Additional Energy and Power Measurements

1 watt (W) = 1 joule/s
1 joule = 0.239 calorie
1 calorie = 4.186 joules
1 W/m^2 = 0.001433 cal/cm^2min
697.8 W/m^2 = 1 cal/cm^2min^{-1}

1 W/m^2 = 2.064 cal/cm^2day^{-1}
1 W/m^2 = 61.91 cal/cm^2month^{-1}
1 W/m^2 = 753.4 cal/cm^2year^{-1}
100 W/m^2 = 75 kcal/cm^2year^{-1}

Solar constant:
1372 W/m^2
2 cal/cm^2min^{-1}

English to Metric

English Measure	Multiply by	Metric Equivalent
Length		
Inches (in.)	2.54	Centimeters (cm)
Feet (ft)	0.3048	Meters (m)
Yards (yd)	0.9144	Meters (m)
Miles (mi)	1.6094	Kilometers (km)
Statute mile	0.8684	Nautical mile
Area		
Square inches (in.^{2})	6.45	Square centimeters (cm^2)
Square feet (ft^2)	0.0929	Square meters (m^2)
Square yards (yd^2)	0.8361	Square meters (m^2)
Square miles (mi^2)	2.5900	Square kilometers (km^2)
Acres (a)	0.4047	Hectare (ha)
Volume		
Cubic inches (in.^{3})	16.39	Cubic centimeters (cm^3)
Cubic feet (ft^3)	0.028	Cubic meters (m^3)
Cubic yards (yd^3)	0.765	Cubic meters (m^3)
Cubic miles (mi^3)	4.17	Cubic kilometers (km^3)
Quarts (qt), U.S.	0.9463	Liters (l)
Quarts (qt), Imperial	1.14	Liters (l)
Gallons (gal), U.S.	3.8	Liters (l)
Gallons (gal), Imperial	4.55	Liters (l)
Mass		
Ounces (oz)	28.3495	Grams (g)
Pounds (lb)	0.4536	Kilograms (kg)
Short ton (tn), U.S.	0.91	Metric ton (tonne) (t)
Velocity		
Miles/hour (mph)	0.448	Meters/second (mps)
Miles/hour (mph)	1.6094	Kilometers/hour (kmph)
Miles/hour (mph)	0.8684	Knots (kn) (nautical mph)
Temperature		
Degrees Fahrenheit (°F)	0.556 (after subtracting 32)	Degrees Celsius (°C)
Fahrenheit degree (F°)	0.556	Celsius degree (C°)
Additional water measurements:		
Gallon (U.S.)	0.833	Gallon (Imperial)
Acre-feet	325,872	Gallons (gal)

Additional Notation

Prefix	Symbol	1000^m	10^n	Decimal	Short scale	Long scale	Since[1]
yotta	Y	1000^8	10^{24}	1 000 000 000 000 000 000 000 000	septillion	quadrillion	1991
zetta	Z	1000^7	10^{21}	1 000 000 000 000 000 000 000	sextillion	trilliard	1991
exa	E	1000^6	10^{18}	1 000 000 000 000 000 000	quintillion	trillion	1975
peta	P	1000^5	10^{15}	1 000 000 000 000 000	quadrillion	billiard	1975
tera	T	1000^4	10^{12}	1 000 000 000 000	trillion	billion	1960
giga	G	1000^3	10^9	1 000 000 000	billion	milliard	1960
mega	M	1000^2	10^6	1 000 000	million		1960
kilo	k	1000^1	10^3	1 000	thousand		1795
hecto	h	$1000^{2/3}$	10^2	100	hundred		1795
deca	da	$1000^{1/3}$	10^1	10	ten		–
		1000^0	10^0		one		
deci	d	$1000^{-1/3}$	10^{-1} 0.1		tenth		1795
centi	c	1000^{-23}	10^{-2} 0.01		hundredth		1795
milli	m	1000^{-1}	10^{-3} 0.001		thousandth		1795
micro	μ	1000^{-2}	10^{-6} 0.000 001		millionth		1960
nano	n	1000^{-3}	10^{-9} 0.000 000 001		billionth	milliardth	1960
pico	p	1000^{-4}	10^{-12} 0.000 000 000 001		trillionth	billionth	1960
femto	f	1000^{-5}	10^{-15} 0.000 000 000 000 001		quadrillionth	billiardth	1964
atto	a	1000^{-6}	10^{-18} 0.000 000 000 000 000 001		quintillionth	trillionth	1964
zepto	z	1000^{-7}	10^{-21} 0.000 000 000 000 000 000 001		sextillionth	trilliardth	1991
yocto	y	1000^{-8}	10^{-24} 0.000 000 000 000 000 000 000 001		septillionth	quadrillionth	1991

[1] The metric system was introduced in 1795 with six prefixes. The other dates relate to recognition by a resolution of the CGPM.

Conversion Equivalents between Units of Energy

To: From:	TJ multiply by:	Gcal	Mtoe	MBtu	GWh
Terajoule (TJ)	1	238.8	2.388×10^{-5}	947.8	0.2778
Gigacalorie	4.1868×10^{-3}	1	10^{-7}	3.968	1.163×10^{-3}
Mtoe*	4.1868×10^4	10^7	1	3.968×10^7	11630
Million Btu	1.0551×10^{-3}	0.252	2.52×10^{-8}	1	2.931×10^{-4}
Gigawatt-hour	3.6	860	8.60×10^{-5}	3412	1

*Million tonnes of oil equivalent.

See also these online tools to convert units:

http://www.iea.org/interenerstat_v2/converter.asp
http://www.iea.org/stats/unit.asp

Glossary

A

Abyssal plain A level area of the deep-ocean floor, usually lying at the foot of the continental rise.

Absolute zero temperature The temperature at which all movement at an atomic level stops.

Active layer The top layer of permafrost soil that thaws during summer and freezes again during autumn.

Adaptation Changes to agricultural, industrial, and social infrastructure that allow a society to cope with environmental changes associated with climate change. See also coping strategy.

Adaptive capacity The ability of a community to make changes to agricultural, industrial, and social infrastructure that allow the community to cope with environmental changes associated with climate change.

Adiabatic A term that describes when a dynamic system evolves without the addition of energy from the external environment or the loss of energy to the external environment. In context this applies to a parcel of air as it rises (expands and cools) and descends (is compressed and warms) through the troposphere.

Adiabatic expansion See Adiabatic.

Adiabatic lapse rate The rate of temperature change with height in the troposphere due to the cooling or warming of air by expansion or compression, not because of the addition or removal of heat from the surrounding environment.

Adjusted Radiative Forcing Radiative forcing after the impact of rapid adjustment of some components of the atmosphere are taken into account.

Advanced Spaceborne Thermal Emission and Reflection Radiometer (ASTER) An imaging instrument that is flying on NASA's *Terra* satellite as part of NASA's Earth Observing System. ASTER is designed to obtain high-resolution global, regional, and local images of Earth in 14 color bands.

Advanced Very High Resolution Radiometer (AVHRR) A radiation-detection imager that can be used to remotely determine cloud cover and surface temperature. Note that the term *surface* can mean the surface of the Earth, the upper surfaces of clouds, or the surface of a body of water.

Aerosols Tiny solid and liquid particles suspended in the atmosphere.

Air mass A distinctive, homogeneous body of air in terms of temperature and humidity that takes on the moisture and temperature characteristics of its source region.

Albedo The reflective quality of a surface, expressed as the percentage of reflected insolation to incoming insolation; a function of surface color, angle of incidence, and surface texture.

Alkenone Highly resistant organic compounds (ketones) produced by phytoplankton. Some phytoplankton such as the coccolithophorids respond to changes in water temperature by altering the production of alkenones in the structure of their cells.

Alliance of Small Island Nations A coalition of small island and low-lying coastal countries that share concerns about the environment, especially their vulnerability to the adverse effects of global climate change. It functions primarily as an ad hoc lobby and negotiating voice within the United Nations.

Alluvial fan A fan-shaped fluvial landform at the mouth of a canyon; generally occurs in arid landscapes where streams are intermittent.

Anaerobic bacteria Bacteria that do not require oxygen for growth and often produce methane and toxic by-products during cellular respiration.

Anaerobic fermentation The breakdown of organic compounds by the action of living anaerobic organisms.

Anopheles A genus of mosquito that is known to spread the malarial parasite.

Anoxic An environment where there is very little free oxygen, typical of deep-ocean floor that is poorly ventilated by currents descending from the surface.

Antarctic Bottom Water (AABW) A water mass in the Southern Ocean surrounding Antarctica with temperatures ranging from –0.8°C to 2°C (33°F to 36°F) and salinities ranging from 34.6 to 34.7 psu.

Antarctic Circumpolar Current An ocean current that flows clockwise from west to east around Antarctica.

Antarctic Coastal Current The southernmost current in the world and the counter-current to the Antarctic Circumpolar Current. On the average, it flows westward and parallel to the Antarctic coastline.

Anthracite A hard, compact variety of coal that has the highest carbon content, the fewest impurities, and the highest calorific content of all types of coal.

Anthropocene An alternative name for the current geological epoch, based on the large impact of humans on the environment.

Anthropogenic Any change in a system that is due to the impact of human activity.

Anticyclonic wind A wind that blows out and flows clockwise about an anticyclone (high-pressure area) in the Northern Hemisphere and counterclockwise about an anticyclone in the Southern Hemisphere.

Aphelion The most distant point in Earth's elliptical orbit about the Sun; reached on July 4 and variable over a 100,000-year cycle. *See also* Perihelion.

Aqueous solution A solution of water and added mineral salt.

Aquifer An underground porous and permeable rock or sediment that contains a reservoir of fresh water.

Aragonite A form of the mineral calcium carbonate that is formed by biological and physical processes in both marine and freshwater environments.

Atlantic Multidecadal Oscillation (AMO) An ongoing series of long-duration changes in the sea surface temperature of the North Atlantic Ocean, with cool and warm phases that may last for 20 to 40 years at a time and a difference of about 06.°C (1°F) between extremes. These changes are natural and have been occurring for at least the past 1,000 years.

Atmospheric window Infrared radiation emitted from Earth's surface at wavelengths between 8 and 11 micrometers and to which the troposphere is transparent.

Atomic radius A measure of the size of the atoms of a chemical element, usually the mean distance from the nucleus to the boundary of the surrounding cloud of electrons.

B

Bali Action Plan A plan adopted after the 2007 United Nations Climate Change Conference on the island Bali in Indonesia. It was a two-year process on the way to finalizing a binding agreement in 2009 in Copenhagen.

Basal sliding The act of a glacier or an ice sheet sliding over the surface beneath due to the lubricating effect of meltwater at the interface. The rate of movement depends on the geothermal gradient, the slope of the glacier, the nature of the bedrock, the amount of meltwater, and the mass of ice.

Base period A long-term average that acts as a point of reference against which climate change can be tracked. To calculate temperature anomalies, scientists compare average temperatures over any given time period—a month or year, for example—to a base period.

BASIC countries A bloc of four large developing countries—Brazil, South Africa, India, and China—that is committed to act jointly in common interest during climate talks at the United Nations.

Belief persistence A tendency of people to favor information that confirms their beliefs, despite clear factual evidence to the contrary.

Benthic The lowest ecological region of a body of water, such as an ocean or a lake, including the sediment surface and some subsurface layers.

Berlin Mandate As part of the United Nations Framework Convention on Climate Change, a 2004 declaration that committed industrialized countries to the legally binding reduction of greenhouse gas emissions. Developing countries were exempted from binding obligations. The Berlin Mandate laid the groundwork for the Kyoto Protocol.

Binge and purge A model for ice sheet growth and decay that involves the slow growth of an ice sheet followed by rapid collapse that starts at the margins with the release of flotillas of icebergs into the surrounding oceans.

Biochar A form of charcoal that can be produced form agricultural waste and used as a soil supplement to sequester carbon.

Biodiesel A diesel fuel derived from vegetable oil or animal fat.

Biodiversity A principle of ecology and biogeography in which the more diverse the species population in an ecosystem (in number of species, quantity of members in each species, and genetic content), the more risk is spread over the entire community. Biodiversity results in greater overall stability, greater productivity, and increased use of soil nutrients than does a monoculture with no diversity.

Bioethanol A form of renewable energy that is produced from agricultural feedstock such as sugar cane or corn.

Biogas Gas produced by the anaerobic digestion or fermentation of biodegradable materials, such as biomass, manure, sewage, municipal waste, green waste, plant material, and crops. Biogas is primarily methane (CH_4) and carbon dioxide (CO_2), a gas produced by the breakdown of organic matter in the absence of oxygen.

Biological pump Biologically mediated processes that transport carbon from the surface zone to the deep ocean and sea floor.

Biomass A renewable energy source derived from biological material.

Biome A large terrestrial ecosystem characterized by specific plant communities and formations; usually named after the predominant vegetation in the region.

Bioreactor Any manufactured or engineered device or system that supports a biologically active environment, commonly used for the production of biogas and liquid biofuels.

Biosphere The totality of life on Earth; the parts of the lithosphere, hydrosphere, and atmosphere in which living organisms can be found.

Biostorage The storage of biomass through the preservation and cultivation of forests, specialized agricultural practices, and the burial or storage of biochar created by the partial burning of biomass.

Bioturbation The reworking of soils and sediments by animals or plants that can destroy primary sedimentary structures.

Bipolar seesaw A model of thermohaline circulation that explains observations from ice cores that record asynchronous warming and cooling in the Arctic and Antarctic; when the northern ocean warms, the southern ocean cools and vice versa.

Black carbon A climate-forcing agent (soot) formed through the incomplete combustion of fossil fuels, biofuel, and biomass.

Bleached Whitened stony reef corals that have expelled their symbiotic zooxanthellae due to elevated ocean temperatures.

Bølling–Allerød interstadial stage A warm and moist interstadial period that occurred during the final stages of the last glacial period from about 14,700 to 12,700 years before the present.

Bottom ocean water Water that flows along or close to the deep-ocean floor due to its relatively high density as a result of elevated salinity and/or low temperature.

C

Calcite A form of the mineral calcium carbonate that often replaces aragonite over time.

Calcrete A fossil soil that is enriched in the mineral calcium carbonate by strong surface evaporation in arid and semiarid environments. Also known as caliche.

Caliche *See* Calcrete.

Cap-and-trade An environmental policy tool that delivers results with a mandatory cap on emissions while providing the owner of those emissions some flexibility in how to comply. Successful cap-and-trade programs reward innovation, efficiency, and early action and provide strict environmental accountability without inhibiting economic growth.

Carbon capture and storage (sequestration) (CCS) The process of capturing waste carbon dioxide (CO_2) at the sources of production, such as coal power plants, and then transporting it to a storage site for disposal in a place such as an underground geological formation where it cannot enter the atmosphere.

Carbon dioxide reduction (CDR) New and existing technologies that could be used to actively reduce the levels of carbon dioxide in the atmosphere.

Carbon fee A fee based on the amount of carbon produced when a fossil fuel is burned. This fee is to be collected at the point of entry or production, and the cost is to be passed on to the public in higher energy prices. The government would then share the revenue generated with consumers, who may decide to use the money to offset some of the higher cost of energy or invest in energy conservation and efficiency and thus reduce their overall energy costs.

Carbon offset A reduction in emissions of carbon dioxide or other harmful greenhouse gases made in one area of production in order to compensate for (offset) emissions made elsewhere.

Carbon sink A natural or artificial reservoir that accumulates and stores some form of carbon for an indefinite period.

Carbon tax A tax levied on the carbon content of fuels.

Carbon wedge An activity that reduces emissions to the atmosphere that starts at zero today and increases to 1 gigatonne per year of reduced carbon emissions in 50 years, a cumulative total of 25 gigatonnes of emission reduction over 50 years.

Carboniferous Period A period of geological time commonly divided into the Mississippian and Pennsylvanian Periods in the United States but generally referred to as the Carboniferous Period in the rest of the world. It extends from the end of the Devonian Period, about 359.2 ± 2.5 million years ago, to the beginning of the Permian Period, about 299.0 ± 0.8 million years ago. The name *Carboniferous* means "coal-bearing" and derives from the Latin words *carbo* (coal) and *ferre* (to carry).

Cellulosic bioethanol A type of biofuel produced from cellulose derived from plant mass.

Cenozoic Era The most recent era of geological history, covering the period from 66 million years ago to the present day.

Charismatic megafauna Attractive large animals with widespread public appeal that environmentalists adopt to engage public support for their conservation goals.

Clean coal technology (CCT) A collection of technologies being developed to mitigate the environmental impact of coal energy generation.

Clean Development Mechanism (CDM) A program under which emission-reduction projects in developing countries can earn certified emission reduction credits. Industrialized countries may use these credits to meet a part of their emission reduction targets under the Kyoto Protocol or may sell them on the carbon market.

Climate forcing Natural and anthropogenic factors such as solar intensity and greenhouse gas emissions that change the balance of incoming and outgoing radiative energy at the top of the atmosphere. Also known as radiative forcing. *See also* Feedback.

Climate model A complex computer-based model that produces generalizations about reality and predictive forecasts of future weather and climate conditions.

Climatic Research Unit Based at the University of East Anglia, a unit that is widely recognized as one of the world's leading institutions concerned with the study of natural and anthropogenic climate change.

Climate sensitivity A measure of how responsive the temperature of the climate system is to a change in radiative forcing. Climate sensitivity is often expressed as the temperature change in °C associated with a doubling of the concentration of carbon dioxide in Earth's atmosphere.

Cloud albedo A measure of the reflectivity of a cloud. Higher values mean that the cloud reflects more solar radiation and thus transmits less radiation to reach Earth's surface.

Cloud condensation nuclei (CCN) Small particles in the atmosphere, typically 0.2 μm in diameter (1/100 the size of a cloud droplet), that are essential for the rapid formation of cloud droplets when air becomes saturated with water vapor.

Coastal aquifer An aquifer adjacent to the coast and vulnerable to invasion by seawater.

Coccolithophores Single-celled phytoplankton that are distinguished by external calcium carbonate plates called coccoliths that can survive the death of the cell and remain as microfossils. When present in abundance, they can accumulate to form a type of limestone known as chalk.

Community-based social marketing Marketing based on research which demonstrates that behavior change is most effectively achieved through initiatives delivered at the community level.

Concentrated solar power (CSP) Systems that use mirrors or lenses to concentrate a large area of sunlight onto a small area. The heat derived through this process is used to drive a steam turbine connected to an electrical power generator.

Conservation of energy Reduction of energy consumption through the use of new and existing technologies.

Continental rifting An early phase of the natural cycle of continental drift that occurs when tectonic stresses start to pull a continent apart. The rift valleys in eastern Africa are a modern example.

Convergence Horizontal inflow of air or water.

Conveyor belt A clearly defined stream of air that is embedded within a cyclonic frontal system. The warm conveyor belt is a stream of warm, moist air that originates in the warm sector of a frontal system.

Coping strategy *See* adaptation.

Coriolis effect The apparent deflection of a moving object from a straight-line path as it moves over the surface of Earth. Deflection is to the right in the Northern Hemisphere and to the left in the Southern Hemisphere. It produces a maximum effect at the poles and zero effect along the equator.

Corporate Average Fuel Economy (CAFE) standard Regulations in the United States that are intended to improve the average fuel economy of cars and light trucks sold in the United States.

Cosmogenic isotope An isotope formed near the top of the atmosphere by the collision between air molecules and high-energy particles (cosmic rays).

Cretaceous Period A period of geological time from about 145 ± 4 to 66 ± 0.3 million years ago (Ma).

Crisis point The point where a community no longer possesses the adaptive capacity to cope with changes in the immediate environment.

Cryosat Launched in 2009 this is the European Space Agency's first spacecraft dedicated to the study of ice.

Cryosphere Parts of Earth's surface that are permanently frozen

Cultural risk theory A conceptual framework that seeks to explain societal conflict about risk.

Cyclonic weather system A dynamic area of low atmospheric pressure with converging and ascending air flows; rotates counterclockwise in the Northern Hemisphere and clockwise in the Southern Hemisphere.

D

Dansgaard–Oeschger (D–O) cycles Rapid fluctuations in climate that occurred 25 times during the last glacial period.

Daughter atom An atom that is produced by the decay of a radioactive (parent) isotope.

Dead zone An area of toxic water where visible life is absent due to the presence of anaerobic bacteria in oxygen-depleted water. Most common on the beds of oceans, lakes, and estuaries where deep water is poorly ventilated by currents originating at the surface.

Decadal Acting over a time period of around 10 years.

Deep-ocean water Water that flows at depth but above the deep-ocean floor due to its intermediate density.

Denudation A general term that refers to all processes that cause degradation of the landscape, such as weathering, mass movement, erosion, and transport.

Desiccation crack A polygonal shrinkage crack that forms when once-wet sediments become desiccated.

Dimethyl sulfide (DMS) A natural organic compound released into the atmosphere by phytoplankton that is oxidized to various sulfur-containing compounds, including sulfuric acid, that have the potential to promote cloud formation and may affect global climate.

Dinoflagellate A large group of flagellate protists (unicellular microorganisms with long tails), most of which are marine plankton, but some of which are also found in fresh water.

Divergence The horizontal outflow of air or water.

Doldrums The equatorial belt of calms or light variable winds that lies between the two trade wind belts.

Drainage basin The total surface area of land that is drained by a connected river system and distinguished from neighboring basins by ridges and highlands that form a divide.

Dropstone Fragments of rock released from melting surface ice that are found within finer-grained water-deposited sedimentary rocks. Dropstones range in size from small pebbles to large boulders and show evidence of having settled vertically through the water column.

E

Earth Summit Also known as the United Nations Conference on Environment and Development (UNCED), a major United Nations conference held in Rio de Janeiro in 1992. In 2012, the United Nations Conference on Sustainable Development was also held in Rio, and it is also commonly called Rio+20 or Rio Earth Summit 2012.

East Antarctic Ice Sheet (EAIS) The larger of the two main ice sheets that cover the continent of Antarctica.

Eccentricity Variation of Earth's orbit from circular to slightly elliptical, which modulates the amount of solar radiation received at the top of the atmosphere over a calendar year.

Ecosystem A self-regulating association of living plants, animals, and their nonliving physical and chemical environment.

Eemian Interglacial Stage An interglacial stage associated with Marine Isotope Stage 5e that began about 130,000 years ago and ended about 114,000 years ago.

Efficiency (energy) The use of new and existing technologies to increase the efficiency of the energy we generate and use.

Effusive volcanism The kind of volcanism most often associated with continental rifting, in which large volumes of lava are rapidly erupted at the surface. Geological examples in India and Siberia released such large volumes of carbon dioxide that the climate was adversely affected.

Ekman transport direction The 90-degree net transport (flow) of the upper layer of the ocean due to the force of the wind on the ocean surface.

El Niño–Southern Oscillation (ENSO) A repeating climate pattern that occurs across the tropical Pacific Ocean every five to seven years. The Southern Oscillation refers to variations in the temperature of the surface of the tropical eastern Pacific Ocean (warming and cooling known as El Niño and La Niña, respectively) and in air surface pressure in the tropical western Pacific.

Electrical Energy Energy associated with the transfer of electrical charge.

Electromagnetic energy Radiant energy produced by the Sun, most of which lies within the visible spectrum.

Emerging Nation An emerging nation is a country that is on its way to becoming an industrialized nation.

Emissions trading scheme A program based on the cap-and-trade principle, in which a cap, or limit, is placed on the total amount of certain greenhouse gases that can be emitted by factories, power plants, and other major sources of greenhouse gas emissions. Within this cap, companies receive emission allowances that they can "trade" (sell) on the carbon market or buy from the market as needed.

Empirical Acquired by means of observation or experimentation.

Energy Information Administration (EIA) The statistical and analytical agency within the U.S. Department of Energy.

Energy poverty A lack of access to modern energy services, such as electricity and clean cooking facilities.

Enhanced greenhouse gas effect The strengthening of the natural greenhouse effect through human activities.

Environmental Lapse Rate The rate of decrease of temperature with height in a stationary atmosphere.

Environmental Protection Agency (EPA) An agency of the U.S. government whose mission is to protect human health and the environment that sets and enforces national pollution-control standards.

Eocene Epoch A major division of geological time lasting from about 56 to 34 million years ago and the second epoch of the Paleogene Period in the Cenozoic Era.

Ephemeral Relating to objects or environments that exist only for a short period of time, such as desert lakes.

Equinox An event that occurs twice a year, when the Sun is directly overhead at the equator. In the Northern Hemisphere the equinoxes occur on March 20 or 21 (the vernal, or spring, equinox) and September 22 or 23 (the autumnal equinox). At each equinox, the lengths of daylight and darkness are equal at all latitudes.

Eukaryotes An organism whose cells contain complex structures such as a nucleus enclosed within membranes.

European Project for Ice Coring in Antarctica (EIPCA) A multinational European project for deep ice core drilling in Antarctica.

Evaporite A term given to rocks comprised of a suite of minerals (mostly halite and sulfate) that precipitate directly from a body of water when it evaporates.

Exhumed landscape The reappearance of ancient topography when erosion removes a mantle of younger sediments.

Exposure In risk theory, a measure of the probable damage to life and property associated with a particular hazard.

Extended Reconstructed Sea Surface Temperature (ERSST) analysis Temperature analysis that is constructed using the most recently available international data set and uses improved statistical methods that allow reconstruction using sparse data.

External factors Physical processes that act on an isolated system from an external source.

F

Faculae Bright areas on the Sun's surface that are slightly hotter than the surrounding photosphere. A sunspot always has an associated facula, though faculae may exist apart from such spots.

Fast breeder reactor A nuclear reactor that is capable of generating more fissile material than it consumes.

Fecal pellet Small pellets, sometimes microscopic, that are the excrement of marine invertebrates and are an important constituent of many fine-grained marine rocks.

Feedback As used in climatic change, any force that augments or diminishes a small initial change in the state of a climate system. *See also* Climate forcing.

Ferrel cell A secondary atmospheric circulation feature that forms at mid-latitudes between the Hadley cell and the Polar cell. It is the result of the transient mixing of air by the passing of high- and low-pressure frontal systems.

Firn A denser recrystallized form of snow that has been left over from past seasons.

Fischer–Tropsch reaction A series of chemical reactions that convert a mixture of carbon monoxide and hydrogen into liquid hydrocarbons.

Fracking (hydraulic fracturing) A technique used to enable the economic extraction of oil and natural gas from reservoir rock formations with low permeability by fracturing the reservoir using a fluid under very high pressure.

Frontal system The complex flow of air associated with moving boundaries between contrasting air masses at the surface and the position of the jet streams at altitude.

Fukushima Disaster A tsunami struck Japan in 2011 and severely damaged a nuclear reactor at Fukushima, resulting in the leakage of radioactive waste into to the environment.

FutureGen Alliance A U.S. government–funded project to construct a near-zero-emissions coal-fueled power plant that produces hydrogen and electricity.

G

Gas hydrate *See* Methane hydrate.

Geocentric Universe An ancient description of the universe where the Earth is at the orbital center of all celestial bodies.

Geoengineering Deliberate large-scale intervention in the Earth's climate system in order to moderate global warming.

Geothermal energy Earth's internally generated energy that produces a positive temperature gradient with increasing depth in the crust. The average is 25°C to 30°C per kilometer (ca. 1°F per 70 feet) in the upper crust.

Geothermal gradient The rate at which the temperature increases with depth in the Earth. The average rate is around 25°C to 30°C per kilometer (ca. 1°F per 70 feet) in the upper crust.

Geothermal power A variety of technologies that use geothermal energy to generate electrical power.

Gigatonne A gigatonne is a billion (10^9) tonnes. A tonne is the metric system unit of mass equal to 1,000 kilograms (2,204.6 pounds) that is used globally for reporting climate related data.

Glacial moraine Glacial sediment left behind by a retreating ice sheet or glacier as it melts.

Glacial Rebound An increase in the elevation of a landmasses when the weight of a large mass of ice is removed from its surface.

Global circulation model (GCM) A large three-dimensional computer model of Earth's air–land–water system that is used to explore climate and climate change. The global wind, pressure, temperature, and moisture fields are computed from basic physical equations. Current GCMs are restricted to regional spatial resolution, largely due to the limited capabilities of today's supercomputers.

Global dimming The decline in sunlight reaching Earth's surface due to pollution, aerosols, and clouds.

Global Historic Climatology Network (GHCN) An integrated database of daily climate summaries from land surface stations around the globe. It is comprised of daily climate records from numerous sources that have been integrated and subjected to a common suite of quality assurance reviews.

Global Warming An increase in average global temperature that can lead to climate change.

Goddard Institute for Space Studies (GISS) A NASA body whose research emphasizes a broad study of global change; an interdisciplinary initiative that addresses natural and human-caused changes in our environment that occur on various time scales.

Gravity Recovery and Climate Experiment (GRACE) A mission whose primary goal is to accurately map variations in the Earth's gravity field over its five-year lifetime. The GRACE mission has two identical spacecraft flying about 220 kilometers apart in a polar orbit 500 kilometers (137 miles) above Earth.

Great ocean conveyor A global pattern of ocean circulation that is driven by density (temperature and salinity) gradients and is responsible for the transfer of large amounts of energy from low to high latitudes. The ocean conveyor gets it "start" in the Norwegian Sea, where warm water from the Gulf Stream heats the atmosphere in the cold northern latitudes. This loss of heat to the atmosphere makes the water cooler and denser, causing it to sink to the bottom of the ocean. As more warm water is transported north, the cooler water sinks and moves south, making room for the incoming warm water. This cold bottom water flows south of the equator all the way down to Antarctica. Eventually, the cold bottom waters are able to warm and rise to the surface, continuing this conveyor belt that encircles the globe.

Greenhouse gas intensity The ratio of greenhouse gas emissions produced to gross domestic product. This is a measure of how efficiently we use energy.

Greenhouse gases Gases such as carbon dioxide, water vapor, methane, and CFCs that absorb and radiate infrared radiation at different wavelengths and delay the loss of infrared wavelengths to space.

Greenland Ice Sheet Project (GISP) A decade-long project to drill ice cores in Greenland that involved scientists and funding agencies from Denmark, Switzerland, and the United States.

Growth ring Internal growth structures of many trees, corals, cave deposits, and some shells that can be used to date the deposits.

Gyre A large, circular surface-current pattern that is found in each ocean and that is created by a combination of surface winds and the Coriolis effect.

H

Hadley cell A thermally driven circulation system of equatorial and tropical latitudes consisting of two convection cells, one in each hemisphere.

HadCRUT *See* Climate Research Unit (CRU).

Half-life The time it takes for half of the parent atoms of a radioactive isotope to decay into daughter atoms; the half-life is a constant for each element.

Halocline A layer in a large body of water in which density changes more rapidly with depth than it does in the layers above or below.

Hazard In risk theory, a situation that poses a threat to life, health, property, or the environment.

Heat capacity A measure of the amount of heat required to increase a substance's temperature by a given amount. In the International System of Units (SI), heat capacity is expressed in units of joule(s) (J) per kelvin (K).

Heat content A term used in thermodynamics to indicate the total energy of a thermodynamic system.

Heinrich Event Periodic events during the Pleistocene Epoch when large volumes of ice broke broke free from boreal ice sheets.

Hockey stick graph A graph in which a sharp increase or decrease occurs over a period of time. The line connecting the data points resembles a hockey stick, with the "blade" formed from data points shifting diagonally and the "shaft" formed from the horizontal data points.

Holocene Epoch A geological epoch that began around 12,000 years ago, at the end of the Pleistocene epoch.

Hothouse A world with no permanent ice caps and global temperatures much higher than they are today.

Human ecology The study of relationships between humans and their natural, social, and built environments.

Hydrogen A chemical element with symbol H and atomic number 1 that burns in the presence of oxygen to produce energy and water with no toxic by-products of combustion.

Hydrogen fuel cell A device that converts the chemical energy from hydrogen into electricity through a chemical reaction with oxygen or another oxidizing agent.

Hydrophilic A molecule or part of a molecule that has a tendency to interact with or be dissolved by water.

I

Ice stream A region of an ice sheet that moves significantly faster than the surrounding ice.

Icehouse A world with permanent ice caps and global temperatures approximately equal to or lower than they are today.

ICESat (Ice, Cloud, and Land Elevation Satellite) The benchmark Earth Observing System mission for measuring ice sheet mass balance, cloud and aerosol heights, and land topography and vegetation characteristics.

Igneous rock A rock formed by the solidification of magma (molten rock).

Ikaite A hydrated form of calcium carbonate that forms within seafloor sediments under near-freezing conditions.

Infrared radiation Radiation emitted by Earth with a wavelength of 0.7 to 200 micrometers.

Infrared radiation The part of the electromagnetic radiation with a wavelength from 0.7 to 200 micrometers.

Insolation The amount of solar energy that strikes a given area over a specific time and varies with latitude or the seasons.

Integrated gasification combined-cycle (IGCC) An efficient technology that turns coal into gas (syngas) that is then burned to drive a gas turbine, with the excess heat generated used to drive a secondary steam turbine. The resulting carbon dioxide gas is compressed and stored, and it can be permanently sequestered.

Interglacial A period of warmer climate conditions between glacial periods.

Intermediate water Major ocean currents that flow at intermediate depth within the oceans above deep and bottom waters and below surface currents.

Internal factors Physical processes that act on an isolated system from an internal source.

International Comprehensive Ocean–Atmosphere Data Set (ICOADS) A set of surface marine temperature data spanning the past three centuries; it is a simple gridded monthly summary of products for 2° latitude × 2° longitude boxes.

Intergovernmental Panel on Climate Change (IPCC) A United Nations group that is the leading international body for the assessment of climate change.

International Energy Agency (IEA) A Paris-based autonomous intergovernmental organization established in the framework of the Organisation for Economic Co-operation and Development (OECD) to act as an adviser on energy policy, energy security, economic development, and environmental protection (including climate change) to its member and associated states.

Intertropical convergence zone (ITCZ) The zone of general convergence between the Northern and Southern Hemisphere trade winds.

Invasive species A non-indigenous species of plants or animals that adversely affect the habitats it invades.

Isotope excursion Sudden changes in the composition of an isotopic system such as carbon in the oceans or oxygen in ice cores.

Isotopes Varieties of a specific element that have the same number of protons and a variable number of neutrons.

J

Jet stream Swift, shifting, intermittent high-altitude winds that travel at 120–240 kilometers per hour (75–150 miles per hour) and remain in relatively narrow belts as they circle the globe.

Joint Implementation Mechanism (JI) One of three flexibility mechanisms set out in the Kyoto Protocol to help countries with binding greenhouse gas emissions targets (so-called Annex B countries) to meet their obligations. An Annex I country can invest in an emission reduction project (referred to as a Joint Implementation Project) in any Annex B country as an alternative to reducing emissions domestically. This may be a cheaper way for a country to meet its emissions goals, as the cost of reducing emissions in another country can be lower.

Joule A unit of energy, work, and heat in the International System of Units. It is equal to the energy used (or work done) in applying a force of 1 newton over a distance of 1 meter—about the same as lifting an apple off the floor.

Jurassic Period A period of geological time that extended from the end of the Triassic Period 145 million years ago to the beginning of the Cretaceous Period about 201 million years ago.

K

K–T boundary A thin boundary layer that contains physical evidence of a catastrophic meteorite impact that marks the transition between the Cretaceous and Tertiary (Paleogene) Periods.

Karst A geological formation shaped by the dissolution of soluble rock layers, usually carbonate rock such as limestone.

Keeling curve A graph that plots the ongoing change in concentration of carbon dioxide in Earth's atmosphere since 1958.

Keystone species A species that has a disproportionately large effect on its environment relative to its abundance. The removal of a keystone species from an area can lead to rapid changes in local ecology.

Kinetic energy The energy of a body in motion, which is a product of its velocity (squared) and mass ($\frac{1}{2}\, mV^2$).

Kyoto Protocol An international agreement linked to the United Nations Framework Convention on Climate Change. The major feature of the Kyoto Protocol is that it sets binding targets for 37 industrialized countries and the European Community for reducing greenhouse gas (GHG) emissions. These targets amount to an average of 5% against 1990 levels over the five-year period 2008–2012.

L

L1 (Lagrange) point A point between Earth and the Sun where the forces of gravitation and orbital motion of a body balance each other so that a satellite placed at this point lies in a direct line between Earth and the Sun at all times during its orbit.

La Niña *See* El Niño–Southern Oscillation (ENSO).

Lake Agassiz An immense glacial lake that formed toward the end of the Pleistocene Epoch and covered parts of Manitoba, Ontario, and Saskatchewan in Canada, as well as North Dakota and Minnesota in the United States.

Last glacial maximum (LGM) A period in Earth's climate history when ice sheets were at their maximum extension, between about 26,500 and 19,000 years ago.

Latent heat Energy that is absorbed or released in the phase change of water and is stored in one of the three states–ice, water, or water vapor. Latent heat includes the latent heat of melting, freezing, vaporization, evaporation, and condensation.

Laterite A type of iron- and aluminum-rich soil that forms under hot and wet tropical conditions.

Laurentide Ice Sheet A thick, massive ice sheet that covered most of Canada and a large portion of the northern United States multiple times during the Quaternary Period.

Lithology The physical characteristics of a rock.

Little Ice Age (LIA) A cold period between A.D. 1550 and A.D.1850. There were three particularly cold intervals: one beginning about 1650, another about 1770, and the last in 1850, separated by intervals of slight warming.

Liverwort Small plants (less than 10 centimeters or 4 inches long) that cover large patches of ground, rocks, and trees. They are distributed globally, most often in humid locations.

Loess A sedimentary deposit formed by the accumulation of wind-blown dust.

Longevity In risk theory, the duration of time over which a specific hazard causes physical and environmental damage.

Luminescence In the context of this book, this refers to the release of visible light from corals that are illuminated by ultraviolet light.

M

Macrofossil Any large fossil that can be examined without the use of a microscope.

Madden–Julian Oscillation (MJO) A large-scale coupling between atmospheric circulation and tropical deep convection that cycles over a period of 30 to 90 days and can have a major impact on global weather.

Magnetic field Earth's magnetic field protects the biosphere from high-energy-charged particles from the Sun and outer space (solar wind and cosmic rays).

Mangrove swamp Small trees and shrubs that grow in saline coastal sediment habitats in the tropics and subtropics and act as a seawall to protect the coast from the effects of major storms such as hurricanes.

Mapping Spectral Variability in Global Climate Project (SPECMAP) A standard chronology for oxygen isotope records that is derived from several isotopic records that help to increase the signal-to-noise ratio of the data.

Marginal desert An area on the margins of a desert that has very poor agricultural potential and that is vulnerable to small changes in precipitation.

Marine atmosphere boundary layer (MABL) The part of the atmosphere that has direct contact with the ocean and where the ocean and atmosphere exchange large amounts of heat, moisture, and momentum, primarily via turbulent transport.

Marine benthic extinction event (MBE) A period of time represented in the fossil record by the extinction of marine organisms that lived near and on the seafloor (benthic) due to sudden changes in their ecosystem.

Marine cloud reflectivity The albedo of highly reflective low-level marine clouds.

Marine Isotope Stages (MIS) The subdivision of the most recent ca. 5 million years of climate history into warm and cool periods, based on the changing isotope composition of shells derived from deep marine sediment cores.

Marker horizon A distinctive layer of rock, dust, or chemical contamination that is rapidly deposited over a wide geographical area and often used to correlate identical moments in time between rock and ice cores.

Marrakesh Accords A set of agreements reached at the 7th Conference of the Parties (COP7) to the United Nations Framework Convention on Climate Change in 2001 that included "flexible mechanisms" (free market–based solutions) as a financial incentive to help developed nations meet their emissions targets and to promote the transfer of technology and human capital in developing nations.

Maunder Minimum A period starting around 1645 and continuing to around 1715 when sunspots were rare and may have contributed to the global cooling associated with the Little Ice Age.

Medieval Climate Anomaly (MCA) A time of warm climate lasting from about A.D. 950 to 1250. Most evidence comes from the North Atlantic region, but there is also evidence of related climate events elsewhere around the world during that time.

Meltwater The water released from a melting body of ice.

Meridional "Along a meridian" or "in the north–south direction."

Methane hydrate A form of ice that combines methane and water and is found under permafrost, in polar shallow marine sediments, and in deeper sedimentary rocks at depth on the continental shelf.

Methanesulphonate (MSA) One of the molecules in the atmosphere that is produced by the oxidation of dimethyl sulfide (DMS).

Methanogenic bacteria Microorganisms that produce methane as a metabolic byproduct in anoxic conditions.

Microwave sounding unit (MSU) A device that has been installed on a polar orbiting satellite to measure, from space, the intensity of microwave radiation emitted by Earth's atmosphere. Different "channels" of the MSU measure different frequencies of radiation that can, in turn, be related to temperature averages of the atmosphere over different vertical regions.

Mid-Pleistocene Transition (MPT) A fundamental change between 1.2 million and 700,000 years ago when the dominant periodicity of climate cycles changed from 41,000 to 100,000 years in the absence of substantial change in orbital forcing.

Migration corridor The intentional preservation of connected areas of natural habitat that enable the unhindered migration of many species.

Miocene Epoch The fourth epoch in the Tertiary Period, extending from 24 to 5 million years ago.

Mitigate To diminish or moderate the impact of (in this case) climate change.

MODIS (Moderate-Resolution Imaging Spectroradiometer) A key instrument aboard NASA's *Terra* and *Aqua* satellites.

Modulate To change the frequency, amplitude, phase, or other characteristic of electromagnetic waves.

Momentum The product of mass and velocity of a moving object.

Monaco Declaration A declaration by 155 scientists from 26 countries expressing concern at the severe threat posed by ocean acidification.

Monsoon Seasonal reversal of wind direction associated with large continents, especially Asia. In winter, the wind blows from land to sea; in summer, from sea to land.

Montreal Action Plan Developed at the UN Framework Convention on Climate Change's 11th Conference of the Parties (COP 11) on December 10, 2005, a plan that is intended to set the scene for extending the Kyoto Protocol and the future inclusion of large developing countries such as India and China in the system of limiting greenhouse gas emissions.

Montreal Protocol An international treaty designed to protect the ozone layer by phasing out the production of numerous substances believed to be responsible for ozone depletion.

Multiyear ice layer A layer of ice that remains frozen year-round and is the oldest and thickest Arctic sea ice.

N

National Aeronautics and Space Administration (NASA) The agency of the U.S. government that is responsible for the civilian space program and for aeronautics and aerospace research.

National allowances The maximum level of greenhouse gas emissions allocated to each participating country under the Kyoto Protocol.

National Climate Data Center (NCDC) A U.S. organization that is the world's largest active archive of weather data.

National Oceanic and Atmospheric Administration (NOAA) A scientific agency within the U.S. Department of Commerce focused on the conditions of the oceans and the atmosphere.

Neogene Period A period of geological time starting 23 million years ago and ending 2.58 million years ago that is subdivided into the older Miocene and the younger Pliocene Epochs.

Neutron A subatomic particle with zero charge that is found in the nucleus of an atom.

Nonlinear system A system where output is not directly proportional to input.

North Atlantic Deep Water (NADW) A cold water mass that flows at depth from its source in the North Atlantic Ocean by the sinking of highly saline water from the Greenland Sea.

North Atlantic Oscillation (NAO) A climatic phenomenon in the North Atlantic Ocean marked by fluctuations in the difference of atmospheric pressure at sea level between the Icelandic low and the

Azores high. The NAO affects the route of the polar jet stream and has a large impact on weather.

Northern Greenland Ice Core Project (NGRIP) A joint project lasting from 1989 to 1994 that used deep ice cores to investigate the climate history of northern Greenland and included investigators from Europe, Japan, and the United States.

O

Obliquity The angle made by a planet's rotation axis with respect to a line drawn through the center of the planet and perpendicular to the plane of the planet's orbit. Also called axial tilt.

Ocean acidification A decrease in the pH of the oceans due to the addition of more carbon dioxide from the atmosphere. More correctly, this is a decrease in alkalinity, as the oceans remain fully alkaline.

Ocean convection The global circulation of major ocean currents that are driven by differences in density arising from changes in salinity and temperature.

Ocean Drilling Program A program funded by the U.S. National Science Foundation and 22 international partners that conducted basic research into the history of the ocean basins and the overall nature of the crust beneath the ocean floor using the scientific drill ship *JOIDES Resolution*.

Ocean fertilization A type of geoengineering based on the idea that adding nutrients to the upper ocean will help enhance the biological pump and remove more carbon dioxide from the atmosphere.

Ocean ridge A continuous elevated zone on the deep-ocean floor of major ocean basins that varies in width from 500 to 5,000 kilometers (300 to 3,000 miles). The rifts at the crests of these ridges, the ridge axes, mark the location of divergent plate boundaries.

Oil crisis Periods of time, especially in the 1970s, when the economies of the major industrial nations of the world were seriously affected by petroleum shortages and elevated fuel prices due to political unrest in the Middle East.

Oil shale A type of fine-grained clay-rich rock that is sufficiently rich in organic material to generate oil.

Oligocene Epoch A subdivision of geological time belonging to the Paleogene (Tertiary) Period that extended from about 34 million to 23 million years before the present.

Opacity A measure of how impenetrable a substance (solid, liquid, or gas) is to electromagnetic radiation, especially visible light.

Optical thickness A measure of the transparency of the atmosphere. Also called optical depth.

Organic carbon Carbon that was once part of living tissue.

Organisation for Economic Co-operation and Development (OECD) An international economic organization of 34 countries founded in 1961 to stimulate economic progress and world trade.

Organization of the Petroleum Exporting Countries (OPEC) An intergovernmental organization of 12 oil-producing countries: Algeria, Angola, Ecuador, Iran, Iraq, Kuwait, Libya, Nigeria, Qatar, Saudi Arabia, the United Arab Emirates, and Venezuela.

Orogenesis A scientific term for the geological process of mountain building.

P

Pacific Decadal Oscillation (PDO) A pattern of climate variability that changes every 20 to 30 years. During a "warm," or "positive," phase, the western Pacific becomes cool, and part of the eastern ocean warms; during a "cool," or "negative," phase, the opposite pattern occurs.

Paleocene–Eocene Thermal Anomaly (PETA) A geological period of rapid global warming that marks the boundary between the Paleocene and Eocene Epochs and is associated with marine extinction and a major excursion in the carbon cycle.

Paleogene Period The first Period of the Cenozoic Era that began 65.5 million years ago and ended 23.03 million years ago.

Pangaea A supercontinent that existed during the late Paleozoic and early Mesozoic eras. 200 million years ago, this proposed supercontinent began to break apart and form the present landmasses.

Particulate matter Tiny pieces of solid or liquid suspended in Earth's atmosphere as atmospheric aerosol.

PDB standard A standard used by the international community for the analysis of carbon isotopes. It has been replaced by a new synthetic standard.

Perihelion The point in Earth's elliptical orbit about the Sun where Earth is closest to the Sun; reached on January 3 and variable over a 100,000-year cycle. *See also* Aphelion.

Permian Period The most recent period of the Paleozoic Era, lasting from 260 to 245 million years ago.

pH A measure of the acidity or alkalinity of a solution, in this case of the oceans, based on the logarithm of the hydrogen ion concentration.

Phanerozoic Eon The current geological eon, which is characterized by the presence of abundant animal life and started about 542 million years ago.

Photon An elementary particle, a quantum of electromagnetic radiation.

Photovoltaic (PV) solar cells Tools for generating electrical power that convert sunlight into direct current electricity using special semiconductors.

Phytoplankton Photosynthesizing microscopic organisms that inhabit the upper sunlit layer of almost all oceans and bodies of fresh water.

Plankton Organisms—often microscopic—that live in the water column and are incapable of swimming against a current.

Pleistocene Epoch The first (and oldest) epoch in the current Quaternary Period, extending from 1.6 million to 10,000 years ago.

Plinian Very powerful volcanic eruptions that are characterized by columns of gas and volcanic ash that extend high into the stratosphere, where they can affect global climate.

Pliocene Epoch A subdivision of geological time that extended from 5.33 million years ago to 2.58 million years before the present. It is the youngest epoch of the Neogene Period, following the Miocene Epoch and before the Pleistocene Epoch.

Plume A tall, high-velocity column of hot volcanic ash and gas emitted into the atmosphere during an explosive volcanic eruption.

Polar front A stormy frontal zone separating air masses of polar origin from air masses of tropical origin.

Positive feedback loop Natural processes that are a consequence of climate forcing and act to augment the direction of climate forcing.

Potable Water that is safe enough to be consumed by humans with low risk of immediate or long-term harm.

Potential energy Energy related to a body's mass and position with respect to gravity. A ball dropped from a height of 10 meters (32 feet) has twice as much potential energy as a ball dropped from 5 meters (16 feet).

Precautionary principle The idea that it's better to be safe than sorry. If an action or a policy may cause harm to the public or to the environment, the burden of proof is on those who claim it is not harmful, not on those who express concern.

Precession A slow motion of Earth's axis that traces out a cone over a period of 26,000 years.

Probability In risk theory, the likelihood of a hazard occurring over a period of time.

Prokaryotes A group of organisms whose cells lack a cell nucleus or any other membrane-bound organelles.

Proterozoic Eon A geological eon that extended from 2,500 to 542 million years ago, representing a time before the proliferation of complex life on Earth but containing evidence of abundant life at the cellular level.

Protists A group of predominantly unicellular microscopic organisms that share certain characteristics with animals or plants or both. The protists now comprise what have traditionally been called protozoa, algae, and lower fungi.

Proton A subatomic particle with single positive charge that is found in the nucleus of an atom.

Proxy data Information gathered from natural recorders of climate variability, such as corals, tree rings, ice cores, and ocean-floor sediments.

Pseudomorph A mineral that replaces a preexisting mineral but retains the original crystal form.

Pycnocline A vertical interval within a body of water where the density gradient is greatest.

Pycnocline A vertical layer within a body of water where the density gradient is greatest.

Q

Quaternary Period The most recent period of the Cenozoic Era, extending from 2.58 million years ago to the present.

R

Radiative forcing *See* Climate forcing.

Radioactive isotope A variety of an element with an unstable nucleus that decays spontaneously at a constant rate into daughter atoms and radioactive particles.

Radiosonde A small radio-device used in weather balloons that transmits a continuous stream of data on the physical conditions of the atmosphere.

Reducing Emissions from Deforestation and Forest Degradation (REDD) A program designed to use market and financial incentives to reduce the emissions of greenhouse gases. Launched in 2008, REDD supports national programs and promotes the informed and meaningful involvement of all stakeholders, including indigenous peoples and other forest-dependent communities.

Regional climate model (RCM) A model that works by increasing the resolution of the global climate model in a small, limited area of interest.

Regional tectonic stress Stress that develops in the lithosphere due to the movement of tectonic plates. Such stresses can lead to the deformation of Earth's crust and affect relative sea level.

Remote Sensing Systems (RSS) A private research company that processes microwave data from a variety of NASA satellites.

Renewable energy Energy that comes from natural resources such as sunlight, wind, rain, tides, waves, running water, and geothermal heat. These resources are considered renewable because they are naturally replenished at a constant rate.

Renewable Portfolio Standard (RPS) A government regulation that requires the increased production of energy from renewable energy sources, such as wind, solar, biomass, and geothermal.

Residency time The period of time a greenhouse gas spends in the atmosphere.

Ridge axis *See* Ocean ridge.

Ripple mark A sedimentary structure that forms naturally on the surface of a layer of sediment under the influence of flowing water.

Risk In risk theory, the combination of the nature of the hazard, the exposure to that hazard, the longevity of the event, and the probability of the event's occurrence.

Rock striations Deep grooves carved into bedrock by sharp rocks frozen at the base of a moving glacier or ice sheet.

Rossby waves Upper-air waves in the middle and upper troposphere of the middle latitudes with wavelengths of from 4,000 to 6,000 kilometers (2,500 to 3,700 miles).

Rudists A group of marine mollusks that first appeared during the Jurassic Period and diversified during the Cretaceous Period, when they became major reef-building organisms.

S

Saffir-Simpson scale A 1-to-5 cyclone rating based on a storm's sustained wind speed.

Sahel A biogeographic transition zone between the Sahara to the north and less arid savannah to the south.

Salinity The proportion of dissolved salts to pure water, usually expressed in parts per thousand (‰) or in practical salinity units (PSU) the conductivity ratio of a seawater sample to a standard KCl solution.

Scleractinia Stony corals that can be either solitary or reef forming. Most schleractinian corals that live in the photic zone have a symbiotic relationship with zooxanthellae.

Schwabe cycle A natural cycle of sunspot intensity and polarity that changes over an approximately 11-year period.

Sea ice volume The volume of sea ice, calculated each day relative to the average over the period 1979–2011. Sea ice volume ranges from about 28,700 km^3 (11,000 square miles) in April to 12,300 km^3 (4,750 square miles) in September and is an important indicator of climate change in the Arctic.

Sea surface temperature (SST) The water temperature close to the ocean's surface, as determined by direct measurement and remotely by satellite.

Second Assessment Report A 1996 assessment published by the Intergovernmental Panel on Climate Change (IPCC), which considered the then-available scientific and socioeconomic information on climate change. It has since been superseded by the *Third Assessment Report* (TAR), published in 2001, and the *Fourth Assessment Report* (FAR), published in 2007.

Short-wavelength flux Radiant energy from the Sun with wavelengths between 0.1 μm and 5.0 μm in the visible (VIS), near-ultraviolet (UV), and near-infrared (NIR) spectra.

Six Americas Survey An ongoing survey of national opinion about climate change in the United States conducted by Yale University.

Skin temperature The temperature at about 5 to 8 kilometers (16,000 to 26,000 feet) high, depending on latitude in the troposphere, where Earth radiates sufficient energy to balance the amount of energy arriving from the Sun.

SMART goals These must be Specific, Measureable, Achievable, Realistic, and deliverable on Time.

Snowpack Layers of snow that accumulate at high altitudes for extended periods during the year. Melting snowpack is an important water resource that sustains many communities over dry summer months.

Social network The ties between individuals in a social structure that can be used to identify local and global social patterns, identify influential entities, and examine network dynamics.

Solar cycle The total energy received from the Sun—often referred to as the total solar irradiance (TSI)—which is not constant but varies over time, especially within the 11-year sunspot cycle.

Solar flare A sudden brightening observed over the Sun's visible surface, associated with the release of a large amount of energy.

Solar radiation management (SRM) Geoengineering projects that seek to increase the amount of reflected sunlight and reduce the impact of greenhouse gas emissions.

Solar updraft tower A renewable-energy power plant for generating electricity from solar power using the energy from rising hot air to drive a turbine.

Solid biomass Any recently living organism or its metabolic by-product, such as wood, sawdust, grass trimmings, domestic refuse, charcoal, agricultural waste, nonfood energy crops, and even dried manure.

Solubility pump The process by which carbon dioxide from the atmosphere first dissolves in cold polar surface water and is then removed (pumped) from the surface to the deep ocean for hundreds of years by thermohaline currents.

Spatial and temporal scales Scales that relate to changes in time and pace.

Specialization When a species becomes so closely adapted to its environment that it is often unable to cope with even small changes in climate.

Special Report: Emissions Scenarios (SRES) A report by the IPCC that outlines how greenhouse gas emissions are likely to change in the future, depending on the actions we take today.

Stable isotope A variety of an atom that has a variable number of neutrons and does not undergo spontaneous radioactive decay.

Stacking Combining two or more data sets from the same data source (for example, an ice or sediment core) to improve the overall signal-to-noise ratio.

Stakeholder An individual or organization that has a special interest in the outcome of climate negotiations.

Stalactite A vertical mineral pillar that grows from the roof of a cave. Stalactites form by the precipitation of calcium carbonate and other minerals from mineralized water solutions.

Stochastic A process whose future behavior is determined both by the process's predictable actions and by a random element.

Stochastic process A process whose behavior is non-deterministic, in that a system's subsequent state is determined both by the process's predictable actions and by a random element.

Stochastic resonance The observed when noise is added to a system and changes that system's behavior.

Stomata Small pores on the surface of a leaf that control the absorption of carbon dioxide and the transpiration of water.

Storm surge A rise in the level of the ocean associated with the approach of a deep low-pressure weather system such as a tropical cyclone (hurricane).

Stratosphere The layer of the atmosphere immediately above the troposphere, characterized by increasing temperatures with height due to the concentration of ozone.

Subtropics Climate zones typical of parts of Earth immediately north and south of the tropical zone, which is bounded by the Tropic of Cancer and the Tropic of Capricorn, at latitudes 23.5° north and 23.5° south.

Sunspot A magnetic disturbance on the surface of the Sun, whose related flares, prominences, and outbreaks produce surges in solar wind. Sunspots occur in an average 11-year cycle.

Supercontinent A large landmass that contains all, or nearly all, of the existing continents.

Sverdrup A unit to measure the flow of water on a large scale. Equal to one millions cubic meters a second.

Swing States States where the outcome of an election is far from certain and can depend on a number of key issues.

T

Taiga A biome characterized by coniferous forests.

Tea Party A political movement in the United States that is partly conservative, partly libertarian, and partly populist.

Tectonic plate A coherent unit of Earth's rigid outer layer that includes the crust and uppermost mantle. Also called a lithospheric plate or simply a plate.

Tectonic subsidence The sinking of Earth's crust on a large scale, associated with the movement of tectonic plates. Tectonic subsidence leads to an apparent rise in sea level.

Teleconnection The strong relation on that changes in the atmosphere at one location can have at another location many thousands of kilometers away.

Terminal moraine Glacial sediment left behind at the snout (end) of a retreating glacier as it melts.

TEX86 An organic paleothermometer based on the membrane lipids (fat) of marine organisms.

Thermal capacity The ability of a body to store internal energy by virtue of a temperature rise when it receives a flow of heat energy.

Thermal expansion The tendency of substances such as the air and oceans to change their volume in response to a change in temperature.

Thermocline A layer in a large body of water in which temperature changes more rapidly with depth than it does in the layers above or below.

Thermohaline circulation Deep-ocean circulation that is driven by changes in surface temperatures and salinity. *See also* Great ocean conveyor.

Total primary energy supply (TPES) The total amount of energy required to meet the global demands of human society. Currently around 568×10^{18} Joules per year, or 568 Exajoules (EJ).

Total solar irradiance (TSI) The amount of radiant energy emitted by the Sun over all wavelengths that falls each second at the top of Earth's atmosphere.

Trace element In geology, a chemical element whose concentration is less than 1,000 ppm, or 0.1% of a rock's composition.

Transform fault A boundary in which two plates slide past one another without creating or destroying lithosphere.

Transition zone The zone of the most rapid temperature change in the oceans, between the surface waters and the more thermally homogeneous deep waters.

Tropical cyclone A storm system that is characterized by a low-pressure center surrounded by a spiral arrangement of thunderstorms that produce strong winds and heavy rain. Also called a hurricane in the Western Hemisphere.

Tropical disturbance A term used by the National Weather Service for a cyclonic wind system in the tropics that is in its formative stages.

Tropical divergence The upper layers of the ocean diverging at the equator and drawing cooler, nutrient-rich water towards the surface because of the impact of surface winds.

Tropopause The boundary between the troposphere and the stratosphere.

Troposphere The lowest layer of the atmosphere, which extends up to the tropopause and is characterized by a decrease in temperature with height. It contains approximately 90% of the total mass of the atmosphere.

Tundra The coldest of all the biomes. Tree growth in the tundra is hindered by the combination of low temperatures, short growing season, low precipitation, and nutrient-poor soils.

Turbulent action The mixing effect of surface wind and waves on the upper layer of the oceans.

Tuvalu A small island in the western Pacific Ocean that is in immediate danger due to rising sea level.

U

U-shaped valley A classic cross-sectional shape of a glacially carved valley.

U.S. Department of Agriculture (USDA) The U.S. federal executive department responsible for developing and executing U.S. federal government policy on farming, agriculture, and food.

U.S. Geological Survey (USGS) The U.S. federal executive department responsible for the study of the landscape of the United States, its natural resources, and the natural hazards that threaten it.

Ultraviolet radiation Solar radiation with a wavelength from 0.2 to 0.4 micrometer that is damaging to living tissue but is significantly reduced at the surface by the presence of ozone in the stratosphere.

Uncertainty Insufficient knowledge of a potential hazard that limits our ability to project the future outcome with any accuracy.

United Nations Framework Convention on Climate Change (UNFCCC) An international environmental treaty that was opened for signature at the Earth Summit held in Rio de Janeiro in 1992 and came into force in 1994.

University of Alabama at Huntsville (UAH) Earth System Science Center (ESSC) A university center that was created to encourage interdisciplinary study of the Earth as an integrated system across traditional boundaries.

Uranium oxide An oxide of the element uranium that is enriched after extraction and used to power nuclear reactors.

V

Varve An annual layer of sediment or sedimentary rock.

Vehicle emission standards *See* Corporate Average Fuel Economy (CAFE) standard.

W

Walker cell circulation Airflow in the tropics that is caused by differences in heat distribution between ocean and land that generate zonal (east–west) and vertical movement in the tropical atmosphere. Walker cell circulation also has considerable motion in the meridional (north–south) direction, as part of the Hadley circulation cell.

Water stress A situation in which the available potable, unpolluted water in a region is less than that region's projected demand.

Weather balloon A balloon used to carry instruments aloft to gather meteorological data in the atmosphere. The balloon contains devices to measure temperature, pressure, and humidity and can reach 35 kilometers (22 miles) in altitude.

Weather satellite A type of satellite that is primarily used to monitor the weather and climate of Earth. Satellites can be polar orbiting, covering the entire Earth asynchronously, or geostationary, hovering over the same spot on the equator.

West African monsoon A monsoon of western Sub-Saharan Africa that results from the seasonal shifts of the intertropical convergence zone that generate great seasonal temperature and humidity differences between the Sahara and the equatorial Atlantic Ocean.

West Antarctic Ice Sheet (WAIS) The smaller of the two main ice sheets that cover the continent of Antarctica.

Western Boundary Current Strong ocean currents that develop on the western side of a large ocean gyre.

World Meteorological Society (WMO) The specialized agency of the United Nations for weather and climate, operational hydrology, and related geophysical sciences.

Y

Yedoma Organic-rich permafrost made of loess with a high ice content from ancient snow.

Younger Dryas event A geologically brief (1,300 ± 70 years) period of cold climatic conditions and drought that occurred between about 12,800 and 11,500 years before present.

Z

Zooxanthellae A group of colored dinoflagellate protists (often still referred to as algae) that live symbiotically within the cells of other organisms such as the stony reef-forming corals.

Credits

Chapter 1

Figure 1.1a and b: Based on "Jan–Dec Global Mean Temperature over Land & Ocean," from *State of the Climate: Annual 2008*, NOAA; Figure 1.1c: "The Physical Science Basis. Working Group I Contribution to the Fourth Assessment Report of the Intergovernmental Panel on Climate Change" from *Climate Change 2007*. Copyright © 2007 by Cambridge University Press. Reprinted with Permission; Figure 1.2b: Based on "Holocene Temperature Variation," from GlobalWarmingArt.com. Copyright © 2007 by GlobalWarmingArt.com. Reprinted with permission; Figure 1.2b: Joy Stein/Shutterstock; Figure 1.3a: "Proxy-Based Reconstructions of Hemispheric and Global Surface Temperature Variations over the Past Two Millennia," by Michael E. Mann, Zhihua Zhang, Malcolm K. Hughes, Raymond S. Bradley, Sonya K. Miller, Scott Rutherford, and Fenbiao Ni, from *Proceedings of the National Academy of Sciences*, September 9, 2008. Copyright © 2008 by National Academy of Sciences. Reprinted with permissions; Figure 1.4: Based on "1000 Year Temperature Comparison," by Robert A. Rohde, from GLOBALWARMINGART.COM. Copyright © by GlobalWarmingArt.com. Reprinted with permission; 7 1.5a: Based on "Jan–Dec Global Mean Temperature over Land & Ocean," from *State of the Climate: Annual 2008*, NOAA; Figure 1.6a: "Build-up in Earth's Total Heat Content," Skeptical Science Graphics, by Skeptical Science. Copyright © by Skeptical Science. Licensed under a Creative Commons Attribution 3.0 Unported License; Figure 1.6b: Planetary Visions Ltd / Science Source/Photo Research Inc.; Figure 1.7: "The Physical Science Basis. Working Group I Contribution to the Fourth Assessment Report of the Intergovernmental Panel on Climate Change" from *Climate Change 2007*. Copyright © 2007 by Cambridge University Press. Reprinted with Permission; Figure 1.9: Based on "Global Warming Predictions Map" by Robert A. Rohde, from GlobalWarmingArt.com. Copyright © by GlobalWarmingArt.com. Reprinted with permission; Figure 1.10: Marek Szumlas/Shutterstock; Figure 1.12: "Source: Based on ""Cenozoic Global Deep-Sea Stable Isotope Data,"" by J. Zachos, et al., IGBP PAGES/World Data Center for Paleoclimatology Data Contribution Series # 2008-098. NOAA/NCDC Paleoclimatology Program, Boulder CO, USA."; Figure 1.13: "The Physical Science Basis. Working Group I Contribution to the Fourth Assessment Report of the Intergovernmental Panel on Climate Change" from *Climate Change 2007*. Copyright © 2007 by Cambridge University Press. Reprinted with Permission; Figure 1.15: Tarbuck, Edward J.; Lutgens, Frederick K.; Tasa, Dennis, *Earth: An Introduction to Physical Geology*, 10th Ed., © 2011. Reprinted and Electronically reproduced by permission of Pearson Education, Inc., Upper Saddle River, New Jersey; Figure 1.16: Tarbuck, Edward J.; Lutgens, Frederick K.; Tasa, Dennis, *Earth: An Introduction to Physical Geology*, 10th Ed., © 2011. Reprinted and Electronically reproduced by permission of Pearson Education, Inc., Upper Saddle River, New Jersey; Figure 1.17: Based on "Milankovitch cycles over the past 1000000 years," by Robert A. Rohde, from GlobalWarmingArt.com. Reprinted with permission;

Figure 1.18: "Synthesis Report. A Contribution of Working Groups I, II and III to the Third Assessment Report of the Intergovernmental Panel on Climate Change," from *Climate Change 2001*. Copyright © 2001 by Cambridge University Press. Reprinted with Permission. Figure 1.19: Richard R. Hansen/Photo Researchers Inc.; Figure 1.20: Leon Neal/AFP/Getty Images; Figure 1.21: Science Source/USGS/Getty Images; Figure 1.22: Based on "Global temperature response to Pinatubo eruption" from *Skeptical Science* website. Reprinted with permission; Figure 1.23: Based on "The NOAA Annual Greenhouse Gas Index (AGGI)," from NOAA Earth System Research Laboratory website; Figure 1.24: "Schematic for Global Atmospheric Model" from NOAA Celebrates 200 Years of Science Service and Stewardship website; p. 18: Excerpt from "IPCC 2000: Special Report on Emissions Scenarios. Prepared by Working Group III of the Intergovernmental Panel on Climage Change," Copyright © 2000 by Cambridge University Press. Reprinted with permission; Figure 1.25: Impacts, Adaptation and Vulnerability. Working Group II Contribution to the Fourth Assesment Report of the Intergovernmental Panel on Climate Change." from *Climate Change 2007*. Copyright © 2007 by Cambridge University Press. Reprinted with Permission; Figure 1.26: "Impacts, Adaptation and Vulnerability. Working Group II Contribution to the Fourth Assesment Report of the Intergovernmental Panel on Climate Change." from *Climate Change 2007*. Copyright © 2007 by Cambridge University Press. Reprinted with Permission.

Chapter 2

Pages 24–25: Elena Elisseeva / Glow Images; John Moore/Getty Images; Figure 2.1a: Based on "Major Biomes Map" (USDA-NRCS, Soil Survey Division, World Soil Resources, Washington D.C.), from National Resources Conservation Service. http://www.nrcs.usda.gov; Figure 2.1b: "21st Century Ecological Sensitivity 1" (NASA/JPL-Caltech), from NASA website; Figure 2.2: Based on "Global Climate Network Temperature Stations," by Robert A. Rohde, from GlobalWarmingArt.com. Copyright © by GlobalWarmingArt.com. Reprinted with permission; Figure 2.3a: Climate Change 2007: The Physical Science Basis. Working Group I Contribution to the Fourth Assessment Report of the Intergovernmental Panel on Climate Change, FAQ 5.1, Figure 1. Copyright © 2007 by Cambridge University Press. Reprinted with permission; Figure 2.4a: Based on "Global Sea Level Fluctuations," by Robert A.Rohde, from globalwarmingart.com. Copyright © by GlobalWarmingArt.com. Reprinted with permission; Figure 2.4b: Martin Shields/Alamy; Figure 2.5a: Based on "Coastline Retreat with Sea-Level Rise" from *Impacts of a Warming Arctic*. Copyright © 2004 by ACIA. Reprinted with permission; Figure 2.6: Source: Based on Intergovernmental Panel on Climate Change Working Group I: The Physical Science Basis of Climate Change, Figure 6.8. 2009; Figure 2.7a: Source: Based on "Estimating Mean Sea Level Change from the TOPEX and Jason Altimeter Missions," by Nerem, R. S., D. Chambers, C. Choe, and G. T. Mitchum, from *Marine Geodesy* 33, no. 1 supp 1 (2010): 435; Figure 2.9a–c: Source: Based on "Climate Impact of Quadrupling CO_2: An Overview of GFDL Climate Model

Results," from the Geophysical Fluid Dynamics Laboratory and the NOAA website, 2012; Figure 2.10: "Minimum extent of Arctic Sea Ice," from NASA Earth Observatory website; Figure 2.11: Source: "Arctic Sea Ice Extent Standardized Anomalies Jan 1953–Oct 2011" by Walt Meier and Julienne Stroeve, from "State of the Cryosphere" on NSIDC website; Figure 2.13: Based on "Greenland's Ice Island Alarm," from NASA Earth Observatory website; Figure 2.14: Jakobshavn Glacier Flow, from NASA GODDARD SPACE FLIGHT CENTER website, September 12, 2009. Figure 2.16: Based on figure by Marco Tedesco, Joint Center for Earth Systems Technology, NASA Goddard Space Flight Center, University of Maryland at Baltimore County. Copyright © 2007 by Marco Tedesco. Reprinted with permission; Figure 2.21: Based on permafrost map from INTERNATIONAL PERMAFROST ASSOCIATION website; Figure 2.22: John E Marriott/Alamy; Figure 2.23: © Accent Alaska.com / Alamy; Figure 2.24a–b: "Projected changes in permafrost (Northern Hemisphere)" by Hugo Ahlenius, from UNEP GRID-Arendal website. Copyright © 2012 by UNEP-GRID Arendal. Reprinted with permission; Figure 2.25: Steve Morgan/Alamy; Figure 2.26: Todd Paris/AP Photo; Figure 2.27a: "Overview on glacier changes since the end of the Little Ice Age" by Hugo Ahlenius, from from UNEP GRID-Arendal website. Copyright © 2012 by UNEP-GRID Arendal. Reprinted with permission; Figure 2.27b: Daniel Beltra/Newscom; Figure 2.28: Based on "Correcting Ocean Cooling: Balancing the Sea Level Budget," from NASA Earth Observatory website, 2008; Figure 2.29a–c: "The Physical Science Basis. Working Group I Contribution to the Fourth Assessment Report of the Intergovernmental Panel on Climate Change" from *Climate Change 2007* Figure 10.19. Copyright © 2007 by Cambridge University Press. Reprinted with Permission; Figure 2.30a–d: "The Physical Science Basis. Working Group I Contribution to the Fourth Assessment Report of the Intergovernmental Panel on Climate Change" from CLIMATE CHANGE 2007. Copyright © 2007 by Cambridge University Press. Reprinted with Permission. Figure 2.31: Based on "Change in Precipitation by the End of 21st Century" (NOAA GFDL), from NOAA Geophysical Fluid Dynamics Laboratory website; Figure 2.32: "The Physical Science Basis. Working Group I Contribution to the Fourth Assessment Report of the Intergovernmental Panel on Climate Change" from *Climate Change 2007*. Copyright © 2007 by Cambridge University Press. Reprinted with Permission; Figure 2.36: Based on "Atlantic Basin Hurricane Counts (1851–2006) 5-year running means," from NOAA website.

Chapter 3

Pages 54–55: NOAA/GSFC/Suomi NPP/VIIRS/Norman Kuring/NASA; Figure 3.1: AGUADO, EDWARD,; BURT, JAMES E., UNDERSTANDING WEATHER AND CLIMATE, 6TH Ed., © 2013. Reprinted and Electronically reproduced by permission of Pearson Education, Inc., Upper Saddle River, New Jersey. Figure 3.2: NASA; Figure 3.4: NASA; Figure 3.5: Based on "Cycle 24 Sunspot Number Prediction (February 2012 Revised)," from NASA website; Figure 3.6: Data from "Unusual activity of the Sun during recent decades compared to the previous 11,000 years," by S.K. Solanki, et al., from *Nature*, vol. 431, pp. 1084–1087, October 28, 2004. Copyright © 2004 by Nature Publishing Group; Figure 3.8: Based on "Milutin Melankovitch (1879–1958)," from NASA website; Figure 3.9: Excerpt from "Climate

Change 2007: The Physical Science Basis. Working Group I Contribution to the Fourth Assessment Report of the Intergovernmental Panel on Climate Change, FAQ 6.1," Copyright © 2000 by Cambridge University Press. Reprinted with permission; Figure 3.12: Based on "Milankovitch Variations," by Robert A. Rohde, from GlobalWarmingArt.com. Copyright © by GlobalWarmingArt.com. Reprinted with permission; Figure 3.13: Source: "Isolation at 65N, Summer Solstice," by Incredio, from Wikipedia website, June 4, 2009; Figure 3.14: Source: "Gas proportions in the Earth's atmosphere," by Mysid, from Wikipedia website, May 30, 2006; Figure 3.15: Dr. Emrys Hall and Allen Jordan, NOAA/OAR/GMD; Figure 3.16: Keller, Edward A.; Devecchio, Duane E., *Natural Hazards: Earth's Processes as Hazards, Disasters, and Catastrophes*, 3rd Ed., © 2012. Reprinted and Electronically reproduced by permission of Pearson education, Inc., Upper Saddle River, New Jersey; Figure 3.18: Based on "Atmospheric CO_2 at Mauna Loa Observatory," from NOAA website; Figure 3.19: Based on "The NOAA Annual Greenhouse Gas Index (AGGI)," from NOAA website; Figure 3.21: Based on "Global Trends in Controlled Ozone Depleting Chemicals," from NOAA website; Adapted from figure from LabSpace website; Figure 3.24: "Mean Annual Global Isolation," from COMET® Website, the University Corporation for Atmospheric Research (UCAR), sponsored in part through cooperative agreement(s) with the National Oceanic and Atmospheric Administration (NOAA), U.S. Department of Commerce (DOC). Copyright © 1997-2012 by University Corporation for Atmospheric Research. All Rights Reserved; Figure 3.25: Based on "Climate and Earth's Energy Budget," from NASA Earth Observatory website; Figure 3.26: "Black Body Emission Curves of the Sun and Earth," from COMET® Website, the University Corporation for Atmospheric Research (UCAR), sponsored in part through cooperative agreement(s) with the National Oceanic and Atmospheric Administration (NOAA), U.S. Department of Commerce (DOC). Copyright © 1997–2012 by University Corporation for Atmospheric Research. All Rights Reserved; Figure 3.27: "Solar Absorbed Minus IR Emitted," from University of Washington, Department of Atmospheric Sciences, http://www.atmos.washington.edu/cgi-bin/erbe/disp.pl?net.ann. Reprinted with permission; Figure 3.28: Based on "NASA Releases Terra's First Global 1-Month Composite Images," from NASA Earth Observatory website; Figure 3.30: Based on "Solar Radiation Spectrum," by Robert A. Rohde, from GlobalWarmingArt.com. Copyright © by GlobalWarmingArt.com. Reprinted with permission; Figure 3.31: Based on "NASA Maps Shed Light on Carbon Dioxide's Global Nature," from NASA website; Figure 3.32: "Earth's Annual Global Mean Energy Budget," from *Bulletin of the American Meteorological Society*. Copyright © 1997 by American Meteorological Society. Reprinted with permission; Figure 3.33: Based on Model output graph generated from web interface by David and Jeremy Archer of University of Chicago. Reprinted with permission; Figure 3.35: Based on "UAH Lower Stratospheric Anomalies for Jan-Dec," from NOAA website; Figure 3.36: Based on "Global Warming Facts and Our Future." Copyright © by Marian Koshland Science Museum. Reprinted with permission; Figure 3.38: Adapted from *Eos*, Vol. 88, No. 45, 6 November 2007. Copyright © 2007 by American Geophysical Union. Reprinted and modified with permission; Figure 3.39:

Chapter 4

Research (UCAR), sponsored in part through cooperative agreement(s) with the National Oceanic and Atmospheric Administration (NOAA), U.S. Department of Commerce (DOC). Copyright © 1997–2012 by University Corporation for Atmospheric Research. Reprinted with permission. All Rights Reserved; Figure 4.37: NOAA; Figure 4.38: Based on illustration by Jack Woods from *Woods Hole Oceanographic Institution*. Copyright © 2012 by WHOI. Reprinted with permission; Figure 4.39: Source: "An enhanced satellite photograph showing a greater concentration of chlorophyll along the equator where divergence . . . " from NASA.ORG; Figure 4.41: NASA; Figure 4.42: Based on "Tracks and Intensity of All Tropical Cyclones," by Robert A. Rohde, from GlobalWarmingArt.com. Copyright © by GlobalWarmingArt.com. Reprinted with permission; Figure 4.43: Source: "The Atlantic Multidecadal Oscillation Index and five-year average counts of tropical cyclones" by Christopher Landsea, et al, from NOAA.GOV; Figure 4.45: Source: "Global Chart of Seasonal Rainfall Patters" from CMAP Dataset, Climate Diagnostics Center, USA; Figure 4.47: NASA; Figure 4.48: Based on illustration by Jack Woods from *Woods Hole Oceanographic Institution*. Copyright © 2012 by WHOI. Reprinted with permission; Figure 4.49: Based on illustration by Jack Woods from *Woods Hole Oceanographic Institution*. Copyright © 2012 by WHOI. Reprinted with permission; Figure 4.51: Source: " . . . January 1997 there is a steep temperature gradient between the western and eastern Pacific Ocean . . . " by Greg Shirah, from NASA.GOV; Figure 4.52: "Global Walker Circulation," from the COMET® Website, the University Corporation for Atmospheric Research (UCAR), sponsored in part through cooperative agreement(s) with the National Oceanic and Atmospheric Administration (NOAA), U.S. Department of Commerce (DOC). Copyright © 1997–2012 by University Corporation for Atmospheric Research. Reprinted with permission. All Rights Reserved; Figure 4.54: Source: " . . . By July 1997 there is a full-blown El Niño and the thermocline is flat." by Greg Shirah, from NASA.GOV; Figure 4.55: Source: "El Niño Southern Oscillation," from NOAA.GOV; Figure 4.56: Source: "Effects of ENSO in the Pacific," from NOAA.GOV; Figure 4.57: Source: "El Niño Southern Oscillation (ENSO)," from NOAA.GOV; Figure 4.58: Source: "Effects of ENSO in the Pacific: El Niño Conditions," from NOAA.GOV; Figure 4.59a–b: Source: "Weather Impacts of ENSO: The Jetstream," from NOAA.GOV; Figure 4.60a–b: Source: "Weather Impacts of ENSO: The Jetstream," from NOAA.GOV; Figure 4.61: Source: "Effects of ENSO in the Pacific: La Niña Conditions," from NOAA.GOV; Figure 4.62: Source: "Monthly Mean SST 2°S to 2°N Average" from NOAA.GOV; Figure 4.63a–c: "Figure 1: Development of a Madden-Julian Oscillation event," from NOAA.GOV; Figure 4.64: Source: based on "200 hecto Pascals Velocity Potential Anomalies Animation," from the NOAA website; Figure 4.65a: "MJO Impacts during Boreal Summer," from the COMET® Website, the University Corporation for Atmospheric Research (UCAR), sponsored in part through cooperative agreement(s) with the National Oceanic and Atmospheric Administration (NOAA), U.S. Department of Commerce (DOC). Copyright © 1997–2012 by University Corporation for Atmospheric Research. Reprinted with permission. All Rights Reserved; Figure 4.65b: "MJO Impacts during Boreal Winter," from the COMET® Website, the University Corporation for Atmospheric Research

(UCAR), sponsored in part through cooperative agreement(s) with the National Oceanic and Atmospheric Administration (NOAA), U.S. Department of Commerce (DOC). Copyright © 1997–2012 by University Corporation for Atmospheric Research. Reprinted with permission. All Rights Reserved; Figure 4.66: Based on "Pacific Decadal Oscillation," by Stepen Hare and Nathan Mantua. Copyright © by University of Washington Department of Atmospheric Sciences. Reprinted with permission; Figure 4.67: Based on "Pacific Decadal Oscillation," by Stepen Hare and Nathan Mantua. Copyright © by University of Washington Department of Atmospheric Sciences. Reprinted with permission; Figure 4.68: Based on "Monthly Values for the PDO Index: 1900–September 2008." Copyright © by University of Washington Department of Atmospheric Sciences. Reprinted with permission; Figure 4.69: Based on figure from CLIMATE WATCH website. NOAA.GOV; Figure 4.70: "Multidecadal Variability In Arctic and North Atlantic Climate System," by Igor Polyakov from International Arctic Research Center, University of Alaska Fairbanks. Reprinted with permission.

Chapter 5

Pages 126–127: MarclSchauer/ Shutterstock; Figure 5.7: Richard Bizley/Science Source; Figure 5.21: "Model of an atom," from NASA website, 2011; Figure 5.23: Source: "Water vapor gradually loses 1^8O as it travels from the equator to the poles . . . ," Illustrated by Robert Simmon, from the NASA website; Figure 5.25: Source: "The concentration of 1^8O in precipitation decreases with temperature . . . " by Jouzel, et al., from the NASA website, 1994; Figure 5.26: Based on figure from "A bi-polar signal recorded in the western tropical Pacific: Northern and Southern Hemisphere climate records from the Pacific warm pool during the last Ice Age" by Reetta Saikku, Lowell Stott, and Robert Thunell, from *Quaternary Science Reviews*, November 2009. Copyright © 2009 by Elsevier. Reprinted with permission; Figure 5.27: Graph from gly.uga.edu. Copyright © by The University of Georgia. Reprinted with permission; Figure 5.31: Climate Chang 2007: The Physical Science Basis. Working Group 1 Contribution to the Fourth Assessment Report of the Intergovernmental Panel on Climate Change, Figure 6.2. Copyright © 2007 by Cambridge University Press. Reprinted with permission; Figure 5.32: Adapted from "Environmental Isotopes in Hydrogeology," by Ian Clark, from University of Ottawa Earth Science Department, based on Veizner, 1989, Ann. Rev. Earth Planet Sci., 17:141–167. Copyright © 2012 by University of Ottawa. Reprinted with permission; Figure 5.35: Source: Data from table 2 in "Using fossil leaves as paleoprecipitation indicators: An Eocene example" by Peter Wilf, Scott L.Wing, David R. Greenwood, Cathy L. Greenwood, from *Geology*, March 1998, Volume 26, 3; Figure 5.36: Based on figure from University of California Museum of Paleontology. Copyright © by UC Regents. Reprinted with permission; Figure 5.38: Data from Table 2 in "Using fossil leaves as paleoprecipitation indicators: An Eocene example," by Peter Wilf, Scott L.Wing, David R. Greenwood, and Cathy L. Greenwood, from *Geology*, March 1998, 26; Figure 5.44: "Bristle Pine - Tree ring widths" from Climate Date Information Website. Copyright © 2012 by www.climatedata.info. Reprinted with permission; Figure 5.53: Santa Cruz Sentinal/ ZUMA Press/Alamy Images.

Chapter 6

Pages 160–161: Catmando/ Shutterstock; Figure 6.5: Publiphoto / Science Source; Figure 6.6b: *Paleoclimatic maps* modified from Boucot, ChenXu and Scotese (2013); Figure 6.8: Francois Gohier / Science Source; Figure 6.10b: BRITISH GEOLOGICAL SOCIETY website. Copyright © 2013 by NERC. Reproduced by permission of the British Geological Survey. All rights reserved; Figure 6.11: Based on "The Story of Global Climate Change in the Upper Cretaceous," from BRITISH GEOLOGICAL SURVEY website. Copyright © 2013 by Natural Environment Research Council. Reprinted with permission. 6.12: Paleographic map from Colorado Plateau Geosystems, Inc. Copyright © 2012 by Ron Blakey. Reprinted with permission; Figure 6.16: Martin Bond/Science Source; Figure 6.17: Source: Massive Terrestrial Strike by Don Davis, from NASA website; Figure 6.18a–b: Climate Change 2007: The Physical Science Basis. Working Group I Contribution to the Fourth Assessment Report of the Intergovernmental Panel on Climate Change, Figure 6.1. Copyright © 2007 by Cambridge University Press. Reprinted with permission; Figure 6.20: *Paleoclimatic maps* modified from Boucot, ChenXu and Scotese (2013); Figure 6.21: Source: Data from "Trends, Rhythms, and Aberrations in Global Climate 65 Ma to Present," by Zachos, J., M. Pagani, L. Sloan, E. Thomas, and K. Billups. 2001, from *Science*, Vol. 292, No. 5517, pp. 686–693, 27 April 2001; Figure 6.23: Data from Zachos et al., SCIENCE 292:686 (2001); Figure 6.24: Dr. Keith Wheeler / Science Source; Figure 6.27b: *Paleoclimatic maps* modified from Boucot, ChenXu and Scotese (2013); Figure 6.28: Based on Figure 3 from "The Early Miocene Onset of a Ventilated Circulation Regime in the Arctic Ocean," by Martin Jakobsson, from *Nature*, vol. 447, p. 986, January 21, 2007. Copyright © 2007 by Nature Publishing Group; Figure 6.29b: *Paleoclimatic maps* modified from Boucot, ChenXu and Scotese (2013); Figure 6.30: Source: "Once the Fram Strait opened, cold water from the Arctic Ocean was able to circulate freely with the Atlantic," from NOAA Ocean Explorer website; Figure 6.31: Based on illustration by Jack Woods from Woods Hole Oceanographic Institution. Copyright © 2012 by WHOI. Reprinted with permission; Figure 6.32: Chase Studio / Science Source; Figure 6.33: "A Phylogeny of the Hominidae," by Richard G. Klein, from NASA website; Figure 6.35: *Paleoclimatic maps* modified from Boucot, ChenXu and Scotese (2013); Figure 6.36: Source: NOAA; Figure 6.37: Mauricio Anton / Science Source; Figure 6.41: "Ice sheets, schematic illustration for Greenland and Antarctica" by Hugo Ahlenius, from UNEP GRID-Arendal website. Copyright © 2012 by UNEP-GRID Arendal. Reprinted with permission; Figure 6.42: Source: Data from "A GIS-based Vegetation Map of the World at the Last Glacial Maximum (25,000–15,000 BP)–(23rd millennium to 13th millennium BC)," Ray, N. and J. M. Adams. 2001, from *Internet Archeology11*; Figure 6.48: Based on ""A Pliocene-Pleistocene Stack of 57 Globally Distributed Benthic D18O Records," by Don Blankenship from *Paleoceanography*, 2005, Volume 20. Copyright © 2005 by University of Texas at Austin. Reprinted with permission; Figure 6.49: "Atlas of the Cryosphere," from the National Snow and Ice Data Center website; Figure 6.50: © Asia Photopress / Alamy; Figure 6.55: Source: Based on *Earth's Climate: Past and Future*, by William F. Ruddiman, 2007; Figure 6.56: Based on figure from "The δ18O record from the GISP2 ice core" adapted from "Comparison of Oxygen Isotope Records from the GISP2 and GRIP Greenland Ice Cores," by P. Grootes, from *Nature*. Copyright © 1993 by Nature Publishing Group; Figure 6.66: Based on "Post-Glacial Sea Level Rise," by Robert A. Rohde, from GlobalWarmingArt.com. Copyright © by GlobalWarmingArt.com. Reprinted with permission; Figure 6.64: Adaptation from "Proxy-Based Reconstructions of Hemispheric and Global Surface Temperature Variations over the Past Two Millennia," by Michael E. Mann, from *Proceedings of the National Academy of Sciences*, September 2008, Volume 105(36). Copyright © 2008 by National Academy of Sciences, U.S.A. Reprinted with permission; Figure 6.65: Source: GISP Core data from NOAA.

Chapter 7

Figure 7.1a–b: Climate Change 2007: The Physical Science Basis. Working Group I Contribution to the Fourth Assessment Report of the Intergovernmental Panel on Climate Change, Figure SPM.6. Copyright © 2007 by Cambridge University Press. Reprinted with permission; Figure 7.2: Climate Change 2007: Impacts, Adaptation and Vulnerability. Working Group II Contribution to the Fourth Assessment Report of the Intergovernmental Panel on Climate Change, Figure TS.2. Cambridge University Press. Reprinted with permission; Figure 7.3a–b: Climate Change 2007: Impacts, Adaptation and Vulnerability. Working Group II Contribution to the Fourth Assessment Report of the Intergovernmental Panel on Climate Change, Figure 11.4. Cambridge University Press. Reprinted with permission; Figure 7.4: Climate Change 2007: Impacts, Adaptation and Vulnerability. Working Group II Contribution to the Fourth Assessment Report of the Intergovernmental Panel on Climate Change, Figure TS.4. Copyright © 2007 by Cambridge University Press. Reprinted with permission; Figure 7.6: Source: "Major Biomes," from Natural Resources Conservation Service website; Figure 7.7: "Biome type in relation to temperature and rainfall" from Annenberg Learner website. Copyright © 2012 by The Annenberg Foundation. Reprinted with permission; Figure 7.8: Climate Change 2007: Impacts, Adaptation and Vulnerability. Working Group II Contribution to the Fourth Assessment Report of the Intergovernmental Panel on Climate Change, Figure 4.3. Copyright © 2007 by Cambridge University Press. Reprinted with permission; Figure 7.10: Based on Housing Density map from Woods Hole Research Center website. Copyright © 2012 by Woods Hole Resource Center. Reprinted with permission; p. 219: Source: David Kitchen; Figure 7.11: "Availability of Freshwater in 200" from UNEP GRID-Arendal website. Copyright © 2012 by UNEP-GRID Arendal. Reprinted with permission; Figure 7.12: Climate Change 2007: Impacts, Adaptation and Vulnerability. Working Group II Contribution to the Fourth Assessment Report of the Intergovernmental Panel on Climate Change, Figure 3.2. Copyright © 2007 by Cambridge University Press. Reprinted with permission; Figure 7.13: "Change in water availability compared with average 1961–1990 (%) 2050 based on IPCC scenario A1," from UNEP GRID-Arendal website. Copyright © 2012 by UNEP-GRID Arendal. Reprinted with permission;

Index